THE SOLAR CHROMOSPHERE AND CORONA: QUIET SUN

ASTROPHYSICS AND SPACE SCIENCE LIBRARY

A SERIES OF BOOKS ON THE RECENT DEVELOPMENTS

OF SPACE SCIENCE AND OF GENERAL GEOPHYSICS AND ASTROPHYSICS

PUBLISHED IN CONNECTION WITH THE JOURNAL

SPACE SCIENCE REVIEWS

VOLUME 53

R. GRANT ATHAY

High Altitude Observatory, National Center for Atmospheric Research
(sponsored by the National Science Foundation)

THE SOLAR
CHROMOSPHERE AND CORONA:
QUIET SUN

D. REIDEL PUBLISHING COMPANY

DORDRECHT-HOLLAND / BOSTON-U.S.A.

Library of Congress Cataloging in Publication Data

Athay, R. Grant.
    The solar chromosphere and corona: Quiet Sun

    (Astrophysics and space science library; 53)
    Bibliography: p.
    1. Solar chromosphere.   2.   Sun – Corona.
I.   Title.   II.   Series.
QB528.A83   523.7′5   75–33385
ISBN 90–277–0244–6

Published by D. Reidel Publishing Company,
P.O. Box 17, Dordrecht, Holland

Sold and distributed in the U.S.A., Canada and Mexico
by D. Reidel Publishing Company, Inc.
Lincoln Building, 160 Old Derby Street, Hingham,
Mass. 02043, U.S.A.

TABLE OF CONTENTS

PREFACE

The widespread tendency in solar physics to divide the solar atmosphere into separate layers and to distinguish phenomena of solar activity from phenomena of the quiet Sun emphasizes the wide ranging diversity of physical conditions and events occurring in the solar atmosphere. This diversity spans the range from a neutral, essentially quiescent atmosphere to a highly ionized, violently convective atmosphere; from a domain in which magnetic field effects are unimportant to a domain in which the magnetic pressure exceeds the gas pressure, and from a domain in which the particle motions are Maxwellian to a domain in which an appreciable fraction of the particles is accelerated to relativistic energies.

It is now widely recognized that the chromosphere and corona have a common origin in the mechanical energy flux generated in the hydrogen convection zone lying beneath the photosphere. Furthermore, magnetic field phenomena appear to be as vital to the structure of the quiet Sun as to the active Sun. For these reasons it appears desirable to present a unified treatment of the entire solar atmosphere, both active and quiet, in a single volume. On the other hand, such a treatise must be very long if it is to avoid being superficial, and it is very difficult for a single author to write authoritatively on such a wide range of topics. In the present volume, I have elected to treat the quiet Sun phenomena of the chromosphere and corona as a unified topic in order to emphasize their common origin and the common problems they pose in terms of energy and momentum balance. The sharply different physical regimes occurring in the corona and the chromosphere are related to the response of the atmosphere to the mechanical energy input rather than to fundamentally different physical mechanisms of energy input. Recognition that this is the case represents one of the milestones in the history of solar physics.

The most widely accepted explanation of the heating that takes place in the chromosphere and corona is through the dissipation of wave energy generated in the hydrogen convection zone. The generation of this wave energy, its relationship to magnetic field phenomena, its passage through the photosphere, its eventual dissipation in the chromosphere and corona, and the response of the chromosphere and corona to the wave energy each form major problems in solar physics which we are only now beginning to understand. These same problems are of major importance to our understanding of stellar atmospheres, and it is the author's hope that both stellar and solar astronomers will benefit from the discussion of these phenomena.

R. GRANT ATHAY

CHAPTER I

INTRODUCTION

1. Chromospheres and Coronas

Solar astronomers are often asked why it is that so much of the effort in astronomy is lavished on a single star — the Sun. If the Sun were just another star among the several billion in our Galaxy, the question could not be answered in a reasonable way. However, the Sun is not just another star. By virtue of its proximity to Earth, the Sun is, to man, the most unique of all stars. It provides the testing ground for our theories of stellar structure and, ultimately for most of our concepts of the nature of the Universe.

The Sun is close enough to Earth that its surface features can be studied in fine detail and its space environment can be probed *in situ*. Its radiations are intense enough that its spectrum can be minutely examined. Thanks largely to the remarkably fortunate circumstances of total solar eclipses, the Sun's outer atmosphere also can be studied in detail. None of this is true for any other star. Stars are too far away, and too faint to be examined, with present day instruments, to a degree that approaches that which is possible for the Sun.

The astronomer's understanding of stars, and star-like objects is based upon a series of assumptions concerning, among other things, the state of matter, the generation and transport of energy through the star, and the chemical composition of the star. In the particular case of the Sun we are able to determine enough about the structure of its outer layers to check, partially, at least, the validity of the assumptions that are normally made about stars. If these assumptions fail for the Sun, they probably fail for the majority of stars. If they predict the wrong structure for the Sun, they undoubtedly predict the wrong structure for most other stars. If our understanding of stellar structure is wrong, our understanding of the Universe is wrong. Thus, in a very real sense a correct understanding of the Universe in which we live depends upon a correct understanding of the Sun.

Had astronomers discovered that the usual theories of stellar structure adequately described the Sun, they could have relaxed and devoted only a normal share of effort to the study of the Sun. In fact, However, the agreement between the theoretical Sun and the actual Sun only serves to tease the astronomer. The agreement is tantalizingly close in many respects but fails badly in certain specific comparisons. Thus, the astronomer proceeds in his analysis of other stars fortified by his successes in largely predicting the solar atmosphere but haunted by the nagging failures. The history of science is filled with examples of theories that 'almost worked' but that ultimately were proven to be totally wrong. We are not suggesting that the same is true of our concepts of stellar structure.

However, our study of the Sun has shown clearly and emphatically that the normal theories of stellar structure are inadequate for many purposes and must someday be revised to bring about better harmony between theory and observation. These facts alone demand that a large amount of attention continue to be devoted to solar astronomy.

In addition to its uniqueness as an astronomical object, the Sun is the source of life giving energy for man. It is also the heat source that drives the Earth's atmosphere in its ever changing patterns of weather and climate. Although the daily weather changes, and likely the climatic changes, are due to terrestrial influences quite apart from any changes in the solar driving energy, the Sun does vary in time and the effect (or lack of it) of these intrinsic solar variations has not been firmly established. The Sun is more directly in control of the space environment of the Earth and of the Earth's ionosphere, magneto-sphere and outermost atmosphere. Here the intrinsic solar variations may entirely govern the local 'weather'. In fact, it is no stretch of truth to say that the Earth is imbedded in the solar atmosphere and that the Earth's immediate space environment represents a sample of the Sun itself.

The influence of the Sun on geophysical phenomena and the control of the Sun over our space environment provide added impetus for intensive study of the Sun, quite apart from the role of the Sun in astronomy. Together these two motivations result in an intensive effort in solar physics, as they properly should.

Just as the Sun is seen out of perspective when viewed simply as another star, the solar chromosphere and corona are seen completely out of perspective when viewed simply as somewhat curious layers of the solar atmosphere. The chromosphere and corona are, in fact, vivid manifestations of the failure of the classical stellar atmosphere to predict the actual Sun. Furthermore, the chromosphere and corona are the solar sources for most of the radiations of interest in geophysical and space phenomena. The classical stellar atmosphere would produce none of the geophysical or space phenomena for which the Sun is, in fact, responsible. Thus, it is precisely the failure of the Sun to conform to the classical model of a stellar atmosphere that makes it of unusual interest to geophysics and space physics as well as to astrophysics. The solar chromosphere and corona represent major manifestations of this failure, and, as such, they represent regions of common and widespread interest in astronomy, geophysics and space physics.

The preceding comments outline the importance of the solar chromosphere and corona to the disciplines of astrophysics, geophysics and space physics. However, both the chromosphere and corona are of great interest as phenomena in and of themselves. They represent large scale plasmas permeated by strong magnetic fields and heated to high temperatures by the dissipation of mechanical energy incident from below the chromosphere in wave modes, which, as yet, are not well understood. The plasma thermal energy density and magnetic energy density in the chromosphere and corona are often comparable in magnitude and complex interactions between the plasma and the magnetic fields are frequently observed.

Within the chromosphere and corona numerous phenomena suggestive of inherent instabilities are evident. Violent solar flares are spawned near regions of strong magnetic fields; the upper chromosphere bristles with geyser-like spicules erupting from the borders of a giant mosaic pattern of supergranule cells; large masses of coronal plasma of abnormally high temperature and density accumulate in regions of strong magnetic fields; the corona condenses into relatively cool dense prominences along neutral lines in the magnetic fields and near the poles of strong fields; the high temperature of the corona forces an outward expansion of the coronal plasma to form the solar wind blowing past the Earth. These phenomena, and several others of similar nature, provide a complex display of plasma pyrotechnics of unparalleled magnitude. They are readily observed, measured and catalogued, but not so readily explained.

Because of the high temperatures of the chromospheric and coronal plasmas much of their radiation falls in the EUV and X-ray regions of the spectrum where observations require rocket and satellite-borne spectrometers. Space programs have added greatly to both our understanding of and our interest in the chromosphere and corona. The added data acquired bring both new understanding of previously known phenomena and discoveries of new phenomena to be explained. This aspect of chromospheric and coronal observing is still in its youth and holds great promise for the future. Already, it has contributed much to our understanding of the chromosphere and corona.

Interactions of the chromospheric and coronal plasmas with the changing magnetic field patterns and with shock waves arising from instabilities provide intense sources of radio noise emission, which, in turn, offer still other means of studying the plasma events. Solar radio astronomy is devoted largely to the study of the chromosphere and corona and of their explosively transient phenomena. Here, too , the added observational data bring invaluable new dimensions and new insight to the phenomena taking place.

Study of the chromosphere and corona encompasses the entire spectrum from wavelengths below 1 Å to wavelengths of several meters. It encompasses, as well, the particle spectrum ranging from a few kev to a few BeV. It involves complex space instrumentation, large radio telescopes, and large optical telescopes located from sea-level to high mountain tops. Solar astronomers anxiously pursue each favourable solar eclipse to take advantage of the unique observing opportunities provided by the circumstances of the eclipse and that can be obtained in no other way. Still others fly telescopes to high altitudes in balloons or carry them aloft in high-flying jet aircraft to escape the murk, turbulence and moisture of the low terrestrial atmosphere. Each type of observation has its rightful place, and each plays an important role in contributing to our attempts to understand the complex phenomena of the chromosphere and corona. With each new observing technique we gain new insight and understanding, but we discover, also, new complexity and sometimes totally new phenomena that previous observations had failed to detect.

The solar chromosphere and corona are marvelously complex phenomena; in my own opinion more complex than any of us are yet prepared to admit. We start always with simplicity and add complexity only as the observations present us with alternatives that are clearly impossible in the simpler framework. Although we have learned a great deal about phenomena of the chromosphere and corona over the past few decades and now feel that we understand the basic properties of the plasma in these layers, few, if any, of the phenomena have been adequately explained. The science of the chromosphere and corona is therefore still relatively youthful in terms of its maturity towards being able to offer an explanation for what is observed.

It would be misleading to end this introduction to chromospheric and coronal physics without acknowledging another notable advance of recent years, viz., large electronic computers. Past efforts to explain the complex phenomena of the chromosphere and corona have been impeded by our inability to handle the problems in mathematical physics that were dictated by our observations and our theories. Thus, even though we were able to write down possibly correct mathematical explanations for what we observed we were often unable to carry them through without committing gross over-simplifications. Computers have greatly relieved this problem and we are now able to solve the mathematical problems more realistically. Perhaps no single advancement in observing techniques has done more for chromospheric and coronal physics than the computer.

Finally, in ending this introductory section, we remind the reader of a few basic facts concerning the Sun that will be helpful in placing the following discussion of the chromosphere and corona in proper perspective.

Solar radius $(R_\odot)$ = 6.96x10^{10} cm = 959.6″.
Surface area = 6.09 x 10^{22} cm^2.
Disk area = 1.52 x 10^{22} cm^2.
Mean Earth-Sun distance = 1.5 x 10^{13} cm = 215 $R_\odot$.
At mean distance 1″ = 725 km.
Surface gravity = 0.274 km s^{-2}.
Escape velocity at surface = 618 km s^{-1}.
Rotational velocity of surface at equator = 2.0 km s^{-1}.
Total radiative flux = 3.86 x 10^{33} erg s^{-1}.
Surface emittance = 6.35 x 10^{10} erg s^{-1}.
Effective temperature = 5785 K.

2. Goals for Solar Physics

It is reasonable for non-solar astronomers, as well as for society in general, to ask of solar astronomers just what goals they have in mind. How far should they carry the study of the Sun before they are content to say that they have done enough — that they have finished the job? It is difficult to obtain a consensus of opinion of such questions. Each solar astronomer justifies his own interest in the disciplines of solar physics in terms that are somewhat different from those developed by his colleagues. Similarly, each has a concept of the importance of solar physics in the overall frame work of science that is peculiarly his own, and each, therefore, would describe the goals of solar physics somewhat differently. Finally, solar physics is a rapidly developing and evolving science and this forces upon each solar physicist a continual reappraisal of where the science is headed and what its goals should be.

The following list of goals for solar physics is representative of the view held by the author at the time this book was written. They will receive, predictably, some support as well as some criticism from my colleagues. The goals of solar physics depend to some extent upon how one views the Sun: as a star, as the driving source for terrestrial and interplanetary phenomena, or as a laboratory for studies in plasma physics. The proposed goals will be stated in these contexts.

When one considers the Sun as a star a number of goals immediately come to mind:

(1) A one-dimensional model of temperature and density as a function of depth.
(2) A quantitative description of the effect and importance of discrete solar features on the one-dimensional model.
(3) A quantitative model of the dynamic Sun.
(4) An understanding of the phenomena of solar activity and of the solar cycle itself.
(5) An understanding of the equilibrium distribution of the chemical elements.
(6) An accurate theory of spectral line formation and a physically meaningful description of line broadening.

Under goal 3 are included such problems as those associated with the convection zone together with the resultant convective modes, wave modes, general circulation, dynamo action and differential rotation. Also included are the problems related to energetic particle production and to the transport of energy, momentum and mass within the chromosphere and corona and the eventual dissipation of the energy as heat. Goal 5 includes an understanding of the processes of stratification and mixing of the chemical elements and a definitive answer to the question as to whether the elements in the Sun are indeed 'well mixed' as is currently assumed. Goal 6 includes a description of line broadening in terms of physically described motions as opposed to the mystical 'turbulence' parameters now used and in terms of established damping constants as opposed to the current practice of simply adopting an 'excess damping'.

These goals are broadly stated and the list is incomplete. However, they provide a convenient scale against which we can measure our understanding of the physical phenomena of the star that is nearest Earth and that, by any measure, is the best understood of all stars. Surely stars, as a family of objects, will not be understood so long as we fail to achieve the above stated goals for solar physics.

When the Sun is viewed as a nearby neighbor who determines the space environment of Earth and who is responsible for a number of terrestrial atmospheric phenomena, new goals emerge. Two such goals are:

(1) The development of forecast skills for predicting solar activity together with its concomitant, short-term changes in XUV radiation, solar wind and energetic particles.
(2) An understanding of the evolutionary changes in solar activity with solar age.

The former goal is necessary for the practical utilization of our knowledge of solar physics in explaining and predicting the wide variety of terrestrial and space phenomena associated with the Sun. Such prediction is necessary for the intelligent use and planning of communication networks and space programs. The second goal is of interest in connection with possible solar influences on climatic change.

Perhaps the longest range goals of solar physics are those associated with the Sun as a plasma physics laboratory. Here, the goal, broadly stated, is to understand what we already observe and to learn new techniques for observing both the known events and possible new types of events. This latter point is perhaps best appreciated in terms of the rate of discovery of new solar phenomena. Major new phenomena are currently being discovered in events related to solar activity at the approximate rate of one per year. In quiet regions of the Sun the current discovery rate of major phenomena declines to approximately one per five years, but even this rate is impressively large. The clever observer can still reshape the direction of solar physics by newly designed observations. Such new observations often unveil a new set of plasma phenomena leading to revised ideas about the Sun itself and to new challenges in plasma physics. Clearly, one goal here is to use our knowledge of solar physics to point the way to observations of new solar phenomena and vice versa.

Having stated, albeit crudely, the goals of solar physics, we should attempt to state where we are relative to those goals. This can be done uniquely and precisely, of course, only after the goals are attained, which is not the case for any of the goals stated. Nevertheless, we can state perhaps with some confidence that we are within about five years of the realization of goal 1 for the Sun as a star. Recent advances in observations from the ground and the existing or planned XUV spectroscopic observations from space are capable of providing much of the data that are needed to fulfill this goal. Also, theory seems to be sufficiently advanced to warrant the speculation that when enough accurate data becomes available reliable models will emerge.

Goal 2 for the Sun as a star is currently held back by the struggle to achieve sufficient angular resolution to permit a proper study of solar features. Most of the known structures on the Sun display marked changes over spatial distances less the $1''$ in size. It seems evident that the final realization of goal 2 will be dependent upon our ability to achieve spatial resolution in both the visual and XUV of the order of $0.1''$. It seems equally evident that this goal will not be achieved within the next decade. Meanwhile, we must continue to do the best we can with the spatial resolution presently attainable.

The achievement of goals 3, 4, 5 and 6 for the Sun as a star is also held back by our current limitations on spatial resolution. Much of the energy, momentum and mass transport processes observable in the solar atmosphere are associated with such structures as granules, filigree, fibrils and spicules whose characteristic sizes are still somewhat beyond our resolving power. Although there is much reason to suppose that major advances in these areas will be made in the next few years, final attainment of our stated goals is dependent upon both ground and space observations with spatial resolution near $0.1''$.

The final realization of goal 4 for the Sun as a star is difficult to predict, partly because the rate of progress is accelerating so rapidly and because the rate of discovery of new phenomena is so rapid. Clearly, we are only beginning to understand the phenomena that are presently known, and, equally clearly, we are not yet observing all of the major phenomena. Progress in this area can be expected for many years to come. Barring some unexpected major breakthrough, there seems little hope of fully attaining this goal within the next decade.

The realization of the goals for the Sun viewed as an Earth neighbor is viewed in widely different terms by solar astronomers. One significant group maintains that predictability of solar activity and its associated radiations does not require understanding of the physical processes involved in producing the active events. They further maintain that such 'analog' prediction may, in fact, lead to better understanding of the events themselves. A second significant group of solar astronomers maintains an opposite view, viz., that it is unwise to expend effort in attempts to predict such events as solar flares and their associated particle and XUV radiations until there is a better understanding of what a flare is. There are elements of truth on both sides, but clearly, the realization of the solar predictability goal does not appear to be within the foreseeable future for either group.

Although it is meaningless to try to evaluate where we are in terms of studying the Sun as a plasma laboratory, it is clear that the future here looks to be richly endowed for many years to come in terms of both better observations of existing phenomena and the discovery of new phenomena. We will be limited only by our imagination and resources.

3. Chromosphere and Coronal Boundaries

The division of a gaseous atmosphere into layers with different names is nearly always a somewhat arbitrary process, the Sun being no exception. In fact, the problem of dividing the solar atmosphere into layers is compounded by the fact that the layers we wish to divide are not layers at all. They are instead assemblages of more or less discrete structures, filaments, loops, spicules, fibrils, condensations, etc. Whether or not the ambient medium in which these structures reside is important to the overall nature of the layers that we shall define is an unresolved question. The answer to the reverse question is clear, i.e., the discrete structures are vitally important in determining the 'average' properties of the layers. Thus, we could with some justification perhaps ignore the ambient medium, but we clearly cannot ignore the discrete structures. We shall, nevertheless, assume that the ambient medium is important and that it makes some sense to construct one dimensional models of temperature and density. Thus, to a given height in the atmosphere we shall assign a given temperature and density and we shall use the resulting temperature curve to define the chromospheric and coronal layers. For this purpose, we shall simply adopt a temperature curve and leave until later the discussion of how such a curve is obtained. We adopt such a curve being fully aware that its meaning is not completely clear but being aware, also, that we need points of reference for future discussion.

Figure I-1 shows a plot of log T vs height in km above the polar limb. The zero point

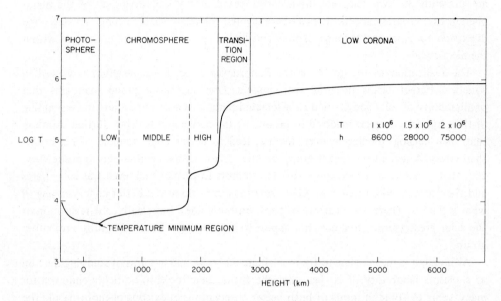

Fig. I-1.   An illustrative temperature model for the chromosphere, transition region and inner corona.

of the height scale is defined here as the point where the tangential optical depth in the continuum at λ 5000 is unity. We define the *radial* optical depth, τ_c, as

$$\tau_c = -\int_\infty^h k_c \, dh \qquad\qquad (I-1)$$

and the tangential optical thickness at the limb, τ_c^t, as

$$\tau_c^t = -\int_\infty^{-\infty} k_c \, dy, \qquad\qquad (I-2)$$

where y is the line of sight coordinate at the limb. The variables y and h are related by (Figure I-2)

$$y^2 = (h+R)^2 - (h_0+R)^2$$
$$= h^2 + 2hR - h_0^2 - 2h_0 R. \qquad\qquad (I-3)$$

For $h_0 \ll R$ and $h \ll R$, Equation (I-3) reduces to

$$y = (2R)^{1/2}(h-h_0)^{1/2}. \qquad\qquad (I-4)$$

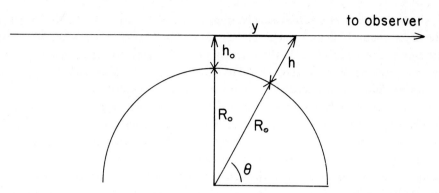

Fig. I-2.   Illustration of the geometrical relationship between the line-of-sight path length, y, at the solar limb, the height, h, and angle θ.

To a good approximation in the solar photosphere

$$k_c = k_0 e^{-h/H}, \tag{I-5}$$

where $H = 60$ km and is approximately constant with depth. Using Equations (I-4) and (I-5) one finds

$$k_c = k_0 \, e^{-y^2/2RH} e^{-h_0/H} \tag{I-6}$$

and from Equation (I-2) one finds

$$\tau_c^t = (2\pi RH)^{\frac{1}{2}} k_0 \, e^{-h_0/H} \tag{I-7}$$

Also, from Equations (I-1) and (I-5) one finds

$$\tau_c = k_0 H \, e^{-h_0/H}. \tag{I-8}$$

Thus,

$$\frac{\tau_c^t}{\tau_c} = \left(\frac{2\pi R}{H}\right)^{\frac{1}{2}} = 2.7 \times 10^2. \tag{I-9}$$

Since we have defined $h = 0$ by the condition $\tau_c^t = 1$, we find $\tau_c = 3.7 \times 10^{-3}$ at $h = 0$. The reader should note that many authors define $h = 0$ by the condition $\tau_c = 1$. According to Equation (I-5) the geometrical distance between $\tau_c = 1$ and $\tau_c = 3.7 \times 10^{-3}$ is 340 km.

The chromosphere, as defined in Figure I-1, extends through a height range of approximately 2000 km beginning near $h = 200$ km. The base of the chromosphere has been placed at the temperature minimum. At this point such an assignment is arbitrary, although we shall later advance physical arguments in support of it. Three regions of the chromosphere are identified in Figure I-1: a low chromospheric region of approximately 400 km extent in which there is an initial temperature rise; a middle chromospheric region in which the temperature rises slowly from about 5500° to about 8500° in a height interval of about 1200 km; and a high chromospheric region in which the temperature first rises sharply then more slowly to about 50 000°. Figure I-1 indicates a thickness for the upper chromosphere of about 500 km. The thickness of this region is open to serious dispute and should be regarded at this stage as being very tentative.

The corona extends upwards (Figure I-1) from a base near 3000 km. Again, the division between the corona and the transition region is largely arbitrary. Because of the high coronal temperature the corona is very extended. The maximum temperature in the corona is expected, for reasons to be given later, to occur at some tens of thousands of kilometers above the solar limb. Thermal conductivity in the corona is very high and prevents the temperature from falling off rapidly in either direction from the maximum.

In a hydrostatic, isothermal atmosphere the density decreases exponentially with a scale height H of

$$H = \frac{2kT}{\mu m_H g},$$
(I-10)

where μ is the mean molecular weight, m_H is the hydrogen mass and g is the gravitational acceleration. Assuming that the corona consists of fully ionized hydrogen and helium with helium being 10% as abundant as hydrogen, one finds that

$$\mu m_H = \frac{10m_H + m_{He}}{20 + 3} = \frac{14}{23} m_H = 0.61 \ m_H.$$

For $g = 2.7 \times 10^4$ cm s^{-1} and $T = 1 \times 10^6$ K, one finds $H = 1.0 \times 10^5$ km. This is one seventh of the solar radius and one expects, therefore, that the corona will extend far beyond the solar surface. For the moment, we will leave the outer limit of the corona unspecified.

Between the chromosphere and corona lies a region of unusual interest known as the transition region. It is defined in Figure I-1 as the region between the temperature limits of approximately 50 000° and 500 000°, but it has a total thickness of only about 500 km. Note that within this region both T and dT/dh change very rapidly with height. Thus, it is somewhat academic to talk about 'a thickness' for the transition region. The transition region, as well as the various chromospheric layers and the corona will be discussed in considerable detail in subsequent chapters. Here we wish only to draw attention to their essential distinguishing properties and to establish the approximate height scale.

The location of the transition region, or, equivalently, the base of the corona has been a subject of much interest and has udergone much evolution. Over the past few decades the supposed base of the corona has moved steadily downward starting from heights of 50 000 to 60 000 km. Its location is still somewhat uncertain and it undoubtedly varies greatly from point to point on the Sun. The curve in Figure I-1 should be regarded, therefore, as a mean temperature profile in which the location in height of the transition region is somewhat uncertain.

The model atmosphere illustrated in Figure 1-1 represents the quiet Sun well removed from centers of activity where such features as sunspots and plages occur. This is in keeping with our intent to discuss primarily the quiet Sun seen as a star. Just as it is somewhat artificial to divide the Sun into layers, it is somewhat artificial to divide it into 'quiet' and 'active' areas. The phenomena of the active Sun are not fundametally different from those of the quiet Sun. The difference is one of degree and of dominance.

Our emphasis on the quiet Sun is done in recognition of the needs of stellar astronomers to understand the more star-like aspects of the Sun without adding unnecessary complexity. Phenomena of the active Sun are perhaps more varied, more complex and less well understood than phenomena of the quiet Sun. To do justice to the active sun would greatly enlarge the scope of the current text without providing a proportionate level of help to the stellar astronomers.

4. Methods of Observation

The solar chromosphere and corona are vividly displayed at total solar eclipse. The lunar shadow on the Earth darkens the sky to such an extent that the corona appears as a large brilliant halo surrounding the dark lunar disk as shown in Figure I-3. In total brightness the corona is very comparable to the full Moon. However, there is a strong gradient in brightness through the corona itself and the inner corona is relatively much brighter. The coronal photograph displayed in Figure I-3 was taken through a neutral filter with a radial (to the center of the Sun) gradient in density such that the brightness of the image is more or less constant.

During an eclipse the lunar limb advances relative to the solar limb at a rather rapid rate. The period from new Moon to new Moon is 29.53 days. Thus, relative to the Sun, the Moon completes an apparent orbit in 29.53 days. This corresponds to an orbital velocity, for an orbit of radius 1 AU, of about 370 km s^{-1}. Thus, the Moon's limb advances over the solar surface during eclipse at an average rate of 370 km s^{-1}. Each eclipse is somewhat different, of course, due to variations in lunar motion and position.

For an eclipse in which totality lasts, say, 3 minutes, the Sun will advance relative to the Moon by about 60 000 km. Thus, the bright inner corona will be covered during a substantial part of totality but the majority of the corona will be visible throughout totality. The chromosphere, however, is visible only briefly. Only about 6 seconds are required for 2000 km of chromosphere to be covered (or uncovered) by the lunar limb. Thus, the lower regions of the chromosphere, which are the brightest, are seen only momentarily. This gives rise to the so-called 'flash-spectrum' just following the onset of totality and just preceding the end of totality.

The brief period during which the chromosphere and corona can be observed at eclipse and the often difficult circumstances under which eclipse observations are carried out makes eclipse observations costly, difficult and hazardous. Expeditions have been ruined

Fig. I-3.   A photograph of the total solar eclipse of June 30, 1973 (courtesy of High Altitude Observatory, a division of the National Center for Atmospheric Research, sponsored by the National Science Foundation).

by minor momentary equipment failures, by human errors committed under the steadily mounting tension as totality approaches and by a single small cloud obscuring the Sun in an otherwise clear sky. In addition to the difficulties inherent in eclipse observing, the fact that the chromosphere and corona are observed at the solar limb where the observed light rays come horizontally through the solar atmosphere adds greatly to the difficulty of interpreting the data. We note from Equation (I-9) that the effective path length of integration at the solar limb is 270 times that at disk center. Thus, if the chromosphere is made up of closely spaced features whose horizontal scale is of the order of the chromospheric thickness observations made at the limb will necessarily average many such features together. Only those features that are small in size and widely separated will be seen as individual features. This makes the knowledge of chromospheric geometry (inhomogeneities) of vital importance in the interpretation of eclipse data. We must know what kinds of features are being averaged in order to define what the average means.

In the corona the situation is not quite so bad because of the relatively large scale height. For a scale height of 10^5 km, which is approximately the coronal value, Equation (I-9) gives the ratio of effective horizontal scale to the vertical scale as 6.6. At large distances from the solar surface the ratio approaches unity. Nevertheless, in the inner corona we must still expect that a given observation represents an average over many discrete structures.

In spite of the difficulties of obtaining and interpreting eclipse observations the quality and quantity of data obtainable at eclipses have sustained eclipse observing as one of the primary sources of coronal and chromospheric data. The advantages of observing through a dark sky as opposed to a bright, sunlit sky and of using the advancing Moon's limb as an absolute reference height far outweigh the disadvantages.

The chromosphere can be (and has been for many years) observed against the solar disk by the simple expedient of isolating the light near the centers of the strongest Fraunhofer absorption lines. At the base of the chromosphere the continuum optical depth at λ 5000 is approximately 10^{-4}. Within a spectral line the shape of the absorption coefficient, ϕ_y, may be approximated by

$$\phi_y = e^{-y^2} + \frac{a}{\sqrt{\pi} y^2},$$ (I-11)

where $y = \Delta\lambda/\Delta\lambda_D$, $\Delta\lambda_D$ is the Doppler width, a is the damping parameter given by

$$a = \frac{\Gamma\lambda^2}{4\pi c \Delta\lambda_D}$$ (I-12)

and Γ is the damping constant due to both radiation and collision damping. (The damping term in Equation (I-11) is to be included only for $y > 1$ to avoid a singularity at $y = 0$.) In the high photosphere, radiation damping dominates over collision damping for strong transitions and a is typically about 10^{-3}. Thus, at three Doppler widths from line center the wing term in Equation (I-11), $a/\sqrt{\pi} y^2$, is beginning to dominate and ϕ_v is of the order of 10^{-4}. If a line has strongly developed wings, it must be that, at the depth $\tau_c = 1$, $\tau_0 a/\sqrt{\pi} y^2 \gg 1$ in the inner wing (τ_0 is the optical depth at line center). Thus, we may infer from the preceding argument that, at $\tau_c = 1$, $\tau_0 \gg 10^4$ in lines with strongly developed wings. Using the simple approximation that the line center is formed near $\tau_0 = 1$ and assuming that $\tau_0/\tau_c = $ const, we conclude that the centers of lines with strongly developed wings are formed at $\tau_c \ll 10^{-4}$, i.e., within the chromosphere. Many such lines are present in the visual solar spectrum.

Disk observations of the chromosphere provide the possibility of obtaining data for single isolated chromospheric features together with their changing aspect from center to limb. Data already obtained using the cores of the Hα line and the H and K lines of Ca II exist in abundance. Figure I-4 shows an Hα image of a portion of the chromospheric disk. Such observations will be further illustrated in Chapter II. Here, however, we encounter another difficulty. Interpretation of spectroscopic data in the cores of strong lines has proven to be exceedingly difficult (see Chapter V). Only since the advent of large computers has significant progress been made in the quantitative use of such data. Thus, the primary use of the disk data in the visual has been to detect, classify and relate the various structural details that are observed.

Fig. I-4.   A photograph of the chromosphere near the solar limb using a 1/4 A birefringent filter center at + 5/8 A from line center. (Courtesy R. B. Dunn, Sacramento Peak Observatory, Air Force Cambridge Research Laboratories.)

The corona is far more transparent than the chromosphere in the visual spectrum and its radiations are weaker. For this reason, disk observations of the corona in the visual spectrum have not been notably successful. In the extreme unltraviolet (EUV) and at

X-ray wavelengths both the chromosphere and corona have appreciable optical thickness in certain strong lines. Also, the photosphere is much fainter than in the visual. These combined effects make it possible to observe the chromosphere and corona against the solar disk at EUV and X-ray wavelengths. An X-ray image of the disk corona is shown in Figure I-5. Observations on the solar disk at these wavelengths from rockets and satellites have added greatly to our knowledge of the temperature and density structure of the

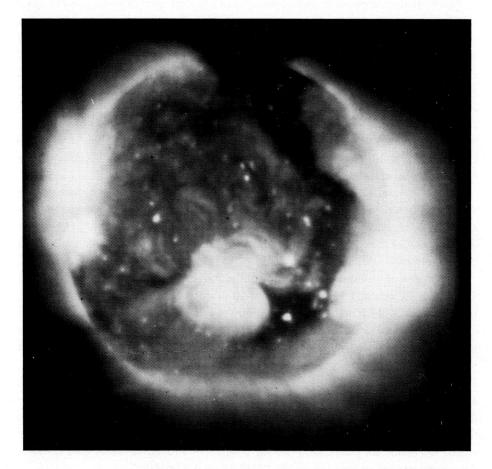

Fig. I-5.   A soft X-ray image of the solar corona at the time of the June 30, 1973 eclipse obtained by the AS&E X-ray Telescope on Skylab. The filter used has a transmission greater than 1% for wavelengths in the range of 3.5–32 Å and 44–54 Å. Active regions are seen as bright, compact loop structures at the limbs of the Sun and near the center of the disk. Small intense bright features, called bright points, are seen distributed over the entire Sun. The 'quiet' corona is seen to consist mainly of diffuse loops which are the remnants of activity. The open region to the right of the picture, which is surrounded by apparently diverging quiet coronal formations, is called a coronal hole.

chromosphere and corona. Although the interpretation of these data bears some of the same difficulties that are present in the case of the strong lines in the visual spectrum, in many cases simpler analyses are justified and are possible. These simple analyses, to be discussed in Chapters VI and VII, have been highly fruitful.

In the opposite end of the spectrum, viz., at radio wavelengths, the chromosphere and corona again become opaque making observations on the solar disk possible. Once again these data have proven very valuable and have contributed much to our understanding of the chromosphere and corona.

Size limitations imposed by radio telescopes and stability limitations imposed by space vehicles have not permitted the attainment of spatial resolution comparable to that obtained in the visual spectrum except in a few isolated cases. Thus, the radio data and XUV and X-ray data have suffered from the same limitations as eclipse data, viz., they exhibit only the coarser structural features and they average over the small discrete features. It is within current capabilities, however, to achieve spatial resolution in the XUV and X-ray regions of the spectrum that equal or exceed the best that has been achieved in the visual spectrum. Assuming the continuation of our space programs, we may anticipate that within the next one or two decades we will have profiles of numerous chromospheric and coronal lines for each type of feature identified on the solar surface with spatial resolution of the order of $1''$.

At XUV wavelengths shortward of $\lambda 1600$ the chromosphere is opaque in the continuum as well as in numerous lines. Thus, at these wavelengths the chromospheric emission is uncontaminated by photospheric light. This simplifies, somewhat, the interpretation of the data, and it makes the XUV data of great value in chromosphere and coronal research.

Near $\lambda 1000$ the continuum brightness of the Sun is about 10^{-6} of the brightness near $\lambda 5000$. The corona at $1.1 R_0$ is also about 10^{-6} as bright as the visual solar disk. Thus, the brightness of the inner corona near $\lambda 5000$ at eclipse and the brightness of the solar disk at $\lambda 1000$ (above the Earth's atmosphere) are about the same. The brighter emission lines of the corona may be 10^2 to 10^3 times as bright as the continuum, per unit wavelength, and are easily observed above the continuum.

On an exceptionally clear day on top a mountain whose elevation is more than 9000 ft above sea level the sky brightness near the Sun may be as low as 10^{-5} as bright as the solar disk. Under such conditions and with a carefully designed telescope the brighter emission lines in the visual spectrum can be observed. Also, the continuum radiation from the corona has significant polarization and the polarization properties can be used to detect the coronal continuum even though the sky itself is much brighter.

Ordinary telescopes introduce a relatively high level of both scattered light and polarization. Attempts to observe the corona outside of eclipse with such telescopes are fruitless. However, by carefully eliminating the scatttered light, as is done in coronagraphs, the spectral lines are readily observed. Also, by carefully accounting for and minimizing the polarization due to the telescope itself the polarized coronal continuum can be successfully detected. Instruments of this type are called white-light coronagraphs. or K-coronameters. They have been successfully operated by the High Altitude Observatory from the tops of high mountains in Hawaii for a number of years.

5. The K- and F-Coronas

The corona we have discussed thus far is that which is an integral part of the Sun itself — the K-corona. There is another aspect of the corona — the F-corona — that arises from the peculiar scattering properties of interplanetary dust grains. Small dust grains whose diameters are of the order of 1 μm in size produce strong forward scattering of visual radiation without pronounced color characteristics.

If we assume that most of the radiation incident on a dust grain is scattered into a narrow cone in the forward direction, a single dust grain of 1 μm diameter in each square cm column from Sun to Earth will produce a sky brightness near the Sun of about 10^{-8} of the disk brightness. Scattering by such particles essentially reproduces the incident radiation including the Fraunhofer absorption lines. The presence of the Fraunhofer lines in this component of the corona gives rise to its designation as the F-corona.

In the K-corona the continuum is produced by electron scattering. The electrons have a mean velocity of

$$\left(\frac{2kT}{m_e}\right)^{1/2} \approx 8 \times 10^8 \text{ cm s}^{-1}$$

corresponding to a mean Doppler shift of the scattered radiation by an amount $\Delta\lambda_D/\lambda \approx 0.027$. Hence, the Fraunhofer lines are broadened some 100 Å and are no longer detectable. An exception occurs in the case of the H and K lines of Ca II whose combined equivalent widths total some 35 Å. These lines have been reported as broad shallow depressions in the K-corona by some authors but others have failed to detect them.

The relative brightnesses of the K- and F-coronas are illustrated in Figure 1-6. Note that beyond $R = 2.2 R_C$ the F-corona is brighter than the mean K-corona. This does not mean that the K-corona becomes invisible at this distance. Indeed, the outer K-corona is composed mostly of filamentary streamers whose brightness is markedly higher than the average K-corona brightness and these brighter features continue to be visible for distances of several radii.

We will not concern ourselves further with the F-corona in this book. Henceforth, we refer to the K-corona simply as the corona, as in the preceding sections.

6. Comments on Discrete Geometrical Features

Suppose, for the sake of argument, that a loop, or arch, structure similar to that illustrated in Figure I-7, exists in the solar atmosphere. The cross sectional diameter of the arch is d and within the arch the particle density differs from the ambient density by Δn and the temperature differs from the ambient temperature by ΔT. Matter is supposed to stream along the arch, rising in one leg and falling in the other.

The chemical composition of the gas within the arch is assumed to be identical to that in the ambient atmosphere. However, because of the differences in temperature and density between the arch and the ambient atmosphere the state of excitation within the

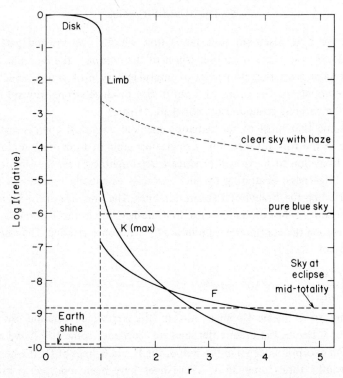

Fig. I-6.   Brightness of the K- and F-continuum coronas and the summed brightness of visual coronal forbidden lines (E) relative to the solar disk and adjacent sky brightness. The abscissa is in units of the solar radius (van de Hulst, 1953, courtesy University of Chicago Press).

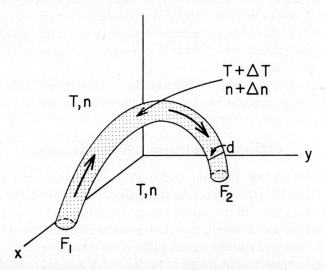

Fig. I-7.   A schematic illustration of an arch of material in the solar atmosphere whose temperature and density differ from the ambient temperature and density by ΔT and Δn, respectively. Material is supposedly flowing along the arch from F_1 to F_2. (See text for discussion.)

arch will differ from that in the ambient atmosphere. This difference, together with the difference in density and the motion of the material, will cause the spectral absorptivity and emissivity in the arch to differ from that of the ambient atmosphere. Thus, the arch will be visible if observed spectroscopically with sufficient care. Several questions about its visibility arise however. 'What will the arch look like?' 'Will it be dark or bright?' 'Will it look the same in all spectral lines?' 'What apparent diameter will it have?'

These questions, and others that could be asked, are difficult to answer quantitatively. A ray of light of intensity I_ν passing diametrically across the arch will be attenuated by an amount $I_\nu\, k_\nu\, d$ and enhanced by an amount $j_\nu\, d$, where k_ν and j_ν are the absorptivity and emissivity, respectively, and where it is assumed that $k_\nu\, d \ll 1$. In general, $I_\nu\, k_\nu$ and j_ν will not be equal in magnitude, but it is possible, in fact, likely, that $I_\nu\, k_\nu$ will exceed j_ν in some lines and be less that j_ν in other lines. Thus, the arch may appear dark in some lines and bright in others. Even within a given spectral line, the same might be true, i.e., the arch could appear dark in, say, the line wings and bright in the line core.

If ΔT and Δn are changing with time, the visibility of the arch will change with time also. It may happen, as is often observed in some solar features, that the arch appears bright against the solar surface during part of its lifetime and appears dark against the surface during other parts of its lifetime even though it is observed in precisely the same manner each time.

It is a common practice in solar observing to observe the chromosphere with narrow band filters, whose width is a fraction of a spectral line width. If such a filter is used to observe the arch in Figure I-7, it is obvious that the 'visibility' of the arch in a given spectral band will vary with position in the arch. The material in the arch near F_1 is rising and will absorb and emit preferentially on the violet side of the line. Just the opposite is true of the material near F_2. Thus, observations made alternatively on opposite sides of the line will very likely show different structural details.

Efforts to observe and identify the arch structure depicted in Figure 1-7 are complicated by factors other than those mentioned above. The diameter d is defined with respect to the locations of discontinuities in temperature and density. Photons are not so restricted, however. They cannot be made to follow discontinuities quite so readily as material particles. If $K_\nu\, d \gg 1$, material within the arch will generate photons, which will diffuse into the ambient atmosphere surrounding the arch. Also, photons that are generated, or scattered, in the ambient atmosphere will diffuse into the arch. These processes will further change the state of excitation in both the arch and in the atmosphere surrounding the arch. If the radiative excitation and de-excitation rates in spectral lines exceed the collisional ones, as they do throughout much of the solar atmosphere in spectral lines of moderate or high strength, the discontinuities in n and T will not necessarily show as discontinuities in the radiative intensity I_ν. Instead, the arch will tend to have a radiative brightness cross-section following something akin to a Gaussian curve of characteristic width, b. If j_ν and k_ν in the ambient medium are of similar magnitude to j_ν and k_ν in the arch, the width b will be determined by the characteristic diffusion length of a photon, i.e., the distance a photon diffuses before it reconverts to thermal energy.

This diffusion length is called, in radiative transfer theory, the thermalization length (cf. Chapter V; also Jefferies, 1968; Athay, 1973). It is characteristically large in the solar chromosphere, sometimes in excess of 2000 km, or 3″. It is also characteristically different in different lines. Hence, if d is less than or near b the 'observed' diameter of the arch will be more nearly equal to b and may differ appreciably from line to line.

The preceding arguments are advanced at this point in order to illustrate the complexity of identifying and labelling discrete geometrical structures observed in the solar atmosphere. The same structure takes on a different appearance when observed in different spectral lines, or with a different pass band within a given line, or at different times. It is perhaps not too surprising, therefore, that solar physicists find it difficult to arrive at a standard set of names and classes for the variety of structures seen in monochromatic images of the Sun. The full list is bewilderingly long, highly redundant, constantly added to and accepted by virtually no one. Unfortunately, it is easy to introduce new names along with new observations but it is almost impossible to discard old names, even though they may be attached to the same basic phenomena. Early solar observers were content with quaint names such as, mottles, rice grains, floculli and burning priaries. Modern observers equipped with vastly improved instruments find such names old fashioned and not sufficiently descriptive so they invent their own set of equally quaint names; rosettes, fibrils, bushes, porcupines and even blobs. The names used in this book will follow the more recent usage in the majority of cases. Before describing the various chromospheric and coronal features in the following chapter, we digress temporarily in order to discuss the problem of identifying the particular depths where spectral lines are formed. The depths of line formation are important, of course, for determining geometrical structures and height gradients of different quantities.

7. Depths of Line Formation

Assignments of heights to features or quantities observed in the solar atmosphere are often done on the basis of a 'contribution function' or on the basis of the Eddington-Barbier relation. To illustrate these assignments, consider a ray of light of intensity I_ν traveling in direction θ with respect to the vertical (see Figure I-8) and which satisfies the relation.

$$\frac{dI_\nu}{ds} = -I_\nu k_\nu + j_\nu. \tag{I-13}$$

Since

$$ds = \frac{dh}{\cos\theta} = \frac{dh}{\mu} \tag{I-14}$$

and since we define $d\tau_\nu$ as

$$d\tau_\nu = -k_\nu \, dh, \tag{I-15}$$

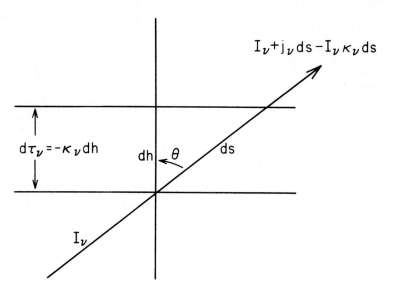

Fig. I-8.   Illustration of the change in specific intensity I_ν by a ray in direction θ traversing an infinitesimal plane parallel layer of thickness dh. Within the layer the absorbtivity is k_ν and the emissivity is j_ν. The optical thickness of the layer is $d\tau_\nu = -k_\nu dh$.

I_ν satisfies the relation

$$\mu \frac{dI_\nu}{d\tau_\nu} = I_\nu - S_\nu,$$

(I-16)

where

$$S_\nu \equiv \frac{j_\nu}{k_\nu}.$$

(I-17)

The solution of Equation (I-17) outside the atmosphere is

$$I_\nu(0) = \int_0^\infty S_\nu e^{-\tau_\nu/\mu} \, d\tau_\nu/\mu.$$

(I-18)

The notation $I_\nu(0)$ denotes $I_\nu(\tau_\nu)$ at $\tau_\nu = 0$.

We here define the contribution function as the integrand in Equation (I-18). For illustration we take S_ν to be of the form

$$S_\nu = a_\nu + b_\nu \tau_\nu.$$

(I-19)

The curves in Figure I-9a show plots of

$$\frac{1}{a_\nu} \frac{dI_\nu(0)}{d\tau_\nu} = \frac{S_\nu}{a_\nu} e^{-\tau_\nu}, \qquad \mu = 1,$$

(I-20)

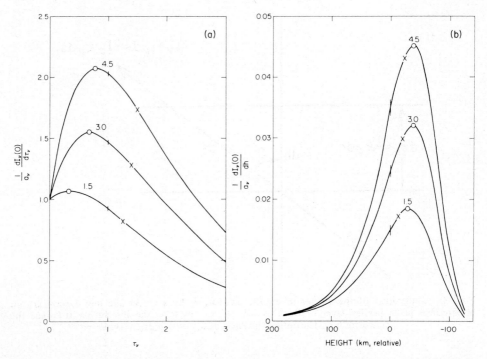

Fig. I-9.   Contribution functions for three values of the ratio b_ν/a_ν defined by Equation (I-19). Part *a* shows the contribution functions on an optical depth scale and part *b* shows the corresponding contribution functions on a geometrical depth scale.

for b_ν/a_ν = 1.5, 3.0 and 4.5. The values of b_ν/a_ν illustrated are representative of the solar continuum at the approximate wavelengths λ 6000 (1.5), λ 4700 (3.0) and λ 4000 (4.5).

Figure 1-9b shows plots of the quantity

$$\frac{1}{a_\nu}\frac{dI_\nu(0)}{dh} = \frac{1}{a_\nu}\frac{dI_\nu(0)}{d\tau_\nu}\frac{d\tau_\nu}{dh} = \frac{S_\nu}{a_\nu}\frac{\tau_\nu}{H}e^{-\tau_\nu}, \qquad \mu = 1 , \tag{I-21}$$

where it is assumed that

$$\tau_\nu = e^{-h/H}.$$

This is the same contribution function as defined by Equation (I-20) except that it is now defined with respect to geometrical height rather than optical height. Note that the reference for height zero is now $\tau_\nu = 1$ rather than the solar limb as used previously. In Figure I-9b H is taken to be 60 km.

Each of the six curves in Figure I-9 is marked at three points: the location of maximum visibility (0); the location of $S_\nu = I_\nu(0)$ (1); and the location above which half of $I_\nu(0)$ is formed (X). As the slope of S_ν increases the depths of maximum visibility on the τ_ν plot move to progressively larger τ_ν, but on the height plot the corresponding effect is not very evident. Note that the strongly visible layers extend from about −100 km to about

80 km, i.e., through about three full scale heights in optical depth or about one-half the total thickness of the photosphere.

The curves in Figure I-9, which may be regarded as 'visibility' curves, are valid for the case where k_ν is constant across the pass band of the observing instrument. One often measures velocity shifts or observes structural features in the solar atmosphere with instruments whose effective spectral band pass is of the order of the Doppler width in the line absorption coefficient. Assume that this is the case and that the band is located in the steep edge of an absorption line. If one edge is at $\Delta\lambda_1 = 0$ and the other at $\Delta\lambda_2 = \Delta\lambda_D$, the line absorption coefficient, which is proportional to $e^{-(\Delta\lambda/\Delta\lambda_D)^2}$ near line center, will change by a factor of e across the spectral band pass. This will broaden the curves in Figure I-9 by a significant amount. Furthermore, if the pass band is set between $\Delta\lambda_1 = \Delta\lambda_D$ and $\Delta\lambda_2 = 2\Delta\lambda_D$, as is more nearly in keeping with observing practices in strong lines, the value of k_ν for the line changes by a factor $e^{-1}/e^{-4} = e^3$. In this case the depth of maximum visibility on one side of the pass band is removed from that on the other side by 180 km (three scale heights), and the curves in Figure I-9 will then show strong visibility over a height interval of approximately 300 km or greater.

In many spectral lines the situation is even worse than is indicated in the preceding discussion. The scale height of 60 km is a good value for the continuum opacity, but in many lines the scale height is considerably larger than 60 km. This is illustrated in Table I-1, which is extracted from the Harvard-Smithsonian Reference Atmosphere (Gingerich et al., 1971). Note that the gas pressure and density each have scale heights that are approximately double that for τ_c.

TABLE I-1
Relative change with depth of continuum optical
depth, pressure and density in the photosphere

Height (km)	τ_{5000}	P_g	P_e	ρ cm^{-3}
−340	1	1.4×10^5	5.7×10^1	3.4×10^{-7}
−205	10^{-1}	5×10^4	3.6	1.5×10^{-7}
− 60	10^{-2}	1.4×10^4	8.0×10^{-1}	4.7×10^{-8}
80	10^{-3}	3.5×10^3	2.4×10^{-1}	1.2×10^{-8}

For an ion that remains at a constant relative abundance throughout the photosphere, which is very nearly true for most singly ionized metal ions, the lines originating near the ground state will show opacity scale heights more nearly like those of total density. Thus, the line opacity will have a scale height approximately double that of the continuum opacity.

Neutral metals are heavily ionized in the photosphere, but because of the outwards decrease of temperature they tend to increase in relative abundance as τ decreases. This effect is strong in some elements and weak in others. In the chromosphere the abundance tends to decrease again because of the temperature rise. These effects are illustrated in Table I-2 for Fe I (Athay and Lites, 1972) and Na I and Mg I (Athay and Canfield, 1969). For $\tau_c \gtrsim 10^{-3}$ the three elements satisfy the conditions of LTE for their ionization. Note that for $\tau_{5000} > 10^{-2}$ the Fe I opacity for ground state transitions decreases less rapidly than ρ. For Na I and Mg I, however, the rate of decrease is nearly the same as for ρ, i.e., about half the continuum rate. For $\tau_{5000} < 10^{-3}$, the opacities for the ground state transitions in Fe I and Mg I decrease a little more rapidly than τ_{5000}, whereas for Na I the rate is again slower.

TABLE I-2
Relative optical depths in lines and continuum

Element	Fe I		Na I		Mg I	
λ	3720	8688	5890	8194	2852	5184
Lower Excit. Pot.	0	2.2	0	2.1	0	2.7
τ_{5000}			τ_0			
1	2.6×10^5	1.3×10^3	1.2×10^4	7.6×10^2	3.0×10^6	4.4×10^3
10^{-1}	1.8×10^5	7.0×10^2	5.1×10^3	1.8×10^2	1.2×10^6	8.6×10^2
10^{-2}	3.7×10^4	1.0×10^2	1.4×10^3	3.8×10^1	2.8×10^5	1.5×10^2
10^{-3}	5.5×10^3	1.5×10^1	3.6×10^2	8.5	5.0×10^4	1.9×10^1
10^{-4}	3.0×10^2	1.5	8.1×10^1	1.4	5.3×10^3	4.5
10^{-5}	1.5×10^1	1.6×10^{-1}	1.3×10^1	1.4×10^{-1}	2.2×10^2	4.7×10^{-1}

Optical depths for lines arising on the excited levels given in Table I-2 behave somewhat differently from those for lines from the ground states. For τ_{5000} between 1 and 10^{-2} the subordinate line opacity decreases more rapidly than the resonance line opacity. This continues to be true for τ_{5000} between 10^{-3} and 10^{-5} for Na I, but for Mg I and Fe I the relative rates between subordinate and resonance lines reverse. The different behaviour is a result of atomic structure and the particular energy levels chosen for illustration. For both Fe I and Mg I the excited levels shown are metastable and are populated relative to the ground state following a Boltzmann distribution. For Na I, however, the excited level is the upper level of the D_2 transition and at small values of τ_{5000} the population of the excited level departs from the Boltzmann distribution. Excited levels of Fe I and Mg I that are radiatively coupled to the ground state, in fact, behave similarly to the excited level for Na I illustrated in Table I-2.

The conclusion to be drawn from Tables I-1 and I-2 is that the contribution functions for metal lines of low excitation potential that are formed in the lower photosphere are very broad and approach the thickness of the photosphere itself. It becomes difficult, therefore, to say that one line is formed near $\tau_{5000} = 1$ another near $\tau_{5000} = 10^{-1}$ and still another near $\tau_{5000} = 10^{-2}$. Two lines in which the optical depth ratio is 10:1 clearly will differ in their mean height of formation, but equally as clearly they will overlap strongly in the layers of the atmosphere that contribute to their emission. Thus, while the lines provide the means of probing the atmosphere in depth, they provide a rather poor discrimination in height. The best we can really hope to do is to discriminate between the low, middle and high photosphere. Similarly we can discriminate between the low, middle and high chromosphere, but any finer division raises strong possibilities of ambiguity in interpretation.

The preceding discussion is applicable to observations made near $\mu = 1$. Since the contribution function contains the factor $e^{-\tau/\mu}$, the width of the contribution function will depend somewhat upon μ. The effect of including $\mu \neq 1$ is equivalent to changing the ratio of b_ν/a_ν to $\mu b_\nu/a_\nu$. It may be seen from inspection of Figure I-9 that the width of the contribution function, in fact, is relatively independent of b_ν/a_0 for the range of values represented. Thus, we expect the μ dependence to be small as long as μ is not too small, say, $\mu \geqslant 0.2$.

In the Eddington-Barbier relation one sets the mean height of formation at $\tau_\nu = \mu(\tau_\nu = 1$ or $h = 0$ in Figure I-9). This follows from integration of Equations (I-18) and (I-19), which gives

$$I_\nu(0) = a_\nu + b_\nu\mu. \tag{I-22}$$

Note that, in general, the depth $\tau_\nu = \mu$ does not coincide with either the depth of maximum visibility or with the depth dividing the atmosphere into two equally contributing halves. The Eddington-Barbier relation ought, therefore, to be used with considerable caution and again is useful only for very coarse assignments of heights.

The illustrations in Figure I-9 are appropriate only for the linear form of $S_\nu(\tau_\nu)$ (Equation (I-19)). This, in turn, is a reasonable approximation only when τ_ν is itself approximately linear in τ_c. It is often the case in spectral lines that in some parts of the line τ_ν may be either a very rapid or a very slow function of τ_c. In these cases, which happen most often in the outer parts of the Doppler core, Equation (I-19) may be a very poor approximation, and the Eddington-Barbier relation may become either grossly in error or improved depending upon the relative behavior of τ_ν and τ_c.

Consider, for example, a case where the Doppler width increases by a factor of two between two depths τ_{c1} and τ_{c2} with $\tau_{c2} = 10\tau_{c1}$. Also, let the ratio τ_0/τ_c be constant with τ_c. Then at some intermediate fixed value of $\Delta\lambda$ (λ_a) the value of y will decrease by a factor of two between τ_{c1} and τ_{c2}. At $y_1 = 2.5$ the value of y_2 will be 1.25. The corresponding values of Φ_y will be $\Phi_{y_1} = 1.9 \times 10^{-3}$ and $\Phi_{y_2} = 0.21$. Thus, $\Phi_{y_2}/\Phi_{y_1} \approx 110$, and we see that Φ_{λ_a} varies approximately as τ_c^2. This means that if the line opacity is large compared to τ_c, $\tau_{\lambda_a} \propto \tau_c^3$ and Equation (I-19) should be replaced by

$$S_{\lambda_a} = a_\lambda + b_\lambda \, \tau_{\lambda_a}^{1/3} \tag{I-23}$$

If we now reverse the situation and let $\Delta\lambda_D$ decrease by a factor of two between τ_{c1} and τ_{c2}, we find $\Phi_{\lambda_a} \propto \tau_c^{-2}$. Thus, there is very little addition to the line opacity between τ_{c1} and τ_{c2} and τ_{λ_a} will be a very slow function of τ_c, if τ_{λ_a} is large compared to τ_c. This means that S_{λ_a} will be a very strong function of τ_{λ_a}, say,

$$S_{\lambda_a} = a_\lambda + b_\lambda \tau_{\lambda_a}^n, \qquad n \gg 1. \tag{I-24}$$

Figure I-10 contains plots of the contribution functions for Equations (I-23) and (I-24) with $n = 3$. The scale height for τ_c is again taken as 60 km for the plots in Figure I-10b, and $b_\nu/a_\nu = 3.0$ in all cases. This gives, for the corresponding scale heights of τ_{λ_a}, 20 km for the $\tau_{\lambda_a}^{1/3}$ case and 180 km for the $\tau_{\lambda_a}^3$ case. It is readily shown that for large n the maxima of the two contribution functions shown occur at $\tau_{\lambda_a} = n$ and at $h = -H \ln (n + 1)$, where H is the scale height of τ_{λ_a}. If we replace H by nH_c, where H_c is the

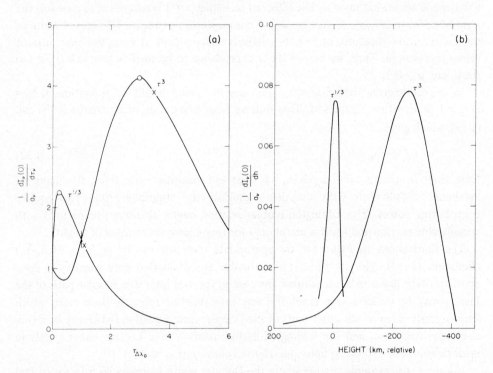

Fig. I-10.   Contribution functions for two values of the parameter n defined by Equation (I-25) for $b_\nu/a_\nu = 3.0$. Part a shows the contribution functions on an optical depth scale and part b shows the corresponding contribution functions on a geometrical height scale.

continuum scale height, the maximum occurs at $h = -nH_c \ln (n + 1)$. For $H_c = 60$ km, this give maxima at $h = -540$ km for $n = 5$ and at $h = -1540$ km for $n = 10$. Such large values of n cannot be immediately dismissed. The opacity in spectral lines can vary markedly with respect to the continuum opacity. Height zero, in this case, is defined as the height where $\tau_{\lambda_a} = 1$, and, in no case, of course, does the height of maximum contribution lie below the height of maximum contribution in the continuum.

Variations in Doppler width are not the only causes for τ_{λ_a} varying much differently than τ_c. A particular absorbing specie such as molecules or neutral metals may exist preferentially in the upper photosphere near the temperature minimum. Similarly, another absorbing process, such as hydrogen Balmer lines or the subordinate helium lines, may exist in appreciable strenth only in the chromosphere and the deep photosphere. Thus, for this latter case, the temperature minimum region may add practically nothing to the opacity and, in this region, τ_{λ_a} is a very weak function of τ_c. In the opposite case where the opacity is heavily concentrated in the temperature minimum region, τ_{λ_a} will be a strong function of τ_c near the temperature minimum and a very weak function of τ_c at other depths.

Note that when τ_{λ_a} is a strong function of τ_c as in Equation (I-23) the contribution function is sharply peaked near $h = 0$ (defined by $\tau_{\lambda_a} = 1$) and becomes increasingly narrow as the scale height of τ_{λ_a} decreases, i.e., as n decreases. By contrast when τ_{λ_a} is a slow function of τ_c (n increasing) the contribution function becomes very broad and maximizes far below $h = 0$. The Eddington-Barbier relation improves, therefore, for small n provided one uses it to define geometrical height. It is still a rather poor tool for picking the value of τ_{λ_a} of maximum contribution. For large n, the Eddington-Barbier relation gives poor results for either the height or optical depth of maximum contribution.

Actually, we have not been completely consistent in substituting $\tau_{\lambda_a}^n$ for τ_c in Equation (I-19) since we should simultaneously change the ratio b_ν/a_ν. In particular, Equation (I-19) with $b_\nu/a_\nu = 3$ implies that S_ν becomes constant for $\tau_c \lesssim 0.03$. Thus, to be consistent we should decrease b_ν/a_ν markedly for the cases we are considering where $\tau_{\lambda_a} \gg \tau_c$. However, Equation (I-19) is a reasonable approximation to the solar case only when τ_c is not too small. For the Sun, for example, The Harvard-Smithsonian Reference Atmosphere (Gingerich et al., 1971) shows the Planck function at λ 5000 decreasing by a factor 1.4 between $\tau_c = 10^{-3}$ and 10^{-4} and by a factor 1.5 between $\tau_c = 10^{-2}$ and 10^{-3}. Also, the fact that the strong Fraunhofer lines are not flat-bottomed and that they continue to show limb-darkening even at line center clearly requires that S_ν is a function of τ even at very small τ_c. Thus it would be unrealistic to decrease b_ν/a_ν by very much. On the other hand, the linear form of S_ν is perhaps equally unrealistic in the high atmosphere, so the results in Figure I-10 should be taken only as being indicative of the kinds of effects that may occur and not as an illustration of a real case.

It should be noted, also, that the large difference in width between the two contribution functions in Figure 1-10b arises primarily from the difference in scale heights. The widths are relatively the same, in fact, in units of the scale height.

The preceding discussion of the contribution function assumes that the line and continuum are formed more or less together in the sense that the ratio of line to continuum opacity is not too strongly varying. This is equivalent to the Milne-Eddington atmosphere, and the line is treated by asking at what depths the observed intensity arises. It is possible to use an entirely different approach, as many authors do (cf. Jefferies, 1968), and consider the line as simply a 'bite' taken out of the continuum. One then asks, 'at what depths is the bite taken out?' This is a quite different question than asking where the observed intensity arises, and it implies that the atmosphere is of the Schuster-Schwarzschild type.

For this second approach, one defines the contribution function as being proportional to $I_c(\tau_c) - I_\nu(\tau_c)$, where the intensities refer to those in the outward direction ($\mu > 0$) at the local value of τ_c. The quantity $I_c(\tau_c) - I_\nu(\tau_c)$ is given, at $\mu = 1$, by

$$I_c(\tau_c) - I_\nu(\tau_c) = e^{\tau_c} \int_{\tau_c}^{\infty} B_\nu e^{-t} dt$$

$$- \exp\left(\int_{1}^{\tau_c} x_\nu \, dt\right) \int_{\tau_c}^{\infty} S_\nu \exp - \left(\int_{0}^{\tau_c} x_\nu \, dt\right) x_\nu \, d\tau \qquad (I\text{-}25)$$

where B_ν is the Planck function and

$$x_\nu = \frac{d\tau_\nu}{d\tau_c}. \qquad (I\text{-}26)$$

For the case $S_\nu = B_\nu = 1 + (b_\nu/a_\nu)\,\tau_c$ and $x_\nu = \mathrm{const}$, Equation (I-25) reduces to

$$I_c(\tau_c) - I_\nu(\tau_c) = 1 + \frac{b_\nu}{a_\nu}(\tau_c + 1) - \left[1 + \frac{b_\nu}{a_\nu}\frac{1}{x_\nu}(\tau_\nu + 1)\right] = \frac{b_\nu}{a_\nu}\left(1 - \frac{1}{x_\nu}\right).$$

$$(I\text{-}27)$$

Note that this latter form is independent of tau, and therefore that Equation (I-25) loses its meaning as a contribution function for this particular case. This provides an illustration, however, where the integrand in Equation (I-18) represents a meaningful contribution function, whereas Equation (I-25) does not. On the other hand, the case where x_ν is identically unity everywhere except in the interval between τ_{c1} and $\tau_{c1} + \Delta\tau_c$, and where x_ν departs from unity within this interval by a small amount and where both τ_{c1} and $\tau_{c1} + \Delta\tau_c$ are small compared to unity, provides another extreme. In this case, Equation (I-25) maximizes near τ_{c1}, i.e., near the upper limit of the line forming layer, whereas the integrand in Equation (I-18) still maximizes near $\tau_c = 1$. Thus, this case provides an illustration where Equation (I-25) is clearly a better definition of the contribution function than the preceding definition. In the intermediate weak line cases in which x_ν differs from unity and is a reasonably strong function of height, it is not at all clear which of the definitions offers a more meaningful assignment of heights to the line forming layers. The two definitions obviously may yield heights that differ by large amounts.

For cases where x_ν is a relatively weak function of τ_c it seems preferable to adopt the first definition of the contribution function. This situation, in fact, covers most of the Fraunhofer lines. There are cases, clearly, where the second definition of the contribution function is preferable, but cases where this can be clearly demonstrated are rare.

For strong lines in which $\tau_\nu \gg \tau_c$ the two alternate definitions of the contribution function lead to very similar results, and it makes relatively little difference which one is used. The problem of choosing one or the other is critical for interpretation of the weaker lines, however.

So far as is currently known most of the weak to medium strong Fraunhofer lines are reformed more or less coexistently with the continuum, i.e., x_ν is a relatively modest function of height. For such lines a crude estimate of where optical depth unity is located can be made from the residual intensity in the line in combination with the limb-darkening coefficient in the continuum. For example, the continuum intensity limb-darkens approximately as

$$\frac{I_c(\mu)}{I_c(1)} = \frac{1 + \dfrac{b_\nu}{a_\nu}\mu}{1 + \dfrac{b_\nu}{a_\nu}} . \qquad (I-28)$$

Assuming the same source function for the line and continuum, we may identify a given residual intensity in a line with a given value of μ. Then, assuming the Eddington-Barbier relation, we equate μ and τ_c. Thus, a residual intensity of $r_{\Delta\lambda}$ corresponds to a value of τ_c given by

$$\tau_c = \mu = r_{\Delta\lambda} - \frac{1 - r_{\Delta\lambda}}{b_\nu/a_\nu} . \qquad (I-29)$$

For $r_{\Delta\lambda} = 0.5$, and $b_\nu/a_\nu = 4.5$, 3 and 1.5, this gives $\tau_c = 0.39$, 0.33 and 0.17, respectively.

Obviously, this suggested interpretation leads to the conclusion that all of the residual intensities in lines greater than, say, 0.5, are formed quite deep in the photosphere. Just as obviously there will be cases where this leads to misleading conclusions. On the other hand, to accurately compute the run of S_ν and τ_ν with depth in the photosphere, and, having done so, to define an unambiguous contribution function is a rather formidable task. The line source functions generally are not equal to the continuum source function and the line opacities are not given accurately by the assumption of local thermodynamic equilibrium. To obtain accurate values of S_ν and τ_ν, it is necessary to solve the coupled radiative transfer and statistical equilibrium equations for the atom giving rise to the spectral line in question (cf. Chapter VI). This has been done only for a few elements using simplified atomic models and, in some case, somewhat uncertain atomic parameters. If there is some clear and unambiguous indication that a weak line is formed high in the photosphere, then, but only then, one should use the contribution function defined by Equation (I-25).

As an illustration of the type of difficulty that may be encountered consider some of the results obtained for lines of Fe I when contribution functions parallel that defined by Equation (I-25) are used to assign heights of formation to the lines and when S_ν and τ_ν are obtained by assuming local thermodynamic equilibrium. Frazier (1968) has considered two such lines, and, on the basis of the resulting height of maximum contribution, has arrived at some very interesting conclusions concerning the nature of the oscillatory motions and convective overshoot in the photosphere. These conclusions are based on the result that one of the Fe I lines, λ 6355, is formed high in the photosphere, $\tau_{5000} > 10^{-2}$, even though the observations are made in the side of the line at a residual intensity of approximately 0.8. This result is purely a consequence of the way in which the contribution function is defined and of the way in which S_ν and τ_ν are related.

The only possible way a line can have a residual intensity of 0.8 if it is formed high in the photosphere in local thermodynamic equilibrium is to have negligible opacity in the line in the low and middle photosphere. This is clearly not true for Fe I, as the results in Table I-2 show. However, as Equation (I-27) shows, the contribution function defined by Equation (I-25) loses its meaning when x_ν is constant. The λ 6355 line arises from a lower level that is 2.83 eV above the ground state and all that the maximum in the contribution function (I-25) in the high photosphere shows, in fact, is that a line of this excitation potential in Fe I, has opacities that are essentially proportional to the continuum opacity everywhere except in the high photosphere. The contribution function that is computed, therefore, has nothing at all to do with where the line is actually formed. It is simply the result of an unfortunate definition that tends to show a maximum in the contribution function at the depths where x_ν is a maximum regardless of the value of τ_ν at that depth.

Perhaps the lesson to be learned from these comments about heights of formation is that the heights should not be taken too seriously. We repeat, for emphasis, the earlier comment that all we can really hope to do is to discriminate between the low, middle and high layers of the photosphere and chromosphere. By working with lines of differing strength from an individual element and from lower levels of common excitation potential, this problem can be minimized. However, when the lines used arise from different elements and have different excitation potentials the relative heights of formation cannot be trusted unless very careful consideration is given to the determination of S_ν, τ_ν and to the definition of the contribution function.

References

Athay, R. G.: 1973, *Radiation Transport in Spectral Lines*, Chap. II, D. Reidel Publishing Co., Dordrecht, Holland.
Athay, R. G. and Canfield, R. C.: 1969, *Astrophys. J.* **156**, 695.
Athay, R. G. and Lites, B. W.: 1972, *Astrophys. J.* **176**, 809.
Frazier, E. N.: 1968, *Astrophys. J.* **152**, 557.
Gingerich, O., Noyes, R. W., Kalfofen, W., and Cuny, Y.: 1971, *Solar Phys.* **18**, 347.
Hulst, H. C. van de: 1953, in G. Kuiper (ed.), *The Sun*, Univ. of Chicago, Chicago.
Jefferies, J. T.: 1968, *Spectral Line Formation*, Blaisdell, Waltham.

CHAPTER II

STRUCTURAL FEATURES

1. Observational Methods

Observations of the brightness of the Sun with moderate spatial resolution in any wavelength band, whether the band be very broad or very narrow, show a highly structured surface. The location and width of the pass-band determine the effective height in the atmosphere that is observed, and, hence, they determine the structural features observed.

Observations in the visual continuum select lower photospheric depths. In the XUV continuum and in the submillimeter and mm continua the depth of formation moves into the chromosphere. Similarly, in the X-ray and meter wavelength continua the depth of formation moves into the corona.

The continuum observations have obvious advantages and disadvantages. The advantages lie in the relative ease of interpretation, in the possibility of defining in an unambiguous way the depth of formation and in the insensitivity of the continuum radiation to velocity and magnetic field effects. The disadvantages, for chromospheric and coronal observations, lie in the relative inaccessibility of the EUV, X-ray, and submillimeter continua and in the difficulty of achieving reasonable spatial resolution at radio wavelengths.

An alternative means of observing the higher layers of the solar atmosphere is to isolate the light in a narrow band within a spectral line. Because of the added opacity in the lines the effective depth of formation of the observed radiation moves higher into the atmosphere. For the strongest lines in the visual spectrum, the effective depth of formation at line center lies in the middle chromosphere. It is possible, in principle, to probe the entire range of heights between the middle chromosphere and low photosphere by stepping selectively in narrow-band steps from the center of a strong line through the line wing to the continuum. There are difficulties with this approach, however.

A fixed, narrow band-pass within a magnetically insensitive ($g = 0$) spectral line may show a change of intensity as a result of: (a) an intrinsic change in solar brightness that is not accompanied by a Doppler shift of the line or, (b) a Doppler shift of the line. If the line is magnetically sensitive ($g \neq 0$), the brightness in the line will show additional changes due to magnetic field structure. Thus, an image of the solar surface made with a fixed band-pass within a magnetically sensitive line will show an apparent structure that is composed of all three effects: (a) intrinsic changes in brightness, (b) Doppler shifts due to motion, and (c) magnetic field changes.

The latter effect can be avoided by using lines with little or no magnetic sensitivity. However, the strongest solar lines all show some magnetic sensitivity.

It is becoming increasingly clear that there are few if any intrinsic changes in solar brightness at local regions on the Sun that are not associated with spatial changes in either macroscopic fluid motion or magnetic field pattern or both. The magnetic field changes may be either in field strength or in orientation. In spite of the obvious association between brightness, velocity and magnetic field it is advantageous to discuss the three aspects of solar structure independently. To do this, it is necessary to use some care in defining what we mean by the brightness features. For purposes of this discussion, we will attempt to distinguish between those brightness features that appear to be bright only because of a velocity effect (Doppler shift) or because of a magnetic field effect (Zeeman splitting) and those that are intrinsically brighter or darker. We include in this latter category, of course, features that have distinctively associated motions and magnetic field.

By defining structural features as those features that have intrinsic brightness that is different from the background, we imply that the structural features represent density and/or temperature perturbations. Thus, it is assumed that the motions and magnetic fields influence the brightness directly only as a result of Doppler shifting or Zeeman splitting, i.e., they do not change the excitation state of the gas. This will not be true in all cases. It is perfectly possible, for example, to change the excitation and ionization state of the solar gases by Doppler shifts that vary differentially with height. An atom whose absorption spectrum is Doppler shifted strongly relative to the Sun as a whole will absorb and emit outside of the normal spectral lines. As such, it is exposed to a different radiation field than is an unshifted atom and, as a result, it will experience different excitation conditions. These effects are frequently small compared to the direct effects of density and temperature fluctuations, however, and we will assume that what we define to be structural features are, in fact, temperature and density fluctuations.

A further difficulty with using the lines to identify structural features lies in the ambiguity in the definition of the contribution function. It is assumed here that it is the residual intensity itself that exhibits the structural feature, i.e., we define the contribution function as the integrand in Equation (I-18). This is important, of course, only for the assignment of heights.

As noted earlier, the radiations that arise primarily from the chromosphere occur mainly in the radio bands at cm and mm wavelengths and in the XUV shortward of λ 1600. At these wavelengths the chromosphere is opaque in the continuum and, in the XUV, in many lines as well. In the visual part of the spectrum only the radiation near the centers of the strongest absorption lines is of chromospheric origin. Also, the helium lines are of chromospheric origin, but in this case it is the depression in the line that is chromospheric and not the radiation itself. The difference is perhaps subtle, but it is nevertheless important.

Although the chromosphere can be observed in several strong lines in the visual as well as in the XUV, most of what is known about chromospheric geometry has been derived from observations made in two spectral lines: Hα and the K-line of Ca II.

Within the K-line of Ca II three regions of the profile are identifiable, as illustrated in Figure II-1. The K_2 emission peaks are located approximately ±0.16 Å from line center and the K_1 minima are located approximately ±0.25 Å from line center. Marked changes in chromospheric structure are evident between observations made in K_3 K_2 and K_1. Although the Hα line of hydrogen does not have emission reversals of the type shown in Figure II-1, it has a distinct core extending to approximately ±0.75 Å through which the

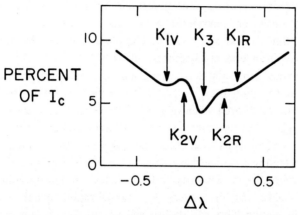

Fig. II-1.   A representative quiet Sun profile for the core and inner wing of the K-line of Ca II. Three distinctive features of the profile are indicated in the figure. These three features are frequently referred to in solar physics literature, as well as in this text. V and R refer to the violet and red sides of the profile, respectively.

observed chromospheric structure again changes markedly. Current evidence indicates that within the middle and low chromosphere where K and Hα are formed the temperature lies below 8500° and the non-thermal broadening motions amount to no more than 8 km s^{-1}. At 8500° the mean thermal velocity of hydrogen atoms is 11.8 km s^{-1}, which, when combined with a non-thermal velocity of 8 km s^{-1}, gives a rms velocity of 14.3 km s^{-1}. Similarly, the mean thermal and rms velocities of calcium are 1.9 and 8.2 km s^{-1}, respectively. These rms velocities produce Doppler widths of 0.31 Å in Hα and 0.11 Å in K. Since we have used an upper limit to the non-thermal motion, these estimates of the Doppler widths are upper limits and are considerably above the true values, as we shall later see. They serve to illustrate, however that the core of Hα and the K_1-K_3 region of the K line are broad compared to the Doppler width. It is to be expected. therefore, that the absorption coefficient in these lines changes markedly across the features under discussion, and it is to be expected, as a result, that the mean height of formation changes markedly also. It is perhaps not too surprising therefore that the observed chromospheric structures change quite dramatically through the Hα core and through the K_1-K_3 region of the K line.

The arguments in the preceding paragraph serve to illustrate another important point: monochromators designed to isolate the radiation in the cores of strong lines in order to identify structures at a particular height need to have a very narrow band pass free from significant side bands. Ideally, they should have a band pass less than one Doppler width, i.e., less than or about ¼ Å in Hα and less than or about $\frac{1}{8}$, Å in K. Of course, if one is interested only in detecting structure, irrespective of the level in the atmosphere at which it occurs, considerably broader pass bands will suffice and will generally show more structural details. The broader band observations are particularly useful for providing an overall view of the geometry whereas the narrow band observations provide more nearly a cut through the geometry on a given optical depth surface. The loop in Figure I-5, for example, might easily be recognized as a loop if observed with a broad band monochromator, but it might be difficult to identify as a loop if observed with a series of narrow band monochromators at adjacent wavelengths.

The chromospheric structural detail that can be seen in a given observation varies from line to line as well as with the pass band. Observations made in Hα characteristically exhibit a wealth of detail even when relatively narrow pass bands are used. Thus, narrow band observations in Hα often show as much detail as broad band observations in K.

There are several reasons why one spectral line may exhibit more details of chromospheric structure than another even when equivalent pass bands are used. Among these are: (a) differences in sensitivity to local changes in density, temperature, motion and magnetic fields, (b) differences in thermalization length (see Chapter V), (c) differences in scale heights of opacity, and (d) differences in ionization properties. The latter two factors are readily identified as favoring structural detail in Hα in the solar chromosphere.

Because of the general rise of temperature through the chromosphere, the population of the second quantum level in hydrogen (the ground state of the Hα transition) first rises to a maximum in the low chromosphere then falls off relatively slowly with height as shown in Figure II-2 (see Chapter V). By comparison, the population of the ground state of the K line (n_1 Ca II) falls off much more rapidly with height. The large scale height in Hα is a consequence of the high excitation energy (10 eV) of the second quantum level in hydrogen. The crosses on the curves in Figure II-2 indicate the approximate depths of $\tau_0 = 1$ in Hα and in K_3 and the open circles indicate the approximate depth of $\tau_0 = e$ and $\tau_0 = e^{-1}$ for the two lines. From the discussion in Chapter I, we concluded that the strongly visible layers extend from approximately $\tau_0 = e^{1.5}$ to $\tau_0 = e^{-1.5}$, and we note from Figure II-2 that the thickness of the layer between $\tau_0 = e$ and e^{-1} is approximately 600 km in Hα and 300 km in K. Thus, even with an infinitely narrow band pass one sees a much thicker layer of atmosphere in Hα than in K.

Effect (d) listed in the preceding discussion plays a further role in making chromospheric features more visible in Hα than in K. At temperatures of the order of 30 000 K or higher there is still an appreciable fraction of hydrogen that is neutral simply because the hydrogen equilibrium is always established between H I and H II. In calcium, however, at these same temperatures the dominant ion stage shifts to Ca IV or higher and

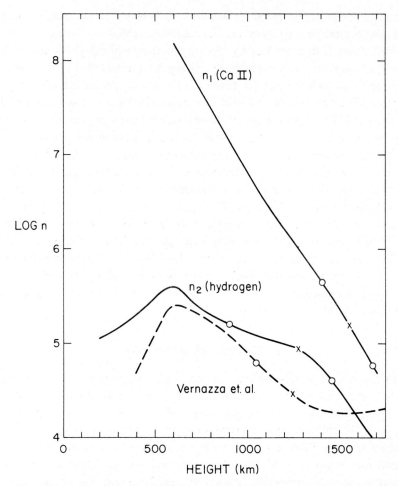

Fig. II-2.   Populations as a function of height for the absorbing levels for the Ca II K-line and Hα for a model chromosphere derived by the author. Values of the $n = 2$ population for hydrogen obtained with the Harvard-Smithsonian Reference Atmosphere (Vernazza *et al.*, 1973) are shown for comparison (dashed line). The heights where $\tau_0 = 1$ are indicated by crosses, and the heights where $\tau_0 = e^{-1}$ and e are indicated by open circles.

the relative concentration of Ca II becomes small compared to the total concentration of calcium. Thus, high temperature features in the upper chromosphere may be invisible in Ca II but visible in Hα.

The Hα line is unique in the visual spectrum as an indicator of chromospheric geometry. It is the only line in the visual spectrum that has the dual characteristics of large opacity and large scale height in the chromosphere. Other Balmer lines have equally large scale heights, but lower capacity. All other lines with high chromospheric opacity in the visual spectrum originate from low lying energy levels and, as a consequence, have relatively small scale heights.

Disk observations of the chromosphere with sufficiently narrow pass band to show detailed chromospheric structure can be obtained either by using narrow band, birefringent filters (filtergrams) or by placing a narrow exit slit at the desired location near the focal plane of a spectrograph. By driving the entrance slits of the spectrograph over the solar disk with the exit slit driven in unison or by driving the solar image across the entrance slit a two-dimensional monochromatic image (spectroheliogram) of the Sun can be produced. The disadvantages of the spectroheliograph are that the image must be built up by scanning, which is time consuming, and that the image quality is dependent upon both the telescopic image and the spectrographic image, i.e., the telescopic image is subject to further deterioration in the spectrograph. Its advantages are found in its flexibility to select almost any desired wavelength within the visual spectrum and to vary the pass band through wide limits. Filtergrams are faster to make, generally have somewhat better image quality, and require less expensive instrumentation but suffer from lack of ability to vary the wavelength and pass band, except through very limited ranges. The apparent advantages of observing in Hα relative to all other lines has made the use of filters practical, and they have found widespread usage. Descriptions of optical systems for solar telescopes, spectrographs, spectroheliographs and birefringent filters are given by Zirin (1966).

2. The Photospheric Structure

Although the subject of this book is the solar chromosphere and corona, these layers cannot be understood without some reference to the underlying photosphere. It is useful at this point, therefore, to remind the reader of some of the basic properties of the photosphere. The chromosphere and corona require an energy supply in some mechanical (non-radiative) form. Such a supply is present in abundance in the subphotospheric convection zone where most of the energy is transported upwards by convection. Some small portion of this convection energy is carried through the photosphere and is dissipated in the chromosphere and corona to produce the temperature rise in these layers. We are particularly interested, therefore, in those aspects of photospheric structure that may be indicative of mechanical energy transport. This will include such phenomena as photospheric structures, motions and magnetic fields. Rather than to devote a separate chapter of this book to a review of these features of the photosphere, it seems more appropriate to discuss photospheric geometry along with chromospheric and coronal geometry and so forth. Thus, we shall review here just those aspects of the photosphere that relate to its structural features and these only in the briefest detail.

The photosphere darkens from center to limb in practically all of the visual spectrum (the only exception being a few spectral lines near the extreme limb). This limb-darkening is readily understood for the continuum in terms of the outwards decrease of temperature in the photosphere. To demonstrate this, consider the solar atmosphere as being plane-parallel and one-dimensional, i.e., free of curvature and free of inhomogeneities except for vertical stratification.

As we noted in Chapter I, the photospheric continuum data can be approximately represented by a source function of the form given by Equation (I-19). However, a more physical representation is obtained with

$$S_\nu = a_\nu + b_\nu\tau_\nu + c_\nu E_2(\tau_\nu), \tag{II-1}$$

where $E_2(\tau_\nu)$ is the second exponential integral,

$$E_2(\tau_\nu) = \int_1^\infty \frac{e^{-\tau_\nu x}}{x^2}\, dx. \tag{II-2}$$

Integration of Equations (I-18) and (II-1) gives

$$I_\nu(\mu) = a_\nu + b_\nu\,\mu + c_\nu\left[1 - \mu\,\ln\left(1 + \frac{1}{\mu}\right)\right]. \tag{II-3}$$

The coefficients a_ν and b_ν are positive and the coefficient c_ν is negative throughout most of the visual spectrum (Pierce and Waddell, 1961). The second exponential integral function, $E_2(\tau_\nu)$ decreases monotonically with τ_ν from a value unity at $\tau_\nu = 0$ to a value 0.148 at $\tau_\nu = 1$. Both a_ν and b_ν are of larger magnitude than c_ν so that S_ν *increases* monotonically with τ_ν.

For the continuum we may identify S_ν with the Planck function B_ν. Hence the monotonic increase of S_ν with τ_ν implies a monotonic increase of T with τ_ν.

As the point of observation moves from center to limb the centroid of the continuum contribution function given by Equations (I-18) and (II-1) lies near $\tau_\nu = \mu$ in similarity with the results in Figure I-9a. Practically speaking, one can observe the limb darkening out to $\mu \approx 0.1$. The geometrical distance between $\tau_\nu = 1$ and $\tau_\nu = 0.1$ obtained from Equation (I-5) with $H = 60$ km is approximately 140 km. Thus, there is still an extensive layer of atmosphere (≈ 200 km) between the depth $\tau_\nu = 0.1$ and the low chromosphere, and we should not necessarily expect any direct carryover of photospheric continuum features into the chromosphere.

Within the stronger Fraunhofer absorption lines the opacity K_ν is several orders of magnitude larger than in the continuum. Thus, the radiation in these stronger lines originates ($\tau_\nu \approx \mu$) well within the chromosphere. This immediately suggests that limb darkening data in the lines could be used to extend the continuum limb darkening analysis through the upper photosphere and well into the chromosphere. However, for the strong lines it is no longer permissible to identify S_ν with the Planck function, and it becomes very difficult to relate an observed $I_\nu(\mu)$ to a temperature distribution. In fact, limb darkening may occur in the lines even when the temperature gradient is reversed, as it is in the chromosphere. This point will be discussed in more detail in Chapter VI.

In spite of the difficulty of interpreting the line spectrum, nearly all that we know about the geometrical structures in the chromosphere is derived from observations made in the stronger Fraunhofer lines. Thus, in this sense, the lines provide an extremely important means of probing the atmospheric structure. Also, spectral line theory is now advanced to the point where the observed intensities within the lines are beginning to provide valuable information on the mean temperature structure (see Chapter V).

Within the photospheric continuum three distinct types of structure are clearly evident: sunspots, granules and faculae, as shown in Figure (II-3). An adequate discussion of sunspots would necessarily carry us far from our main purpose of studying the quiet chromosphere and corona. Suffice it to say here that the spots define centers of activity where the magnetic fields are strong and where much transient activity takes place. The phenomena associated with sunspots in the chromosphere and corona are of considerable importance and will be discussed in some detail. However, the sunspots, as such, will remain of interest only as the most easily recognized symptom of the activity.

Photospheric faculae are observed preferentially in proximity to sunspots and near the solar limb. They appear typically as chains of small bright points ($\approx 1''$ in size). The fact that they are seen near the limb but not at disk center suggests that they are phenomena of the higher photosphere. Rather similar phenomena (plages) are observed in the chromosphere near active regions and the two phenomena may, in fact, be closely related. On the other hand, there is a substantial class of faculae that is unassociated with other visible forms of activity and it seems to have a counterpart in the chromospheric network. This latter class of faculae shows a polar maximum near the minimum of the sunspot cycle. They are clearly associated with magnetic field phenomena, but, apart from this, little is known about them.

Facular areas are long lived, often outliving the sunspots themselves. Individual bright grains within a facular area are shorter lived, but their statistics are poorly known.

Granules cover the sun more or less uniformly in a pattern suggestive of non-stationary convective cells. Under good seeing conditions they are readily observed in the continuum, even near the limb, but they are not seen in the spectral lines except at intensity levels relatively near the continuum. This suggests that they are phenomena of the deeper photosphere. Both theory and observation are consistent with the conclusion that the granules are associated with convection in the upper layers of the sub-photospheric convection zone. Whether the granules are themselves convective cells or a by-product of them is not absolutely clear, but there is no strong reason to suppose that they are not the convective cells. The granule lifetime is of the order of 5 min and their sizes range from about $2''$ to less than $\frac{1}{2}''$. The mean appears to be near $1''$. It is estimated that the total number of granules on the Sun at a given time is a few million, but the actual number estimated depends, obviously, upon the definition of what constitutes an individual granule. Granules often occur as close associations of several bright features closely packed together but separated by narrow lanes of intermediate brightness. It is not clear whether such associations should be considered as several independent cells or as a single cell.

The brightness contrast between granules and the average background is about 10 to 15% near λ 5000. The contrast increases toward the violet and decreases toward the red. Brightness contrast, of course, gives only the contrast averaged over the contribution

Fig. II-3a. The solar photosphere as observed at 2 A from the center of the Hα line. Enough absorption is present in the line to show increased facular (bright) structure on the disk and dark curved arches of material overlying one of the sunspots. A system of bright network structures bordering supergranule cells can be seen in the figure (R. B. Dunn, Sacramento Peak Observatory, Air Force Cambridge Research Laboratories).

function. It may be inferred from the existence of the granule structure that a surface of constant optical depth in the photosphere, when referred to physical depth, is somewhat

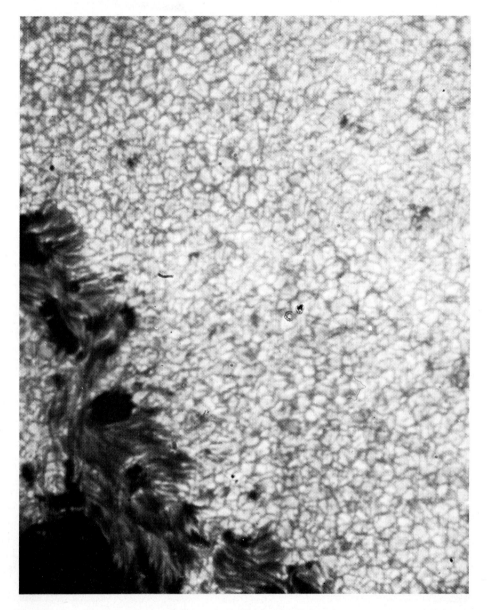

Fig. II-3b   A stratascope image showing photospheric granulation near a sunspot group (Princeton University).

like that of an egg carton — alternate mountains and valleys. The relative elevations of the mountains and valleys are difficult to determine, but there is no evidence that they are excessively large.

If the height amplitude of the peaks and valleys in the optical depth surfaces were large relative to the opacity scale height, the center-limb opacity would not scale in proportion to μ^{-1}. In such a case the analysis of center-limb data assuming the μ^{-1} dependence would inevitably lead to marked inconsistencies. The fact that center-limb analyses seem to work rather well in the photosphere suggests that the roughness or 'rugosity' effect is not very pronounced.

To a first approximation the bright granules and dark lanes separating them are more or less symmetrical in brightness about the mean. In other words, the granulation pattern is composed about equally of negative and positive excursions from the mean. Closer inspection shows, however, that the dark lanes tend to be somewhat narrower than the granules and that it is possible to distinguish between the negative and positive features.

Spectroheliograms made in lines whose contribution function maximizes in the upper photosphere or in the temperature minimum region show evidence of fine granulation type features as well as the faculae and sunspots. However, new features begin to appear,

Fig. II-4.   Strips of photospheric images as seen at regularly spaced intervals from the center of the Ca II K-line. Image e is at K_3 (line center) and subsequent images are spaced 0.11 A to the red. The K_2 region lies between images f and g, and K_1 lies between images g and h. The network is seen as broken chains outlining the supergranule cells. About 50 such cells can be counted along a solar diameter. Note that the network have maximum visibility in K_2 and K_3 and, at these wavelengths, are quite distinct over the entire disk. Beyond about +0.44 A (image i), however, the network structure is distinct only near active centers (Hale Observatories, California Institute of Technology).

also. One of the more interesting features showing up in the high photosphere is a network of narrow veins apparently outlining large cells of the order of 30 000 km in diameter. These cells form a mosaic over the entire Sun. They are most distinctive in areas contiguous with active centers, but often become indistinct within the active centers themselves. The network becomes even more pronounced in the chromosphere.

Figure II-4 shows a series of spectroheliograms made in the K-line of Ca II. Image e is at line center and subsequent images are stepped 0.11 Å from line center ending at +0.77 Å in image L. The average K_2 position lies between images f and g and the average K_1 position lies between images g and h. The K_1 region of the profile is formed near the temperature minimum and K_2 is formed in the low chromosphere. Note that network structure is evident over most of the disk through image h, i.e., in the chromosphere and temperature minimum region, but becomes relatively indistinct over much of the disk beyond image h, i.e., in the photosphere. There seems little doubt that the bright network contiguous with the active centers is identical with the photospheric faculae.

Little evidence for the same widespread mosaic of network exists in the deeper layers of the photosphere. Exceptions occur in the case of photospheric faculae and the recently discovered *filigree*, to be discussed shortly. However, Doppler spectroheliograms in which two images of the Sun made in a weak Fraunhofer line at equally spaced intervals either side of line center are subtracted photographically show a low photospheric cellular flow pattern (see Figure II-5) that coincides with the cells seen in the high photosphere and in the chromosphere. The flow in the photosphere is predominantly from the cell center to the borders where the network lies. These 'supergranule' cells will be discussed in more detail in Chapter III.

The network in the upper photosphere is brighter than the ambient background in all lines in which it has been observed except in the wings of Hα and in the higher Balmer lines where it is relatively dark. It has an apparent lifetime, in the chromosphere and presumably in the photosphere as well, of approximately 20 h. The fact that the lifetime is about one terrestrial day means, however, that the lifetime is difficult to determine accurately.

It is difficult to assign a depth in the atmosphere at which the network first exists as a physical entity. Observations in the wings of the Ca II K-line show the most pronounced regions of network clearly out to $\Delta\lambda \approx 6$ Å from line center, beyond which it disappears. Assuming that Equation (I-19) is approximately correct in the wings of the line and that $b_\nu/a_\nu = 4.5$ (the value for the continuum at λ 4000), we find that the residual intensity in the K-1 line at $\Delta\lambda = 6$ Å $(I = 0.5\,I_c)$ corresponds to S_λ at $\tau_c \approx 0.4$. If we further assume that the contribution function extends an optical distance of $e^{1.5}$ either side of the depth where $I_\lambda = S_\lambda$ (see Chapter I), we conclude that the network first appears near $\tau_c = 0.1$. It is clear from the general disappearance of most of the network just outside of K_1, however, as shown in Figure II-4, that the network over most of the Sun is not pronounced in the middle photosphere.

R. B. Dunn at Sacramento Peak Observatory has recently discovered a new feature of the photosphere associated with both the network and granules. These features, which are

Fig. II-5.   A Doppler image of the Sun obtained by photographically subtracting two images made in opposite sides of a spectral line. Rotation of the Sun shows a dark hemisphere and a bright hemisphere. Supergranule cells show as elongated features consisting of parallel bright and dark bands. They are more visible near the limb and least visible near disk center. These effects are consistent with a horizontal outflow of matter from the cell center to the cell border. The network appears to lie over the cell borders (Hale Observatories, California Institute of Technology).

called *filigree* (Dunn *et al.*, 1974), are shown photographically in Figure II-6. In the upper right hand panel of this figure, which is a filtergram made 2 Å to the red side of Hα, the filigree are seen as bright chains of very small features whose characteristic size is observed to be near ¼″. Since this is near the resolution limit of the telescope, the actual filigree size could be much smaller. It is readily seen from this panel that the characteristic size of the filigree is considerably smaller than that of the granules seen in the same image.

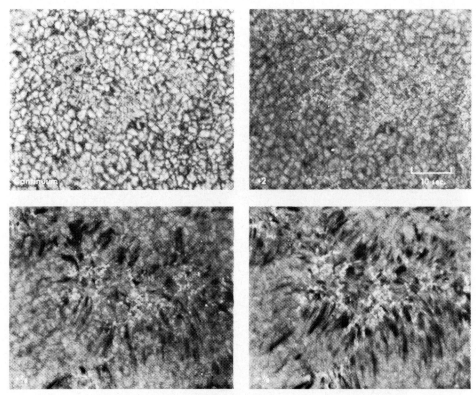

Solar Filigree

Fig. II-6.   Solar filigree shown in the continuum (upper left), +2 Å from the center of Hα (upper right), +7/8 Å from the center of Hα (lower left) and −7/8 Å from the center of Hα. The filigree are the bright, narrow chains of points (Sacramento Peak Observatory, Air Force Cambridge Research Laboratories).

The upper left hand panel in Figure II-6 was photographed in the continuum outside of the Hα line. The filigree are not prominently visible, as such, but the granules underlying the filigree regions are seen to have a fragmented appearance. According to Dunn *et al.* (1974), this appears to represent a real change in the granule structure rather than simply a superposition of granules and filigree. Since the filigree show more prominently at 2 Å from the center of Hα, which is still within the line wings, than they do in the continuum, we infer that the filigree are more pronounced in the middle photosphere than in the deep photosphere.

The lower two panels in Figure II-6 were photographed at ±7/8 Å from Hα line center. These panels show pronounced filigree together with faint granules in the background and elongated dark features of chromospheric origin. At these wavelengths in the Hα line most of the radiation comes from the middle photosphere. However, some chromospheric features of higher than average opacity or of greater than average Doppler width show in projection against the photospheric image. There is virtually no emission or absorption in the Balmer lines in the high photosphere and temperature minimum region because of the low temperatures occurring at these heights. Thus, Figure II-6 is consistent with the idea that the filigree are bright in the middle photosphere but does not rule out the possibility that they extend into the upper photosphere as well.

Studies of filigree are in their rudimentary stage only and their statistical properties are not established. However, Dunn *et al.* (1974) conclude that the filigree are most easily seen near active centers underlying aggregates of chromospheric rosettes. The rosettes, to be discussed in Section 4 of this chapter, are elements of the chromospheric network. Thus, the filigree seem to lie in the mid-photosphere as the lower extensions of the brighter net-work. Lifetimes and evolutionary characteristics of the filigree are unknown, but individual elements of the filigree appear to move about with transverse velocities of the order of 1.5 km s^{-1} as though they were being jostled about by the circulation associated with granule cells (Dunn *et al.*, 1974).

As we shall see in the following chapter, the network outlining the supergranule cells is the locus of many interesting chromospheric phenomena suggesting that the supergranule cell is a remarkably important component of the solar atmospheric structure.

Little is known about the fine scale features of the upper photosphere and their specific relationship to the granules except that they are of comparable size. Theoretical models of the photosphere clearly indicate that the primary convection ceases near $\tau_c \geqslant 1$. On the other hand the supergranule cells show characteristics that are suggestive of convection cells. The network structure, therefore, may be directly associated with convection in the convection zone. Nevertheless, there is not much reason to suppose that the fine structure in the upper photosphere can be tied in any one-to-one fashion back to the granules. There is, in fact, good reason to suppose that this is not the case.

3. Chromospheric Network

The supergranule cells, which are evidenced in the low photosphere by their internal flow patterns, become visible in the upper photosphere and chromosphere as a mosaic of polygonal network chains outlining the borders of the cells. The chromospheric network is similar in appearance to the network in the upper photosphere. However, in the chromosphere the network chains are somewhat broader than in the photosphere and the fractional surface area covered by the network chains is correspondingly larger. Also, a number of distinctive features can be identified in association with the network chains in the chromosphere. The broadening of the network in the chromosphere is illustrated by the comparison of the K_2 and CN spectroheliograms in Figure II-7. The CN molecule is believed to occur in greatest abundance in the high photosphere where the reduced temperature favors molecular associations. Thus, the CN spectroheliogram is representa-tive of the high photosphere whereas K_2 is representative of the middle chromosphere. Bright features showing in the CN image are seen to be considerably finer in scale than the corresponding features shown in the K_2 image. Some of this effect is probably

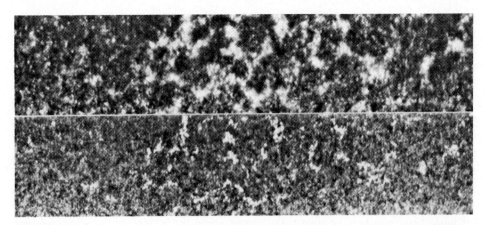

Fig. II-7.   Network structure seen in K_{2V} (top) and at the CN band head near λ 3883 (Kitt Peak
National Observatory)

illusory, and results simply from the fact that more features are visible in K_2 than in CN.
However, many of the individual bright grains in the two images are larger in K_2 than in
CN.

When observed with moderate spatial resolution, the network is often broken or
indistinct but can be crudely traced over most of the Sun. It is more evident in some lines
than in others and its appearance may change markedly with the wavelength position
within the lines. In some lines the network is bright against a darker background of cells
whereas in others it is dark against a brighter background of cells. Lines in which the
network appears brighter than the background include: H, K, λ 8498 and λ 8542 of Ca II,
λ 4227 of Ca I, D_1 (λ 5896) of Na I, b_2(λ 5173) of Mg I, λ 304 of He II, Lyman-α of
hydrogen, and Hα, as well as many others. In the case of Hα the network is bright only
near line center. In the outer Doppler core the Hα network is dark. Other lines in which
the network appears dark include: Hβ, Hγ and λ 10 830 of He I. The network structures
seen in Hα-center Hβ-center, K_3, λ 8542-center (Ca II) and λ 5896-center (Na I) are
shown in Figure II-8.

One of the most remarkable properties of the network structure lies in the persistence
of the structure throughout the entire chromosphere and the transition region. The
network structure in the transition region is illustrated in Figure II-9, which shows solar
images in the λ 304 line of He II and the λ 465 line of Ne VII. These lines are formed at
temperatures near 80 000° and 500 000°, respectively. Note that the network structure is
not remarkably more diffuse in Ne VII than in CN (Figure II-7). Between the mean
heights where these two radiations originate the matter density decreases by seven orders
of magnitude and the temperature increases by over two orders of magnitude!

Network structure appears to disappear almost entirely for lines formed in the corona
where the temperature exceeds about 1×10^6 K. However Reeves et al. (1974) find that
some of the smaller bright features in Mg X images of the Sun are traceable back to the

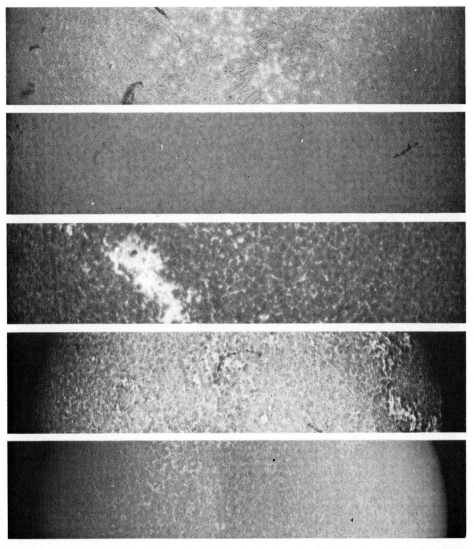

Fig. II-8. Representative examples of network structure as seen in the centers of five lines. From top to bottom the lines are: Hα, Hβ, K₃ λ 8542 Ca II, and λ 5896 Na I (Hale Observatories, Calif. Inst. of Technology).

chromospheric network. Mg X is most abundant in the corona for temperatures near 1.5×10^6 K. Thus, it seems clear that some remnant of the network structure persists into the corona but that, for the most part, it disappears at temperatures below 1×10^6 K.

As we shall see in Chapter IV the network structure in the photosphere and low chromosphere is closely associated with concentrations of vertical magnetic flux.

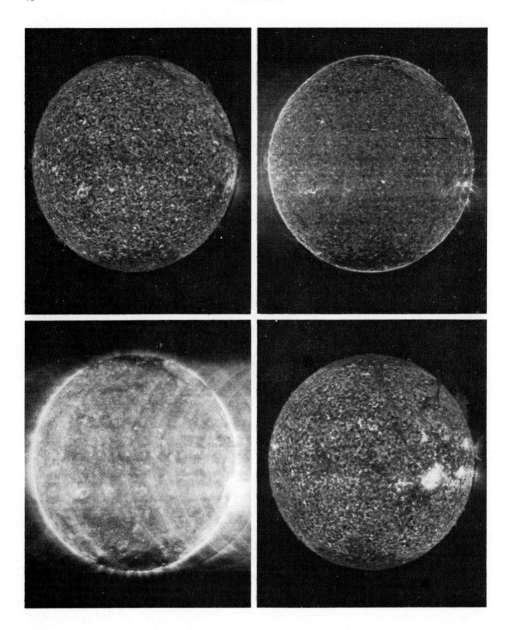

Fig. II-9.   Network structure seen in lines formed in the chromosphere-corona transition region. The lines shown are, beginning in the upper left, λ 304, He II; λ 465, Ne VII; λ 386, Mg IX; and λ 304, He II. Note the absence of pronounced network structure in the Mα IX image, which is formed at a temperature near 1×10^6 K. Also, note the coronal holes at both polar regions. The first three images were made simultaneously on December 30, 1973. The fourth was made 30 days earlier on November 30, 1973. North is at the top (Naval Research Laboratory, Tousey *et al.*, 1973).

Presumably, the same is true of the upper photosphere and the transition region. The absence of network structure in the corona could arise either from a basic change in the character of the magnetic field structure or in the long radial extent of the contribution function which always tends to suppress the contrast of small features.

The statistical properties of the network pattern are very similar to those of the supergranule cells. The network encloses cells of supergranule size ($\sim$30 000 km diameter) and Simon and Leighton (1964) have demonstrated that the network strands do, in fact, overlie the supergranule borders. Lifetimes of the network pattern are estimated at about 20 h. The cells appear to break up and dissolve as new cells form rather than to migrate to new positions.

4. Chromospheric Fine Structure on the Disk

When observed with high spatial resolution both the network and their enclosed cells break up into a collection of small scale features. Studies of the small scale features are complicated by the fact that a given feature takes on a different appearance in different spectral lines, or at different wavelengths in the same line or with different widths of the pass band when the central wavelength of the pass band is held constant. This often leads to ambiguity when attempts are made to compare one set of observations to another. One observer may recognize a characteristic pattern of features with a given set of observations, whereas a different observer, with different observing equipment, may not be able to recognize the same pattern at all. This difficulty undoubtedly contributes to the proliferation of names for small scale solar features. It is likely that some structures in the solar atmosphere are parading under two or more names depending upon how they are observed and by whom.

A fundamental question arises in the classification of solar phenomena as to how far such classification should be carried. If, for example, one views a forest of trees from far above the forest it may seem to be arranged in distinct large clumps of trees. Closer inspection will first show individual trees, then different species of trees, then branches on individual trees, then leaves on individual branches, etc.

At some point in solar physics we must decide when we have left the trees and have started examining the leaves, so to speak. There is no obvious guideline to use in such a decision at this time. However, as a general rule one might argue that those features of the solar atmosphere whose horizontal dimensions are of the order of the vertical density scale height ought still to be classed as trees. Since the density scale heights in the solar chromosphere are a fraction of an arc second, this rule would argue that the solar trees are probably less than an arc second in size and that we need not worry about whether we are resolving the tress into individual leaves until we achieve a resolution of less than about $^1/_{10}$". The best observations available to date give about ¼" resolution. Thus, we can perhaps safely assume that even the smallest features thus far seen on the Sun fall into the category of trees rather than leaves and should, therefore, be studied as individual structures.

Within the cells enclosed by the network the surface appears mottled with bright and dark features. This structure is generally described in Hα as dark 'grains' against a brighter background. The grains are readily seen in a number of lines. In Hα they are more visible in the violet wing. Their lifetime is of the order of 75 s, and their size of the order of one arc s. Similar features are observed in K_3 and more vividly in K_{2V}. Here, however, the features are bright and it is not known whether the bright K grains are identical with the dark Hα grains.

The appearance of the grains, or points as they are often called, of the supergranule cells in the K line is illustrated in Figure II-10. This unusual image of the Sun is made by

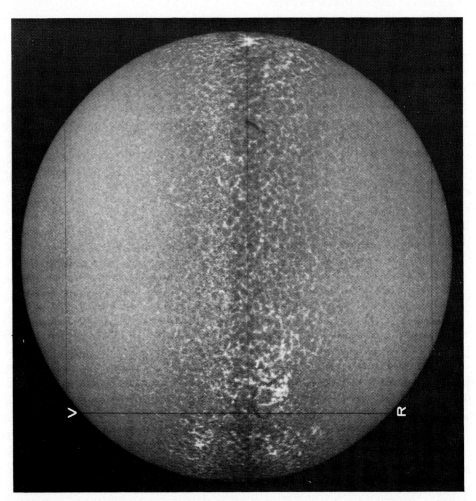

Fig. II-10.   A spectroheliogram of the Sun in which the wavelength changes progressively from left to right. The vertical line at the left is −1.1 Å from the center of the K-line and the vertical line at the right is +1.1 Å. K_3 is marked by a vertical line near disk center. The K_2 regions show as bright bands either side of K_3 and the K_1 regions show as faint dark bands just outside of K_2 (Hale Observatories, Calif. Institute of Technology).

scanning the solar disk in wavelength simultaneously with the position scan. The three vertical lines in the figure are made at ±1.1 Å and at line center. K_3 is visible as a dark lane near the central vertical line and the K_2 peaks are visible as two bright bands either side of K_3. Both K_2 and K_3 broaden considerably near the two solar limbs. Several conclusions may be reached from images of this type. Two conclusions of interest at the moment are: (a) the network is about equally visible in both K_2 peaks and (b) in the violet K_2 region a number of bright points are found in addition to the network. The bright points are scattered more or less randomly and, therefore, lie predominantly within the cells. There is little evidence of a counterpart of the points in the red K_2 region. It seems clear from these features of the K line image in Figure II-10 that the network are bright in both K_{2V} and K_{2R}, but that the bright points are, in most cases, bright in K_{2V} only. In other words, the K line has bright emission peaks on both sides of the line for the network elements but has only a violet K_2 peak in the majority of bright points. These conclusions are consistent with a statistical study of K_2 features by Liu (1972).

High resolution studies of the chromosphere indicate that the pattern within the cells is more complex than that of a uniform background with a sprinkling of bright or dark points. Figure II-11 shows two images of the chromosphere made at −0.3 Å and +0.63 Å from the center of Hα. The image nearer line center shows the solar surface to be covered more or less completely with bright and dark mottles, some of which are point-like and some of which are elongated. So complex is the structure that it is difficult for an inexperienced observer to even distinguish between network and cells in such an image. An experienced observer, however, recognizes a network of features from which the elongated features appear to radiate. These elongated features, which we shall refer to as *fibrils*, often extend over most of the cells enclosed by the network and intermingle with the bright and dark points.

The network itself consists of a variety of features. Beginning with features whose characteristic size is comparable to the width of a network wall, we find bright coarse mottles and dark coarse mottles. The lifetimes of these features are of the order of 20 h, i.e., the same as the network itself. Their characteristic sizes are 5 to 10″. Dark coarse mottles are also called *rosettes* or *bushes*. The former name is most often used for the feature when viewed well onto the solar disk and the latter is used for the same features seen near the limb. The name 'rosette' derives from observations with high spatial resolution and what some observers describe as a typical pattern of fibrils (or elongated dark fine mottles) emanating outward from a central core. Examples of rosettes are indicated in Figure II-11 by the encircled areas. (Note that the two images in Figure II-11 are not simultaneous in time nor are they cospatial. They were, in fact, made at different observatories on different dates.)

Near disk center the rosettes have a bright core in Hα center and in K. In these same lines the bright core regions observed with high spatial resolution are seen to extend outwards between the dark fibrils. However, the relative sizes of the bright and dark features are different in different lines and the rosette pattern is therefore different in different lines. The K line shows the central bright features to be brighter and larger than

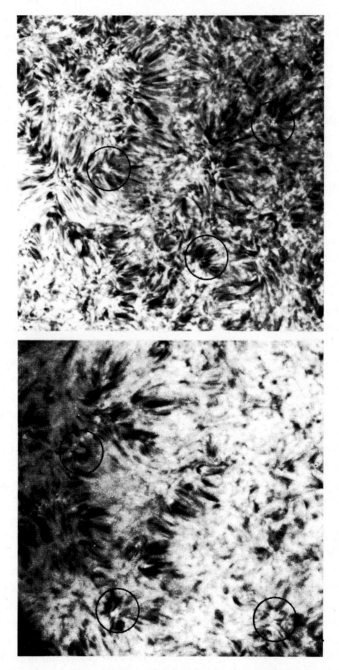

Fig. II-11.   Filtergrams at −0.3 Å (top) from the center of Hα and +0.63 Å (bottom) from the center of Hα. The encircled areas are examples of dark coarse mottles, also called rosettes by some observers. The two images were not made at the same time and represent different areas of the Sun (Sacramento Peak Observatory, Air Force Cambridge Research Laboratories and Hale Observatories, Calif. Institute of Technology).

does Hα. This central bright region together with its bright elongation makes up the bright coarse mottles.

Near the limb where the dark structure is described as bushes of fine dark mottles, the dark features are seen to be more elongated toward the limb and to lie nearer the limb than the bright features. This structure is most easily seen in the wings of Hα, as

Fig. II-12. An Hα image of the solar limb +1 Å from line center (top), and an Hα image of the disk near the limb +7/8 Å from the center (bottom). The elongated dark features on the disk outline the network and show in projection at the limb as spicules. These same dark features make up the bush structure of the rosettes (Sacramento Peak Observatory, Air Force Cambridge Research Laboratories).

illustrated in Figure II-12, and is interpreted to mean that the dark features of the bush structure either have greater vertical extent than or lie above the bright features. Some observers, such as Bray (1969), conclude that the dark and bright features of the rosette are distinctly different phenomena, whereas one observer (Dunn, 1974) has suggested that the bright points near the limb are the foot points of the dark fibrils of the bush. The tendency for a single structure to appear bright at low elevations and dark at high elevations is a common phenomenon in surge prominences seen against the disk. Thus, Dunn's suggestion should be taken seriously even though it is difficult to establish a one to one correspondence between a dark bush fibril and a bright foot point.

The question as to whether one should consider the bright fine mottles of the rosette pattern as the lower extension of the dark fine mottles or as a different class of structures is, of course, fundamental. Their lifetimes are of similar duration but not firmly established. Bray (1968) finds a mean lifetime of 5 min for dark mottles and, in a later paper (Bray, 1969), he finds a mean lifetime of 11 min for the bright fine mottles. Beckers (1968), however, gives 10 min as the mean lifetime of the dark fine mottles.

It seems clear that near the core of the rosette the bright and dark fine mottles (or fibrils) are interlaced as alternate structures. Such an appearance could result, however, simply from a partial obscuration of the bright rosette core by the dark fibrils. Thus, one could assume that the bright core of the rosette is more or less circular in shape and with a diameter near disk center given approximately by the outer extension of the bright fine mottles. If the dark fine mottles were then supposed to come mainly from the central part of the bright region and to lie above this region, the outer parts of the bright region would be most visible between the dark fine mottles. This would give the appearance of alternate dark and bright mottles with similar lifetimes. It is not clear that such a picture is not acceptable.

Figure II-13 is a schematic illustration of three (a, b and c) alternate models of rosettes. The distinction between spicules and fibrils will be made shortly. For the moment we are concerned with the structure of the bright core of the rosette. In panel *a* of Figure II-13 the bright mottle is separate from and underlies the dark mottles. In panel *b* the bright mottle underlies the dark mottles but is a physical extension of the dark mottles, i.e., the bright and dark mottles are parts of the same fine structures. A less likely, but still possible, alternative is that the bright and dark mottles are separate but similar features with the bright mottles lying statistically lower than the dark ones.

The third three panels of Figure II-13 represent alternate associations between dark fibrils of the rosettes and the bright points of the interior supergranule cell. The two features may be unassociated as in *d* and *e* with the fibril either reconnecting to the lower layers as in *e* or remaining elevated above the level of the points as in *d*. A third possibility is that the bright points are the outer feet of the fibrils. It could be, for example, that the fibril has bright foot points at either end, one being a cell point and the other being a bright mottle of a rosette.

There are nine possible combinations of models *a*, *b* and *c* with *d*, *e* and *f* and all nine

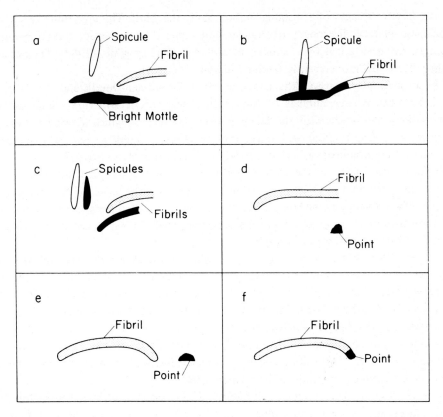

Fig. II-13.   Three alternative proposals illustrating the possible relations between spicules, fibrils and bright coarse mottles (a, b and c), and three alternative proposals illustrating the possible relations between the outer ends of fibrils and the cell bright points.

would perhaps find some advocates. A definitive choice between the nine alternatives cannot be made at this time.

Perhaps the most important distinction among the various alternatives shown in Figure II-13 is that between *d* on the one hand and *e* and *f* on the other. If the fibrils do not reconnect to the surface (*d*), it implies that whatever determines the fibril shape does not reconnect. For example, if the fibril follows a magnetic flux tube, which is a highly likely and plausible possibility, then model *d* suggests that the magnetic field radiates out from the rosette but does not reconnect to the surface presumably because the cell interior is either field free or of the same polarity as the rosette field. A reconnection of the fibril to the cell interior, however, clearly implies the presence of magnetic polarity of opposite sign to that in the rosette. This point will be discussed in more detail in Chapter IV.

In all six of the illustrations in Figure II-13 the fibril is drawn as a distinct feature arching above the general level of the Hα chromosphere. This picture itself is not an obvious one. The fact that the fibrils are predominantly horizontal in character prompts the question as to whether they are not fully emersed in the Hα chromosphere and made visible largely because of streaming motion along the fibril. This question, i.e., is the brightness contrast due mostly to Doppler shift rather than to density and temperature changes, has been asked by a number of authors including White (1962), Thomas and Athay (1961) and more recently Gebbie and Steinitz (1974).

In an attempt to answer the above question, Grossmann-Doerth and von Uexküll (1971) studied a large number of fine mottles near rosette structures. Simultaneous observations were made with Hα filtergrams and Hα spectra, which allowed a study of both the geometrical nature of the structure and its spectroscopic properties. Two classes of models were assumed: (a) a cloud model of the type shown in Figure II-13, and (b) an imbedded stream flow. In the former model, the line source function, the opacity and stream flow within the cloud were allowed to be different than in the background chromosphere whereas in the latter model the only distinctive property of the feature was its vertical motion. A range of profiles was computed for each class of models and compared with the observational results.

Of more than 400 features studied by Grossmann-Doerth and von Uexküll, the vast majority (88%) were well represented by the moving cloud model but were poorly represented by the pure stream model. No clear distinction was found between the motions of the bright and dark features, but a statistical preference for downward motion was present in the features studied. The streaming velocities of the features studied fell predominantly in the range -4 to $+8$ km s^{-1}, with extremes of -24 to $+20$ km s^{-1} and an average of $+3.9$ km s^{-1}. (Positive velocity is associated with positive wavelength displacement, hence, downward motion in the solar atmosphere).

The analysis by Grossmann-Doerth and von Uexküll shows, of course, only that the source function, opacity and streaming motion are different from the ambient medium. It does not prove unambiguously that the changes in source function and opacity are due primarily to density and/or temperature changes rather than to the effects of the motion itself (cf. Chapter VI). However, the logical deduction from their analysis is that the fine mottles of the rosettes are best represented, in fact, by a combination of density and temperature perturbations combined with streaming motion.

Bhavilai (1965) has identified a further class of network features, which he refers to as tunnels. These are similar to greatly elongated rosettes in that they show alternating bright and dark fibrils emanating from a bright core (in Hα center). The core in this case, however, elongates along the network chain for some distance rather than clumping into the more nearly circular rosette pattern. Structures of this type are clearly evident in Figure II-11. There seems to be no compelling reason, however, to consider these features as being basically different from rosettes.

It is most likely that the bright appearance of the network in certain lines results from the prominence of the bright coarse mottles, and we should expect to find prominent bright coarse mottles in all lines in which the network is bright. Conversely, we should not expect to find prominent bright coarse mottles in lines such as He I λ 10 830, Hβ or the wings of Hα. In these lines the dark coarse mottles should be dominant.

It is clear from observations made of the limb that most of the Hα fibrils have limited vertical extent. Relatively few extend as much as 5″ above the base of the chromosphere. The characteristic length of the fibrils seen near disk center is greater than 5″ when viewed with the wider of the Hα filters (¼ to ½ Å). Thus, the fibril structure is predominantly of a low lying, horizontal character. Most authors (cf. Beckers, 1964; Bhavilai, 1965) have assumed that the fibrils are predominantly closed arches with both ends terminating in the low chromosphere or the photosphere, but the evidence for true arch structure appears to be weak. The open ended banner model of Figure II-13 is an alternative. In either picture the dominant fibril characteristic is its large horizontal extent.

In their narrower dimensions the dark fibrils are typically about 1–3″ in size, though some appear to be larger. Their long dimension is very sensitive to the manner in which they are observed. They are longest when viewed in broad band (centered ½ Å off Hα). Typical lengths, when viewed in this way, are 5–10″, but some are considerably longer. Some authors (Beckers, 1968) report that the bright fine mottles are characteristically larger in dimension than the dark fine mottles but others (Bray, 1969) give the opposite result. Thus, it seems safest to conclude, at this time, that the bright and dark fine mottles have similar statistics.

At the limb of the Sun the extreme upper chromosphere consists entirely of fine fibril-like features known as spicules. The spicules appear bright in projection against the corona as shown in Figure II-12. Their average lifetime is 5 to 10 min. They have an average instantaneous vertical extent of 5–10″, and typically reach a maximum extent of about 15″. Their width is of the order of 1–2″. Thus, in some respects they have statistical properties similar to those of the dark fine mottles and bright fine mottles.

It seems clear from photographs made in the side of the Hα line core, as in Figure II-12, that the spicules at the limb are identical with the dark fibrils of the bushes seen on the disk. The spicules at the limb have a predominantly radial orientation (average inclination to radial direction is 22°, Beckers (1968)), however, and should show little elongation near disk center. Thus, the bushes cannot be identified both with the spicules and with the dark fine mottles near disk center. For this reason one cannot conclude that the average fibril necessarily lies above the bright fine mottles. Observations made near the limb show the higher features preferentially and thus select the more nearly vertical features. It seems clear, in fact, that the majority of the dark fine mottles do not show at the limb and that the 'average' fibril lies well below the spicules which make up the bush structure.

Besides the spicule-fibril distinction in terms of predominantly vertical structure as opposed to predominantly horizontal structure there is a further distinction that arises from the material motion within the spicules and fibrils. Spicules shoot out of the rosettes in the form of high speed, jet-like eruptions. Fibril material, on the other hand, appears to flow from the cell interior toward the rosette and has a much lower velocity amplitude than the material motion in spicules. This point will be discussed further in Chapter III.

It is important for the subsequent discussion of chromospheric structure and models to recognize that the fibrils and spicules appear to be quite separate and distinct phenomena and, as such, may have very different physical properties. It seems equally clear that the bright features of the rosettes probably have still different physical characteristics. This is true irrespective of whether the bright features are physically connected to the spicules and fibrils or are separate from them. Whatever the geometrical relationships, the change from bright feature to dark feature heralds a change in physical conditions that must be taken into account. Thus, it seems that within the network itself we have at least two separate classes of discrete features, spicules and fibrils, and that the bright mottles represent either changes in the physical properties of either spicules or fibrils (or both) with height or they represent a third class of discrete features.

The fibrils, of course, extend into the cell structure and are properties of both the cells and the network. In the upper chromosphere and transition region, as we shall see in the following section, spicules cover only a small percentage of the network structure. Nevertheless, the network structure remains well defined and more or less continuous in character. It seems inescapable, therefore, that we picture the network as a more or less continuous chain of material and not merely as a collection of spicules.

5. Spicule Structure

As noted in the preceding section the bushes of the network extend beyond the solar limb as spicules. At any one time the number of spicules seen varies with height above the solar limb and, more slowly, with latitude on the Sun. Hα filtergrams of spicules at the limb and on the disk are shown in Figure II-12. The number of spicules counted as a function of height and latitude is shown in Figure II-14. These latter data were obtained from much earlier observations made at the center of Hα (see Thomas and Athay, 1961) and which tend to show fewer spicules than can be observed in the Hα wings as in Figure II-12. Spicule counts from the newer observations by Lynch *et al.* (1973), show a maximum of 110 spicule features per 12° of solar latitude. Lynch *et al.* (1973) argue that even these numbers should be doubled in order to include the spicules seen in the opposite wing. It is well known, however, that many spicules have Hα profiles whose width at half intensity exceeds 1.5 Å and most of these spicules will be seen in both wings of the Hα line. Doubling the spicule counts seems unjustified, therefore.

Spicules seen at the limb are distributed over a relatively wide range in solar longitude. A spicule that sits off the limb but that extends quite far above the solar surface will project against the sky as though it were located at the limb with a reduced height. The reduced height will depend upon the angular distance from the limb. If a spicule of height h is at position angle α (measured from the limb), its apparent height r' above the limb is given by (see Figure I-2)

$$\cos \alpha = \frac{r' + R_\odot}{h + R_\odot}$$

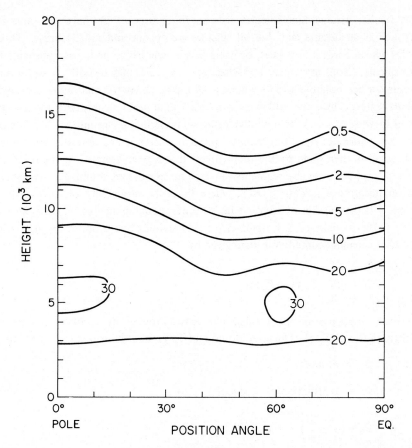

Fig. II-14. A contour map of the number of spicules counted in a 12° arc of the solar limb as functions of solar latitude and height above the visual limb (after Athay, 1959).

For small values of α, this reduces to

$$h - r' = 105\,\alpha^2,\qquad\qquad\text{(II-4)}$$

where α is expressed in degrees and h and r' are in kilometers. Thus, for $h - r' = 10\,000$ km $\alpha = 9.7°$ and we see that many of the observed spicules, particularly at low heights, are quite far from the true limb position. Under these conditions one spicule will often lie in front of another and thus be lost in the spicule counts.

To correct for these projection effects (blending and foreshortening), we assume that the spicules are randomly distributed over the solar surface. We know from the association of spicules with the network that this is not true. However, the network cell size subtends a heliocentric angle of only about 2.5° and since this angle produces a foreshortening at the limb of only about 650 km, we expect any given line of sight through the spicule field to include spicules from several adjacent network chains.

We further assume that spicules seen in the center of Hα have the same surface brightness at all heights and that all spicules are opaque and equally bright. Thus, two spicules whose borders just meet, or overlap, are assumed to be indistinguishable from a single spicule. (Note that in the line wings where spicules may be both transparent and of differing surface brightness there will be a natural tendency to count more 'features' than at line center. Thus, the following analysis would need appropriate modification if applied to wing counts.) We now divide the solar circumference into m elements each of which has a diameter equal to the mean spicule diameter d. The condition for a spicule to be observed is then given by the requirement that a given segment of the solar limb be bordered, on both sides, by segments in which no spicules are found. The probability for such an occurrence is $P_0^2(1 - P_0)^j$, where P_0 is the probability of finding an empty element, j is the number of adjacent limb elements containing one or more spicules and $1 - P_0$ is the probability of a single element containing one or more spicules. The total number of spicules counted, S, is then given by

$$S = mP_0^2 \sum_{j=1}^{m-2} (1 - P_0)^j/j,$$ (II-5)

since each configuration of j filled elements bordered by empty elements occurs $(m/j)P_0^2(1 - P_0)^j$ times. For large m, Equation (II-5) is approximately equivalent to

$$S = m P_0^2 \ln P_0,$$ (II-6)

and for a random distribution of N' spicules

$$P_0 = \left(1 - \frac{1}{m}\right)^N \approx e^{-N'/m}$$ (II-7)

Thus,

$$S = N'e^{-2N'/m}.$$ (II-8)

Equation (II-8) is sufficient to determine the true number of spicules, N', from the counted number S provided we know m, the mean spicule diameter. Note that Equation (II-8) has a maximum, for a given m, at $N' = m/2$ or $S_{max} = m/2e$. Thus, the maximum number of spicules counted depends only on the mean spicule diameter. For a mean diameter of $1''$, $m = 6000$ and $S_m = 1100$, which corresponds to approximately 3 spicules per degree of solar latitude.

Because of the assumption of uniform brightness for all spicules, Equation (II-8) will tend to underestimate the number of spicules counted. Thus, the counts in Figure II-14, which shows a maximum of 3.3 spicules per degree, suggest a mean spicule diameter of $0.92''$, which is probably an underestimate. Note that the spicule diameter indicated by this method is the observed diameter including the effects of telescopic and atmospheric resolution and is not the true spicule diameter.

The N' given by Equation (II-8) still contains the foreshortened spicules lying off the limb. Since the spicules are counted at intervals of 1000 km, we divide the Sun into longitudinal sectors $\Delta\alpha$, $\Delta\alpha_2$, $\Delta\alpha_3$, ... each of which reduces the projected height by 1000 km. Values of $\Delta\alpha_i$ are readily determined from Equation (II-4). We then assume that the spicules in the largest height group, N'_1, , lie within $\Delta\alpha_1$. Thus, $N'_1 = N_1$ and the true number in the next height interval is given by

$$N_2 = N'_2 - N_1 \frac{\Delta\alpha_2}{\Delta\alpha_1}.$$

similarly, the number in the third interval is given by

$$N_3 = N'_3 - N_1 \frac{\Delta\alpha_3}{\Delta\alpha_1} - N_2 \frac{\Delta\alpha_2}{\Delta\alpha_1}.$$

Using this procedure, we find successively each value of N_j. The results are given in Table II-1.

The distribution N_j gives the number of spicules in a zone of width (in the line-of-sight) $2\Delta\alpha_1 = 6.2°$ and length $1°$. This area subtends a fraction of the solar disk given by $(6.2 \times 1)/4\pi(57.3)^2 = 1/6660$. Thus, the number of spicules on the whole Sun extending to any height interval is $6660 N_j$. The values of N in Table II-1 at heights of 3000 and 4000 km exceed the total number of spicules (≈ 9 deg^{-1}) counted by Lynch *et al.* (1973), but not by a factor of two. It must be admitted, however, that the uncertainty in N is at least of the order of a factor of two, at these heights, and the results in Table II-1 should not be interpreted as being inconsistent with the results of Lynch *et al.*

The spicule numbers under discussion represent an average condition, or a 'snapshot' of the Sun at a given moment. Spicules are dynamic phenomena with large vertical velocities (see Chapter III). Thus, it should not be concluded from the preceding study that most spicules terminate at low height. In fact the successive differences in N_j are approximately constant to 10 000 km. It may be concluded from this that spicules move upward at an approximately uniform speed and that most of the spicules eventually reach to approximately 10 000 km. This is consistent with their observed motions.

It is of interest to determine the density of spicules within the network itself. The area of a given supergranule is approximately $(2.5°)^2$. Since this is closely equal to $1° \times 6.2°$, the values of N_j we have determined are closely equal to the number of spicules per supergranule. The fractional area of the Sun covered by spicules extending to 3000 km or higher is given by

$$6660 N_{3000} \frac{\pi d^2}{4} / 4\pi R_\odot^2 = 0.02$$

for $d = 1000$m and $N_{3000} = 25$. Thus, we see that spicules cover only a small fraction of the solar surface. It seems evident from Figure II-12 that network cells are highly varied in the numbers of spicules they contain and that some have many more than 25 and some

TABLE II-1

Spicule counts and computed values of N and N'. All counts are per degree of solar latitude

Latitude height (10 km)	6°			30°			54°			78°			Average
	S	N'	N	S	N'	N	S	N'	N	S	N'	N	N
3	1.4	27	13	1.6	24	13	1.5	23	14	1.7	19	12	13
4	2.7	17	11	2.8	15	11	2.9	14	9.8	2.8	11	8.0	10
5	3.1	14	10	3.2	10	.1	3.2	9.6	7.7	1.9	7.2	5.9	7.9
6	3.3	9.6	8.0	3.2	7.2	6.2	3.1	6.8	6.1	2.6	4.6	4.2	6.1
7	3.3	7.2	6.2	2.8	5.1	4.6	2.4	3.8	3.4	1.9	2.7	2.5	4.2
8	2.9	5.0	4.6	2.0	2.8	2.4	1.6	2.0	1.8	1.2	1.5	1.4	2.6
9	2.3	3.3	3.0	1.4	1.8	1.7	0.92	1.1	1.0	0.64	0.72	0.70	1.6
10	1.5	1.9	1.8	0.80	0.88	0.80	0.46	0.48	0.44	0.31	0.31	0.30	0.84
11	0.93	1.1	1.0	0.46	0.46	0.22	0.22	0.22	0.22	0.14	0.14	0.14	0.45
12	0.49	0.49	0.46	0.22	0.22	0.22	0.10	0.10	0.10	0.08	0.08	0.08	0.22
13	0.27	0.28	0.27	0.10	0.10	0.10	0.05	0.05	0.05	0.04	0.04	0.04	0.11
14	0.14	0.14	0.14	0.06	0.06	0.06	0.02	0.02	0.02	0.03	0.03	0.03	0.06
15	0.07	0.07	0.07	0.02	0.02	0.02	0.01	0.01	0.01	0.02	0.02	0.02	0.03
16	0.05	0.05	0.05	0.01	0.01	0.01	–	–	–	0.01	0.01	0.01	0.02
17	0.02	0.02	0.02	–	–	–	–	–	–	–	–	–	0.01

many fewer. Also, of course, the estimate of 25 spicules per supergranule cell at a height of 3000 km is, itself, only a rough estimate. A count of the spicule features in Figure II-12 suggests that this average number is not grossly in error, however.

At heights where spicules are observed the fractional area of the Sun covered by the network is much larger than the value 0.02 estimated for spicules. Thus, it appears that even within the network features the spicules are relatively widely separated.

On the other hand the dark Hα fibrils appear to cover at least half of the solar surface (Figure II-11), and we are again forced to conclude that the fibril structure is predominantly horizontal in character and that it is confined to heights below about 3000 km.

Spicules seen at the limb often show systematic orientations with respect to each other. The so-called 'porcupine' features are of the order of half the supergranule size and show steadily increasing inclination of spicules away from the vertical with increasing distance from the porcupine center. The lifetime is of the order of 5 min (Beckers, 1964). A phenomenon known as a 'wheat field' shows a more or less simultaneous eruption of spicules all with similar orientation along an arc length of about 12° on the solar limb, i.e., about 5 supergranule diameters. Again the lifetimes are of the order of the spicule lifetime of 5 min. Porcupines and wheatfields occur mainly in high solar latitudes (> 60°) (Beckers, 1964). Aside from their existence and lifetimes little is known about these phenomena.

It is clear from Figure II-14 that the polar regions of the Sun are somewhat richer in spicules than the equatorial regions. The reasons for this are not known, but possible reasons will become evident when we discuss magnetic field phenomena in Chapter IV.

As noted earlier, observations in the wings of Hα show that the spicules project onto the disk as dark absorption features. However, it is of interest to know what the spicules look like on the disk at line center in Hα, and in other lines as well. Figure II-15 shows a limb picture at line center in Hα, which illustrates.the difficulty of tracing the spicules onto the disk. The wealth of detail in Hα both on the disk and at the limb makes any one-to-one association very difficult, and, needless to say, there is difference of opinion as to what the associations are.

Prior to 1965 most observers associated spicules with the dark fine mottles emanating from the rosette structure, though frequently the rosette structure itself was not recognized. Much of the evidence for this association was statistical in nature. In 1965, Bhavilai pointed out that the statistical properties of the bright fine mottles of the rosettes are very similar to those of the dark fine mottles, and he noted that bright fine mottles near the limb were often traceable to spicules at the limb whereas the dark fine mottles were more often traceable to dark features in the chromosphere rather than to spicules. Bhavilai's conclusions have not succeeded in converting the majority of solar observers to his point of view, but they have shaken the belief that the spicules are exclusively the dark features.

There is, in fact, no firm reason for supposing that spicules should all be bright or all be dark in projection on the disk, or even that a given spicule should have the same

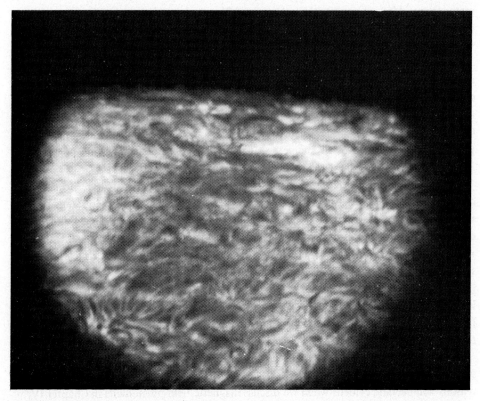

Fig. II-15.   The disk and limb regions as observed ¼ Å from line center in Hα. Note the difficulty of following individual features from the limb to the disk (Sacramento Peak Observatory, Air Force Cambridge Research Laboratories).

appearance throughout its entire length. As we shall later show the spicule model is still sufficiently flexible to allow either bright or dark spicules. Most observers seem to regard the question as still unanswered.

6. Transition Region Structure

The chromosphere-corona transition region has been successfully observed only on the solar disk at cm and XUV wavelengths. The radio data have insufficient resolution to exhibit other than the coarse structure such as active regions and center-limb brightness changes. A few observations are available in the XUV with moderate spatial resolution, but such observations are still rather limited.

Figure II-9 shows spectroheliograms made in the λ 304 line of He II and λ 465 line of Ne VII. The network structure is clearly present and would undoubtedly show more clearly with improved spatial resolution. It is not altogether clear whether the λ 304 line forms in the upper chromosphere or in the lower transition region. Computations of the expected He II line intensities as a function of temperature (Athay, 1965) indicate that the line should be formed near 80 000 K if it is assumed that the solar atmosphere has a uniform temperature gradient. It is known, however, that the temperature gradient is much steeper near 10^5 K than at temperatures near 3×10^4 K and this tends to give more of the emission at the lower temperatures. In any case it is clear that the λ 304 line is formed either in the extreme upper chromosphere or in the lower part of the transition region. Ne VII forms predominantly at temperatures near 5×10^5 K (Jordan, 1969), which places it relatively high in the transition region so far as temperature is concerned.

The contrast between network and cell brightness reaches its maximum value in the transition region. Thus, Reeves *et al.* (1974) give a network: cell contrast ratio ranging from about 15:1 to 10:1 for O IV and somewhat smaller ratios for C III and O VI, which are formed at lower and greater heights, respectively. The contrast ratio has essentially reached unity at Mg X and is of the order of 2:1 for the brighter network features in K_3 and K_2 often dropping to unity in areas where the network is undisturbed. Accurate contrast ratios require observations with high spatial resolution and it is quite possible that the reported contrast ratios are still seriously degraded by lack of spatial resolution.

Existing observations of the network at transition region heights, such as those shown in Figure II-9, suggest that the network width is typically $\frac{1}{4}$ to $\frac{1}{5}$ of the total cell width. For a system of square cells of dimension l uniformly separated by network of width xl, the ratio of network area to cell area is given by $x^2 + 2x$. If the network width, xl is $1/n$ of $xl + l$, then $x = 1/(n-1)$ and for $n \gg 1$ the ratio of network area to cell area is approximately $2/(n-1)$. This would give typical network to cell area ratios of $\frac{2}{3}$ to $\frac{1}{2}$, i.e., the network would cover 40% to 33% of the surface area. However, this estimate treats the network as covering the entire sun uniformly, which is not the case. Some observers estimate the network area as being about 10% of the total surface area at the level of the transition region.

The transition region is punctured by spicules in the sense that the spicules extend well above the base of the transition region. To a reasonable approximation, the mean thickness of the transition region may be estimated from the equation (see Chapter VII)

$$h = 2[(10^{-5} \, T)^{7/2} - 1], \tag{II-9}$$

where Δh is the thickness in km between the layers where the temperature is T and 10^5 K. For $T = 10^6$ K, Equation (II-9) gives $\Delta h = 6300$ km. Since the base of the transition region is near 2000 km, the level where $T = 10^6$ K is near 8000 km. Most spicules reach a height of 10 000 km, and, hence, most of them penetrate through the layer where $T = 10^6$ K.

We noted earlier that a given network cell has a number of spicules that is approximately equal to N_j in Table II-1. Thus at a mean height of 3000 km a network cell has an average of approximately 25 spicules. At 5000 km this average spicule population is reduced to about 10 per cell. This is not a large enough number to give the cell border a conspicuous appearance near the center of the disk.

Ten spicules of average height 5000 km and average diameter 1000 km have a total cylindrical surface area of 1.5×10^8 km^2. By comparison a typical network cell has a surface area of approximately 9×10^8 km^2. Thus, the total surface area of the spicules in the transition region is considerably smaller than the surface area of the Sun, and even if the spicules were themselves sheathed in a transition layer similar to that covering the rest of the Sun this would add only about 20% to the total volume of transition region material.

The relative importance of the spicule phenomenon becomes more important in some respects as the limb is approached. If, for example, the spicules are opaque at some wavelengths, they may obscure much of the transition region radiation near the limb. At 5000 km height, one can count about 4 spicules per degree. Using a mean spicule diameter of 1000 km and the linear distance along the limb corresponding to $1°$ of solar latitude (12 200 km), we find that approximately a third of the limb is obscured by spicules. Near 3000 km the limb becomes almost completely obscured. Thus any radiation that arises near to or below 5000 km and for which the spicules are opaque would be strongly modified by the spicules near the limb if the spicules were the only discrete structures present.

Observations of transition region emission lines that lie within the wavelengths of the Lyman continuum of hydrogen exhibit anomalous limb-brightening when compared to similar lines that do not fall within the Lyman continuum (Withbroe, 1970). This effect is illustrated in Figure II-16. Such observations show clearly that features of the chromosphere that are opaque in the Lyman continuum extend into the transition region, an effect which could easily be due to spicules. However, we caution that spicules are not the only source of roughening for the chromosphere and the transition region. In the lower levels of the transition region, fibrils may occupy considerably more volume than spicules.

7. Non-Spherical Modeling

Spicule statistics show unambiguously that the fractional surface area of the Sun covered by spicules in the upper chromosphere and in the transition region is far less than the surface area of the network. The simplest complete picture of the upper transition region thus appears to consist of the cells, the general network walls and the spicules with fractional areas of the components decreasing in that order. Foukal (1974) has proposed a qualitative model of this type. Within the chromosphere itself, and possibly in the lower extremities of the transition region as well, fibrils make up an important fourth component.

Extant quantitative models of the chromosphere and transition region have not reached a level of complexity sufficient to include a multi-component network. Indeed, much of what is known about quantitative network modeling as opposed to the average chromosphere model derives from the center-limb behavior of mm and cm radio data and

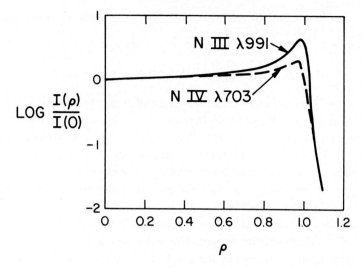

Fig. II-16.   Differential center-limb behavior for two transition region lines formed either side of the Lyman Series limit (λ 912). The absence of marked limb brightening in λ 703 is attributed to absorption by spicules or fibrils.

XUV line and continuum data such as illustrated by Figure II-16. Such centre-limb studies lead, almost universally, to the conclusion that the chromospheric and transition region surfaces are rough and that relatively cool components of the atmosphere obscure the hotter components at and near the limb. Thus, one pictures discrete, cool extensions of the lower layers into the overlying hotter layers. Unfortunately, one then immediately jumps to the conclusion that these cool upward extensions are, in fact, spicules. This has tended to create an image of spicules that is truly schizophrenic in character. We have tried to make spicules the scapegoat for all of the non-spherical structure of the chromosphere and transition region in spite of the readily verified fact that the fibrils cover a much larger area of the Sun and in spite of the fact that there is strong evidence for a general network wall many times more extensive in horizontal cross section than the spicules. Perhaps one could escape this latter conclusion by encasing the spicules in a thick sheath of network material so that the spicule area became much greater. All this would accomplish, however, is to give the spicules major horizontal structure as well as vertical structure and this in no way simplifies the overall model.

Center-limb studies of chromospheric radiation clearly reflect the properties of the cells, fibrils, spicules and network wall combined. It seems equally clear that for radiation originating at heights above 3000 km the spicules and the network wall dominate the center-limb effects. Possibly the spicules produce the dominant effect because of their extreme density and temperature characteristics.

For radiation arising below 3000 km, such as the Lyman continuum and transition region radiation characteristic of temperatures below about 5×10^5 K, the fibrils probably play the dominant role in center-limb behavior simply because of their strong tendency for a nearly horizontal orientation and their resultant overwhelming cross-sectional area.

8. Chromospheric Active Region Structure

Active regions on the Sun show rich structural detail in the chromosphere as is illustrated by the Hα filtergrams in Figure II-17. Perhaps more is known about these structures in terms of their morphology and patterns than is known about the structure of the quiet chromosphere. To do justice to these structures, together with their motions and magnetic fields, would require the addition of a large body of material without markedly increasing our understanding of the 'average' chromosphere. It is clear that the chromosphere exists independently of active regions and that the Sun observed as a star would show only moderate effects from chromospheric activity. Thus, in the interest of brevity we will comment only briefly on active region structure.

Several features of active regions stand out in contrast to quiet solar regions. Within the close confines of the active region the network structure is not pronounced. The bright and dark fibrils are both more elongated and more organized on a large scale and have a longer lifetime. Bright plages appear near the sunspots and occasionally flare for brief periods. In the outlying areas surrounding the active regions the network structure appears to be enhanced.

For the most part the typical features of the quiet chromosphere, e.g., network cells, rosettes and fibrils, occur within a region of predominantly one magnetic polarity. In other words, the scale of the general magnetic polarity reversals is large compared to the chromospheric features. In active regions this is no longer true. The magnetic polarity reverses, often several times, within a distance scale that is small compared to the extent of the active region itself. In addition, the average magnetic fields in active regions are 10^2 to 10^3 as strong as the average fields in the quiet chromospheric region. This has the consequence that the average magnetic energy density exceeds the thermal energy density in the active regions whereas the reverse is true in the quiet chromosphere. Thus, in the active region one sees structure that is dictated by the magnetic field configuration, and in the quiet chromosphere one sees magnetic field configurations that are dictated by the supergranule. Clearly this sharp dichotomy is not true for all of the fine features, but it accounts for the highly ordered structure of the active region and for the lack of network structure within the interior of the region.

Again, in the quiet chromosphere the structures observed are often associated with a single magnetic polarity. In active regions the structures often connect regions of opposite polarity. A characteristic chromospheric structure seen in regions of field reversal is the so-called 'field transition arch' (Prata, 1971) also referred to as 'arch filament systems' by earlier observers (Bruzek, 1967). These arches present a pattern of dark fibrils similar to the other dark fibrils of the active regions except that the field transition arches are shorter, more sharply curved and higher in elevation than the fibrils. The orientation of the dark fibrils, including the field transition arches, is consistent with the supposition

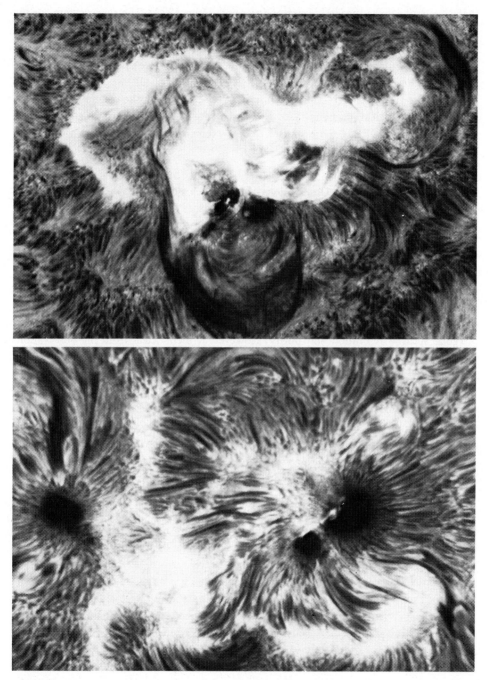

Fig. II-17.   Hα photographs of active regions. The top image shows the late stage of a large, double-ribbon chromospheric flare overlying a sunspot group. This flare (August 7, 1972) produced energetic protons at Earth. The lower image shows a bi-polar spot group (May 21, 1972) (Hale Observatories, Calif. Institute of Technology).

that they parallel the magnetic lines of force (cf. Zirin, 1972). The same is true for the bright and dark fibrils of the rosettes away from active centers (cf. Chapter IV). However, because of the polarity reversals in the active regions the magnetic field pattern is complex and the field transition arches sometimes appear to have orientations opposed to those of the longer dark fibrils.

The plage regions surrounding sunspot areas appear to be regions of strong, predominantly vertical magnetic field. As we shall see in Chapter IV, the same is true of the network regions. Thus, there appears to be a close analogy between the plages and the network.

9. Observations of Coronal Structures

Geometrical structures observed in the corona differ in fundamental ways from those observed in the chromosphere. The coronal structures have much greater vertical extent and larger horizontal dimensions and the corona is dominated more completely by active region structures. No evidence for a general network structure similar to that found in the chromosphere has been observed in the corona. Similarly, no fine scale structure on a scale comparable to rosettes and fibrils has yet been identified in the corona. Admittedly, observations of sufficient quality to reveal such structure are limited to a few cases of marginal spectral resolution and purity. However, it seems clear from the observations that do exist that the dominant geometrical structures of the corona are much coarser than those found in the chromosphere.

There are three major reasons for supposing that observed coronal structures should be coarser than those in the chromosphere: (1) the coronal scale height is much greater, (2) the corona is further removed from the poles of the local magnetic fields, and (3) the corona is optically thin at visual and most XUV wavelengths where the structure is most readily observed. The temperature of the corona is approximately 2×10^6 K, which gives an isothermal, hydrostatic scale height in the lower corona of 10^5 km. This is to be compared with a corresponding scale height of approximately 200 km in the chromosphere. Because of the large scale height in the corona, the mean level at which features are observed against the disk, say, 20 000 km, lies far above the chromosphere. Thus, the coronal features, on the average, lie at least 10 times as far above the photosphere as does the 'average' chromosphere feature. Around active regions where the photospheric magnetic field strength and polarity change rapidly with position on the solar surface the 'poles' of the magnetic field evidently lie close to the photosphere. This means that the average coronal feature is much further from the magnetic poles than is the average chromospheric feature. As a result, we expect the average coronal magnetic fields near active regions to be considerably weaker and to show smaller gradients than those in the chromosphere. Geometrical features associated with magnetic field structure, therefore, should generally be more diffuse in the corona than in the chromosphere. Also, the optical thinness of the corona tends to obscure the individual structures by superposing the emission from all of the structures in the line of sight.

At the solar limb, a line drawn tangent to the limb intersects the surface whose height is 10^5 km above the limb at a heliocentric angle of $(\pi - \theta) = 29°$ (see Figure I-2). The length of the chord, y, from $+29°$ to $-29°$ is $1.1\ R_\odot$, or 7.7×10^5 km. An observation of the corona at the limb, therefore, exhibits a superposition of features along about $\frac{1}{6}$ of the solar circumference. Any given feature whose dimensions are small compared to this scale will be observed as a relatively weak 'signal' against a large background of 'noise' and will appear with greatly reduced contrast over what it would have if the background were eliminated. An example of the maximum contrast expected for a cylindrical feature whose brightness is 10 times the mean brightness is given in Table II-2 as a function of the diameter of the cylinder. The background corona is assumed to be uniformly bright for a distance of 7.7×10^5 km and to have zero brightness outside this interval. Thus, the location of the cylinder with respect to the limb is unimportant. A second result is given in Table II-2 for the case of a cylinder of zero brightness within the 7.7×10^5 km strip of the corona.

TABLE II-2
Brightness contrast of cylindrical coronal features

Diameter of cylinder	$(0.01\ R_\odot)$	$(0.03\ R_\odot)$	$(0.1\ R_\odot)$
$\frac{\Delta I}{I}$ (brighter x 10)	0.085	0.25	0.85
$\frac{\Delta I}{I}$ (void0	−0.0094	−0.028	−0.094

The results in Table II-2 illustrate two important aspects of coronal features observed at the limb: (1) features as small as $10''$ are not likely to be observed unless they are very bright, and (2) features of reduced emission are very inconspicuous unless they are of very large dimension.

Observations of coronal structure on the disk have the advantage that the effective path length of integration is smaller by a factor of about 7 near disk center than it is at the limb. Hence, the predicted contrasts in Table II-2 are increased by about a factor 7 for disk center observations and by about a factor 3.5 at $\mu = 0.5$ $(r = 0.87\ R_\odot)$. The disadvantage of observations of coronal structure on the disk is that height relationships are largely lost.

One of the primary means of studying coronal structure is by means of white light photographic images of the corona taken at total eclipse. The corona is readily observed in the continuum and a wealth of structural detail is evident. Continuum observations of the corona have the advantage of being insensitive to velocity, magnetic or temperature effects. The structures that are observed, therefore, can be related unambiguously to the density structure of the corona.

Because of the high coronal temperature both of the major constituents of the atmosphere (hydrogen and helium) are fully ionized. This means that the electron density of the corona is not directly dependent upon the temperature (the election density may be indirectly related to temperature, of course, through the gas pressure). The continuum spectrum of the corona mimics that of the photosphere which gives two important clues to the origin of the continuum; (i) the coronal continuum is scattered photospheric radiation, and (ii) the scattering is independent of wavelength. A third clue comes from the fact that the Fraunhofer lines are lost in the scattering process, i.e., they are broadened so much that they can no longer be recognized. All three clues point to electron scattering as the source of the continuum. If we define a scattering thickness, τ_S, for the inner corona by the relationship

$$I_c = \tau_S I_p, \tag{II-10}$$

where I_c and I_p are the coronal and photospheric brightness, respectively, it is clear from Figure I-6 that τ_S is the order of $10^{-5.5}$ to 10^{-6}. Further out in the corona it is necessary to take into account the dilution of the photospheric radiation.

The scattering thickness, τ_S, as defined by Equation (II-10) is directly proportional to electron density (see Chapter VI). Thus, features observed in the continuum are directly related to the electron density structure, which, in turn, is directly related to the total density structure.

Eclipse observations of the continuum have the strong disadvantage that they do not readily permit studies of lifetimes and motions. By placing a series of instruments along the path of totality the observing period can sometimes be extended to a few hours. However, this is a costly and uncertain procedure and difficult to carry out.

Because the coronal continuum arises from electron scattering, the radiation is plane polarized with the electric vector parallel to the solar limb (see Chapter VI). The polarization properties of the coronal radiation can be used to discriminate against the normal sky brightness outside of eclipse. Such observations have been made now for a number of years with the High Altitude Observatory's K-coronameter at Hawaii. These observations give only coarse spatial resolution and therefore permit studies of only the coarser coronal features at the limb.

Externally occulted coronagraphs flown on the OSO-7 satellite and on the manned ATM mission have provided additional opportunities to study the white light corona. Observations of this type provide the opportunity to observe the corona frequently over a sustained period of time with relatively good spatial resolution. Hence, they are ideally suited for studying the evolution of small scale coronal structures and transient events as well as the large scale features. Because the satellite instruments are externally occulted by a disk that is relatively near the objective lens of the coronagraph, these instruments do not see the inner corona.

Spectral line radiation from the corona, in visual regions of the spectrum, may be up to 100 times brighter than the continuum, per unit wavelength interval, and the coronal line emission can be observed routinely outside of eclipse with a suitably located coronagraph. The most common method employs the use of spectra taken near the limb at selected locations. Narrow band filters designed to isolate the green (λ 5303) line of Fe XIV have been used successfully also. These latter observations are, of course, more useful for studying structural details and their morphology. They are much more difficult to obtain, however, and only limited observations have been made.

The use of spectral lines for studying coronal structures introduces complications in the interpretations of the data. Velocity and magnetic field effects are readily controlled by looking at the entire line rather than some narrow band within the line. However, the line radiation in the inner corona is excited by collisions, and, as a result is proportional to the square of the density. It is, in addition, a strong function of the local temperature, since this determines the density of the atoms in the particular state of ionization that produces the line in question (see Chapter VI). Because of these effects structural details of the corona are generally of higher contrast in the lines than in the continuum. It is not always clear, however, whether the contrast is due to density changes, temperature changes, or both. Usually, it is both.

The best observations of coronal structure on the disk have been achieved in the XUV and X-ray regions of the spectrum. Observations with sufficient spatial resolution to exhibit the finer features of the corona have been made from both rockets and satellites.

10. Fine Structure of Inner Corona

The results in Table II-2 suggest that any small scale ($<30''$) bright feature observed prominently in the coronal continuum represents a large local density increase and that a small scale local density decrease, no matter of what amplitude, will be very inconspicuous. In view of this, it seems somewhat surprising that continuum photographs of the corona do, in fact, exhibit many small scale features. This can only be interpreted to mean that the geometrical structure of the corona is very pronounced.

Figures I-3, II-18 and II-19 are eclipse photographs of the white light corona taken through a neutral density filter whose transmission varies radially from the center of the filter outward in a cylindrically symmetric manner. The solar image is centered on the filter and the filter transmission is designed to vary inversely with the mean coronal brightness. Hence, the product of coronal brightness and filter transmission is approximately independent of the radial distance from the Sun (within limits, of course). This provides an image in which the overall radial gradients are suppressed and, as a consequence, in which structural details are more readily apparent.

In the inner corona ($r < 1.2\,R_\odot$) many small scale features can be seen whose characteristic dimensions are $10''$ and for which $\Delta I/I \approx 0.1$ (order of magnitude estimates only). These structures imply local density increases of approximately a factor of 10 (Table II-2). Furthermore, they imply that the apparently darker fine scale features are nothing more than 'normal' corona showing between regions of density enhancement, i.e., they do not represent voids in the corona. This follows from the results in Table II-2, which show that a void of $10''$ size will show with a 1% contrast rather than 10%. It does

Fig. II-18.   An eclipse photograph of the white light corona made with a radial gradient filter in order to reduce the apparent brightness gradient. Eclipse date November 12. 1966. The bright feature to the Northwest of the Sun is Venus (High Altitude Observatory, National Center for Atmospheric Research).

Fig. II-19.   The same as Figure II-18 except that the eclipse date was March 7, 1970.

not follow, of course, that small regions of abnormally low density do not exist in the corona. They are simply unobservable.

We conclude from the above arguments that the inner corona is highly structured on a scale of $10''$ with frequent local density fluctuations in excess of a factor of 10. Our reason for bringing this point out quantitatively at this point is to emphasize again that coronal geometry is of utmost importance. It has often been assumed from the appearance of coronal eclipse photographs that the small scale geometry is not of much importance and that to a first approximation, the corona is spherically symmetric. While this may be a valid approximation for some purposes, it is grossly invalid as a general rule.

Because of the long integration path through the corona, it is difficult to avoid active regions in observations at low solar latitudes. A single bright active center may more than double the local surface brightness of the corona. For a region of $100''$ diameter, this implies (Table II-2) an enhanced intrinsic brightness in the active region in excess of 10 fold. Even an embryo or dying region may have such properties and mature regions may be still brighter. It is not too surprising, therefore, that a path around the Sun at constant r frequently encounters intensity variations, in the continuum, of about a factor 2 and occasional variations of a factor 10. Changes of this amplitude, however, are associated with structures of the order of $100''$ in size, or larger. On the scale of $10''$ one rarely sees brightness variations much in excess of 10 to 20%. It appears, therefore, that the common density variations about the mean are more nearly a factor of 10 than a factor of 100. Two-basic classes of small scale structures can be recognized in the inner corona. These are: arches, or systems of arches, and vertical rays. Individual arches have cross-section diameters of the order of $10''$ though clearly some are larger and some, perhaps many, are smaller. The rays are often grouped in bunches and the arches commonly occur in groups. Thus, these small scale structures are clearly associated with larger structures. They are most pronounced over active centers, but are not restricted to these centers.

Separation of coronal structures into quiet and active classes is both difficult and of questionable merit. Coronal forms associated with active centers are much more extensive, spatially, than are chromospheric forms, and it is often difficult to draw a boundary between the active and quiet features. In both the active and quiet Sun the coronal features seem to be dictated by the magnetic field structures. The fields in active centers are more intense and perhaps more complex in topology than in quiet areas. This is reflected in a more complex array of features within some active centers than one would normally expect to find in quiet coronal areas. Nevertheless, both the active and quiet corona exhibit systems of rays and arches of rather similar character.

Because the spectral line radiation from the inner corona is proportional to the square of electron density, features in the inner corona are more readily apparent in the spectral lines than in the continuum. By the same token, the line observations preferentially favor the active regions because of the generally enhanced densities in these regions. Nevertheless, we choose the line spectrum data to illustrate the nature of the structures in the inner corona.

Figure II-20 contains three photographs of coronal regions in the green (λ 5303) line

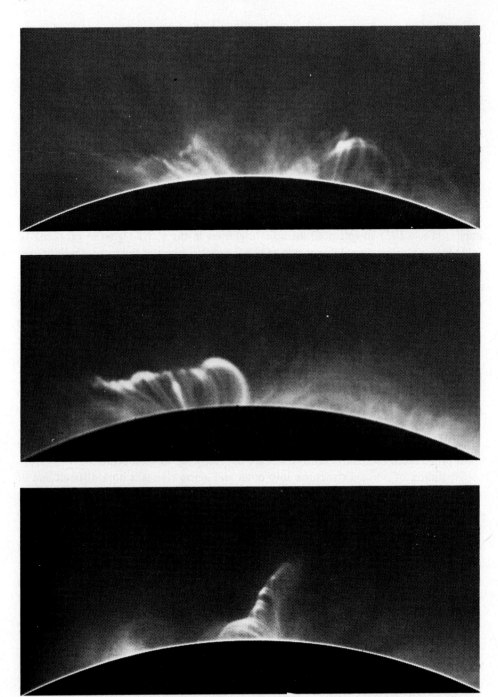

Fig. II-20.   Examples of arch structures observed in the corona in the λ 5303 line of Fe XIV. Arches of this type are frequently evolving on a relative short time scale (minutes) (Sacramento Peak Observatory, Air Force Cambridge Research Laboratories).

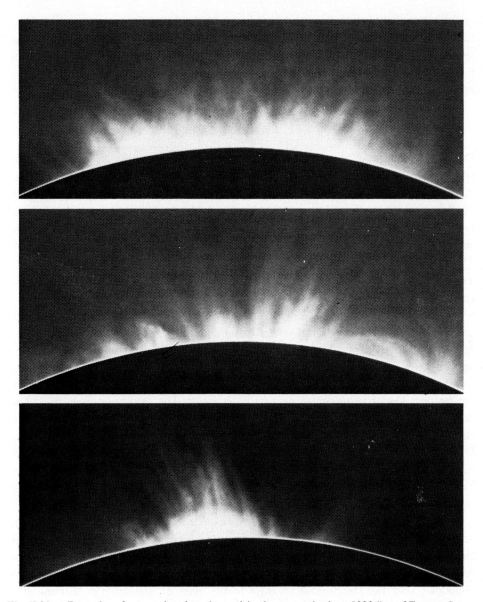

Fig. II-21. Examples of rays and arches observed in the corona in the λ 5303 line of Fe XIV. Some of what appears to be rays may be one leg of an arch or an arch viewed edge on (Sacramento Peak Observatory, Air Force Cambridge Research Laboratories).

of Fe XIV obtained with a birefringent filter at the Sacramento Peak Observatory. The three photographs were selected to show regions dominated in the inner corona by systems of closed arches. Figure II-21 shows three similar photographs in which a more open structure of diverging rays seems to dominate, at least in the higher levels of the regions observed. Little is known about the lifetimes and morphology of the individual features in such structures due to a paucity of observations. The overall structural forms undoubtedly have lifetimes typical of the active regions themselves, viz., of the order of two weeks or more. It seems highly probable, however, that a given arch or ray has a lifetime that is measured in hours rather than weeks.

11. Coronal Streamers

One of the more prominent large scale features of the corona is the streamer. several of which are visible in Figures I-3, II-18 and II-19. These streamers have a characteristic bulbous appearance of an ellipsoidal dome topped with a long tapering spire, as is illustrated schematically in Figure II-22. The ellipsoidal dome at the base of the streamer often terminates below $r = 1.5\,R_\odot$ and rarely, if ever, exceeds $r = 2.0\,R_\odot$. At the base of and near the center of the streamer, there is often a bright prominence (see Section 13). These are generally less than $100''$ in height, but may extend to $1000''$ ($1\,R_\odot$) in length along the solar surface. However, such long prominences are usually elongated in longitude and the latitudinal extent is typically much smaller. Surrounding the prominence, when it is present, and at the center of the dome when no prominence is present, there is often a region of reduced continuum brightness with dimensions of the order of $100-300''$. Around this region of reduced brightness, and concentric with it, one sees bright arches more or less equally spaced with successively increasing radii. The arches seem to extend throughout the remainder of the ellipsoidal base of the streamer, but become harder to recognize near its outer limits. The spacing between successive bright arches appears to be typically of the order of 50 to $200''$, although there is considerable variation. Above the arch system of the ellipsoidal base the streamer filaments stretch out into long parallel fibers extending many solar radii into space. Beyond about $1.5\,R_\odot$ these streamer extensions dominate much of the coronal structure. The base of the streamers may vary in latitudinal extent from about $300''$ ($0.3\,R_\odot$) to about $1000''$ ($1\,R_\odot$).

The preceding description of a helmet streamer is somewhat idealized and does not accurately describe the appearance of all streamers. A fundamental question regarding the idealized streamer described is whether it is axially symmetric, as an ellipsoid of revolution, or planarly symmetric, as in a long arcade. If it is the latter, then, we should sometimes view the structure at an angle to the plane of symmetry so that the arch structure of the arcade would not be concentric with the apparent center of the streamer and would sometimes intersect each other. The fact that the prominences associated with streamers are often very elongated along the solar surface suggests the arcade model. Additional suggestions for the arcade model are provided by modeling of the coronal magnetic fields (Chapter IV), by X-ray images of the disk corona and by the asymmetric appearance of many streamers. It appears to be consistent with the observations, therefore, to visualize the streamer as an arcade of arches enshrouded by the rays that tend first to follow the curvature of the arches out to their crest then sweep outwards

Fig. II-22.   A schematic drawing of the corona emphasizing features of the corona described in the text.

into a fan of parallel rays in the outer parts of the streamer. In such a picture, it is clearly necessary that the axis of the arcade has a preferential east-west orientation. Otherwise, the arcade would be observed as often from the side as along the axis and a 'typical' streamer pattern would be hard to recognize. Clearly, however, there is a more or less typical pattern and thus a preferred orientation.

Figure II-23 shows intensity contours for the corona at $1.17\,R_\odot$ and $1.50\,R_\odot$ as observed with the white light coronagraph of Mauna Loa for the period 3–11 March, 1970. The ordinate gives the product of polarization times brightness and is thus not a direct brightness measure. Heliocentric position angles in the figure are measured

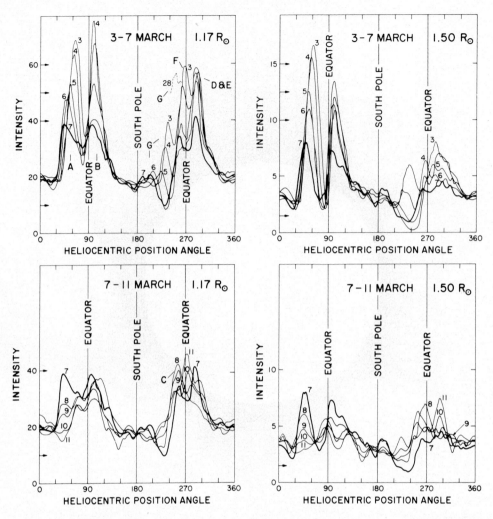

Fig. II-23.   K-coronameter contours of the inner corona at $R_\odot = 1.17$ and 1.50 for the period
surrounding the eclipse of March 7, 1970 (Fig. II-19). Coronal features identified with the contour
features are indicated in part b of the figure (after Hansen *et al.*, 1970).

eastward from the north solar pole at $0°$. The feature marked A is the bright helmet
streamer located in the northeast quadrant in Figure II-19. This feature is seen to have
had maximum brightness on 3 March, 4 days prior to the eclipse, and to remain visible at
the east limb until about 9 or 10 March. Also, the point of maximum brightness in the
limb projection of the streamer is seen to move steadily poleward as time progresses. A
model of this streamer constructed by Hansen *et al.* (1970) shows an elongated feature
extending upwards of $120°$ in longitude along an axis tilted from near $90°$ heliocentric at
its leading extremity to near $45°$ heliocentric latitude at its trailing extremity. It is

Fig. II-23b.

therefore highly suggestive of the arcade structure described above. Other features in the photograph in Figure II-19 tend to show much more limited longitudinal extend. Hence, feature *A* is perhaps somewhat atypical.

From a summary analysis of three years of data from the white light coronagraph, Hansen *et al.* (1972) have described the life cycle of two types of idealized coronal streamers. Since their idealized model is intimately associated with the behavior of photospheric magnetic fields, it is necessary to digress for a moment in order to describe the magnetic field patterns of interest in connection with streamers. The description of the magnetic fields given here is cursory and will be given in more detail in Chapter IV.

Active regions on the Sun are the sources of new magnetic flux much of which remains on the Sun in a dispersed configuration long after the active region itself has lost identity. Although active regions tend toward a balance between fields of opposite polarity, the polarity of the leader sunspots lies, statistically, nearer the solar equator than does the polarity of the follower spots, i.e., the axis of the approximate magnetic dipole is tilted toward the equator at its leading edge. As the active region decays the follower polarity drifts poleward and becomes stretched out in longitude by differential solar angular velocity, which is faster at the equator and slower at the poles. This results in a band of unipolar magnetic field trailing the site of the original active region and arching poleward along a path that is concave towards the equator. This unipolar region supposedly reconnects to opposite polarity across the equator from its lower latitude side and to polar magnetic fields from its high latitude magnetic fields from its high latitude side. Thus, in this idealized picture proposed, in part, by Babcock (1961) the unipolar region is envisioned to form two magnetic arcades: one spanning the equator and one reaching toward the polar crown.

Hansen *et al*. (1972) find evidence for two systems of long lived coronal streamers whose characteristics coincide with the idealized magnetic picture of unipolar regions. Six prominent coronal features were selected from the three year period of the study because of their appearance at high solar latitude, poleward of 45°, their long duration, at least three solar rotation periods, and their isolation from other major features. Each of the six features were found to have originated over active regions and thence moved steadily poleward with succeeding solar rotations. In their final configuration, they were found to be stretched out in longitude astraddle polar crown prominence filaments, which themselves were lying along the poleward borders of unipolar magnetic regions. Thus, one imagines that this group of streamers traces out the magnetic field pattern over the prominence filament. In cross section, the streamer-prominence relationship is more or less as shown in Figure II-22. In projection on the disk, however, the prominence filament extends serpent-like over the solar surface and is encased by the coronal arcade.

Hansen *et al*. (1972) found a second set of five distinctive, long-lived coronal features that appeared to be consistent with the picture of an equatorial arcade (of shorter longitudinal extent) connecting fields of opposite polarities in the two hemispheres. However, the detailed histories and associations of these features were not presented.

The models of streamers presented by Hansen *et al*. (1972) are perhaps overly idealized to be representative of the majority of streamers. Nevertheless, they do show clearly that many of the prominent streamers are fan-like in structure as opposed to the pencil-like structures often postulated.

Arch structures in the lower corona that are associated with helmet streamers are relatively faint features when seen at the limb in white light. Brighter more prominent arch systems are often seen in association with active regions near the limb. It should not be assumed, therefore, that all arch structures, such as those shown in Figure II-20, are indicative of streamers. Helmet streamer structure is most pronounced in middle solar latitudes on the poleward side of active centers and is more pronounced when the Sun is active than when it is quiet.

As was mentioned in the preceding discussion some streamers are directly associated with active regions on the Sun. Such streamers do not usually show prominences at their centers and the centers are enhanced in brightness rather than depleted. Arch structures may be present, but they are not usually so simply arranged as are the arches over prominences in the higher latitude streamers. Some authors (cf. Newkirk, 1967; Hansen *et al*., 1972) have suggested that streamers over active regions and helmet streamers over prominences are simply different evolutionary stages of a single phenomenon. Active region streamers exist only for a week or two during the average 'mature' lifetime of the active regions and it is suggested that these streamers gradually evolve into the longer lived prominence streamers. This picture is largely based on speculation and plausibility arguments derived from the fact that both types of streamers are dependent upon the active region and its magnetic fields. It remains conjectural.

Within the inner corona the brightness contrast between a streamer and its immediate neighborhood has a value near two. For a long arcade structure, this would imply a factor two increase in density within the streamer, but for a cylindrically symmetric streamer it would imply a density increase of about a factor 5. The lower value is more in keeping with the proposed model of streamers. Far out in the corona, say, $r = 3\,R_\odot$, the streamer contrast is much larger and approaches a value of 10 for the brighter streamers. Assuming that these brighter cases represent those where the arcades are aligned with the line of sight, we conclude that a characteristic density increase for the streamers at these distances is about a factor of 10. It was noted in the preceding section that the arch structures within the streamers implied local density increases of a factor of 10. This, however, assumed that the arches were of circular cross section. If, instead, they represent an alignment of individual arches within an arcade the required density increase would be correspondingly smaller. However, this seems like a remote possibility because of the small cross section of the arches and the long path length through the arcade. It seems more plausible to consider the arches as individual tubes of unusual density enhancement. Thus, the arches in which the density increases by a factor 10 are to be regarded as the extremes rather than the average. It should be noted, however, that any inferences about density increases presupposes a particular geometry and illustrates the need to develop proper geometrical models.

It is readily apparent from the photographs in Figures II-18, II-19 and I-3 that the axis of the streamers is not always radial to the Sun. More often than not the streamer axis in the inner corona is curved rather sharply as if drawn towards some preferred direction or, conversely, as if avoiding some forbidden region. In the outer corona the streamers are more gently curved, and one streamer studied rather extensively with balloon borne coronagraphs, eclipse observations and K-coronameter observations (Bohlin, 1970) exhibited the Archimedes spiral structure characteristic of matter flowing radially out of the Sun with constant radial velocity and with the tangential velocity of the Sun at $r = 2\,R_\odot$. In other words, the streamer trails the Sun beyond $r = 2\,R_\odot$ rather than rotates as a rigid body. This type of effect does not account for the more sharply defined curvature in the inner corona.

Streamer curvature in the inner corona is sometimes directed towards the equator and sometimes toward the poles. It appears sometimes to be directed away from nearby active centers and sometimes towards them. Thus, no clear pattern is evident. The statistics of such relationships are still very poor, however. Satellite observations of the white light corona with coronagraphs provide hope of obtaining sufficient observations of streamers to clarify their relationships to other coronal and solar features.

An illustration of the potential for studying the structure and evolution of the corona from satellite borne coronagraphs is provided by Figure II-24. This photograph obtained with the ATM Skylab clearly reveals much of the structural detail of the corona

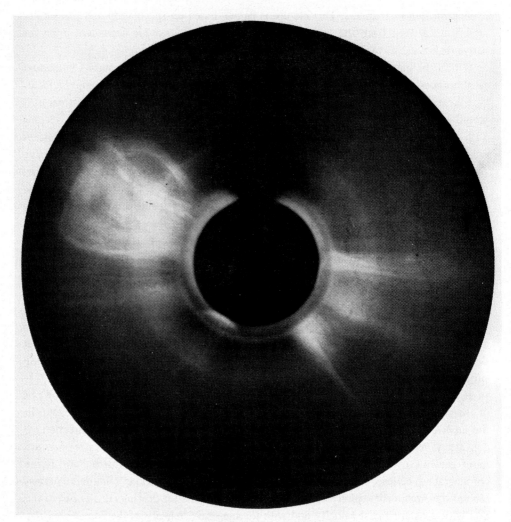

Fig. II-24.   A photograph of a large transient event in the corona observed by the ATM Skylab. Observations were made in white light with an externally occulted coronagraph. The solar diameter is approximately half of the centrally occulted area. Vignetting due to the coronagraph occulting disk reduces the apparent radial brightness gradient. The dark feature at the top is formed by the occultation produced by the support post for the occulting disk. A bright feature near the limb at about 7 o'clock position is produced by a small piece of lint on the occulting disk, later removed by one of the Skylab astronauts. Transient features of the type shown were observed by ATM Skylab at an average interval of about 3 days (High Altitude Observatory, National Center for Atmospheric Research).

beyond $r = 2 R_{\odot}$ at the time of a major transient event in the corona. Transient events of such magnitude were unknown to visual observations prior to the ATM experiment even though their existence had been suspected on the basis of radio noise events. The ATM data are too new at the time of this writing to allow a detailed description of the results.

One result that is clear and of obvious importance, however, is that pronounced changes in coronal form occur relatively frequently on a time scale that is less than the solar rotation period. This casts some doubt on the practice of constructing three-dimensional coronal models from observations of the coronal structure at the limb over a full rotation period.

12. Coronal Rays and Plumes

A second characteristic large scale feature of the corona is the system of more or less vertical rays shown near the south and north polar regions in Figure I-3 and illustrated schematically in Figure II-22. Such structures can occur at any latitude although they are seen most often near the poles during periods near sunspot minimum.

A question that immediately arises is whether the ray structures referred to are not simply side views of streamer arcades. We noted in the preceding section that in the outer corona the streamers are much brighter than the ambient corona and that they often curve poleward. It follows from this and the added fact that polar corona is generally fainter than the corona at lower latitudes that streamer arcades whose base is quite some distance from the solar limb could easily appear as relatively conspicuous features in projection above the polar limb. Some of the ray structure undoubtedly arises from this effect. However, both Figure II-21 and X-ray images of the disk corona (see Figure I-5) show clearly that open ray structures are not at all uncommon in the corona, particularly near the poles.

Near sunspot minimum the polar corona has both reduced brightness and an increased gradient in brightness. Hence the minimum polar corona is much less conspicuous than either the maximum polar corona or the minimum equatorial corona. This is apparent from a comparison of Figures I-3 and II-18 with Figure II-19.

In the polar regions shown in Figure I-3 and in the illustration in Figure II-22, there are plume structures of diverging rays that are characteristic of the minimum polar corona. As we shall note in Chapter IV, the polar regions of the Sun tend, during times of sunspot minimum, to show a more or less uniform magnetic polarity of a given sign. It has been commonly assumed that the polar plumes outline the magnetic lines of force of the polar fields. However, it seems clear that some of the polar plume structure does arise from projection of lower latitude streamers as well.

A typical plume filament has a diameter of approximately $10''$. The line of sight integration through the corona at the limb has an effective path length of about 7.7×10^5 km, or approximately $1100''$. Thus, if the polar corona were uniformly broken up into alternate plumes and voids of $10''$ size, a given line of sight would intersect approximately 50 plumes and it would be difficult indeed to recognize a plume structure. The most one could hope for would be to observe something akin to the ray structure with an occasional, and random, appearance of an unusually bright plume. This is not what is observed, however. The plume structure is distinctly visible and the plumes are spaced more or less regularly along the limb. So regular is the plume structure, in fact, that several observers have traced the diverging plume structure back to a common focal point. These focal points occur near $r = \frac{2}{3} R_\odot$ but their exact location varies during the solar cycle and does not usually lie on the rotation axis. However one interprets this evidence of a common focal point, it dramatizes the regularity of the plume structure.

The clear regularity of the plume structure has led some observers (Saito, 1965) to the speculation that the plumes are concentrated in well defined latitude zones near 75–80° (measured from the equator). Thus, one envisions the plumes as a crown of regularly spaced features girdling the poles somewhat askew and showing in projection above the poles. Disk observations tend to support such a picture, and it is difficult to see how the observed regularities permit any other interpretation. Such a picture is not inconsistent with the interpretation of plumes as part of the streamer arcade in projection, but other pictures are possible and perhaps equally plausible. Thus, one can envision the plumes as simply a crown of fine rays encircling the polar regions without identifying them in any particular way with helmet streamers.

The brightness contrast in the plumes is again of the order of 10%, which together with their size, implies density enhancements of about a factor of ten. Lifetimes of individual plumes are difficult to establish, but they have been reported as being approximately 15 h (cf. Newkirk, 1967). The occurrence of plume structure itself obviously has a lifetime of at least a few years.

Some observers associate polar plumes with polar photospheric faculae, and others have attempted to associate them with spicules. Such associations are exceedingly difficult to establish and no firm conclusions can be drawn. The fact that the polar faculae and the plumes are most evident during solar minimum favors such an association. Further credence to this association is given by the evident association of both the faculae and the plumes with the polar magnetic field.

13. Coronal Disk Structures and Coronal Holes

The study of coronal structures on the solar disk holds obvious advantages over limb studies. Until relatively recently, however, disk observations in the EUV and X-ray regions of the spectrum with sufficient spatial resolution to exhibit the coronal structure have not been available. Photographic X-ray observations of the solar disk obtained with the ATM Skylab satellite are now available. These data are of excellent quality and cover a few solar rotations during the latter half of 1973. They will clearly add a great deal to our understanding of coronal structure and morphology. Unfortunately, only preliminary results are available at this time.

Figures I-5 and II-25 exhibit two images of the solar disk in soft X-rays (Vaiana *et al.*, 1973a). The two images are separated by 33 days so that some of the features appearing in the western hemisphere in Figure I-5 were in the eastern hemisphere in Figure II-25. The large bright active region below disk center in Figure I-5, for example, was near the east limb in Figure II-25.

Many features of interest are immediately evident in these figures (cf. Vaiana *et al.*,

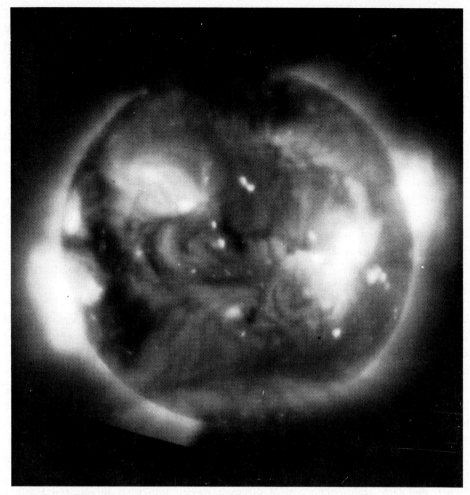

Fig. II-25.   A soft X-ray (3–32 Å and 43–54 Å) image of the Sun obtained by ATM Skylab May 28, 1973. Prominent features of special interest include: (a) arches connecting different active centers, (b) bright points, and (c) coronal holes (American Science and Engineering).

1973a). A general limb brightening occurs except in the polar regions. This is expected from the low optical thickness of the corona to X-rays at these wavelengths. Active regions at the limb, as well as on the disk, show strong enhancements of the X-ray intensity. Again, this behavior is expected on the basis that the active regions, on the average, are hotter and denser than the ambient corona.

The diffuse X-ray emission from the non-active regions shows unmistakable evidence of large scale patterns within which considerable fine structure is observed (Vaiana *et al.*, 1973a; van Speybroeck *et al.*, 1970). Both arches and open ray structure appear to be typical. Also, the diffuse structure on the disk in the southeast quadrant in Figure II-25 is suggestive of the arcade structure postulated for some helmet streamers. Many of the finer filaments seen in Figure I-5 and II-25 are in the form of loops connecting two points on the solar surface that are relatively far from each other. Loops whose foot points are separated by more than $0.3\,R_\odot$ (300") appear to be common. Also, arches frequently bridge across the solar equator connecting the northern and southern hemisphere (Vaiana *et al.*, 1973; van Speybroeck *et al.*, 1970).

Two additional features of interest in Figures I-5 and II-25 are the small bright points and large, dark coronal holes. Dark holes can be seen in Figure II-25 at both poles and on the disk just inside the east limb. This latter dark hole shows in the western hemisphere in Figure I-5 made one and a quarter rotations later. We will return to a discussion of these features after discussing the relationship between the limb and disk features.

Figure II-26 is a montage of an X-ray image of the solar disk in the combined pass bands 3–36 Å and 44–64 Å and a white light photograph of the corona at the limb. The coronal photograph is the same as is shown in Figure II-19 and was made at the 1970 eclipse. The X-ray photograph was made from a rocket flight near the end of the eclipse. Note that the X-ray image is printed slightly smaller than the white light image so as to preserve the X-ray limb structure.

At first glance it appears that the two bright streamers near the east solar limb extend back onto the disk to join with the two bright bands of X-ray emission. A more careful study shows, however, that this is not the case and that the two streamers lie above regions that are behind the limb (van Speybroeck *et al.*, 1970). Similarly the streamer near the south solar pole extends to the disk behind the limb. On the other hand, most of the white light features near the west limb join with X-ray features on the disk near the limb. Some of the larger bright areas on the disk are remnants of active regions that have lost their visible signs of activity in the chromosphere and are likely the bases of streamers. Again we note that arches in the diffuse X-ray pattern rather frequently bridge over distances of the order of 0.2 to $0.3\,R_\odot$, and, in some cases, cross the equator to join the leading polarities of active regions in the two hemispheres (van Speybroeck *et al.*, 1970). This large scale connection of different regions into what Bumba and Howard (1965) refer to as a 'complex of activity' plus the long lived remnants of active regions illustrates the meaninglessness of attempts to separate coronal structures into 'quiet' and 'active' components.

The X-ray image in Figure I-5 was made from the ATM Skylab near the time of the eclipse photograph shown in Figure I-3. Also, the X-ray image in Figure II-25 preceded the eclipse photograph by about one and a quarter solar rotations. The arcade structure seen below disk center and to the east in Figure II-25 is near the west limb at the time of the photograph in Figure I-3. Helmet streamers appear at the limb at this latitude. It is tempting to identify the streamer in the northwest quadrant in Figure I-3 with the bright X-ray region just northeast of disk center in Figure II-25. This apparent association could be misleading, as we have seen from Figure II-26. Nevertheless, there appears to be a close association between bright X-ray regions near the limb in Figure I-5 and the brighter regions of the white light corona shown in Figure I-3. A feature of particular

Fig. II-26.   A montage of a soft X-ray image of the Sun (American Science and Engineering) with an eclipse photo of the white light corona (High Altitude Observatory, National Center for Atmospheric Research) for the March 7, 1970 eclipse. The X-ray photo was taken a little after the eclipse picture. A separate soft X-ray image (upper left) and a harder X-ray image (upper right) show the X-ray structure somewhat more clearly.

interest is that the south polar plumes appear, in the X-ray photograph, to originate on the disk inside the limb.

The dark coronal holes that are so evident in the X-ray photographs show also in the λ 304 line of He II, which ·originates in the lower transition region or upper chromosphere, and the λ 584 line of He I, which is of chromospheric origin (see Figure II-9; also Munro and Withbroe, 1972). Harvey (1974) has discovered that the coronal holes are visible also in the λ 10 830 and λ 5876 lines of He I which are of even lower origin in the chromosphere. On the other hand, Munro and Withbroe (1972) were unable to locate the coronal holes in lines formed in the transition region (with the exception of λ 304 of He II).

Altschuler and Perry (1972) have used time series of K-coronameter data to construct three-dimensional models of coronal electron densities. At the location of coronal holes as identified by the XUV data, they find markedly reduced electron densities. Munro and Withbroe (1972) reach the same conclusion from a comparison of relative line intensities. One property of the coronal holes, therefore, is that the electron densities are abnormally low.

Altschuler *et al.* (1972) have superposed three dimensional electron density models of the corona obtained from K-coronameter data and coronal magnetic fields computed from the observed photospheric fields (see Chapter IV). They find a general correspondence between coronal holes and magnetic regions where the fields are weak and diverging. In other words, the magnetic field appears to open into interplanetary space and to be relatively weak in the vicinity of the holes. This picture is consistent with the frequent appearance of coronal holes at the polar positions. Altschuler *et al.* (1972) have traced the coronal holes out to $2\,R_\odot$.

Lifetimes of individual holes are not well established, but many holes clearly last more than one solar rotation.

The absence of the hole in the transition region lines is interpreted by Munro and Withbroe (1972) to mean that both the electron density and the temperature gradient are reduced in the transition region. The reduced temperature gradient effectively increases the thickness of the transition region to compensate for the reduced density.

The small bright points observed in the X-ray photographs, which are especially notable in Figure I-5 have been found to be associated with small bipolar magnetic regions in the photosphere (Vaiana *et al.*, 1973b). The bright points occur over the entire Sun, including the regions of coronal holes and the polar regions. They appear, therefore, to be associated with the low altitude magnetic geometry rather than with the larger scale properties of the corona. Further study of the bright points may lead to important clues concerning the heating of the chromosphere and corona and the generation of the magnetic cycle as suggested by Vaiana *et al.* (1973b).

14. Prominences

Up to this point we have said little about solar prominences as part of the corona, except as they relate to the streamer structure. It is customary, in fact, to treat prominences as an entirely separate class of phenomena. This approach tends to ignore the obvious association between coronal streamers and certain types of prominences. An association, of course, does not necessarily imply causal relationships, but it certainly suggests them.

A simple demonstration that prominences may be of fundamental importance in the corona can be made from estimates of the fraction of total coronal matter that is contained in the prominences. The mean hydrogen (proton) density at the base of the corona is approximately 3×10^8 cm^{-3}. When multiplied by the area of the Sun ($\approx 6 \times 10^{22}$ cm^2) and the scale height of the corona ($\approx 10^{10}$ cm), this gives a total of 2×10^{41} hydrogen atoms (protons) in the corona. Maps of the solar surface during periods of high sunspot activity give for the total length of prominence filaments on the solar surface values of 10^7 km and higher. If we take an average height of 3×10^4 km and an average thickness of 6×10^3 km, the volume of prominence material in the corona is 1.8×10^{15} km^3 or 1.8×10^{30} cm^3. To obtain an estimate of density in the prominence, we assume that the gas pressure is the same inside and outside the prominence and that the mean prominence temperature is 6000 K. We then find the hydrogen density from the equation

$$[(n_p + n_e)T]_{\text{prom.}} = [(np + n_e)T]_{\text{cor.}}$$

or, assuming $n_p \approx n_e$,

$$n_p = \frac{3 \times 10^8 \times 2 \times 10^6}{6000} = 1 \times 10^{11} \text{ cm}^{-3}.$$

Thus, the total number of hydrogen atoms in the prominences is $1 \times 10^{11} \times 1.8 \times 10^{30} \approx 2 \times 10^{41}$. This just equals the estimate of the total number of hydrogen atoms in the corona, exclusive of prominences. The numbers are, of course, very crude, so they should not be taken too seriously. Nevertheless, they demonstrate clearly that the prominences are very probably of fundamental importance as geometrical structures of the corona.

A further, and even more impressive, argument for including prominence structures as a fundamental part of the corona comes from considering the coronal mass balance. Matter in prominences is observed to flow gradually downward towards the chromosphere. Even if one takes a mean downward velocity of only 1 km s^{-1}, which is lower than most observers estimate, the average prominence would drain back into the chromosphere in a few hours time. Thus, it seems clear that if the corona supplies the prominence matter the prominences provide a very rapid sink for coronal material.

Prominence phenomena are too widely varied to permit a detailed discussion of them here. Many of these phenomena are of highly transitory nature and because of this they are not of as much interest in terms of the overall coronal structure as are the longer lived prominences. Thus, we shall attempt to single out those aspects of long lived prominences that do seem germain to our discussion of the corona.

Prominences whose lifetimes are a few hours, or even a few days, will influence mainly the local coronal region around the prominence. These prominences, which include the majority of prominence types. such as surges, sprays, loops, coronal rain and funnels, are all closely associated with active regions and we shall not consider them further in this text. The interested reader will find an extensive discussion of these phenomena in a recent book by Tandberg-Hanssen (1974).

The prominences of interest to us here are the long-lived, or the so-called 'quiescent', prominences. These are the most commonly seen prominences on the Sun (excluding spicules, which are sometimes classified as prominences) both because of their long lifetimes and their number. They are visible on the solar disk in Hα and in K_3 as dark filaments. When seen at the limb they average some 40 000 to 50 000 km in height and some 100 000 km in length. Their width is much smaller and is difficult to determine

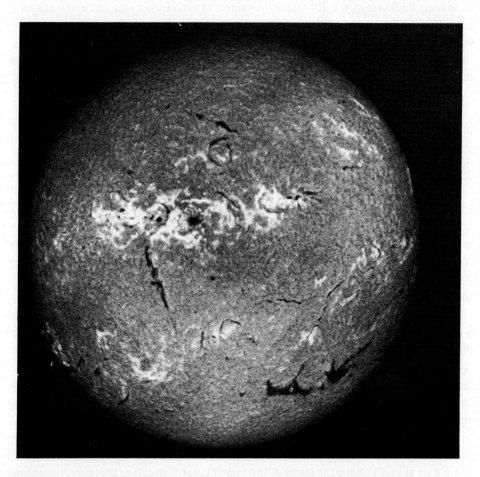

Fig. II-27.   A full disk Hα image of the Sun during a period of relatively high activity. The dark, irregular patches and ribbons are quiescent prominences (or filaments) (Sacramento Peak Observatory, Air Force Cambridge Research Laboratories).

because their extended length and height always gives them an observed cross section that is larger than their true cross section. The best observations suggest thicknesses of less

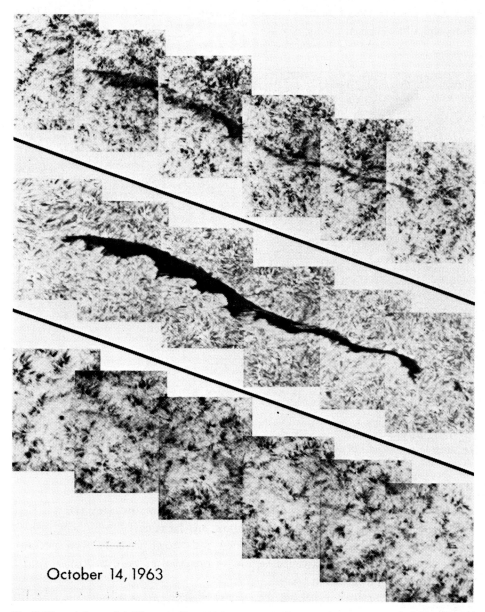

October 14, 1963

Fig. II-28.   A large disk filament observed in Hα at +δ Å (center image), (-0.5 + δ) Å (top image), (+ 0.5 + δ) Å (bottom image). The quantity δ is the result of a small mistuning of the filter. This prominence appears to be optically thick near line center. In the case of some prominences, however, the fine structure of the chromosphere can be seen through the prominence even near line center in Hα. Each of the horizontal steps between adjacent frames is approximately 120″ long.

than 10^4 km but greater than 10^3 km. Most observers adopt values between 10^4 km and 5×10^3 km. Prominences of this type show their greatest variation in their lengths, which often exceed a solar radius and often fall below a tenth of a radius. Figures II-27 and II-28 show photographs of quiescent prominences on the solar surface. Figure II-29 is a photograph of a quiescent prominence at the limb.

Fig. II-29.   A high resolution image of a quiescent prominence at the limb. Of particular interest are the sharp lower boundaries and the filamentary fine structure (Sacramento Peak Observatory, Air Force Cambridge Research Laboratories).

Lifetimes of quiescent prominences vary from a few weeks to many months. They occur both in close association with active regions and far removed from them. Like all other prominences they are closely associated with magnetic fields. The quiescent prominences form between extended regions of opposite magnetic polarity, wherever (but not whenever) such regions occur.

Quiescent prominences associated with active regions follow the migrations of the active regions (statistically) during the sunspot cycle. Early in the cycle they occur near latitudes of 30°–40° and they move progressively nearer the equator as the cycle progresses. Those that are not immediately associated with active regions tend to lie poleward of the activity and to migrate closer to the poles as the sunspot cycle progresses.

They typically reach the pole itself about 3 years following sunspot maximum, i.e., 2 to 3 years before sunspot minimum. As we shall see in Chapter IV, this poleward migration follows the parallel migration of the remnant magnetic fields of active regions. The poleward group of filaments associated with the remnant magnetic fields of active regions constitutes the prominences seen at the base of prominence streamers. They are typically longer lived and of greater length than the active region filaments.

When viewed broadside, as in Figure II-29, the quiescent prominences show intricate structure. Their borders are sharply defined and, internally, they consist essentially of numerous, closely spaced vertical threads. The threads have a typical diameter of approximately $1''$, or smaller.

Filaments on the disk and at the limb tend to show a large scale quasi-periodic structure visible as pillars, or feet, extending from the main filament body to the chromosphere somewhat resembling the feet of a caterpillar. Nakagawa and Malville (1969) find from a study of approximately 90 filaments that the mean separation of the prominence feet is 6×10^4 km but that distribution of sizes is approximately flat between 3×10^4 km and 9×10^4 km. It is tempting to identify the foot points with network structure, since the average diameter of a network cell is just 3×10^4 km. However, Nakagawa and Malville point out that for some filaments, at least, the lifetime of the foot points is several times the lifetime of the network structure, and they use this as an argument against the supposed association with the network. It could be, of course, that the network structure under the filaments is longer lived than the average network, so the lifetime argument is not conclusive. Studies of the network structure under the filaments in lines that do not show the filament itself would be of interest in this connection. However, such studies have not yet been carried out.

Nakagawa and Malville seek an explanation of the foot points in terms of instabilities resulting from the interaction between the prominence and the magnetic field in which it is imbedded. The association between filaments and magnetic fields will be discussed further in Chapter IV.

The high latitude filaments lie underneath the prominent coronal streamers and are surrounded by a region of reduced brightness in the white light continuum. This reduced brightness is clearly evident around some of the prominences seen in Figure II-18. It arises from a reduction in density in the corona surrounding the prominence and suggests that the prominence is formed from a condensation of coronal material.

The mean-free-path of a particle in the corona is much larger than in a prominence because of the enhanced density in the prominence. Thus, coronal particles striking the prominence as a result of their thermal motion will be trapped by the prominence. An estimate of the extent of this effect can be obtained from the mean-free-path length in the corona.

The mean-free-path length, λ, is given by

$$\lambda = \frac{1}{ns} ,$$

(II-11)

where s is the collision cross section and n is the particle density. Since the corona is ionized, the collisions are largely Coulomb collisions. The approximate cross section for this type of collision can be obtained by equating the Coulomb potential energy of the particles to a fraction γ of the mean kinetic energy of the particles. The Coulomb potential energy, Ω, of particle 1 in the field of particle 2 is given by

$$\Omega = \frac{Z_1 Z_2 e^2}{x},$$

where x is the distance between particles and $Z_1 e$ and $Z_2 e$ are the charges on the particles. The mean kinetic energy, U, is

$$U = \frac{1}{2} m_1 (\bar{v})^2 = \frac{1}{2} m_1 \frac{3kT}{m_1} = \frac{3}{2} kT,$$

and the cross section S is

$$S = \pi x^2. \tag{II-12}$$

Equating Ω to γU, we find

$$x = \frac{2}{3} \frac{Z_1 Z_2 e^2}{\gamma kT} \tag{II-13}$$

and from Equation (II-12) we find

$$S = \frac{4\pi Z_1^2 Z_2^2 e^4}{9 \gamma^2 (kT)^2}. \tag{II-14}$$

For protons colliding with protons, Equation (II-14) gives

$$S = \frac{3.9 \times 10^{-6}}{\gamma^2 T^2}. \tag{II-15}$$

The quantity γ varies slowly with temperature and density, but for order of magnitude estimates an average value of 0.4 is reasonable (cf. Alfvén, 1950). For $T = 2 \times 10^6$ and $n = 3 \times 10^8$, Equations (II-11) and (II-15) then give $\lambda \approx 5000$ km. We designate this value as λ_0.

Protons within a distance λ_0 of the prominence will be captured in a time of approximately

$$t = \frac{\lambda_0}{\bar{v}} = \frac{\lambda}{(\frac{3}{2} kT/m)^{1/2}}$$

$$\approx 30 \text{ s}.$$

for $T = 2 \times 10^6$ and $\lambda_0 = 5000$ km. These protons will be replaced, of course, by other protons flowing in from outlying regions. Eventually, a steady state will be reached in which the density is depleted for some distance around the prominence. Since the mean free path decreases as n^{-1}, we expect the region of density depletion to be large compared to λ_0.

Mechanisms other than entrapment of thermal particles may play a role in the corona-prominence mass transfer. Pikel'ner (1971) suggests, for example, that a systematic stream flow from the corona to the prominence occurs as a result of the magnetic field configuration around the prominence. This point will be discussed further in Chapter IX.

15. Solar Cycle Effects

The equatorward migration of active regions and the poleward migration of filaments are reflected in the coronal structure. In addition to, but perhaps as a result of, these particular migrations, there is a distinct difference in coronal shapes from sunspot minimum to sunspot maximum. The minimum corona is strongly elliptical in shape, as shown in Figure II-18. The maximum corona, on the other hand, is more nearly spherical in gross outline. Comparisons of visual continuum brightness (mean electron density) indicate that the ellipticity of the minimum corona results from a reduction in brightness and a steepened gradient in brightness at the poles. Near sunspot maximum the polar zone of coronal streamers associated with the polar filaments enhances the coronal brightness near the poles. This is clearly not the only effect present, however. Waldmeier (1950) has shown that intensity isophotes of the green (λ 5303) coronal line show a poleward maximum during years near sunspot maximum as well as the active region maxima. The poleward maximum of the green line follows the same migration pattern as the filaments, but precedes the filaments by about $10°$ of latitude. As a result of preceding the filaments, the green line maximum reaches the polar regions about 2 years ahead of the filaments, i.e., about 1 year after sunspot maximum.

Since the line emission decreases as the square of the local density, the line is usually confined to the low corona. Thus, the line emission observed at the poles must be explained in terms of features lying within about $15°$ of the poles.

It was indicated in Section 11 of this Chapter that some prominence filaments migrate poleward along with the coronal streamers and that these filaments and streamers lie on the poleward side of the unipolar magnetic regions. This is not meant, however, to imply that all prominence filaments follow this pattern. The major poleward migration of filaments, as noted by Waldmeier (1950), seems to lag behind the migration of green line emission features. It is tempting to identify the green line emission regions with the unipolar regions, since the initial coronal enhancement is closely identified with the magnetic field enhancement. This picture would then require two zones of quiescent prominences, one either side of the unipolar region, with the low latitude zone being the more important. Such a picture is not firmly established, however.

References

Alfvén, H.: 1950, *Cosmical Electrodynamics*, Clarendon Press, Oxford.
Altschuler, M. D. and Perry, R. M.: 1972, *Solar Phys.* **23**, 410.
Altschuler, M. D., Trotter, D. E., and Orrall, F. Q.: 1972, *Solar Phys.* **26**, 354.
Athay, R. G.: 1959, *Astrophys. J.* **129**, 164.
Athay, R. G.: 1965, *Astrophys. J.* **142**, 755.
Babcock, H. W.: 1961, *Astrophys. J.* **133**, 572.
Beckers, J. M.: 1964, 'A Study of the Fine Structure of the Solar Chromosphere', Thesis University of Utrecht.
Beckers, J. M.: 1968, *Solar Phys.* **3**, 367.
Bhavilai, R.: 1965, *Monthly Notices Roy. Astron. Soc.* **130**, 411.
Bohlin, J. D.: 1970, *Solar Phys.*, **12**, 240.
Bray, R. J.: 1968, *Solar Phys.* **4**, 318.
Bray, R. J.: 1969, *Solar Phys.* **10**, 63.
Bruzek, A.: 1967, *Solar Phys.* **2**, 451.
Bumba, V. and Howard, R.: 1965, *Astrophys. J.* **141**, 1492.
Dunn, R. B.: 1974, private communication.
Dunn, R. B., Zirker, J. B., and Beckers, J. M.: 1974, in R. G. Athay (ed.), *Fine Structure of the Solar Chromosphere*, D. Reidel Publishing Co., Dordrecht, Holland, p. 45.
Foukal, P. V.: 1974, *Solar Phys.*, in press.
Gebbie, K. B. and Steinitz, R.: 1974, in R. G. Athay (ed.), *Fine Structure of the Solar Chromosphere*, D. Reidel Publishing Co., Dordrecht, Holland, p. 55.
Grossmann-Doerth, U. and von Uexküll, M.: 1971, *Solar Phys.* **20**, 31.
Hansen, R. T., Hansen, S. F., and Garcia, C. J.: 1970, *Solar Phys.* **15**, 387.
Hansen, S. F., Hansen, R. T., and Garcia, C. J.: 1972, *Solar Phys.* **26**, 202.
Harvey, J. W.: 1974, private communication.
Jordan, C.: 1969, *Monthly Notices Roy. Astron. Soc.* **142**, 501.
Liu, S. Y.: 1972, 'Fine Structure in the Solar Chromosphere', Thesis Univ. of Maryland.
Lynch, D. K., Beckers, J. M., and Dunn, R. B.: 1973, *Solar Phys.* **30**, 63.
Munro, R. H. and Withbroe, G. L.: 1972, *Astrophys. J.* **176**, 511.
Nakagawa, Y. and Malville, M.: 1969, *Solar Phys.* **9**, 102.
Newkirk, G., Jr.: 1967, *Ann. Rev. Astron. Astrophys.* **5**, 213.
Pierce, A. K. and Waddell, J. H.: 1961, *Mem. Roy. Astron. Soc.* **68**, 89.
Pikel'ner, S. B.: 1971, *Solar Phys.* **17**, 44.
Prata, S. W.: 1971, *Solar Phys.* **20**, 310.
Reeves, E. M., Foukal, P. V., Huber, M. C. E., Noyes, R. W., Schmahl, E. J., Timothy, J. G., Vernazza, J. E., and Withbroe, G. L.: 1974, *Astrophys. J.* **188**, L27.
Saito, K.: 1965, *Publ. Astron. Soc. Japan* **17**, 1.
Simon, G. W. and Leighton, R. B.: 1964, *Astrophys. J.* **140**, 1120.
Tandberg-Hanssen, E.: 1974, *Solar Prominences*, D. Reidel Publishing Co., Dordrecht, Holland.
Thomas, R. N. and Athay, R. G.: 1961, *Physics of the Solar Chromosphere*, Interscience, New York.
Tousey, R., Bartoe, J. D. F., Bohlin, J. D., Brueckner, G. E., Purcell, J. D., Scherrer, V. E., Sheeley, N. R., Jr., Schumacher, R. J., and Van Hoosier, M. E.: 1973, *Solar Phys.* **33**, 265.
Vaiana, G. S., Davis, J. M., Giacconi, R., Krieger, A. S., Silk, J. K., Timothy, A. F., and Zombeck, M.: 1973a, *Astrophys. J.* **185**, L47.
Vaiana, G. S., Krieger, A. S., and Timothy, A. F.: 1973b, *Solar phys.* **32**, 81.
Van Speybroeck, L. P., Krieger, A. S., and Vaiana, A. S.: 1970, *Nature* **227**, 818.
Vernazza, J. E., Avrett, E. H., and Loeser, R.: 1973, *Astrophys. J.* **184**, 605.
Waldmeier, M.: 1950, *Z. Astrophys.* **27**, 237.
White, O. R.: 1962, *Astrophys. J.* **137**, 1210.
Withbroe, G. L.: 1970, *Solar Phys.* **11**, 208.
Zirin, H.: 1966, *The Solar Atmosphere*, Ginn and Blaisdell, Waltham.
Zirin, H.: 1972, *Solar Phys.* **22**, 34.

CHAPTER III

MACROSCOPIC MOTIONS

1. Measuring Systematic Motions

Astronomers have long recognized that many of the structural features of the Sun's atmosphere are related to and associated with the dynamic properties of the atmosphere. Thus, granules, and spicules have long been associated, respectively, with convection in the photosphere and jets of matter ejected from the upper chromosphere. More recently, it has been recognized that the network is related to the flow in supergranules and that matter flows along the fibrils as if channelled by their structure. These and other clear associations between organized motion and structural features point to a close association, and very likely a one-to-one association, in which each class of structural features represents a distinct class of flow phenomena.

It is also becoming more universally accepted by astronomers that the state of motion of the atmospheres of stars and the thermodynamic structure of the atmospheres are mutually dependent — each affects the other. This represents a considerable reversal from the once popular notion that one could determine the temperature structure of a stellar atmosphere purely from radiative and hydrostatic equilibrium, and that one could then deduce the average properties of the state of motion from the known thermodynamic properties and observed profiles of spectral lines. Stellar astronomers seemed for some time to overlook the fact that solar astronomers who followed this approach were completely unable to arrive at any agreed upon thermodynamic and fluid dynamic model. It is now abundantly clear that this older approach is inconsistent. The thermodynamic structure of the solar atmosphere, as well as the geometrical structure, is closely related to the state of motion and vice versa. The same is probably true for all stars with convection zones, and we cannot immediately dismiss the idea that a similar inter-dependence exists in stars without convection zones.

Streaming motions in the solar atmosphere can be detected in two basic ways: by time-lapse photography and by the displacement of a spectral line from its rest position. Time lapse photography is useful for features that are sharply defined and are either long-lived or rapidly moving. For ill defined or short-lived features, however, the measurement of motions by time lapse photography requires a combination of meticulous care and unusual observing conditions that are difficult to realize. In any high resolution observations made from the ground the effects of atmospheric seeing cause distortions of the solar image. Such distortions cause small features on the solar surface to appear to wander about. An excursion by $1''$ in a period of 5 s produces an apparent velocity of 145 km s^{-1}. This is a good deal larger than typical velocities found in the chromosphere, and it is difficult, therefore, to identify a given motion as being a true solar motion rather than simply a result of seeing. Nevertheless, such observations have been very useful in specific cases and will undoubtedly find more usage as solar telescopes improve and as better observing sites are discovered.

When motion is observed by the time-lapse method, there is still the question of whether the feature that moved represents a bodily transport of material or whether the feature itself is a pattern moving through the material. A campfire flame, for example, may dance about above the fire as a result of the gases in a given location alternately becoming incandescent then invisible. The motion of the gas itself is primarily vertical, which is quite different from the motion of the flame. Similarly, gravity waves on the surface of the sea appear to show a streaming of the sea surface whereas they are actually traveling perturbations whose motion is more or less independent of the true streaming motion. Such ambiguity can sometimes be resolved in the solar case by observing associated phenomena or by simultaneously observing both the lateral displacement and the Doppler shift of spectral lines. However, in most cases, particularly those that pertain to short-lived, low-velocity features, the ambiguity remains.

Motion detected by the Doppler displacement of a spectral line is unambiguous insofar as the existence of true material motion is concerned. On the other hand the location of the moving material, the magnitude of the velocity and even the direction of the motion remain ambiguous in many cases. If the entire gaseous layer in which a spectral line forms is moving with a uniform, constant velocity v in the line of sight, a spectral line will shift by an amount

$$\Delta\lambda = \lambda\frac{v}{c}, \qquad\qquad\qquad\qquad\qquad\qquad\qquad\qquad (\text{III-1})$$

where v is measured positively in the direction away from the observer. When observing an individual bright feature beyond the solar limb, such as a knot of prominence material, there is little question but that the motion is uniform throughout the feature. Thus, Equation (III-1) gives a valid relationship between $\Delta\lambda$ and v. Similarly, if one is measuring the motion of the Sun itself, as is the case in spectroscopic measurement of solar rotation, for example, Equation (III-1) is valid.

When one is measuring the Doppler displacement in a spectral line at a localized area on the solar disk, however, it must be remembered that the radiation is built up by integration along the line of sight and that the layers contributing strongly to the observed intensity extend through a layer of the atmosphere that is thick compared to the opacity scale height. In most lines the opacity scale height is comparable to or greater than the density scale height. Thus, the layer of line formation, even with perfect spectral resolution, is thick compared to the density scale height. With less than perfect spectral resolution it is even thicker. In a gaseous atmosphere it is simply unrealistic to expect the motions of the gas to remain uniform for more than about one density scale height. Continuity conditions alone make such motions the exception rather than the rule. One is faced, therefore, with the difficulty that most of the Doppler shifting of spectral lines on the solar disk will be caused by motions in which the gradient of velocity in the line of sight is important in determining the resultant Doppler shift. Under such conditions Equation (III-1) is not valid.

When strong gradients of motion are present Equation (III-1) in some cases may give a simple average of v in the layers of line formation. In most cases, however, the v determined from Equation (III-1) will not be a simple average of the true v. It is easy, in fact, to imagine situations in which even the sign of v is given incorrectly. Suppose, for example, that the loop depicted in Figure I-7 is in emission relative to the background radiation and is semi-transparent in a given line. When the slit of a spectrograph is placed over the leg of the loop in which material is rising the resultant line profile will consist of the sum of the normal undisplaced absorption line plus the net emission from the loop, which will be displaced to the violet. If the displacement is not large, the sum of the two lines will result in a filling in of the normal absorption line on its violet side with a resultant shift in the center of gravity of the line to the red. Without any other evidence to go on an observer knows only that the absorption line is apparently shifted to the red and he is likely to conclude that the material, at the level of line formation, is moving downward, which is completely erroneous.

The type of difficulty illustrated in the preceding paragraph is perhaps an infrequent one. Nevertheless, it must be carefully considered in specific cases. Simon and Leighton (1964) find, for example, that over the borders of the supergranules the wings of $H\alpha$ indicate 'funnels' of downflowing material with velocities of 1.2 km s^{-1}. Could these be interpreted alternatively as upward moving bright features in the chromosphere? Unfortunately, without further information to go on the answer is yes. Observations of similarly descending motions, though of lower amplitude, at the borders of supergranules in lines of iron, sodium and magnesium by Tanenbaum (1971) support the actual existence of the downward motions. However, without such supporting evidence the conclusions by Simon and Leighton would remain ambiguous.

A type of ambiguity similar to that discussed above is particularly important in interpretations of the K_2 and K_3 structure in the K line. The K_2 and K_3 structure varies markedly from point to point on the solar disk. One often observes the K_{2V} component to be brighter than K_{2R}, as in Figure II-1. Thus, if one imagines the line to consist of a normal absorption line with a superposed emission line whose width extends from K_{2V} to K_{2R} and which arises from a bright cloud elevated above the layers where the absorption line forms, then the 'emission' line appears to be shifted to the violet relative to the absorption line. A typical shift corresponds to an upward velocity for the emission line region of 1 or 2 km s^{-1}. On the other hand, one may interpret the K_2, K_3 structure as a single line formed in a continuous layer. The asymmetry of the emission component then means only that the motion is such that K_{2V} is more visible than K_{2R}. It has been found by numerical modeling of the profile (Athay, 1970) that the visibility of K_2V can be enhanced over K_{2R} in a manner similar to what is observed by permitting either *upward* motions in the region where K_1 is formed or *downward* motions in the region where K_3 is formed. The amplitude of the motion depends upon the particular velocity model adopted and may exceed 10 km s^{-1}. Thus, a wide range of interpretations is possible, illustrating again the ambiguity of 'velocity measurements'. This effect will be discussed in more detail in a later chapter.

Perhaps a common type of motion to be expected in the solar atmosphere is that of a gravity wave propagating more or less horizontally in the solar atmosphere and confined to a layer whose thickness is of the order of one or two scale heights. Such a wave will produce vertical displacements of material with a considerable vertical gradient in velocity. Figure III-1 shows computed profiles for the Hα line for crude models of such motion. In the models the vertical velocity amplitude is constant for about 500 km in the chromosphere and decreases linearly to zero at the borders of the layer, i.e., continuity is not preserved. The profile labeled 0 in Figure III-1 is the profile for the stationary atmosphere. For the profile labeled 1 the moving layer lies between 500 km and 1000 km, and for the profile labeled 2 the moving layer lies between 1500 km and 1900 km. It can be seen from Figure 11-2 that for the profile labeled 2 all of the motion lies above the surface $\tau_0 = 1$. Nevertheless the profile shows a 'Doppler shift' even out to 1 Å from line center where the maximum of the contribution function lies well within the photosphere. Again, an observer would likely conclude from profile 2 that motion exists throughout the Hα chromosphere, i.e., at all depths below about 1500 km and that the amplitude of the velocity is approximately 9 km s^{-1} whereas in the model the motion is confined to layers above 1500 km and has an amplitude of 23 km s^{-1}.

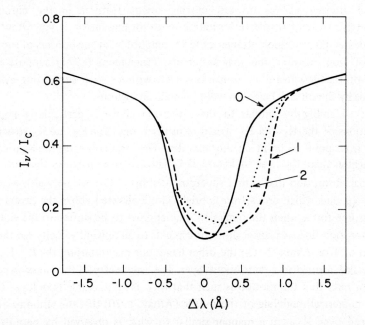

Fig. III-1   Profiles of Hα computed for vertical motion in a restricted layer of the chromosphere. The moving layer lies between 1900 and 1500 km in case 2 and between 1000 and 500 km in case 1. Case 0 represents the static atmosphere. The velocity amplitude is 23 km s^{-1} in both cases 1 and 2.

The preceding illustrations are given not to discredit observations of velocities by the Doppler shift method but to point out that the interpretation of such observations should not be made naively. An observer faced with an observed line shift has little recourse other than to interpret it directly as a velocity, assuming that the velocity is in fact uniform in depth and constant in time. However, when different lines show different velocities, or when the observed line profiles are asymmetric, the observer should recognize immediately that gradients are present and should thereafter exercise caution in analyzing the data and reporting the results.

In practice, several different methods are used for measuring Doppler shifts. A wiggly line spectrum, such as shown in Figure III-2, obtained with high spatial, temporal and spectral resolution may be used to determine the shift in center of gravity or center of symmetry of the line as a function of position along the slit. This is a simple direct method requiring a minimum of instrumentation. Alternatively one may equip a spectrograph with two closely spaced exit slits of equal width each of which feeds into balanced photoelectric detectors. The two detectors are combined to produce a null signal when the two slits receive equal amounts of light, i.e., when they are located symmetrically on the line profile, and the location of the slits with respect to the average line position then gives directly the Doppler shift.

A third method makes use of the photographic subtraction processes used for producing the Doppler image of the Sun shown in Figure II-5. This method is useful for detecting patterns of motion but is not useful for quantitative measures.

The most indirect method used is that of comparing features seen in different portions of a line profile. The image of the Sun shown in Figure II-10 was made by scanning the spectrohelioscope in wavelength across the central regions of the K line simultaneously with a scan across the solar disk. Thus, one side of the solar image shows the features seen in one side of the line and the other side of the disk shows the features seen on the opposite side of the line. It is readily apparent in this way that the structure seen in K_{2V} is different in detail from that seen in K_{2R}. One infers from this, correctly or incorrectly, that a feature seen in K_{2V} but not in K_{2R} is moving toward the observer and vice versa. This interpretation is subject to all of the uncertainties discussed in the preceding paragraphs of this section. As mentioned, it may be totally misleading, particularly in the case of the K line. We will return to this point in later Chapters when we discuss spectral line formation in more detail.

2. Photospheric Motions

So-called Doppler spectroheliograms are obtained by superposing two images of the Sun made at two wavelengths within a spectral line, $\Delta\lambda_+$ and $\Delta\lambda_-$, equally and oppositely spaced from line center. In principle, then, a Doppler shift of the line in a local area on the Sun will result in a difference in intensity at $\Delta\lambda_+$ and $\Delta\lambda_-$. If no motion is present, the line should be unshifted and the intensities at $\Delta\lambda_+$ and $\Delta\lambda_-$ should be equal. Now, if one of the two images that are superposed is a positive and the other is a negative and if the

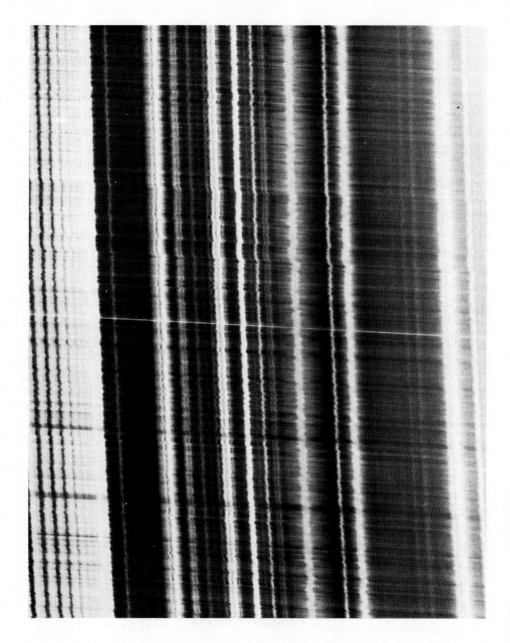

Fig. III-2.   A high spatial and high spectral resolution image of the Sun. The spatial distance from top to bottom is approximately $440''$, or about ten supergranule diameters. Along the fiducial line the wavelength range from left to right is λ 3882.3 – 3889.0. The CN bandhead at λ 3883.29 is prominent at the left hand side of the spectrum, and two strong Fe I lines at λ 3886.29 and λ 3887.06 are prominent just right of center (Sacramento Peak Observatory, Air Force Cambridge Research Laboratories).

photographic film is controlled to have the transmission, t_{ν_+} proportional to I_{ν_+} on the positive and the transmission t_{ν_-} proportional to I_ν^{-1} on the negative (photographic gamma of unity), the superposed images will have a combined transmission of

$$t_{\nu_0} = t_{\nu_+} t_{\nu_-} = \text{const} \frac{I_{\nu_+}}{I_{\nu_-}} . \tag{III-2}$$

Thus, if no velocity displacement is present t_{ν_0} will be constant with position on the Sun, even if I_{ν_+} and I_{ν_-} themselves are fluctuating due to a mottled brightness pattern. On the other hand, a velocity displacement in one direction will increase t_{ν_0} and a displacement in the opposite direction will decrease t_{ν_0}. A spectrogeliogram made in this manner, therefore, will show a uniform grey pattern where no motions are present and a relatively bright or dark pattern where motions exist.

Alternatively, one may carefully guide the slit of a spectrograph on the solar surface and measure the actual displacement of the spectral lines. This method has the advantage of giving more useful quantitative measures of the line shift but suffers the disadvantage of being restricted to a one-dimensional sample of the solar surface, i.e., the line defined by the spectrographic slit on the solar image. The former method can be used with whole Sun images.

Using the former technique, Leighton (1962) discovered small vertical velocity cells oscillating with a regular period of 300 s and larger horizontal velocity cells of steady flow and long lifetime. The larger cells are the supergranules shown in Figure II-5. The two types of cells coexist in the same physical layers and cover essentially the entire solar surface away from active centers. Thus, within a given supergranule cell, which is defined primarily by a horizontal outflow of material from center to edge, there are smaller cells in which the velocity is mainly vertical and of relatively short oscillatory period.

Supergranule cells are characteristically of approximately 30 000 km diameter, have lifetimes of about 20 h and have maximum horizontal outflow velocities of about 0.4 km s^{-1} (cf. Noyes 1967). There is an upward velocity of about 0.065 km s^{-1} at the cell center and a downward velocity of similar magnitude at the cell borders (Tanenbaum, 1971). The horizontal motions within the cells are observed in medium strong Fraunhofer lines, but they have not been observed either in weak lines or in very strong lines. On the other hand, Tanenbaum (1971) has observed the vertical velocities at the centers and borders of the cells in both weak and very strong lines. Thus, it appears on this basis that the cell exists as an organized motion throughout the photosphere and into the low chromosphere. The strongest horizontal flow is evidently found in the middle photosphere. Although the horizontal flow has not been observed in either weak or strong lines, a horizontal flow of magnitude comparable to the vertical flow would be difficult to detect and could easily be present.

The persistence of the network structure throughout the upper photosphere and the entire chromosphere is difficult to understand unless the supergranule flow pattern also persists throughout these layers. This will become even more evident in Chapter IV when the magnetic field configuration is discussed. The network is the site of strong vertical magnetic field, and it is presumably the supergranule flow that concentrates the magnetic field in the network.

The second type of systematic motion observed in the photosphere, viz., the oscillatory cells are also of major interest in connection with chromospheric stuctures and in connection with possible heating mechanisms for the chromosphere and the corona. Early observations of these oscillatory features led to rather simple descriptions of the phenomena. More refined observations have tended to add complexity to these descriptions as well as to confuse the overall picture. Because of the relative simplicity of the early descriptions, it is convenient to retain them as an initial reference in discussing the later results.

The granulation in the photosphere presents a varying brightness pattern on a small scale, with a characteristic lifetime of 5 to 10 min and a characteristic size of $1''$. If an area on the surface of the Sun of the order of $5''$ in diameter is observed in the continuum, for a period of time, the brightness fluctuates in a periodic manner (at least for one or two periods) with a period of approximately 300 s. The amplitude of this periodic component is approximately one tenth of the amplitude of the smaller scale granule fluctuations.

Similar observations made within spectral lines reveal similar periodic brightness changes plus periodic motion that is predominantly vertical. The amplitude of the motion is 'measured' to be of the order of 0.3 to 0.4 km s^{-1}. To a first approximation both the brightness and velocity oscillate with periods near 300 s.

Early observations of these oscillations gave the following overall properties:

(a)   The onset of the oscillation in velocity is progressively later at increasing heights and the velocity oscillations lag the continuum brightness pulses.

(b)   The velocity amplitude and the frequency of oscillation both increase slowly with height.

(c)   The size of the oscillatory cell increases rapidly with height from slightly more than $2''$ in the low photosphere to about $5''$ in the low chromosphere (cf. Evans and Michard, 1962b; Evans et al., 1963; Noyes, 1967).

A. WAVE FREQUENCY

The wave frequency was given in the early observations as 3.3×10^{-3} s^{-1} in the low photosphere increasing to about 3.5×10^{-3} s^{-1} in the middle photosphere. For a given wave train, therefore, there is a continuing change of phase with height. After two cycles of oscillation in the low photosphere, the oscillation in the middle photosphere has advanced in phase, relatively, by about one-eighth of a period, or about 40 s. Thus, a delay in onset of the oscillations by 40 s between the low and middle photosphere is quickly overcome.

Power spectra of the oscillations, as shown in Figure III-3 show a characteristic strong resonance peak near 3.3×10^{-3} s^{-1}. Weak Fraunhofer lines (those near the bottom in Figure III-3) formed in the low photosphere show an additional strong peak near zero

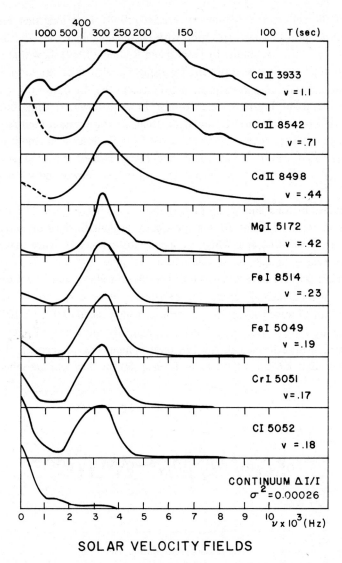

SOLAR VELOCITY FIELDS

Fig. III-3.   Power spectra for the periodic component of motion in the solar photosphere and chromosphere. The mean height of formation of the spectral lines used to obtain the power spectra increases from the next to bottom panel (C I) to the top panel (K line of CA II). The bottom panel shows the power spectrum of the continuum brightness changes in the low photosphere (after Noyes, 1967).

frequency. This component is identified with convective overshoot of the granulation and is non-periodic. In the upper photosphere and low chromosphere, observed by means of the very strong lines (those near the top in Figure III-3), a higher frequency component near 6×10^{-3} s^{-1} becomes strongly evident (cf. Jensen and Orrall, 1963; Orrall, 1966).

Studies of the velocity oscillations by Frazier (1968) using long time series of data showed a double peaked power spectrum with maxima at fequencies of 3.8×10^{-3} s^{-1} and 2.9×10^{-3} s^{-1}. Similar studies by Gonczi and Roddier (1969) show that the 'typical' power spectrum for a long time series and for a single small area on the surface of the Sun consists of several peaks that are separated from each other by $0.85 \pm 0.2 \times 10^{-3}$ s^{-1} as are the two peaks in Frazier's data. Unlike Frazier, however, they find that the 3.3×10^{-3} s^{-1} component is nearly always dominant. They interpret their result in terms of a fundamental carrier frequency of 3.3×10^{-3} s^{-1} modulated in amplitude by wave packets at intervals of about 20 min. The different wave packets remain in phase in excess of 40 min. This modulation produces the subsidiary power peaks separated by 0.85×10^{-3} s^{-1}.

Similar studies by Cha and White (1973) show a similar pattern. The wave packets separated by intervals of about 20 min are clearly evident in the time series shown in Figure III-4. However, Cha and White are unable to find any evidence that subsequent wave packets bear a unique phase relationship to preceding ones. They suggest that subsequent wave packets have a random distribution in phase, which at times gives the appearance of a fundamental carrier that remains in phase over two or more packets. Within the individual wave packets the power is strongly peaked near 3.3×10^{-3} s^{-1}.

The studies by Frazier, Gonczi and Roddier, and Cha and White all refer to weak to moderately strong lines formed in the low photosphere. Frazier claims that one of the lines in his study, $\lambda 6355$ of Fe I, is formed in the low chromosphere or high

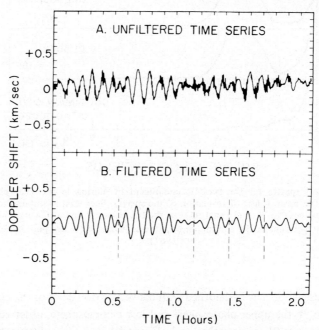

Fig. III-4. A time series of Doppler displacements for a photospheric line. Different wave packets are indicated by vertical dashed lines (after Cha and White, 1973).

photosphere. However, this is clearly a misinterpretation, and all of the lines he used are of low photospheric origin (see Chapter I).

B. PHASE RELATIONS

Evans and Michard (1962a, b) found a correlation between a pulse of continuum brightness (granule) and the onset of a velocity oscillation in the photosphere. The time delay between the continuum brightening and the onset of velocity oscillation in the λ 5172 (b_2) line of Mg I was 40 s, whereas a weaker line of Ti II commenced oscillations 25 s after the continuum brightening. They interpreted this delay in terms of a disturbance moving upward with a velocity near the sound velocity (≈ 7 km s^{-1}). A re-examination of their data, however, led to the conclusion (Athay, 1963) that the 'heights' assigned to the lines as observed by Evans and Michard were too great and that a much lower disturbance velocity was indicated. (This question of assigning a 'height' to give spectral band, $\Delta\lambda$, within a line is crucial to much of the discussion of height relationships and is dealt with in Chapter I. Following that discussion, we will restrict the height classifications to low and upper photosphere and low chromosphere.)

Evans *et al.* (1963) find that between the upper photosphere and low chromosphere (as observed in λ 8514 of Fe I and λ 8542 of Ca II) the phase lag in velocity varies strongly across the 'resonance' peak in the power spectrum. The phase lag is zero at the low frequency side and increases to about 30° (25 s) at the high frequency side.

Tanenbaum (1971) finds velocity phase lags of between 10 and 14 s between the low photosphere and low chromosphere, which is broadly consistent with the phase lags found by earlier workers. However, in Tanenbaum's observations care has been used to stay near line center in the strong lines so that the supposed relatively large height separation between the very strong and weak lines is more firmly established. For the lines he uses (C I, λ 5380.2 and Na D$_1$), the heights of formation are separated by 500 to 800 km. Thus, the phase lag of 10 to 14 s clearly implies phase velocities much in excess of the velocity of sound.

Tanenbaum (1971) also studied the relationship between brightness oscillations and velocity oscillations. The brightness oscillations are observed both in the continuum and in the lines, the latter being at the same wavelengths as the observations of velocity oscillations. For the latter case, the velocity oscillation lags the brightness by an amount that increases with height. For a weak line (C I, λ 5380.2) the lag is 34 ± 13 s (40° ± 15°) and for a strong line (D$_1$) it is 78 ± 7 s (98° ± 8°). The weak line brightness is closely in phase with continuum brightness. Thus, Tanenbaum's results are consistent with those of Evans and Michard (1962b). It is clear from Tanenbaum's results, however, that the brightness-velocity phase lag is a property of the wave form itself and is not the result of a 'traveling disturbance.'

Tanenbaum concludes from the wave properties, i.e., a brightness-velocity lag of less than 90° in the weak lines and in excess of 90° in the strong lines, that the waves are traveling waves in the photosphere and standing waves in the chromosphere. Such an interpretation seems unwarranted, however, until we have a more positive identification of the wave mode and its properties.

Tanenbaum's results, if taken literally, suggest that the brightening increases in phase (leads) with increasing height, i.e., the chromosphere brightens before the photosphere. This follows from the fact that the brightness-velocity lag increases more rapidly with height than the velocity-velocity lag. If this is a valid result, it is an extremely important one for establishing the nature of the brightness and velocity fluctuations. A commonly accepted picture is that of a piston (granule) compressing the low photosphere and a subsequent resonant 'ringing' of the photosphere and chromosphere. If the chromospheric brightening truly precedes the photospheric one, such a picture is clearly untenable.

Within the convective component of power at low frequencies the phase properties are not clearly defined. Edmonds *et al.* (1965) find that the velocity phase leads the brightness phase by about 100 s. However, Frazier (1968) finds only a slight tendency for the velocity phase to lead the brightness. Thus, there is no general agreement on the results.

Frazier makes the claim that the near equality of phase between brightness and velocity in the convective component persists even into the upper photosphere. He argues, on this basis, that the convective over-shoot penetrates, in large measure, the entire photosphere. However, this conclusion again results from an incorrect assignment of heights and his results are to be taken as applying only to the low photosphere.

C. VELOCITY AMPLITUDE AND CELL SIZE

The velocity amplitude of the oscillatory cells can be properly measured only if the cells are resolved individually and if the velocity gradient is small. Thus, cell size and velocity amplitude are interrelated. Measurement of cell size has led to some disagreement though most observers agree that the cells are considerably larger than the granules. Early observations indicated cell sizes of approximately 1600 km in the low photosphere increasing to approximately 3500 km in the upper photosphere (Noyes, 1963). These cell sizes were defined in terms of the half width of the spatial autocorrelation curves of Doppler velocity.

One of the surprising features of the oscillatory motion is that it is observed even with very coarse apertures. In fact, there are claims (not universally accepted) that it has been observed in the radio spectrum at a wavelength of 3 cm with very low spatial resolution. However, the velocity amplitude diminishes as the spatial resolution becomes coarser. This is illustrated in Figure III-5, taken from Tanenbaum (1971). The aperture size given is the width of a square aperture of the same effective area as the observing aperture, which is often not square. For apertures larger than about 3000 km, the velocity amplitude is inversely proportional to aperture size, as is expected if the number of oscillatory elements is proportional to the aperture *area*. On this basis, Tanenbaum argues that the cell size is near 3000 km. If this were really true, however, it is difficult to see why the velocity amplitude continues to increase for aperture sizes even less than

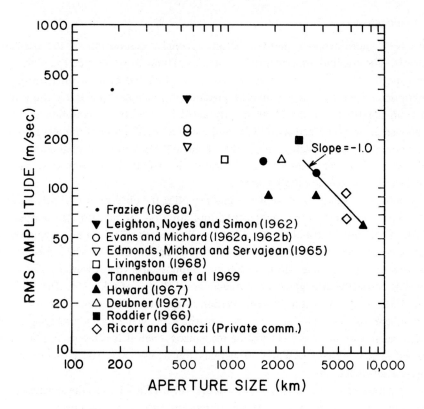

Fig. III-5.   Velocity amplitude of the 300 s oscillation as a function of observing aperture diameter
(Tanenbaum, 1971).

one-tenth of this amount. The statistics used by Tanenbaum are probably invalid when the number of cells included becomes small, and the plot in Figure III-5 may be consistent with a cell size much less than 3000 km.

Figure III-5 is complicated by the fact that both the cell· size and the velocity amplitude show an overall increase with height and not all of the results plotted refer to the same physical heights in the solar atmosphere. The results in Figure III-4 indicate a growth in velocity amplitude from 0.18 km s^{-1} for the C I line at λ 5052 to 0.71 km s^{-1} for the Ca II line at λ 8542.

The gradient of cell size and velocity amplitude with height may not be of the same sign throughout the atmosphere. There is some indication in the work of Frazier (1968), and in the results in Figure III-4, that the velocity amplitude may decrease with height in the low photosphere before it finally increases with height.

It is perhaps safest, at this stage to regard both the velocity amplitude and cell size as being relatively uncertain. Certainly more observations are needed in order to adequately evaluate these parameters and their height variations.

D. DISCUSSION

Obviously, more work is needed before a concise description of the oscillations is possible. The oscillations evidently are manifestations of some type (possibly types) of wave motion. A variety of wave modes is possible, and given a sufficiently accurate description of the motion it would be possible, in principle, to identify the wave mode. Such an identification would be greatly aided by a correct description of the height relationships, in phase and amplitude, and a correct description of the relative horizontal and vertical motions, such as might be obtained from center to limb studies. It is not possible at this time, however, to uniquely identify the dominant wave mode or to conclusively rule out any particular mode.

A point that needs to be kept in mind regarding the observations of periodic waves in the solar atmosphere is the fundamental limitation on the observations imposed by the width of the line contribution function. For illustration of this point, suppose that a slit placed in the side of a spectral line 'sees' a wavelength interval in which the contribution function has an effective width of 200 km. Suppose also that a wave of 100 s period is traveling vertically through the atmosphere at a velocity of 6 km s^{-1}. The wavelength is then 600 km. An instantaneous observation in the line will average over 200 km of the atmosphere, or over one-third of a wavelength. With such a coarse averaging it is highly questionable whether the wave will be recognized. Even if it is recognized, its properties will be largely obscured. Waves of periods much less than 100 s clearly will not be observed as oscillations.

It is pertinent to ask, at this point, whether the oscillations and the supergranule flows represent the only prominent types of motion within the photosphere, or, indeed, whether they are even the most prominent. Other evidence of motion in the photosphere comes from the broadening of the Fraunhofer lines. It appears to be impossible to account for the observed widths of the lines without invoking motions, in the low photosphere at least, that, for the heavy elements, exceed the thermal motions. Iron atoms, for example, have a mean thermal velocity at 6000 K of 1.3 km s^{-1}. This would produce a Doppler width at λ 5000 of

$$\Delta\lambda_D = \lambda\frac{v}{c} = 5000\,\frac{1.3}{3 \times 10^5}$$

$$= 0.022 \text{ Å}$$

If this were the only source of broadening, an unsaturated iron line at λ 5000 would therefore have a full width at half intensity of approximately

$$W_2 = 2\,\Delta\lambda_2,$$

where $\Delta\lambda_2$ is defined by $e^{-(\Delta\lambda_2/\Delta\lambda_D)^2} = \frac{1}{2}$ or

$$\Delta\lambda_2 = (\ln 2)^{\frac{1}{2}}\Delta\lambda_D.$$

The above value of $\Delta\lambda_D$ gives $W_2 = 0.037$ Å. By comparison, however, the observed widths of the weak iron lines near λ 5000 are more than twice this amount. More careful analyses of the lines (Holweger, 1967) yield so-called microturbulent velocities in the low photosphere of about 2.8 km s^{-1}. The word microturbulence, as used here, means only that the motion produces a line broadening similar to, but in excess of, the thermal motions. It could just as well be a small scale ordered motion as a random motion, and it could, of course, be a result of short period waves.

Neither the oscillatory motion nor the supergranule motion has an observed velocity amplitude approaching 2 km s^{-1}. Thus, we seem forced to conclude that we have not yet resolved the non-thermal components of motion in the low photosphere in which most of the mechanical energy resides. The nature of these motions remains largely in the realm of complete mystery.

Perhaps the conclusion just stated is not too surprising. The smallest cell size reported for the oscillatory motion is 1600 km, more than twice the average granule size. There is no *a priori* reason to expect the smallest velocity cells to be larger, necessarily, than the brightness cells. In fact, there is reason to suppose that within a given brightness cell velocity fluctuations may exist.

In the upper photosphere the situation appears to be somewhat different. Fine granular structures are still present at these heights but the network pattern is now clearly visible (see Chapter II). Also, the required microturbulence for lines observed near disk center, in some models (Holweger, 1967; Athay and Canfield, 1969; Athay and Lites, 1972) is of the order of 0.5 km s^{-1}, or less. All Fraunhofer lines broaden towards the solar limb, however, and the low 'microturbulent' velocity found at disk center is not sufficient to produce the broadening at the limb. This phenomena leads to the concept of 'anistropic microturbulence' in which the horizontal component of motion is considerably larger than the vertical component. Typical values for the horizontal component of microturbulence in the upper photosphere are 1.5 to 2.0 km s^{-1}.

Alternative models of the 'microturbulence' in which the motion is assumed to be isotropic require that the microturbulent velocity increases steadily outward in the solar atmosphere. This result follows immediately from the observational fact that *all* lines broaden towards the solar limb. Thus, if the motion is isotropic its amplitude must increase with height in all layers of the atmosphere. Models of this type lead to velocity amplitudes exceeding 2 km s^{-1} in the upper photosphere.

It again seems clear that most of the kinetic energy in the upper photosphere does not reside in the vertical oscillations, at least as the oscillations are currently pictured. Either the dominant energy is contained in predominantly horizontal motions or it is contained in unresolved features.

Solar physicists have not yet been able to resolve the ambiguity between anisotropic and isotropic microturbulence. All of the recent interpretations of strong Fraunhofer lines whose Doppler cores are formed in the upper photosphere and temperature minimum region, however, show a clear preference for the anisotropic model. This is a question of fundamental importance and should be resolved.

3. Chromospheric Motions

The photospheric velocity field discussed in Section 2 consists of a complex pattern of superposed large scale and small scale motions. Large scale motions are dominated by the supergranule flow and the oscillatory component. We remind the reader that the supergranule flow consists of an outflow from center to edge that is apparently strongest in the middle photosphere and of an amplitude of approximately 0.5 km s^{-1}. At the borders of the supergranules there is a downflow of material and at the centers of the supergranules there is an upflow. The upflow is weak with an amplitude of approximately 0.05 km s^{-1}. The downflow at the borders, under low spatial resolution, averages about 0.05 km s^{-1} when observed in the λ 5250 line of Fe I (Tanenbaum, 1971), but when observed in the outer core of the Hα line ($\Delta\lambda = 0.7$ Å) with high spatial resolution the downflow is apparently confined mainly to narrow 'funnels' where the flow velocity is about 1.2 km s^{-1} (Simon and Leighton, 1964).

This circulation model for the supergranule cell is at photospheric depths and does not necessarily reflect the circulation model in the chromosphere. Attempts to 'measure' the supergranule flow pattern in the chromosphere immediately encounter the difficulty that the lines of chromospheric origin are heavily saturated lines whose profiles have very low slopes (in the cores) and which, as a result, are not well suited for velocity measurements. The problem is further compounded by the complexity of geometrical structures observed in the chromosphere, each of which has its own characteristic motion. Because of these difficulties no clear picture of the circulation associated with the network (the chromospheric outline of the supergranules) exists.

A. SPICULES AND FIBRILS

The spicules, which are located at the network (supergranule borders), have lifetimes of 5 to 10 min and move predominantly upward at velocities of the order of 20 to 25 km s^{-1}. However, spicules evidently originate in the middle or high chromosphere (see Chapter IX) where the matter density is several orders of magnitude lower than in the photosphere. Thus, the spicule flow is not expected to be related in any simple way to the photospheric flow. Downward motions are also observed in spicules, sometimes as reversals of an original upwards motion and sometimes as the first manifestation of motion. About $\frac{1}{3}$ of the spicules show downward motion, whereas considerably more than $\frac{2}{3}$ show upward motion during some part of their life. Spicules observed in Hα at the limb are illustrated in Figure II-12.

It has been inferred from the fact that the ends of the Hα fibrils near the network are most readily seen in the red wing of Hα whereas the opposite ends (away from the network) are most readily seen in the violet wing of Hα that the flow along the Hα fibril is upwards near the center of the network cell and downwards at the network. Since the fibrils are low lying features the flow throughout most of the fibril length must be predominantly horizontal. Also, the very nature of the fibril structure suggests that the flow is again concentrated into narrow channels or 'funnels'. This suggested model in which the flow in the low lying fibrils is directed oppositely from the flow in spicules again indicates that the low lying fibrils are distinctly different phenomena from the spicules.

According to Bhavailai (1965) the ends or 'foot points' of the fibrils are most clearly visible near ± 0.5 Å from line center in Hα. Using the naive interpretation that the increased contrast results from a Doppler shift of 0.5 Å, we infer a flow velocity of approximately 23 km s^{-1} in the fibril. This is of the same order as the spicule velocity, but it is not to be taken too seriously. The solar Hα profile has maximum slope near ± 0.5 Å and this fact alone will tend to produce maximum contrast near ± 0.5 Å (≈ 23 km s^{-1}) for any small displacement of the profile.

Studies of streaming motion in the fine scale rosette features (mixture of bright and dark mottles) by Grossmann-Doerth and von Uexküll (1973) give an average downward motion in the rosette of + 3.9 km s^{-1}. A few features were observed with velocities as large as −24 km s^{-1} (upward) and +20 km s^{-1}. However, the large majority of features had streaming velocities between −4 and +8 km s^{-1}. Since this study (see Section 4, Chapter II) incorporated an actual synthesis of line profiles using a discrete moving feature projected against a background chromosphere at rest, it provides a realistic model from which the line-of-sight velocity should be reasonably well determined. The fact that most of the velocities lie in the range −4 to +8 km s^{-1} with an average of +3.9 km s^{-1} again argues strongly that the predominant motion is downward in the rosettes and that the bright and dark mottles are not universally to be identified as spicules.

It is probably true that the velocities determined by Grossmann-Doerth and von Uexküll suffer from the dual facts that the models are oversimplified and that the observed features are not perfectly resolved. These effects could lead to serious underestimates for the true streaming motions. On the other hand, there is ample evidence that the dark fine mottles themselves are too numerous and predominantly too horizontal in character to be classed entirely as spicules. Thus, there is no particular reason to suspect that the true streaming motions are only those associated with spicules.

Although the pattern of flow suggested for the fibrils, viz., from the cell interior to the network, is consistent with the photospheric flow pattern, the evidence for such flow is not absolutely conclusive. The existence of such a flow is inferred mainly from the observation that the ends of the fibrils towards the cell interior show highest contrast in the blue wing of Hα whereas the ends of the fibrils near the network show highest contrast in the red wing of Hα. Such evidence is relatively conclusive but not absolutely conclusive. Also, it gives little indication as to the magnitude of the velocity.

The matter density in the chromosphere at a height of, say, 500 km above its base is about two orders of magnitude smaller than the matter density of the middle photosphere. Thus, the equation of continuity can be satisfied by low velocities in the photosphere even though the chromospheric velocities may be quite large. The flow of 0.5 km s^{-1} toward the supergranule border observed in the middle photosphere is much more than sufficient to sustain the spicule flow out of the network. Similarly, an average downflow of 0.05 km s^{-1} at the supergranule borders is much more than sufficient to sustain a flow of 10 km s^{-1} in the fibrils (assuming a mean density for the fibrils of 10^{-13} g cm^{-3}, and assuming that the fibrils cover half the surface area). Thus, the possibility of a rather large flow velocity in the fibrils cannot be ruled out on the basis of continuity.

In addition to flow along the fibrils there is some indication that the entire fibril structure may move transversely back and forth across the Sun in what has been described as flagellant motion. Such motion is difficult to separate from the effects of atmospheric seeing, but, under good seeing conditions, one frequently sees such motion on all of the Sun except in the immediate vicinity of sunspots where the magnetic field strength is large. The lack of motion near the spots suggests that the motion away from the spots is of solar origin and not of terrestrial atmospheric origin. Spicules observed at the limb, according to some observers, show a similar lateral motion. Nikolsky and Platova (1971) report a quasiperiodic lateral motion of approximately 1″ amplitude and a period of approximately 1 min. The mean lateral velocity is 10 to 15 km s^{-1}, i.e., about half of the vertical growth velocity. Spicule spectra show a quasiperiodic oscillation in Doppler shift that is consistent in amplitude and period with the motion in the plane of the sky. Although no quantitative studies of flagellant motion exist, the spicule results appear to be consistent with the flagellation phenomena.

The sound velocity in the middle chromosphere is approximately 7 km s^{-1}. A feature moving with sonic velocity will be displaced only about 400 km ($^4/_7$″) after 1 min of time. Thus, any quasiperiodic motion with a period near 1 min must have a velocity comparable to or greater than the sound velocity in order to have been detected in ground based observations. As we shall later show the chromosphere requires a considerable heat source to sustain its radiation. The most widely accepted mechanism for heating the chromosphere is the dissipation of mechanical wave modes. We should not be too surprised, therefore, to discover motions that are comparable in magnitude to the sound speed. By the same token, the detection of such motions represents one of the primary challenges in observational solar physics.

B. CELL BRIGHT POINTS

The increased visibility of the bright points within the supergranule cells (see Figure II-10) in the violet emission peak of the K-line suggests that the points are upward moving features even though we noted in Section 1 of this chapter that a violet emission peak in the K-line could be produced either by upward motion in the low chromosphere or *downward* motion in the high chromosphere. Time sequence studies of the cell points by Liu (1974) and Cram (1974) clearly suggest the upward moving model.

Liu (1974) has identified spectroscopic features in the K-line that he identifies with the cell points and has studied their spectroscopic evolution in time. Similar studies have been reported by Cram (1974). It appears from Liu's study that the cell points are associated with the 180 second oscillations in the chromosphere. According to Liu, some 90% of the cell interior is experiencing brightness oscillations at any one time. The oscillating cells are of the order of 3 to 6″ in size with a given cell undergoing typically two or three periods of oscillation. However, the brightness oscillation described by Liu consists of a more or less steady background brightness with superposed, quasiperiodic bright pulses, i.e., diminutions of intensity occur only as a bright pulse decays back to the normal intensity.   During a typical cycle, the regions near K_1, as well as those lying between K_1 and about 2 Å from line center brighten for about half the period. K_{2V} and K_{2R}, however, brighten appreciably for only about one-third of the period. The brightening in K_2 follows after the brightening in K_1. Near the middle of the pulse in K_{2V} and K_{2R} and lasting for about one-sixth of the period, the K_{2V} brightness exceeds considerably the K_{2R} brightness. This suggests quite clearly that the K_{2V} bright points represent a short-lived evolutionary stage in the oscillations. One troublesome aspect of this picture, however, is that the bright cell points in K_{2V} are considerably smaller than the 3 to 6″ size of the oscillating feature. It may be, therefore, that the bright point is restricted to only a portion of the oscillating region.

Brightening in the K_2 regions statistically lags behind the brightening in K_1, and, similarly, the K_1 brightening lags behind the wing brightening. This is illustrated in the time sequence of difference profiles shown in Figure III-6. In the example shown, the time delay from peak brightness at ± 2 Å to peak brightness at ± 0.44 Å is approximately 20 s. The delay in brightness peak between ± 0.44 Å and K_2 is approximately 40 s. Very crudely, the heights associated with these regions of the profile are 0, 100 and 600 km, respectively. Thus, for the wavelength pair ± 2 Å and ± 0.44 Å the inferred upward phase velocity is of the order of 5 km s^{-1} and for the pair ± 0.44 Å and K_2 it is of the order of 12 km s^{-1}. The sound velocity in these layers of the atmosphere is near 6 km s^{-1}. As we noted earlier in this chapter, brightening of the K_{2V} peak relative to K_{2R} is produced by upward motion in the K_1 region more effectively than by upward motion in the K_2 region itself. Thus, the height of the disturbance at the time of the K_{2V} brightening should be taken as being intermediate to the heights where K_1 and K_2 are formed. The brightness pattern is not necessarily inconsistent, therefore, with a constant vertical velocity near 6 km s^{-1}.

Although the pattern of brightening seen in Figure III-6 is somewhat typical, the phase relations are not. For approximately half of the brightenings studied out of a total of 459 the K_2 brightening appeared in phase with the wing brightening to within ± 5 s. For the entire sample of events the maximum brightening at ± 0.44 Å preceded the maximum K_{2V} brightening by an average of about 9.3 s. Thus, for the half of the sample showing a definite phase lead at ± 0.44 Å the average delay from maximum brightness at ± 0.44 Å to maximum brightness at K_{2V} is consistent with a propagation velocity of 20 km s^{-1}.

In 3% of the cases studied by Liu, the brightening at K_2 preceded the brightening at K_1 by more than 5 s. This was interpreted by Liu as a downward moving wave, possibly resulting from reflection of an earlier wave moving upwards. Other interpretations are possible. The large average phase velocity together with the large dispersion in phase velocity for the sample of events studied emphasizes the fact that the propagation characteristics of the waves are complex and not well determined.

No counterpart of these upward moving features seen in the cell interiors have been

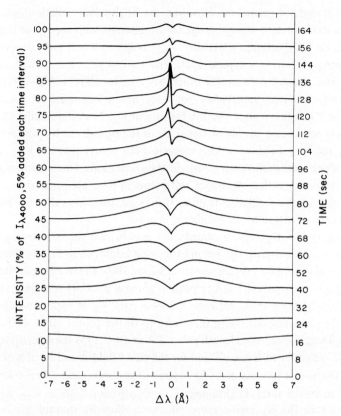

Fig. III-6.   A time sequence of observations in the K-line showing a bright feature moving from the wings toward line center. Each curve represents the residual profile after a mean profile has been subtracted and is shifted 5% in intensity relative to the preceding profile (Liu, 1973; Courtesy *Solar Physics*).

seen at the limb. It is conceivable, however, that the bright cell points are the outer foot-points of the rosette fibrils and that the upward motion observed at the cell point continues along the fibril becoming downward motion at the inner foot-point in the rosette. Such a picture is purely speculative at this time, however.

C. CHROMOSPHERIC OSCILLATIONS

Vertical oscillatory motion analogous to that in the photosphere has been detected in the chromosphere. The power spectra of the oscillatory component of motion are shown in Figure III-3. The λ 3933 (K line) line of Ca II and the λ 8542 line of Ca II are of chromospheric origin. Power spectra for these lines show a component at a period near 180 s (frequency 5.5×10^{-3} s^{-1}) that dominates the power in λ 3933 and that is a prominent feature of the power in λ 8542. Since λ 3933 is formed considerably above λ 8542, the 180 s period seems clearly to be dominant in the chromosphere. It is of interest to note in this connection that power spectra of radio noise observations of the Sun at 3.3 and 3.5 mm wavelengths show a concentration of power in a periodic component with a period of 180 s (Simon and Shimabukuro, 1971). Similar results were previously obtained by Yudin (1968) at a wavelength of 3 cm. The 3 mm radiation arises from the low chromosphere at heights between those where the K_2 and K_3 emission arises. The 3 cm radiation arises in the upper chromosphere somewhat above the K_3 region. Thus, the 180 s period observed in the radio spectrum is consistent with the results for the Ca II lines.

The amplitude of the chromospheric oscillations in the K line is given as 1.1 km s^{-1} (Orrall, 1966). Again, however, we emphasize that the 'measured' velocity is actually the average measured shift in the two sides of the line profile at a fixed reference intensity. As noted earlier in this chapter this may be interpreted as a true velocity amplitude only if the entire domain of line formation is moving with constant velocity. If sizable velocity gradients are present, the measured Doppler shift may reflect the actual velocities in a very artificial way.

The velocity amplitude of 1.1 km s^{-1} reported by Orrall (1966) corresponds to a wavelength shift at the K line of $\Delta\lambda = 0.014$ Å. Statistical studies of the K_2 and K_3 features of the profile show that these features vary in position about their average positions with rms amplitude of approximately 0.04 Å. In the discussion in Section 2 we noted that such shifts could be explained in a variety of ways and with velocities ranging from 3 km s^{-1} to in excess of 20 km s^{-1} depending upon the velocity gradients and the location of the moving material. The same ambiguity is inherently present in the interpretation of the oscillatory motion. Since the 180 s component appears to be restricted to the chromosphere, it may be associated with an appreciable velocity gradient, and the actual velocity amplitude may be much in excess of 1.1 km s^{-1} in the high chromosphere.

D. DOPPLER BROADENING OF CHROMOSPHERIC LINES

A further indication of non-thermal motion in the chromosphere comes from the widths of chromospheric lines. Chromospheric line widths can be studied for the strong disk absorption lines or for the emission lines seen beyond the solar limb at total eclipse. The latter method has the advantage of being able to select an unsaturated line, which gives more reliable Doppler widths, and of having the height of the lunar limb known absolutely relative to the solar limb. Its disadvantages are that the observed profiles are integrated over a transverse path through the chromosphere as well as over the height interval covered by the spectrograph slit. In most cases, this latter interval is comparable to the total height of the chromosphere.

Studies of strong Fraunhofer lines necessarily require that the Doppler widths be obtained from saturated profiles. In addition, the depth of line formation must be deduced from a model atmosphere together with a theory of line formation. In either method, therefore, there is considerable ambiguity in the assignment of a height scale.

Since there is considerable evidence that the line broadening motion in the photosphere is anisotrpic, the possibility of anisotropy in the chromospheric line broadening should not be overlooked. This presents the possibility that analyses of absorption line on the disk may give different results from emission line studies at the limb. In the limb studies the vertical and horizontal components of motion are necessarily averaged in some complex way and it would be extremely difficult, if not impossible, to separate the two components. Disk studies at different μ positions provide the opportunity of separating the horizontal and vertical components, but this has not been attempted for the strong chromospheric lines. Thus, it should be kept in mind that the absorption line studies probably give preference to the vertical component whereas the emission line studies at eclipse give preference to the horizontal component.

Results for the broadening velocity as a function of height in the chromosphere are given in Figure III-7. The velocity plotted is the quantity ξ defined by,

$$v^2 = v_{th}^2 + \xi^2 \tag{III-3}$$

where v is the total Doppler velocity, v_{th} is the mean thermal velocity given by

$$v_{th} = \left(\frac{2kT}{m_i}\right)^{1/2} \tag{III-4}$$

and m_i is the mass of the atom giving rise to the line. To determine v_{th} it is necessary to know the temperature, T. Different authors use different model atmospheres and, hence, somewhat different values of v_{th}. However, the results in Figure III-7 are restricted to elements for which $v_{th} \ll \xi$. In the low chromosphere where $\xi \lesssim 2$ km s^{-1}, however, v_{th} is not negligible even for the heavy elements and some scatter among the results by different authors is introduced as a result.

The results in Figure III-7 from Redman and Suemoto (1954) and from Hirayama (1971) were obtained at eclipse. The points in the low chromosphere from their results represent the averages from many spectral lines, mainly of iron. Note that they do not fall below about 3 km s^{-1}. The remaining results in Figure III-7 are obtained from attempts to represent the center-limb variations of the stronger Fraunhofer lines as well as some of the emission lines observed on the disk in the XUV spectrum. The disk lines used are Na I and Mg I (Athay and Canfield, 1969), O I (XUV) (Athay and Canfield, 1970), Ca II and Mg II (Athay and Skumanich, 1968; Linsky and Avrett, 1970) and O I and Mg II (Chipman, 1971).

While there is considerable spread in the values of ξ in the chromosphere used by different authors, it is clear that ξ increases rapidly with height and apparently approaches the sound speed somewhere near 1000 km to 1500 km. These results are based, however, on spherically symmetric models. There is ample evidence that the chromosphere is not spherically symmetric, particularly about 1000 km, and it is not clear how the results should be modified to account for this.

There is undoubtedly a relationship between ξ and the flagellant motion of fibrils, the systematic flow in fibrils, and the oscillatory motion. It is not clear at this time, however,

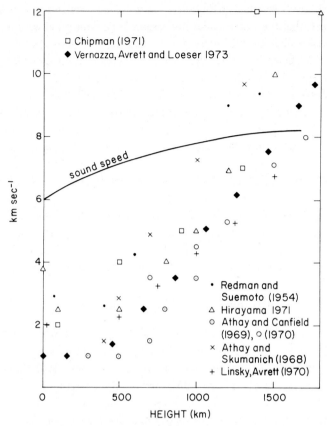

Fig. III-7.   A compilation of microturbulent velocities deduced from the widths of chromospheric
emission lines at the limb (●,Δ) and from the widths of strong Fraunhofer lines on the disk.

just what the relationship is. It is of interest to note that ξ is much larger than the
reported amplitude of the oscillatory motion, which again indicates that most of the
macroscopic kinetic energy is not in the long period oscillatory mode. As previously
noted, however, the reported amplitude of the oscillatory motion may be misleading.

Although there is little reason to assume that line broadening motions in the solar
chromosphere are isotropic, little or no attempt has been made to include possible effects
of anisotropy. This is due largely to lack of sufficient data and the complications of
chromospheric inhomogeneities. The differential center-limb effects of the inhomogeneities already introduce a major unknown into line profile analyses and the addition of
a second major unknown arising from angular dependence in ξ is more than present data
require.

As in the deeper layers of the photosphere, it appears that the great predominance of
mechanical energy in mass motions is associated with the unresolved elements of motion
that produce the line broadening. The rms velocity amplitude of this motion is about an

order of magnitude larger than is found for the resolved oscillatory elements. Although we have assumed that the line broadening motion is randomly and isotropically directed and has a Gaussian distribution function, we really have no basis for these assumptions. There are not sufficient data, at this time, to rule out the possibility of ordered anisotropic motion. The resolution of these questions must await our acquisition and successful analyses of sufficiently accurate line profiles, with their center-limb variations, for a number of spectral lines formed in the chromosphere. Such profile data currently exists only for the Ca II H, K and λ 8452 lines, Hα, the Na D lines, the Mg b lines, and a few Fe I lines. Many of these lines are not formed sufficiently high in the chromosphere to be of great help and none of them have been sufficiently studied to answer meaningful questions about directional and other ordered effects in the broadening motions. The richest reservoir of unused data lies in the numerous far uv emission lines and in the strong uv absorption lines of Mg, Fe and Si shortward of λ 3000. None of these lines have been sufficiently well observed to permit the type of detailed analysis required in order to answer questions concerning ordered effects in the mass motions. The OSO-I satellite scheduled for launch in 1975 was designed to provide much of this badly needed data.

4. Small Scale Coronal Motions

In this section, as in the preceding two, we shall be concerned with the motions in the quiet solar atmosphere. The higher one looks in the solar atmosphere the more one sees a predominance of transitory events and the less one knows about the so-called quiet atmosphere. Transitory motions will be discussed briefly in Section 6.

In spite of the fact that profiles of coronal emission lines have been observed for several decades, we still know very little about line broadening motions in the corona. We have been faced with the difficulty that the coronal kinetic temperature has not been determined to an accuracy much better than a factor of two. Thus, the thermal velocities are known only to accuracies of the order of $2^{\frac{1}{2}}$, and the observed line widths do not differ from the predicted widths by more than this amount. Attempts to separate thermal motions from microturbulent motions by using ions of different mass have been frustrated by the fact that the lines in the visual spectrum come mainly from ions of nearly equal mass or from ions formed under different excitation conditions. The former difficulty places too stringent conditions on the accuracy to which the profiles must be observed, and the latter invalidates the comparison of line widths.

The fact that the corona is optically thin and has such a large scale height makes any comparison of line profiles difficult. At the limb, the effective path length of integration in the corona is of the order of one solar radius. Any conclusions based upon small differences in profiles are always suspect in that the observed emissions in two lines may not have the same spatial contributions, and, hence, may not refer equally to the same volumes of space.

Photospheric and chromospheric lines, when observed with high spatial, temporal and spectral resolution, show a characteristic wiggly line structure. This wiggly line structure, as noted earlier in this chapter, arises from the Doppler shift associated with cells of matter moving in an organized way. No counterpart of the wiggly line structure has been observed in the corona. This is due, in part, to the long path length of integration in the corona. A given line of sight evidently intersects many individual moving cells. However, part of the reason for not observing wiggly lines in the corona may be simply that the only observations made with sufficient spatial and spectral resolution to show such structure are limb observations where the path length of integration is a maximum. Observations of coronal lines on the solar disk have a much shorter effective path length (approximately 10^5 km) and have much better prospect for exhibiting the wiggly line character. This is particularly true for transition region lines where the effective path length is greatly reduced.

Coronal lines can be observed on the solar disk only in the XUV and X-ray wavelengths. Observations in these spectral regions have not yet been made with sufficient spatial and spectral resolution to reveal such structure. However, experiments now waiting for launch on OSO-I will provide data of the necessary quality.

Limb spectra of the corona in the visual spectral region do occasionally show Doppler shifted features. However, these are invariably associated with active regions and motions of relatively large scale features and are not analogous to the photospheric and chromospheric oscillations. the lack of evidence for either non-thermal broadening motions of coronal lines or for Doppler displacements of coronal lines in quiet regions is not usually interpreted to mean that such phenomena are absent. On the contrary, it is more generally believed that these phenomena are present, as they are everywhere else in the solar atmosphere, but that we do not yet have sufficiently good observations to reveal them. The equally general belief that the mechanical energy that heats the chromosphere and corona is carried by propagating waves places a high priority on obtaining coronal line observations on the solar disk with sufficiently high spatial ($\lesssim 5''$), high temporal ($\Delta t \lesssim 20$ s) and high spectral ($\Delta\lambda/\lambda < 3 \times 10^{-5}$) resolution to exhibit the line profiles in the necessary detail.

The recent discovery by Gabriel et al. (1971) that neutral hydrogen in the corona scatters the chromospheric Lyman-α line with considerable intensity provides the possibility for determining sensitively the non-thermal component of line broadening. From the coronally scattered profile together with the obseved chromospheric profile one can deduce the mean velocity of the coronal hydrogen atoms. Similarly, from the observed profile of, say, a coronal iron emission line one can deduce the mean velocity of coronal iron atoms. The relation given by Equations (III-3) and (III-4) should allow an evaluation of both T and ξ. Note that the values of V_{th}^2 for hydrogen and iron differ by a factor $m_{Fe}/m_H = 56$. Thus, even values of ξ that are only a fraction of $V_{th}(H)$ may appreciably increase $V_{th}(Fe)$. Equation (III-3) is strictly valid only if the motions represented by ξ have a Maxwellian distribution, which they probably do not have. Nevertheless, Equation (III-3) should give a clear idea as to whether or not the ξ component is important in the corona.

In spite of the fact that we do not know the temperature of the corona well enough to accurately evaluate V_{th}, we can at least place upper limits on ξ by setting $V_{th} \ll \xi$. The coronal green line (λ 5303) of Fe XIV has an average half width corresponding to a broadening velocity of 27 km s^{-1}. This provides a firm upper limit to ξ. Since the sound speed in the corona is near 200 km s^{-1}, the non-thermal broadening velocities are small compared to the sound speed. This is in sharp contrast to the upper chromosphere where the macroscopic broadening velocity is near the sound speed.

5. Coronal Expansion: The Solar Wind

Ejections of streams or clouds of particles from the Sun have long been postulated as the source of auroral and geomagnetic phenomena associated with solar activity. The concept that particles might be continuously flowing outwards from the Sun was first advanced by Biermann (1951) to account for the acceleration away from the Sun of matter in comet tails. A theoretical argument leading to the existence of a solar wind was developed by Parker in 1958.

Direct observations of the solar wind in interplanetary space by spacecraft outside the Earth's magnetosphere have now been carried out for over a decade. These observations reveal that the solar wind has a steady background component at the orbit of the Earth of about 6 protons per cm^3 moving at a speed of about 400 km s^{-1}. The interplanetary medium is permeated by magnetic field lines carried out from the Sun by the solar wind.

At times of solar flares a complex variety of phenomena occurs that influence the solar wind. These include the creation of shock waves that propagate in the magnetic field and that are followed by enhanced velocities and particle densities in the solar wind.

Theoretical and observational studies of the solar wind have advanced to the point where they now form the basis for separate books (cf. Parker, 1963; Hundhausen, 1972). We can do little more, here, therefore, than to indicate the basic nature of and the importance of the phenomena. Additional details will be given in Chapter IX.

Imagine, for the sake of argument, that a mass of unconfined gas is heated to some temperature T and that internal gravitational forces are negligible. Under such circumstances the gas will, of course, expand. If, now, the force of gravity is increased within the gas, the expansion will slow down and will eventually stop when the gravitational force becomes strong enough. The gravitational force required to stop the expansion will depend upon the mean molecular weight, μm, of the gas and upon T. If m is too small or if T is too large, the expansion will continue.

To illustrate this effect for the corona, consider the equation of hydrostatic equilibrium

$$dP = - n\mu m_H g(r) \, dr, \tag{III-5}$$

where g is the gravitational acceleration and r is the radial distance from the center of gravity. We let

$$P = nkT, \tag{III-6}$$

$$g(r) = g_0 \left(\frac{R_\odot}{r} \right)^2 \tag{III-7}$$

and

$$T = T_0 \left(\frac{R_\odot}{r} \right)^\alpha \tag{III-8}$$

Equation (III-5) then becomes

$$\frac{dP}{P} = \frac{\mu m_H g_0}{k T_0} R_\odot^{2-\alpha} \frac{dr}{r^{2-\alpha}}, \tag{III-9}$$

which integrates to

$$P = P_0 \exp -\frac{\mu m_H g_0 R_\odot^{2\alpha}}{k T_0 (1-\alpha)} \left(\frac{1}{R_\odot^{1-\alpha}} - \frac{1}{r^{1-\alpha}} \right) \tag{III-10}$$

At $r = \infty$, Equation (III-10) gives

$$P_\infty = P_0 \exp -\frac{\mu m_H g_0 R_\odot}{k T_0 (1-\alpha)}, \qquad \alpha < 1,$$

and

$$P_\infty = 0, \qquad \alpha > 1.$$

Note that for $\alpha < 1$, P_∞ is finite. In fact, for solar values of g_0 and $R_\odot$ and for $\mu = 0.6$ and $T_0 = 1.6 \times 10^6$ K, $P_\infty = 10^{-3.8/(1-\alpha)} P_0$.

The thermal conduction model of the hydrostatic corona proposed by Chapman (1957) gives $\alpha = 2/7$. This value is too small to stop expansion and gives $P_\infty = 10^{-5.3} P_0$. By comparison, the pressure in the interstellar medium is several orders of magnitude smaller than this estimate of P_∞, and, as a consequence, the corona expands.

As a first order approach to the solar wind one solves the equation of motion,

$$\frac{dP}{dn} = -\mu N m_H g_0 \left(\frac{R_\odot}{r} \right)^2 - \mu N m_H v \frac{dv}{dr}, \tag{III-11}$$

together with the equation of continuity,

$$r^2 vN = R_\odot^2 v_0 N_0, \tag{III-12}$$

and the energy equation,

$$\frac{d}{dn} K(T) \frac{dT}{dn} = -kTv \frac{dN}{dr} - 3/2\, Nkv \frac{dT}{dr} \tag{III-13}$$

This latter equation equates the divergence of conductive flux to the work done in expansion and the energy used in heating the gas. The conduction coefficient, $K(T)$ is proportional to $T^{5/2}$.

In this form, the equations neglect several important factors, including viscosity, magnetic fields and the coronal heating term. The solution to Equations (III-11) to (III-13) give a reasonable approximation to the solar wind observed at the orbit of Earth provided T_0 is chosen properly and provided that T remains near T_0 out to a value of r near 5 to 10 $R_\odot$. More will be said about this point in Chapter IX. These restrictions on T_0 are conditioned by the neglect of viscosity and magnetic fields. Although attempts have been made to include each of these effects, a completely satisfactory model of the solar wind, including all of these effects, has not been developed.

The Solar wind observed at Earth orbit ($r = 215 R_\odot$, $v = 400$ km s^{-1} and $n = 6$) implies an expansion velocity at the base of the corona ($r = 1\, R_\odot$ and $n = 3 \times 10^8$) of 350 m s^{-1} and at $r = 2\, R_\odot$ ($n \approx 2 \times 10^6$) of 14 km s^{-1}. Velocities of this magnitude are too low to appreciably broaden the coronal emission lines observed in the inner corona.

6. Prominence Motions

We discussed in some detail in Chapter II the importance of quiescent filaments in the coronal mass balance. Of interest in that connection is the slow downward motion of matter within the prominences. In this section, we shall discuss, briefly, the more spectacular motions of the prominence body itself, or some portion thereof.

Prominence material is observed to move in all directions in the solar atmosphere: up, down and laterally. In a given prominence, however, there is usually a dominance of either upward or downward motions. Classes of prominences in which upward motions dominate include spicules, surges, sprays and eruptive filaments. The first three classes of prominences are short-lived, impulsive events in which upward motion dominates throughout most of the lifetime of the prominence. The lifetimes of these events extend from a few minutes up to an hour or so. In the latter case, however, the prominence pattern is stable most of it lifetime with an internal, slow downward motion of matter. The eruptive phase is short-lived (an hour or so) and the velocities are large. Downward motions predominate in filaments, coronal rain, and loop prominences. The first class contains a variety of relatively stable features with lifetimes up to weeks or months. The latter two are short-lived (lifetimes of an hour or so) and occur in close association with sunspots.

In discussing prominence motions it is of interest to consider three velocity classes: $v \approx 20$ km s^{-1}, $v \approx 200$ km s^{-1}, and $v > 615$ km s^{-1}. The former is the approximate sound velocity in the upper chromosphere, the second is the approximate sound velocity in the corona and the third is the escape velocity from the solar surface, or, equivalently, the free-fall velocity from infinity. It is perhaps not surprising that the velocities observed for certain classes of prominences tend to cluster (this is an inference from miscellaneous measurements and is not based on a systematic study) around the two sound velocities.

Spicules, for example, are ejected from the upper chromosphere with typical velocities of 20 to 25 km s^{-1} whereas surges are ejected from the upper chromosphere with typical velocities of about 200 km s^{-1}. Velocities of 100 to 200 km s^{-1} are typical, also, of knots of falling material in coronal rain, loop prominences, and in arches connecting to active region filaments.

The classes of events in which v is less than the excape velocity may be regarded as an internal transport process in which the material is confined to the solar atmosphere. An association of the velocity of such features with the sound velocity implies that the moving material produces shock waves which then exert a 'braking' action against further increases in velocity.

Prominence classes in which the material velocity may greatly exceed the sound velocity include sprays and other fast ejecta and eruptive prominences. In sprays and fast ejecta the outward velocities often exceed the excape velocity. Features of this type have either a fragmented or filamental structure (sprays) or a compact cloud structure (fast ejecta). Their outward motion does not typically show large acceleration or deceleration.

Eruptive prominences move upwards with velocities ranging upwards from 200 km s^{-1} to well in excess of the escape velocity. In many features of this type there is a pronounced upwards acceleration of the matter. The eruptive phase is typically preceded by an 'activation' of the filament in which the internal motions become much more violent (20 to 50 km s^{-1}) than is typical of the quiet phases (5 to 10 km s^{-1}) of the filaments life. Approximately two-thirds of eruptive prominences reform after two to three days in a pattern very similar to that existing before the eruption. Evidently the 'mold' in which the prominence forms is more stable than the prominence itself.

A vivid illustration of prominence motion and of a potentially powerful new technique of observing is shown in Figure III-8. The lower images are Hα filtergrams crossed by a series of spectrograph slits. The upper images in this figure show the spectral region near Hα for each of the slit positions. Sunspots show up as broad dark absorption features in the left three slits. Prominences show as dark features with Doppler shifted profiles. Examples are clearly evident in the right hand image. All of the spectra are made simultaneously by a technique developed by Title, 1974.

This discussion of prominence motions is both brief and superficial. Our interest in prominences here is only with regard to their relationship to the overall properties of the chromosphere and corona. For a more complete discussion of prominence phenomena and motions the reader is referred to a recent book by Tandberg-Hanssen (1974).

As was noted in Chapter II, the prominences contain a very significant fraction of the total coronal mass. The preceding discussion of motions of prominence material should be sufficient to indicate that prominences are characterized by their transitory, unstable nature and that strong forces other than gravitation (magnetic) act on the prominence material. It should also be apparent to the reader that prominences are not rare phenomena. They are common, everyday occurrences in the corona and ought to be considered as an intergral part of the corona as a phenomenon.

We have little reason to suppose that there is a complete dichotomy of coronal

Fig. III-8.   Three examples of simultaneous Hα images of active regions (bottom panels) with Hα spectra taken through a series of slits in a single spectrogram. The vertical lines in the bottom panels show the slit positions and the dark lanes in the upper panels are the cores of the Hα lines. Sunspots produce the very wide line cores and plages produce the brighter line cores. Prominence can be seen as Doppler shifted absorption features (Lockheed Solar Observatory, Rye Canyon).

material that can be classed as prominence and non-prominence in the sense that the non-prominence material sits quiescently in hydrostatic equilibrium while the prominence material is hurled about by magnetic forces. The so-called quiet corona undoubtedly experiences the same types of magnetic forces to some degree.

7. Impulsive Motions

Since the emphasis in this book is on the quiet Sun as opposed to the active Sun, it is not appropriate to include an extensive discussion on motions associated with phenomenon of solar activity. There is a variety of such motions that are of an impulsive nature. These will be mentioned only briefly.

The birth of a solar flare occurs on a time scale of several minutes. Brief periods of unusually rapid growth or brightening occur on time scales of approximately 1 min or less. This rapid growth, often referred to as the flash phase, usually occurs in some preferred direction. Associated with the rapid expansion of the flare, there is frequently a propagating disturbance having the appearance of a wave emanating from the flare in the direction of growth. Such a disturbance is illustrated in Figure III-9. The wave character of the disturbance is evidenced in Hα pictures of the chromosphere which show the leading edge of the disturbance as a violet shift of the line and the trailing edge as a red shift. Propagation velocities for these disturbances, commonly called 'blast waves' or 'Morton-waves', are typically 1000 to 2000 km s^{-1}.

Flare associated blast waves do not spread circularly from their point of origin in the flare. Instead they follow well defined channels, or cones, that often connect one active

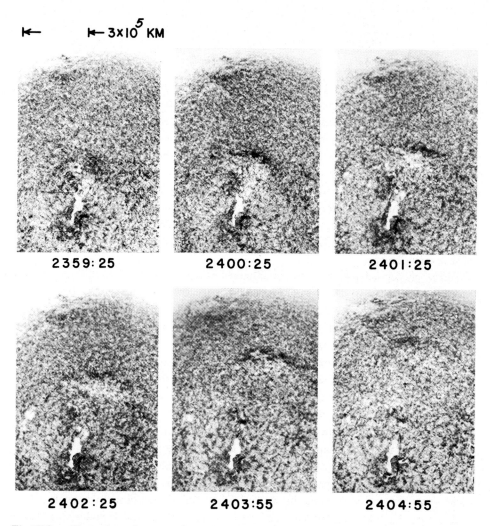

$\leftarrow$ $\leftarrow$ 3×10^5 KM

| 2359:25 | 2400:25 | 2401:25 |

| 2402:25 | 2403:55 | 2404:55 |

Fig. III-9.   Illustration of a wave front (narrow dark band followed by a broader bright band) propagating laterally across the solar surface (upward in the figure) in association with a solar flare (bright feature). The images are made by photographically subtracting two images of the Sun made in opposite sides of the Hα core (Lockheed Solar Observatory, Rye Canyon).

region to another. Blast waves are observed to propagate for distances across the solar disk that, when converted to distance along the spherical surface, correspond to about one solar diameter. Since the disturbance is often visible in the chromosphere over much of the path, the propagation evidently follows the solar curvature.

The velocities of 1000 to 2000 km s^{-1} observed for flare blast waves are much in excess of the sound velocity, which suggests that the waves are of the Alfvén type. Alfvén waves have velocities given by

$$V_A = \frac{B}{(4\pi\rho)^{1/2}} \, , \tag{III-14}$$

where B is the magnetic field strength and ρ is the matter density. In the upper chromosphere ρ is approximately 3×10^{-14} and a velocity of 1000 km s^{-1} corresponds to $B \approx 60$ G. Similarly, in the low coronaa $\rho \approx 5 \times 10^{-16}$, which, for $V_A = 1000$ km s^{-1}, corresponds to $B \approx 8$ G. These are reasonable results for active regions. Thus, the Alfvén mode appears reasonable and accounts quite naturally for the channeling of the waves (cf. Uchida *et al.*, 1973).

Slower propagating disturbances initiated by flares are also evidenced in prominence phenomena. The 'activation' of filament prominences preceding their eruption phase is flare associated. Time lags between flare and filament activation often suggest propagation velocities of only a few km s^{-1}. Relatively little is known about the nature of such disturbances.

Direct photographs of the corona obtained with 'green line' (λ 5303) filters show occasional events in which coronal arches expand and break apart in a whiplike action very reminiscent of eruptive prominences. Also, surge-like phenomena are observed. The detailed association between such phenomena and prominences has not been established, but their essential similarity suggests a close analogy if not a detailed correspondence.

Observations recently obtained using the coronagraph aboard the Skylab satellite have greatly extended the total amount of high resolution coronal observations available. Initial inspection of these data reveal a variety of large scale coronal transient events occurring in association with solar activity. One such event is illustrated by the time sequence in Figure III-10. In this particular event a large arch is seen to expand outwards with a velocity of approximately 400 km s^{-1} at the leading edge.

The statistics of such events are still not well known, although Skylab data should give a reasonable sample for a particular epoch of the sunspot cycle. It is clear already that such events are not rare. Also, the spectral properties of the coronal transients are not well known. The Skylab instrument accepts a broad spectral band including the Hα line and the coronal emission lines. Much of the radiation observed could be Hα radiation from prominences, although there is reason to believe that continuum radiation forms an important part of what is observed. Resolution of this problem must await a more thorough analysis of the data.

Radio noise bursts give further evidence of large scale mass motions in the corona. Type II radio bursts are flare associated and are generally understood to be excited by surge ejections. The type II source moves upwards in the corona at a characteristic surge velocity and has been demonstrated, for a few cases, to coincide in time and position with the surge.

Type III radio bursts show characteristic source velocities of one-quarter to one-third the velocity of light. The sources have been observed with the aid of satellite borne radio detectors (Fainberg *et al.*, 1972) to travel beyond the orbit of the Earth. It is believed

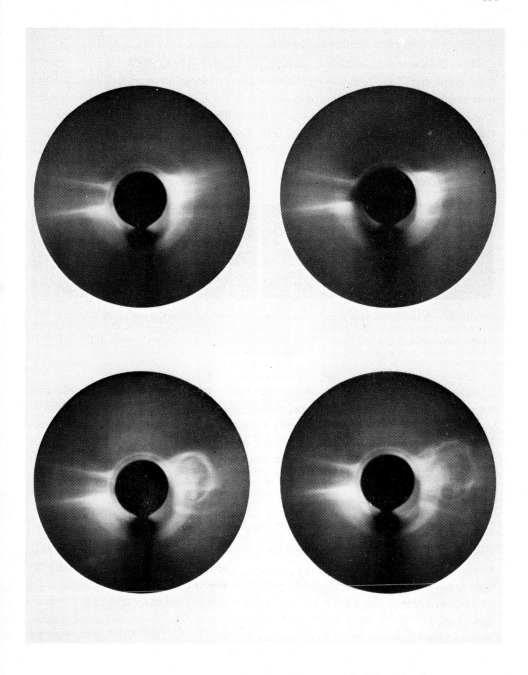

Fig. III-10. A time sequence of white light images of the corona observed by ATM Skylab, August 10, 1973. The four photographs were taken at 1332 UT, 1343 UT, 1424 UT and 1447 UT beginning with the upper left image and ending with the lower right image (High Altitude Observatory, National Center for Atmospheric Research).

that these are excited by relativistic particles ejected from the Sun, although there is not general agreement as to whether the particles are protons or electrons or both.

Other indications of relativistic particles in the solar atmosphere come from type IV radio bursts and X-ray bursts as well as from direct observation of relativistic particles of solar origin from instruments flown in space near the Earth and in the planetary system.

An example of a moving radio source region observed with the radio-heliograph at Culgoora, Australia, is shown in Figure III-11. The radio source in this case is type IV and is moving at a speed of approximately 270 km s^{-1}. In the last frame of this figure, the radio source is located at $5\frac{1}{2}R_\odot$.

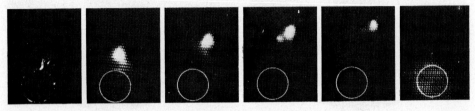

Fig. III-11.   A rapidly moving type IV radio event observed with the Culgoora radioheliograph. The left hand image shows an Hα flare spray observed at 2158 UT, March 1, 1969. The successive radio images were made at 2314, 2350, 0009, 0031 and 0119 UT, respectively (Division of Radio Physics, CSIRO, Australia).

Again, we mention these phenomena only briefly in order to remind the reader that the solar corona, and perhaps the upper chromosphere, is an area in which acceleration of particles to relativistic and sub-relativistic velocities is a relatively common occurrence. MeV particles accompany many of the major flares as well as some of the minor ones. A review of the observational evidence for solar injected relativistic particles in space is given by Fichtel and McDonald (1967).

8. Mass Balance and Energy Transport in Mass Flow

Mass exchange processes in the chromosphere appear to be of such a long time scale that they can be largely ignored with the possible exception of the upper chromosphere. In these layers the mass flow in spicules is probably important.

As an estimate of the mass in the upper chromosphere, we adopt a hydrogen number density of 10^{11} cm^{-3} and a scale height of 5×10^7 cm. The product of density, scale height and surface area of the Sun then gives a total of 3×10^{41} hydrogen atoms. By comparison, the number extimated in Chapter II for the corona is 2×10^{41}. The total mass of the low chromosphere is several orders of magnitude larger.

Spicules carry matter upwards from the upper chromosphere to the corona. The rate at which they transport matter is given by

$$n_\mathrm{H} v_s a_s N_s,$$

where v_s is the mean spicule velocity and a_s is the average surface area covered by each spicule and N_s is the total number of spicules. For the product $a_s N_s$, we take one percent of the total surface area of the Sun, or 6×10^{20} cm^2, and for n_H we take 6×10^{10} cm^{-3}. Then, for $v_s = 2.5 \times 10^6$ cm s^{-1} we find a total transport of about 1×10^{38} hydrogen atoms s^{-1}. It follows that the matter transport in spicules is sufficient to both replace the entire corona and to remove the entire upper chromosphere in approximately 2×10^3 s.

Most of the matter carried upwards by spicules must return to the chromosphere. This is immediately evident from a comparison of the matter flow in spicules to the matter flow in the solar wind. Using $v = 4 \times 10^7$ cm s^{-1} and $n = 6$ at the orbit of the Earth and assuming an isotropic solar wind, we find the total matter flow in the solar wind to be

$$nv4\pi(\text{AU})^2 = 6 \times 4 \times 10^7 \times 4\pi \times (1.5 \times 10^{13})^2 = 6.8 \times 10^{35}$$

hydrogen atoms s^{-1}. This is only about 0.7% of the estimated spicule flow but, nevertheless, represents a rapid depletion of coronal matter sufficient to remove the corona in about 80 h.

We noted in Chapter II that approximately half of the coronal mass may be in quiescent filaments and that the matter in filaments flows downward at a velocity of approximately 5 km s^{-1}. A system of filaments 50 000 km high and containing half the coronal mass would drain the coronal mass back to the chromosphere in approximately 10^4 s, of 3 h. Hence, even at times of heavy filament activity the primary mass cycle could not be represented by balancing the upflow in spicules against the downflow in filaments. Also, the filaments are often not numerous enough to represent more than a few percent of the coronal mass, but there is no evidence for a corresponding decrease in spicule density. It must be concluded, therefore, that much of the spicule upflow simply falls back into the chromosphere. Spicule statistics (cf. Beckers, 1972) indicate that the down flowing matter visible in Hα is not sufficient to balance the outflow visible in Hα. Hence, the primary downflow must be in a form that is not easily visible in Hα.

A single large surge prominence may reach upwards some 200 000 km and have a cross sectional area of 10^8 km^2. The corresponding volume is 2×10^{28} cm^3, and, for a mean hydrogen density of 5×10^{10} cm^{-3}, we find that it contains 1×10^{39} hydrogen atoms. This represents about 0.5% of the total matter in the upper chromosphere (or in the corona). Since the surge covers only about 0.006% of the solar surface area, however, the surge represents an enormous local perturbation. This is made even more evident by considering the energy flow in the surge. The rate of transport of kinetic energy density plus the rate of gain of potential energy is given by

$$F_{\text{transport}} = \tfrac{1}{2} n_H m_H v^3 + n_H m_H g v R_0 = 3.5 \times 10^9 \text{ erg cm}^{-2} \text{ s}^{-1}, \quad \text{(III-15)}$$

where we have used a surge velocity of 200 km s^{-1}. Thus, the energy flow in the surge is seen to be as much as 6% of the total rate of energy flow in radiation from the photosphere. Most of the energy flux in the surge is in the potential energy term. The kinetic energy transport does not become dominant until v exceeds the escape velocity of 615 km s^{-1}. The thermal energy of the corona in a 1 cm^2 column is approximately

$$nkTH = 6.7 \times 10^8 \, \text{erg cm}^{-2},$$

and the energy flux in the surge would be sufficient to regenerate the coronal (locally) in a fraction of a second.

Similar reasoning applied to the rate of energy flow in spicules is equally impressive. For a mean spicule density of 6×10^{10} cm^{-3} and mean spicule velocity of 25 km s^{-1} the energy flux due to mass transport given by Equation (III-15) is

$$F_{\text{transport}}(\text{spicules}) = 4.8 \times 10^8 \, \text{erg cm}^{-2} \, \text{s}^{-1}.$$

This is the local value within the spicule, and, again, almost all of the energy flux is in the potential energy term.

Although the spicule energy flux is only about a tenth of the surge energy flow, spicules are always present on the Sun and cover some one percent of the solar surface area at a given time. As we shall later show, the spicule energy flux, when averaged over the Sun, is comparable to the total energy flux required to heat the chromosphere and corona. It is a much larger energy flux than is needed for the corona itself.

The foregoing considerations of mass cycling and the realization of the magnitude of the role played by prominences and spicules in the corona and upper chromosphere again lend support to the concept that the upper chromosphere and the corona are dynamic regions in which the flow of matter itself and the energy transported by the flow play major roles. As we shall see in the next chapter, and as has already been implicit in much of the earlier discussion, the magnetic field structure also plays a major role.

References

Athay, R. G.: 1963, *Astrophys. J.* **138**, 680.
Athay, R. G.: 1970, *Solar Phys.* **11**, 347.
Athay, R. G. and Canfield, R. C.: 1969, *Astrophys. J.* **156**, 695.
Athay, R. G. and Canfield, R. C.: 1970, in H. G. Groth and P. Wellmann (eds.), *Extended Atmosphere Stars*, NBS Special Publ. 332, Washington.
Athay, R. G. and Lites, B. W.: 1972, *Astrophys. J.* **176**, 809.
Athay, R. G. and Skumanich, A.: 1968, *Solar Phys.* **4**, 176.
Beckers, J. W.: 1972, *Ann. Rev. Astron. Astrophys.* **10**, 73.
Bhavilai, R.: 1965, *Monthly Notices Roy. Astron. Soc.* **130**, 411.
Biermann, L.: 1951, *Z. Astrophys.* **29**, 274.
Cha, M. Y. and White, O. R.: 1973, *Solar Phys.* **31**, 55.
Chapman, S.: 1957, Smithsonian Contr. to *Astrophys.* **2**, 1.
Chipman, E.: 1971, Smithsonian Astrophys. Obs., Special Report 338.
Cram, L.: 1974, in R. G. Athay (ed.), *Chromospheric Fine Structure*, D. Reidel Publishing Co., Dordrecht, Holland, p. 51.
Deubner, F. L.: 1967, *Solar Phys.* **2**, 133.
Edmonds, F. N., Michard, R., and Servajean, R.: 1965, *Ann. Astrophys.* **28**, 534.
Evans, J. W. and Michard, R.: 1962a, *Astrophys. J.* **135**, 812.
Evans, J. W. and Michard, R.: 1962b, *Astrophys. J.* **136**, 493.
Evans, J. W., Michard, R., and Servajean, R.: 1963, *Ann. Astrophys.* **26**, 368.
Fainberg, J., Evans, L. G., and Stone, R.: 1972, *Science* **178**, 743.
Fichtel, C. E. and McDonald, F. B.: 1967, *Ann. Rev. Astron Astrophys.* **5**, 351.
Frazier, E. N.: 1968, *Astrophys. J.* **152**, 557.

Gabriel, A. H., Garton, W. R. S., Goldberg, L., Jones, T. J. L., Jordan, C., Morgan, F. J., Nicholls, R. W., Parkinson, W. J., Paxton, H. J. B., Reeves, E. M., Shenton, C. B., Speer, R. J., and Wilson, R.: 1971, *Astrophys. J.* **169**, 595.
Gonczi, G. and Roddier, F.: 1969, *Solar Phys.* **8**, 255.
Grossmann-Doerth, U. and von Uexküll, M.: 1973, *Solar Phys.* **28**, 319.
Hirayama, T.: 1971, *Solar Phys.* **19**, 384.
Holweger, H.: 1967, *Z. Astrophys.* **65**, 365.
Howard, R.: 1967, *Solar Phys.* **2**, 3.
Hundhausen, A. J.: 1972, *Coronal Expansion and Solar Wind*, Springer-Verlag, New York.
Jensen, E. and Orrall, F. Q.: 1963, *Astrophys. J.* **138**, 252.
Leighton, R. B., Noyes, R. W., and Simon, G. W.: 1962, *Astrophys. J.* **135**, 474.
Linsky, J. L. and Avrett, E. H.: 1970, *Publ. Astron. Soc. Pacific* **82**, 485.
Liu, S. Y.: 1973, *Solar Phys.* **31**, 127.
Liu, S. Y.: 1974, *Astrophys. J.* **189**, 359.
Livingston, W. C.: 1968, *Astrophys. J.* **153**, 929.
Nikolsky, G. M. and Platova, A. G.: 1971, *Solar Phys.* **18**, 403.
Noyes, R. W.: 1963, Thesis, California Inst. of Tech., Pasadena.
Noyes, R. W.: 1967, in R. N. Thomas (ed.), *Aerodynamic Phenomena in Stellar Atmospheres*, Academic Press, New York.
Orrall, F. Q.: 1966, *Astrophys. J.* **143**, 917.
Parker, E. N.: 1958, *Astrophys. J.* **128**, 664.
Parker, E. N.: 1963, *Interplanetary Dynamical Processes*, Interscience, New York.
Redman, R. O. and Suemoto, Z.: 1954, *Monthly Notices Roy. Astron. Soc.* **114**, 238.
Roddier, F.: 1966, *Ann Astrophysic* **29**, 639.
Simon, G. W. and Leighton, R. B.: 1964, *Astrophys. J.* **140**, 1120.
Simon, M. and Shimabukuro, F. I.: 1971, *Astrophys. J.* **168**, 525.
Tandberg-Hanssen, E.: 1974, *Solar Prominences,* D. Reidel Publishing Co., Dordrecht, Holland.
Tanenbaum, A. S.: 1971, Thesis Univ. of California, Berkeley.
Tanenbaum, A. S., Wilcox, J. M., Frazier, E. N., and Howard, R.: 1969, *Solar Phys.* **9**, 328.
Title, A.: 1974, private communication.
Uchida, Y., Altschuler, M., and Newkirk, G.: 1973, *Solar Phys.* **28**, 495.
Yudin, O. I.: 1968, *Soviet Phys. Doklady* **13**, 503.
Vernazza, J. E., Avrett, E. H., and Loeser, R.: 1973, *Astrophys. J.* **184**, 605.

CHAPTER IV

MAGNETIC FIELDS

1. Method of Observation

Solar magnetism is a topic of great importance and of strong current interest in solar physics. Magnetic field phenomena on the Sun are complex, highly variable and of interest in their own right. It seems clear, for example, that the surface phenomena of the active Sun, in almost all their aspects are primarily manifestations of the variable nature of solar magnetism. Thus, if we were to discuss solar activity in any detail we would necessarily delve deeply into magnetic field phenomena.

Our interest in this text is primarily in the chromosphere and corona as phenomena which reflect upon the overall character of the solar atmosphere and, hence, upon the nature of stellar atmospheres. A detailed study of solar activity would lead us necessarily into consideration of the solar interior, convection, rotation, dynamo action, etc., which is beyond the intent of this book. Thus, our discussion of magnetic fields will necessarily be somewhat limited and will relate primarily to the 'quiet Sun' structure.

Once again, however, we note the artificialness of this distinction. There is clear evidence suggesting, for example, that the magnetic field emerging through the photosphere in active regions migrate far from their place of origin, in both latitude and longitude, before they dissipate. Thus, rather than say that we will discuss the quiet Sun as distinct from the active Sun, we should more accurately say that we will discuss the magnetic structure of the average Sun as distinct from disturbed solar areas immediately identifiable as active regions. The average magnetic properties of the Sun are not determined by the strong local fields of active regions but by the more diffuse areas of magnetic field removed from the active centers.

We begin with an elementary review of how observations of magnetic field strength are carried out.

An atomic energy level with a given angular momentum quantum number, J, is made up of $2J+1$ states of different magnetic quantum number, M. In the absence of a magnetic field the different M states each have the same energy and the J level is degenerate. When a magnetic field is present, however, the M states separate and the corresponding spectral lines arising from different J levels separate into a Zeeman pattern of three or more components.

The separation between the different line components in the Zeeman pattern depends upon the Lande 'g' factor. In L-S coupling

$$g = \frac{3}{2} + \frac{1}{2} \frac{S(S+1) - L(L+1)}{J(J+1)} \qquad \text{(IV-1)}$$

If $J = 0$ $(S = L = 0)$, Equation (IV-1) is indeterminate, but, in this case, $g = 0$ since there is only one M level $(M = 0)$. Also, $g = 0$ whenever

$$\frac{S(S + 1) - L(L + 1)}{J(J + 1)} = -3.$$

This is satisfied, for example, with $S = \frac{3}{2}$, $L = 2$ and $J = \frac{1}{2}$ or with $S = 2$, $L = 3$ and $J = 1$. Thus, some lines show no Zeeman splitting.

Zeeman triplets occur whenever $J = 0$ $(g = 0)$ for one level or whenever g is the same for the initial and final levels. Singlet lines occur when $S = 0$. Since this leads to $L = J$, we note from Equation (IV-1) that $g = 1$ for both levels. Thus, all lines of a singlet spectrum produce Zeeman triplets and, since $g = 1$, this is defined as the 'normal' Zeeman effect. If g differs from unity for a triplet Zeeman pattern or if the Zeeman pattern has higher than triplet multiplicity, the pattern is described as 'anomalous'.

For the weak field case, the spacing of the different Zeeman components from the normal line position in zero magnetic field is given by

$$\Delta\lambda = (g, M' - g''M'') \frac{\lambda^2 eH}{4\pi mc^2}$$

$$= 4.67 \times 10^{-13} (g'M' - g''M'')\lambda^2 H, \tag{IV-2}$$

where both $\Delta\lambda$ and λ are in ångstroms. The prime and double prime denote initial and final states and the magnetic quantum number M is subject to the selection rules $M'' - M' = 0$ or ± 1. M may take on any of the $2J + 1$ quantized values $-J < M < J$. For a normal pattern $g'M' - g''M'' = \pm 1$ and at $\lambda 5000$ a field of $1000\,\text{G}$ produces a separation of $\Delta\lambda = \pm 0.0117\,\text{Å}$. Many spectral lines exist, however, for which the factor $g'M' - g''M''$ (known as the 'effective g' factor) exceeds unity by a considerable amount. Since increasing the effective g increases $\Delta\lambda$ for given λ and H, one customarily measures magnetic field strength using lines whose effective g is large. The combined requirements for large effective g, for unblended lines and for relatively simple Zeeman patterns usually result in the use of lines with effective g values in the range 1.5 to 3. The often used lines of Fe I at $\lambda 5250$ and Ca I at $\lambda 6103$, for example, have effective g values of 3.0 and 2.0, respectively. For these lines, the separation of the two circularly polarized components in a field of $1000\,\text{G}$ is $0.077\,\text{Å}$ and $0.070\,\text{Å}$ respectively.

There is still further advantage to using lines of long wavelength. A line of Fe I at $\lambda 15648.6$ has been used by Harvey and Hall (1974). The spectroscopic terms from which this line arises have not been identified, but by comparison with the Zeeman splitting of other lines in sunspot spectra Harvey and Hall conclude that this line has an effective g value of 3.0 and has a simple Zeeman pattern. The full separation between the two circularly polarized components for this line in a field of $1000\,\text{G}$ is $\Delta\lambda = 0.22\,\text{Å}$, which is considerably in excess of the line width. Not all of the apparent advantages of lines of longer wavelength suggested by Equation (IV-2) are realized, of course, since the line widths increase in proportion to λ. Thus, the real advantage from using longer wavelength is proportional to λ rather than λ^2. Nevertheless, the combination of longer wavelength and large effective g values for the λ 15 648.6 line leads to a strong advantage in the use of this line for measuring solar fields. Harvey and Hall find, for example, that for many areas of locally strong magnetic fields the Zeeman components of the line are clearly resolved and their separation can be measured directly.

Zeeman lines arising from $\Delta M = 0$ are linearly polarized in the direction of the magnetic field (π lines) and those arising from $\Delta M = \pm 1$ are circularly polarized (σ lines) with the plane of polarization at right angles to the field and with opposite senses of polarization for $\Delta M = +1$ and -1. The circular polarization is used to measure the longitudinal component of magnetic field strength and the linear polarization is used to measure the transverse component of the field strength.

Nearly all magnetic field measurements of the Sun thus far have used longitudinal field magnetographs employing two exit slits placed in opposite wings of a selected spectral line in the focal plane of a spectrograph. When no field is present there will be no circular polarization. Thus, if the slits are equally spaced from line center, they will each see steady signals of equal intensity when the line is viewed alternately in left hand and right hand circular polarization. In the presence of a magnetic field, the two σ components are now circularly polarized and shifted in $\Delta\lambda$. Thus, the two slits will now see fluctuating signals as the line is viewed alternately in the two senses of circular polarization. The amplitude of the signal will depend upon both the slope of the intensity profile in the line at the location of the slit and upon the magnetic field strength, i.e., upon the amount of Zeeman shift.

For the sake of illustration, let us consider just one of the exit slits. We assume that the half-profile of the line can be represented, as illustrated in Figure IV-1a, by two straight lines. For $\Delta\lambda < \Delta\lambda_1$, the intensity is constant at I_0. Between $\Delta\lambda_1$ and $\Delta\lambda_2$, the intensity is given by

$$I = I_0 + \left(\frac{I_c - I_0}{\Delta\lambda_2 - \Delta\lambda_1} \right) (\Delta\lambda - \Delta\lambda_1),$$

where I_c is the continuum intensity. We denote the slope $(I_c - I_0)/(\Delta\lambda_2 - \Delta\lambda_1)$, for this part of the profile by b. A shift of this hypothesized line by an amount $\delta\lambda$ produces a change of intensity at $\Delta\lambda(\Delta\lambda > \Delta\lambda_1)$ by an amount $b\delta\lambda$ for $\delta\lambda < \Delta\lambda - \Delta\lambda_1$. However, if $\delta\lambda > \Delta\lambda - \Delta\lambda_1$ the change of intensity is limited to $b(\Delta\lambda - \Delta\lambda_1)$. For this hypothetical case, therefore, there is a clear saturation effect arising from the change in slope of the line profile. Real line profiles do not change slope so abruptly as the profile illustrated in Figure IV-1a. On the other hand, real profiles do have zero slope at line center and in the far wings and they have a maximum slope at some intermediate point. They are subject, therefore, to the same saturation effect as illustrated.

It is clear from the above example that magnetographs employing exit slits a fixed distance apart require both an accurate knowledge of the slope of the line profile in the solar surface areas where the magnetic fields are located and that the line splitting is not too large. To the contrary, however, it has been found that the line profiles in the areas

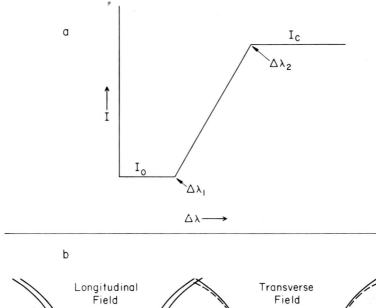

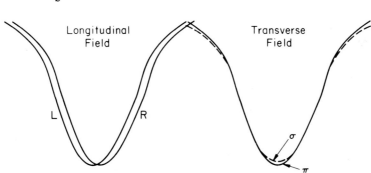

Fig. IV-1.   (a) An idealized line profile used for discussing the technique for measuring solar magnetic field. (b) Comparison of the effects on the line profile for magnetic fields viewed along the field lines (longitudinal) and at right angles to the field lines (transverse).

where most of the magnetic flux is located change markedly from the average Sun profile, for some lines, and that the Zeeman splitting in these same areas is large enough to lead to saturation effects. Both of these effects will be discussed in more detail in Section 5 of this chapter.

Although longitudinal field magnetographs provide a great deal of valuable and interesting material, it would be useful to know the transverse (relative to line-of-sight) components as well. However, transverse field magnetographs are severely handicapped by the fact that the σ components are now linearly polarized in the same direction and the alternating signal must now be made between the sum of the π component and the sum of the σ components. Because the σ components are symmetrically spaced about the π components, the sum of the various σ components gives a net profile that is undisplaced from the sum of the π components. Also, the total line strength from the σ components must exactly equal that from the π components in the optically thin case. Since the σ

components separate further than the π components the combined profile of the σ components will be somewhat broader than the combined profile of the π component. Thus, in the optically thin case the only difference in signal obtained by observing first the π then the σ components arises from slightly different widths and slightly different central intensities for the two components. Radiative transfer effects will induce further small changes as the opacity increases. Such differences as exist, however, are very much smaller than for the corresponding case using the circularly polarized light. These differences are illustrated schematically in Figure IV-1b. For further details the reader should consult Evans (1966).

An additional complication of the transverse field measurements is that linear polarization in spectral lines may be induced by resonance scattering processes when no magnetic fields are present. This additional polarization is difficult to evaluate and causes uncertainty in the interpretation of the linear polarization data. The magnetic field and resonance scattering polarizations are solar induced. Optical instruments also induce linear polarizations that can be removed only with exceeding care. This further complicates the transverse field measurements.

Although transverse field magnetographs now exist, it is somewhat doubtful that any completely reliable results have yet been obtained.

Interpretation of magnetographic observations suffers, from the same type of difficulties as the interpretation of line shift measurements. Thus, to extract the magnetic field strength directly it is necessary to assume that the fields are uniform over the surface area observed and constant with depth. It is further assumed that the spectral lines are symmetric. Any Doppler displacements that are present must be carefully compensated for to insure that the magnetograph is centered on the actual solar line at its 'local' wavelength rather than on some average wavelength. For this reason most magnetographs also serve very usefully as velocity spectrographs as well.

Unlike velocity fields, however, we expect the magnetic field to vary more coherently with depth. Thus, the lack of height discrimination that is inherent in all observations using spectral lines is perhaps not so critical in the case of the magnetic field. Effects of line asymmetries have not been carefully studied. On the other hand, solar magnetic fields are now known to be highly localized with the field strongly concentrated into small bundles whose typical size appears to be small compared to the sampling aperture of most magnetographs. Since it cannot be assumed safely that the spectral line is itself uniform over the sampling aperture, the 'measured' field obtained with modest resolution instruments is some average of the fields over areas in which the field gradients are large. The exact nature of the averaging is not clear.

2. Energy Considerations

It is of interest to ask whether the magnetic field structure of the Sun could be of importance in the overall energy balance of the solar atmosphere. The magnetic field strength averaged over the solar surface is of the order of 2 G. However, this gives a very misleading impression of the fields and their energy. As we shall later see, the solar fields appear to be clustered into small patches whose field strength is of the order of 1000 to 2000 G. Thus, we should more properly describe the average field as being composed of elements of strength B_1 occupying a fractional area a_1 of the solar disk. For the sake of illustration, we choose $B_1 = 2000$ and $a_1 = 10^{-3}$.

The energy content of the solar magnetic fields is then given by

$$\frac{a_1 B_1^2}{8\pi} L_1 = 160 L_1 \text{ erg cm}^{-2},$$

where L_1 is the effective vertical scale length of the field. By comparison, the mechanical energy required to heat the chromosphere and corona (see Chapter IX) is about 4×10^6 erg cm^{-2} s^{-1}. Thus, the mechanical flux could be supplied by the magnetic field if the field energy were converted to heat energy in a time period, T_B, given by

$$T_B = \frac{160 L_1}{4 \times 10^6} = 4 \times 10^{-5} L_1 \text{ s.}$$

The scale length L_1 is unknown, but is probably of the order of the density scale height, viz., 10^7 cm. Thus, we estimate T_B to be of the order of 400 s.

The lifetimes of individual field clusters on the Sun are unknown; however, the fields are associated with brightness patterns whose identity persists for periods of 10^4 to 10^5 s. It seems clear, therefore, that the generation and decay of solar magnetic fields are not major factors in the energy balance in the solar atmosphere.

The preceding conclusions are altered, of course, in the areas where the field concentrations occur. A field of 2000 G with scale length 10^7 cm has an energy of 1.6×10^{12} erg cm^{-2}. If fields of this strength were converted to heat energy in a period of one day (10^5 s), the local average mechanical energy flux resulting from the field annihilation would be 1.6×10^7 erg cm^{-2} s^{-1}. Since this is well above the average mechanical energy flux for the whole Sun, the local effects may be important.

This does not prove, of course, that the presence of strong magnetic fields strongly modifies the mechanical energy supply, but it does show that the magnetic energy changes in strong field areas may be important in the local energy balance.

A second way of illustrating the role of magnetic field phenomena in the solar atmosphere is by comparing the potential energy of the fields with the internal kinetic energy of the atmospheric gases. This comparison is given in Table IV-1. The internal

TABLE IV-1
Comparison of magnetic potential energy to
internal thermal energy

	Base of photosphere	Base of chromosphere	Middle chromosphere ($n_H \approx 10^{12}$)	Base of corona	$R = 1.5\,R_\odot$
U	1.3×10^5	800	1.2	0.18	0.001
B_0	1800	140	6	2	0.15

kinetic energy density of the gas is given to sufficiently good approximation by

$$U = nkT \qquad\qquad (IV\text{-}3)$$

and the magnetic energy density is given by

$$W = B^2 / 8\pi \cdot \qquad\qquad (IV\text{-}4)$$

Entries in the table give values of U at the base of the photosphere ($\tau_c = 1$), base of the chromosphere, upper chromosphere, base of the corona and corona at $R = 1.5R_\odot$. Values of B_0 given in the table are the values at which $U = W$.

Two conclusions may be reached from a comparison of the observed solar fields to the results in Table II-2: (1) the average observed fields are less than B_0 and (2) in local regions the greatest observed field strengths at photospheric depths are near B_0. It is not known whether this latter conclusion is true at other levels indicated in the table.

Throughout the solar atmosphere the electrical conductivity is high enough that, for many purposes, the magnetic field is 'frozen' to the plasma. Then, since the solar atmosphere is a dynamic medium in which large scale mass motions exist, it is not unreasonable to expect that in the quiet Sun regions the magnetic field will be carried about by the fluid motions and compressed in local areas of fluid convergence until W is of the order of the energy in the mass motions. In the quiet Sun, therefore, the magnetic field configuration will be locally concentrated by the fluid flow and we expect to see little evidence of homogeniety or small scale regularity in the field patterns. In strong field areas, on the other hand, the field strength is sufficient to resist the flow and the flow is expected to conform to the more or less regular pattern of the magnetic field. Broadly speaking, these inferences are consistent with observations. However, it is surprising to find fields whose energy is of the same order as the thermal energy density.

3. Polar Fields and UM Regions

The appearance of the sunspot minimum corona at total solar eclipse led to the early speculation that the Sun was a magnetic sphere with magnetic poles near the rotation poles in similarity with the Earth. This belief was strengthened by the discovery by Hale in 1908 that sunspots exhibited the Zeeman effect and therefore had magnetic fields. Hale's measurement in 1913 (near sunspot minimum) of polar Zeeman splitting resulted in his conclusion that the polar field strength in the photosphere was of the order of 50 G and reaffirmed the concept of a solar dipole field.

Although there was much debate about the polar field strength following Hale's initial observations (as well as additional measurements), and even some doubt expressed as to the reality of the field itself, the general idea of a solar dipole field persisted for several decades. The Babcock magnetograph developed in 1940 demonstrated clearly that solar magnetism was more complex than had been supposed. The polar fields were found to be generally much weaker than had been reported by Hale, although occasionally field strengths of the order of 50 G were observed. Both the Babcock and the Hale magnetographs record only the line-of-sight component of the field. Thus, the polar fields are inferred from the observation of a small component directed towards the observer and are subject to large errors, particularly when the fields are weak.

At the sunspot maximum of 1958 the average magnetic polarity of the Sun reversed in the polar regions. Strangely enough the two poles did not reverse simultaneously but had a phase lag of a year and a half (the southern hemisphere field reversed first) during which both poles had the same polarity. During this period and during the years immediately adjacent the polar fields were weak and often of mixed polarity. It has become clear during the intervening period that the polar field tends to be strongest near sunspot minimum and weakest near sunspot maximum. A similar reversal of the polar magnetic fields occurred during the current solar cycle. In mid-1971, the average field in the north polar regions reversed from predominantly negative polarity to predominantly positive (Howard, 1972). The reversal at the south polar regions, however, lagged about one year behind the reversal at the north polar regions (Gillespie et al., 1973). Both polar reversals occurred well after the epoch of maximum sunspot activity in 1968.

An extensive catalog of the photospheric magnetic field observed at Mt. Wilson Observatory (now part of the Hale Observatories) during the period August, 1959 to June 1966 has been published by Howard et al. (1967). This catalog contains a wealth of data on the large scale features of solar magnetism.

Synoptic maps of the solar magnetic field, such as those published by Howard et al. show that the predominant features of the field are generally at lower latitudes associated with centers of activity rather than with a general dipole field. The magnetic fields associated with active centers on the Sun generally outlive the active centers themselves. As the fields increase in age, they drift poleward. This produces large areas of enhanced magnetic field that have evolved from the active centers.

Figure IV-2 exhibits a magnetic map of the solar surface obtained by J. Harvey and W. Livingston at Kitt Peak Observatory on June 1, 1970. Several characteristic features of the solar magnetic field are immediately evident from Figure IV-2, which was made about 1 yr after sunspot maximum: (1) The polar fields are relatively weak. (2) Large areas of predominantly unipolar character (UM regions) trail eastward (heliocentric) and poleward from the active latitudes. (3) Much intermingling of opposite polarities exists, particularly near active centers. (4) The network structure is clearly evident. In the sunspot cycle at the time Figure IV-2 was made, the northern solar hemisphere (top of the figure) had negative magnetic polarity (shown as dark) for the leader sunspots. Just the opposite is true for the southern hemisphere.

It is not clear from Figure IV-2 just what polarity the extreme polar regions had at that time. This is due, in part, to the fact that the polar fields were near their lowest ebb. The polar fields observed at Kitt Peak some 2 yr later on June 19, 1972, are shown in Figure IV-3. Here it is clear that a mixture of polarities occurs at both poles but that positive (light) magnetic polarity dominates in the north and negative magnetic polarity dominates in the south. As at lower latitudes, the polar fields continue to show the patchy outline of the network structure.

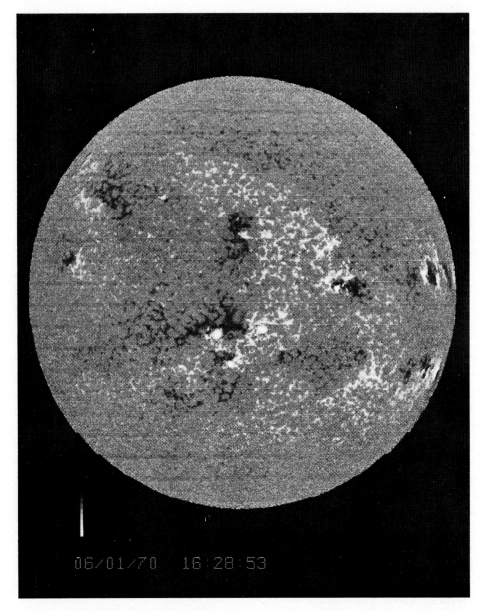

Fig. IV-2.   A magnetic map of the solar surface, June 1, 1970. North is at the top and west is at the
right. Black indicates negative (south) magnetic polarity (Kitt Peak National Observatory).

The large, predominantly unipolar areas (UM regions) on the Sun shown in Figure IV-2
were first reported by H. W. and H. D. Babcock in 1955. Extensive studies of the
evolution of UM regions have been carried out by Bumba and Howard (1965), and by
Stenflo (1972). The axes of the UM regions correspond roughly to the lines of solar

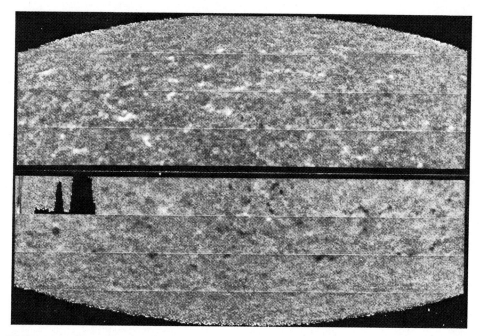

Fig. IV-3.   Magnetic fields observed at the two polar regions June 19, 1972. North is at the top and
              is dominated by positive (North) polarity (Kitt Peak National Observatory).

longitude as distorted by differential rotation starting from the birth of the active region
with which they are identified. With each solar rotation the axis of a given UM region
moves poleward and assumes a more nearly east-west orientation.

Although it is not immediately evident from Figure IV-2, there is a marked statistical
tendency for the dominant UM regions in a given hemisphere to have the magnetic
polarity of the *follower* spots in that hemisphere. For the current sunspot cycle, which
includes the date of the observation in Figure IV-2, the follower spots are of positive
(light) polarity in the northern hemisphere and of negative polarity in the southern
hemisphere.

In both of the polar field reversals observed thus far, the polar fields prior to sunspot
maximum are dominantly of the same polarity as the follower spots of the preceding
cycle (which is the same as for the leader spots of the cycle in progress). After the field
reversal occurs, the polar fields have the dominant polarity of the follower spots of the
cycle in progress, i.e., they have the same polarity as the dominant UM regions of that
cycle.

The magnetic field pattern near active centers is often very complex, as is illustrated
by Figures IV-4 and IV-5. Complex patterns of opposing polarities almost seem to be the
rule rather than the exception. Also, the patterns are typically dynamic rather than
typically static. Large changes in field configuration occur during the lifetime of a given
active center.

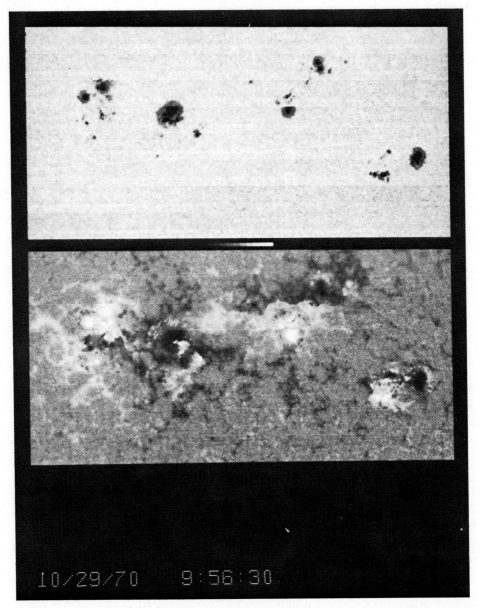

Fig. IV-4.   A magnetic map of an active region. The top image is a white light photograph of the
same region (Kitt Peak National Observatory).

At its inception, a new active region has relatively simple magnetic geometry and the
magnetic field lines most likely close within the region itself. As the region grows in
complexity over a period of time, however, magnetic reconnection evidently occurs. Such

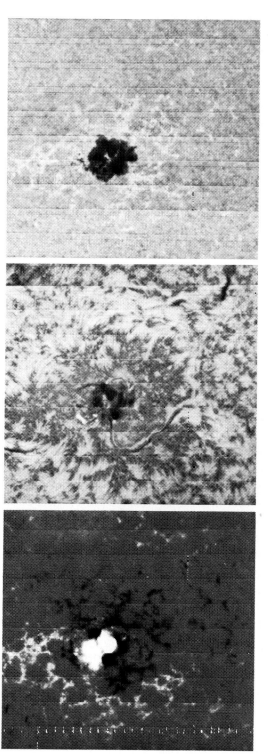

Fig. IV-5.   The magnetic field (top) and photospheric network structure (bottom) near an active region. The center image was made in Hβ (Kitt Peak National Observatory).

reconnection is suggested by coronal structures (see Chapter II) interconnecting different active centers, sometimes even between opposite solar hemispheres, and by coronal arcades apparently straddling from one UM region to another. These coronal structures clearly imply magnetic couplings that apparently did not exist during the early appearance of the magnetic regions. Further evidence of magnetic connection from active centers to distant sites is furnished by the phenomena of sympathetic flares, in which a flare in one active center evidently triggers a flare in a distant active center (cf. Zirin, 1968) and by the clearly 'channeled' propagation of blast waves associated with flares (see Chapter III) into well defined corridors.

In still another sense, the very phenomena of a single UM region growing out of a given active center means that one magnetic polarity has become dominant over the other in that particular region. Clearly, this cannot happen without major magnetic reconnection. Direct measures of field strength in a given active region generally show a net imbalance of the flux, i.e., the net flux is not zero.

4. Evolution of Large Scale Fields

Some observers have suggested that all photospheric solar magnetic fields may originate in active centers near sunspots, i.e., that active centers are the only locations where photospheric fields are produced. As evidence of this they cite the following observations: (1) the evident association of the dominant UM regions with the follower fields of active regions; (2) the poleward migration of UM regions; (3) the tendency for polar fields to be stronger at sunspot minimum than at maximum; and (4) the reversal of the polar fields near sunspot maximum. The phenomenological picture that emerges (Babcock, 1961; Leighton, 1964) is one in which pieces of the follower field are carried stochastically by the supergranule circulation step by step towards the poles. This process favors the follower fields over the leader fields because of the well-known statistical tendency for the follower fields of active regions to lie nearer the poles than do the leader fields. The pieces of follower field carried from the active regions coalesce to form the dominant UM regions and gradually migrate to the poles. During a given sunspot cycle the polar field collects increasingly large amounts of magnetic flux of a given polarity. The following cycle starts the process anew but with dominant UM regions of opposite polarity. As this new polarity reaches the poles it gradually erodes the fields of the previous cycle and eventually dominates. Approximately the first half of the cycle is required to cancel the field of the preceding cycle and the latter half builds the polar field back to its sunspot minimum value. Field reversal occurs near, but somewhat after, solar maximum.

The stochastic model suggested by Leighton has many attractions. It accounts in a quite natural way for most of the observed phenomena and seems quantitatively compatible with the observations (see Leighton, 1964; 1969, for details). Perhaps of equal importance it allows for considerable irregularity: one cycle need not duplicate another and even one hemisphere may exhibit a very different detailed behavior from the other depending upon the particular distribution of sunspot activity and upon the level of activity in both the current and the preceding cycles in the two hemispheres. Although the most consistent feature of the solar activity cycle is the stability of the average cycle over long periods of time, there are marked differences in activity between individual cycles and between the two hemispheres during a given cycle.

The preceding picture of the magnetic evolution of the Sun has many attractions, but it leaves several questions unanswered. Solar magnetism is almost certainly a result of dynamo action involving both convection and differential rotation in the sun (cf. Parker, 1970). Both poloidal (polar) and toroidal fields are essential parts of the dynamo theory. The toroidal fields girdling the lower solar latitudes are formed from the poloidal fields by differential rotation and the poloidal fields are formed, in turn, from the toroidal field by the convective motions (including those due to magnetic buoyancy). In such theories, photospheric fields in active regions result from segments of the toroidal field floating to the surface and erupting as sunspot regions. The theory necessarily requires that each active region has zero net magnetic flux and that all field lines reconnect within the active region when the region is young. The equatorward migration of sunspot zones through each solar cycle arises, according to the theory, from a circulation within the toroidal field region. The details of this circulation depend upon the sign of $d\omega/dR$ where ω is the angular velocity, upon the depth of penetration of $d\omega/dr$ and upon the sign of the vorticity in the convection. If $d\omega/dr$ falls to zero just below the solar surface, i.e., the differential rotation is just a surface effect, then it is $d\omega/d\phi$ that determines the direction of drift of sunspot zones. In Leighton's theory the convection is identified specifically with magnetic buoyancy and the supergranulation. The equat or-pole tilt of sunspot regions with follower spots nearer the poles corresponds to cyclonic vorticity and gives the pole to equator drift only if $d\omega/dR < 0$. If $d\omega/dR > 0$, as some authors prefer, then the convection must have anticyclonic vorticity. At present, the sign of $d\omega/dr$ and its depth of penetration are still completely open to speculation.

Dicke's (1970, 1973) observations of solar equatorial oblateness suggested that $d\omega/dR < 0$. However, Dicke's results are still highly controversial and are subject to other interpretations even if the observations are correct. New, more definitive, observations are in progess by H. Hill and his collaborators at the University of Arizona. Preliminary results indicate that such oblateness as does occur is time dependent, and that definite periods exist when the oblateness is much smaller than reported by Dicke.

It is probable that both the stochastic drift models and dynamo models are still too grossly oversimplified to be reasonable representations of the true magnetic picture. Recent observations of newly emerging magnetic flux on the sun are beginning to shed light on this problem.

The emergence of new magnetic flux on a scale that subsequently grows into active centers, according to Zirin (1974), occurs at the average rate of about 2 per day near sunspot maximum. Near sunspot minimum, of course, tne corresponding rate is near zero. If we adopt an average birth rate of 1 active region per day for the whole sunspot cycle and an average lifetime for active regions of 2 weeks, the implied average, steady-state population of active regions is 14. This is a reasonable number, for the average, but is perhaps a little larger than actually exists.

A typical active region has a total magnetic flux of approximately 10^{21} Mx (Vrabec, 1974). Thus, if there were an average of 14 active centers on the Sun, and no other fields were present, the total magnetic flux would be only 1.4×10^{22} Mx. This corresponds to an average field strength of only 0.2 G, which is too small by a factor of about ten. However, the Figure 14 for the number of active centers is arrived at by assuming an average lifetime of 14 days for active centers. As previously noted, the magnetic lifetime is much longer than 14 days. Typical UM regions last several solar rotations. This suggests that the lifetime of the magnetic region should be increased by a factor of the order of 5 to 10. On the other hand, it is known that the UM regions do not maintain a flux balance. By inference, then, some of the field from the active center decays. Thus, one cannot argue conclusively, at this point, that the active centers, in fact, are the primary sources of solar magnetic flux. As we shall see in the next section, there is good evidence for substantial emergence of new flux in non-active areas.

The magnetic fields immediately associated with active regions are both complex in topology and evolving in time. It has long been known that small sunspots tend to coalesce with large spots of the same magnetic polarity thereby building up the total flux in the large spot. Such a coalescence has suggested to some (cf. Piddington, 1974; Vrabec 1975) that the separate spots of the same polarity are simply branches from a common trunk of magnetic field buried beneath the photosphere and that coalescence results from upward motion of the trunk. The rate of drift of sunspots towards each other during coalescence is about 1 km s^{-1}. If the average branching angle is θ, the average upward velocity of the magnetic trunk is given by $v_c/\tan \theta$, where v_c is the relative velocity of the coalescing spots. Thus, for $\theta = 45°$ the average upward velocity would be 1 km s^{-1}, also, but for $\theta = 75°$ the average upward velocity would be only 0.27 km s^{-1}.

In addition to the magnetic inflow to large sunspots represented by the coalescence of smaller sunspots and pores surrounding large spots of the same polarity, there is often a similar inflow of magnetic features of the same polarity as the spot but not associated with visible spots or pores. These latter features are generally smaller in size than pores, but more numerous.

In addition to the inflow of magnetic flux, there is often an outflow of flux of perhaps even greater consequence. Both the inflow and outflow may be in progress simultaneously at different points on the periphery of a spot. The outflowing flux contains both polarities in almost equal amounts. Flux outflow is observed most prominently in spots where the magnetic fields adjacent to the spot are weak and disordered near the spot but becomes stronger and more ordered further from the spot. Some observers (cf. Vrabec 1974) have referred to this configuration as a magnetic 'moat'. Many spots show such moats over at least part of their periphery. An example of a magnetic moat is shown in Figure IV-6.

Magnetic field observations made with high spatial and time resolution show a predominance of outflow of magnetic flux through the moat. The outer edge of the moat

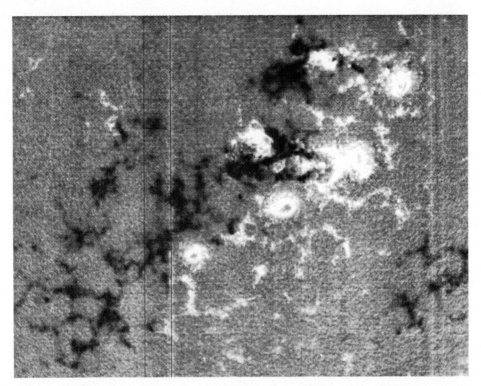

Fig. IV-6.   A magnetic map of an active region illustrating a magnetic moat around a sunspot. The spot of particular interest is just right of center and shows an area (moat) of weak field immediately surrounding the spot (San Fernando Observatry, Aerospace Corp).

usually shows the normal network structure. This network structure often seems to have been influenced by the moat in that the borders of the network nearest the moat follow the general curvature of the moat, i.e., the curvature tends to be concave towards the spot. The mean lifetime of the outflowing flux elements is given by Vrabec (1974) as 2 to 3 h. Most of the moving features cannot be followed to the outer edge of the moat. Those that do reach the outer boundary seem to lose their identity quickly without appreciably disturbing the existing network structure.

Vrabec (1974) describes the typical outflowing features as being about 2.5 $''$ in size, having a velocity of about 1 km s^{-1} and having a total magnetic flux of about 10^{19} Mx. He further describes the pattern of outflowing field as being highly fragmented and having a 'salt and pepper' appearance of opposite polarities, closely intermingled in about equal amounts. He counts some 50 to 200 such features in a moat at a given time and estimates the total rate of outflow of magnetic flux to be 4×10^{14} to 1.5×10^{20} Mx h^{-1} Harvey and Harvey (1973) have estimated the *net* flux outflow to be 1×10^{19} Mx h^{-1}, which is roughly one tenth of the total flux outflow. The net flux outflow has the same sign as the spot flux and is generally consistent with observed rates of decay of sunspot fields. For example, a spot of area 10^{18} cm^2 and field strength 3×10^3 G would decay, as a result of typical magnetic outflow in some $3 \times 10^{21}/10^{19} = 300$ h. This is a reasonable average decay time for such a spot.

5. Structure and Evolution of Small Scale Fields

The large scale properties of the solar magnetic field are important to our understanding of the field generation and its evolution. For other purposes, however, it is the small scale structure of the field that is important. In quiet areas of the Sun the regions in which a given magnetic polarity dominate are large compared to the size of supergranules. Thus, it is normally to be expected that the borders of a given supergranule will have a dominance of one magnetic polarity. Figures IV-2, IV-4 and IV-5 show clearly that this is the case.

Figures IV-7 and IV-8 show magnetic maps of individual solar areas made with an angular resolution of approximately 2.5″. At this resolution, the field is seen to be concentrated into a few small knots scattered along the network features rather than to

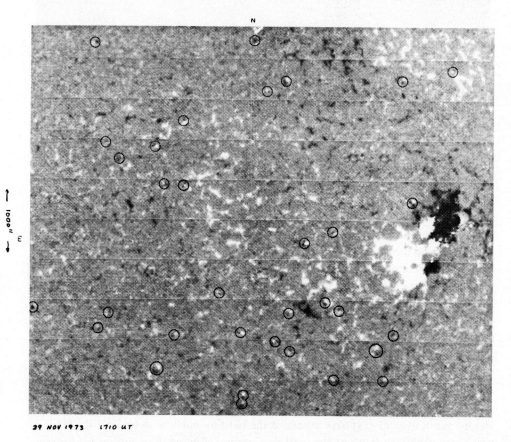

29 NOV 1973   1710 UT

Fig. IV-7.   Fine scale magnetic field structure on the Sun showing closely spaced fields of opposite polarity. The circles enclose some of these bipolar regions. The vertical scale in this image is 1000″ (Kitt Peak National Observatory).

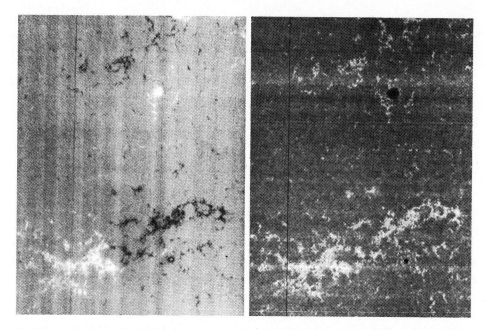

Fig. IV-8.   A high resolution magnetic map (left) together with a spectroheliogram showing network structure in the wing (+2.5 A) of the CA II K ling. The vertical scale is 600″ (San Fernando Observatory, Aerospace Corp).

be uniformly spread throughout the network. Thus, a given supergranule cell seems to have most of its associated magnetic flux concentrated in a few magnetic knots at the cell borders. We note, also, from Figure IV-7, particularly, that the dominance of one field polarity seems to be a statistical effect rather than an exclusive one. Several areas of closely spaced fields of opposite polarity are encircled in this figure. Many other such areas are evident.

There is considerable evidence to suggest that the tendency of solar magnetic fields to concentrate into small areas of strong field is even more pronounced than is indicated by direct observations with 2.5″ resolution. It appears, in fact, that the magnetic knots have a typical diameter of only about ¼ to ½″. Since there are magnetographs that approximate this resolution, it is worthwhile to consider the evidence leading to the conclusion that the magnetic knots are very small. The inferred field strength, of course, is inversely proportional to the square of the diameter of the magnetic knots. For knots of the size mentioned, the implied field strengths are in excess of 1000 G.

Since the solar magnetic fields appear to be made up mainly of small areas of high field strength, the saturation effects discussed earlier in this chapter will very likely influence the measurements. One of the manifestations of saturation will be an incorrect measure of the mean field strength. It follows that lines of different Zeeman splitting, hence different saturation levels, will exhibit different mean field strengths. This would be true even if the line profiles themselves retained the same values of $dI/d\lambda$ in both the magnetic and non-magnetic regions. It is known, however, that certain lines, such as λ 5250.2 of Fe I, show marked changes in the shapes of the profiles at the locations of the magnetic knots. This adds further error to the measurement of mean field strength and provides a second source of disagreement between the mean field strengths measured in different spectral lines. Still a third source of differences in the measured mean magnetic field strength between different spectral lines lies in the different levels of the solar atmosphere where the lines are formed, i.e., in height gradients in the magnetic field itself.

Severny (1966) found from a comparison of field strengths measured in lines of λ 5250.2 (Fe I) and λ 6102.7 (Ca I) that considerable discrepancy existed. A later study of the λ 5250.2 and λ 5233.0 lines of Fe I by Harvey and Livingston (1969) showed a systematic difference in the field strengths measured in the two lines. The λ 5233.0 line gave field strengths more than double those observed in λ 5250.2. Sheeley (1967) had shown earlier that the λ 5250.2 line of Fe I weakened markedly in small local areas where the field strength exceeded 350 G. The λ 5233.0 line does not show a similar weakening of the profile. This led Harvey and Livingston (1969) to attribute most of the difference in the mean field strengths to the weakening of the λ 5250.2 line.

We noted in Section 1 of this chapter that the total splitting in λ 5250.2 in a field of 1000 G is 0.077 Å. This is comparable with the width of the line itself, and saturation effects will clearly be present in fields as high as 1000 G. The Zeeman splitting in the

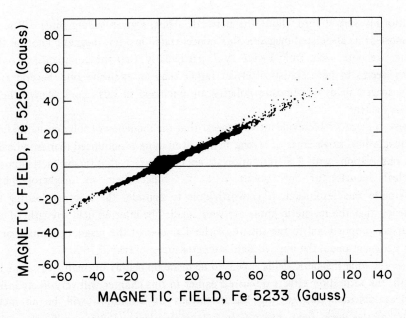

Fig. IV-9.   Correlation between the field strengths observed in two different spectral lines (after Frazier, 1970).

λ 5233.0 line of Fe I, however, is only 0.033 Å in a field of 1000 G. This line would show only a small saturation effect for fields near 1000 G, or less, but would show clear saturation effects for fields near 2000 G.

An example of the observed ration of field strengths in λ 5250.2 and λ 5233.0 of Fe I obtained by Frazier (1970) is shown in Figure IV-9. In this example, the average fields measured in λ 5233.0 have almost exactly twice the strength of the average fields measured in λ 5250.2. However, this ratio is dependent upon the widths of the magnetograph exit slits and how far they are spaced from line center.

The plot shown in Figure IV-9 is remarkable in that the correlation appears to be perfectly linear, at least for mean fields under 1000 G. A possible explanation for the linear relationship is that the fields are confined to small areas in which both the magnetic field strength and the profile of λ 5250.2 are essentially invariant. In other words, the magnetic field exists only in 'magnetic quanta' of a given field strength and the line profile has only two characteristic shapes — one for the magnetic areas and one for the non-magnetic areas. If this were true, then the mean field strength is simply a measure of the number of magnetic knots within the aperture of the magnetograph (Frazier, 1974). A slope different from unity would result from the combined effects of the profile changes and saturation.

An alternative explanation is that the different mean magnetic field strengths seen in

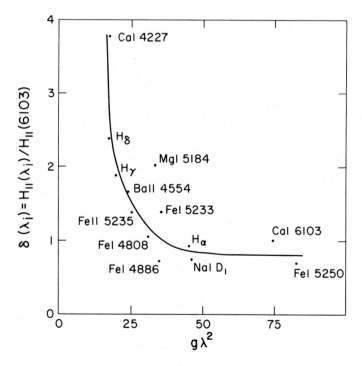

Fig. IV-10.   An illustration of the dependence of observed field strength on the magnetic sensitivity ($g\lambda^2$) of the line (after Gopasyuk *et al.*, 1973).

the two lines is primarily a result of just one of these effects, i.e., either the saturation dominates or the profile changes dominate. Again, however, it seems necessary to invoke the idea that the actual magnetic field strength has only a narrow range of values. Otherwise, we would be left with the very strange suggestion that the line profile changed markedly when the field strength was non-zero but was otherwise independent of the field strength.

Evidence that the primary effect in Figure IV-9 is due, in fact, to saturation has been presented by Gopasyuk et al., (1973). By comparing field strengths determined in a number of lines in a manner similar to that displayed in Figure IV-9, these authors found what appears to be a systematic dependence of the mean field strength on $g\lambda^2$, where g is the effective value. Their results are shown in Figure IV-10.

The plot in Figure IV-10 seems to show conclusively that saturation is a dominant effect in determining the mean field strength. On the other hand, it must be remembered that the relative values of the field strengths measured in two lines depends upon where the magnetograph slits are placed. Thus, the reported ratios of field strengths measured in 5233.0 to those measured in 5250.2 vary from 1.5 (Howard and Stenflo, 1972) to about 3.0 (Harvey and Livingston, 1969). It follows that the points plotted in Figure IV-10 are subject to some arbitrariness, depending upon the characteristics of the magnetograph. Nevertheless, the evidence for a dependence of the measured mean field strength on $g\lambda^2$ seems clearly established.

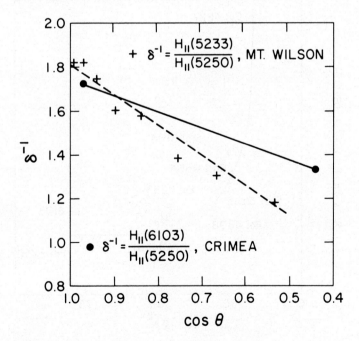

Fig. IV-11.   An illustration of the dependence of δ on position angle on the solar disk (after Howard and Stenflo, 1972).

Howard and Stenflo (1972) and Gopasyuk *et al.* (1973) have shown that the ratio of measured mean field strength in two different spectral lines varies systematically with the μ value on the solar disk. Their results are shown in Figure IV-11. The μ dependence near $\mu = 1$ appears to be too strong to be accounted for by changes in the shapes of the profiles. Thus, the observed changes must result from height gradients in the field that lead to a reduction of the saturation effect. Most authors have adopted the intrinsic height gradient of the field as the explanation (cf. Gopasyuk *et al.*, 1973). However, the longitudinal component of the fields, assuming they are primarily radial, varies directly as μ, and this may well represent a stronger component of the μ dependence than the actual intrinsic height gradient.

Evidence that the field strength may indeed be 'quantized' has been given by ·

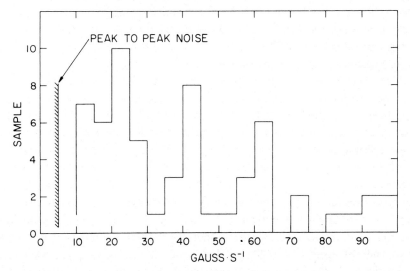

Fig. IV-12.   A histogram of observed field strength distribution using a 5 × 5″ aperture (after Livingston and Harvey, 1969).

Livingston and Harvey (1969). Figure IV-12 shows their plot of the number of areas sampled vs the measured mean field strength. The histogram shows a tendency for maxima near 20, 40 and 60 G. This is consistent with the assumption that each magnetic knot contributes a mean field strength of 20 G and that the magnetograph aperture tends to see either 1, 2 or 3 such magnetic knots. The magnetograph aperture, in this case, was 5″ × 5″. Thus, a mean field of 20 G would correspond to an actual field strength of 2000 G if it were confined to an area measuring 0.5″ × 0.5″.

Stenflo (1973) has used two lines of Fe I, λ 5250.2 and λ 5247.1, from the same multiplet to separate the effects on the measured field strength due to changes in line profile and saturation. These two lines have similar oscillator strengths and arise from lower energy levels lying very close together. Thus, changes in profiles due to temperature and density changes should be essentially the same in both lines. However, the two lines have different effective g values and will therefore have different Zeeman splitting. Stenflo concludes from his study that most of the change in the λ 5250.2 profile in magnetic regions is due, indeed, to the Zeeman splitting. He further concludes that the field strengths are strongly clustered near 2000 G in areas with diameters of 100 to 300 km ($\frac{1}{7}$ to $\frac{3}{7}$"). The above arguments for very strong fields concentrated into small magnetic knots are all somewhat indirect. Simon and Zirker (1974) and Harvey and Hall (1974) have recently demonstrated by very direct and convincing observations that this picture of the magnetic fine structure is the correct one. Harvey and Hall (1974) have reasoned that if the local field strengths are as large as 1000 G some spectral lines should have their different Zeeman components almost completely resolved. If this is true and if the different components can be identified, the field strength will be given directly by the Zeeman splitting even though the magnetic features themselves are not resolved. For example, if the magnetic area occupies 10% of the aperture area, the Zeeman components in a simple triplet pattern will be approximately a tenth as strong as the unsplit line from the remaining area of zero or low field strength.

Fully resolved Zeeman components are most likely to be found for lines with a simple Zeeman pattern located as far to the red as possible. After searching the spectrum for suitable lines Harvey and Hall selected the λ 15648.6 line of Fe I. They identified the Zeeman pattern as a simple triplet with an effective g value of 3.0 by comparison of the observed splitting in large, stable sunspots with the splitting given by lines of known g value. This was necessary since the term structure in Fe I giving rise to the λ 15648.6 line is unknown.

The procedure followed by Harvey and Hall (1974) was to first locate small areas of the Sun that were bright in the Mg b lines. Such regions are known to be identified with magnetic knots in the network. Once the spot was located, the telescope was guided on the spot and the spectra in right and left hand circular polarization were recorded photoelectrically. In the dozen or so cases studied, the results were always the same, viz., the right and left hand Zeeman components could be readily identified as distortions of the line wings. By deconvolving the spectra with a set of Voigt functions, the positions of the right and left hand components could be accurately determined. In all cases, their separation corresponded to field strengths of 1500 to 2000 G. The strength of this method lies in the fact that the measured field strength is independent of the assumed model of the magnetic region. The method still depends somewhat, but only slightly, on the assumed shapes of the profiles. By comparison with other methods of inferring the field strengths, however, the method of Harvey and Hall (1974) is both direct and unambiguous.

Simon and Zirker (1974) have utilized the Sacramento Peak tower telescope during moments of good seeing to achieve magnetic field measurements with a spatial resolution of approximately 0.75". With such resolution they have observed field strengths up to 1300 G in magnetic knots. The results of Stenflo (1973) indicate that the magnetic knots are still unresolved at 0.75″ resolution. Thus, the results of Simon and Zirker are consistent with those of Harvey and Hall and with those of Stenflo.

The conclusion that solar magnetic fields are confined largely to small knots of high field strength is, of course, of fundamental importance to solar physics. Undoubtedly, the field configuration is related to convective phenomena somewhat below the photosphere as well as in the photosphere itself. Thus, the field configuration carries valuable information about the nature of the internal convective modes that may help in our eventual understanding of these phenomena.

In this brief survey of the recent work establishing the idea that the fields are confined to small knots we have overlooked the very important earlier work of several authors who established the concept and offered convincing supporting evidence.

The conclusion that much of the magnetic flux in quiet areas of the solar disk is localized in small knots of 1500 to 2000 G field strength seems inescapable. However, Livingston and Harvey (1971) have noted that there is some evidence of a weak field background of up to 2 to 3 G covering the Sun more or less uniformly. This evidence was obtained from observations with 5″ resolution and a magnetic field sensitivity of 0.4 G. The magnetic knots are much smaller than 5″ in size, but they are also separated, on the average, by much more than 5″ Thus, the weak background fields detected by Livingston and Harvey do not appear to be explainable as just unresolved knots of the type discussed above. If all the magnetic flux were located in the network, for example, observations with 5″ resolution should show essentially zero field strength at the centers of the supergranule cells. Instead, Livingston and Harvey find field strengths of 2 to 3 G.

Since a typical supergranule area is about 10^3 sq arc s, a field of 2000 G in a single magnetic knot of $\frac{1}{3}$″ diam would contain only $\frac{1}{9}$ as much total flux as a field of 2 G spread over the area of the supergranule.

It seems clear from studying magnetic maps of the whole Sun, that shown in Figure IV-2 being somewhat typical, that the average magnetic field strength of the Sun varies quite markedly across the solar surface even well away from active centers. It seems perfectly possible, therefore that a given supergranule cell in a location of high field strength may have a general field of some 2 to 3 G plus several magnetic knots at its borders with field strengths of the order of 1500 to 2000 G, whereas a supergranule cell located in a weaker field area may have both a weaker background field and as few as one magnetic knots at its border. The cell in the stronger field area, assuming a background field of 2 G and 10 magnetic knots of $\frac{1}{3}$″ diameter and field strength of 2000 G, would have a total magnetic flux of 2.3×10^{19} Mx, nearly half of which is in the knots located in the network. A supergranule cell in a weak field area, having a background field of 1 G and a single magnetic knot with the same flux as in a knot in the strong field area, would have a total flux of 0.8×10^{19} Mx, only an eighth of which is in the network knot.

The total surface area of the Sun corresponds to approximately 10^4 supergranule cells, and the total magnetic flux of the Sun is estimated at about $0.3 - 1 \times 10^{23}$ Mx. On this basis, we would conclude that the weak field supergranule with only one magnetic knot is more typical than the strong field supergranule having several such magnetic knots and a stronger background field. On the other hand, the total magnetic flux of the Sun can be derived only from direct observation of the surface fields or from observations of the fields in the solar wind. The former method is unreliable for the reasons cited earlier in this chapter. The latter method is quite sensitive to the detailed magnetic field configuration near the Sun which is not well known. Thus, neither method establishes a firm value for the total solar flux. The difficulty of locating a clearly defined spectroscopic network over much of the Sun (see Chapter II) as well as the appearance of large regions of the Sun where the magnetic field network is poorly defined, or absent entirely (see Figures IV-2 and IV-3) tends to confirm the conclusion that the average supergranule cell together with its associated network does not contain a large number of magnetic knots with field strengths as high as 1500 to 2000 G.

It was mentioned earlier in this section that the field in a given supergranule cell tends to have the same magnetic polarity. On the other hand, it was suggested in Chapter II and III that the fibril structure of the network may be indicative of the presence of the opposite polarity. In Figure IV-7, a number of areas are encircled in which fields of both polarity are observed in close proximity. Numerous other regions in which field reversals occur on a scale that is small compared to supergranule size are evident in this figure. At very high spatial resolution, one sees more and more evidence of opposite field polarities on a small scale. It seems clear, therefore, that the UM regions are only statistically dominated by a single polarity and that a given supergranule cell has a high probability for having one or more regions of magnetic polarity opposite to the dominant polarity of the cell.

Closely spaced bipolar magnetic regions (ephemeral active regions) have been studied by Harvey and Martin (1973) and more recently by Harvey and Harvey (1973). From a study of some 4400 such regions observed during a period of 17 months divided between 1970 and 1973, Harvey and Harvey conclude that about 200 ephemeral regions are present on the solar disk at a given time. The ephemeral regions were found to have a mean lifetime of about 12 h and to occur more or less randomly with respect to the cell-network strcture.

From the number of ephemeral regions quoted by Harvey and Harvey, we would conclude that only about 5% of the supergranule cells contain such regions at a given time. However, the ability to detect closely spaced field reversals is limited by both instrumental and observational restraints. Thus, the occurrence of ephemeral regions could be considerably more common than current observations suggest.

The ephemeral regions studied by Harvey and Harvey were more common during periods of high sunspot activity. However, the ephemeral regions are much more uniformly distributed in latitude and longitude than are the active sunspot centers. The average rate of magnetic flux emergence due to the ephemeral regions observed by Harvey and Harvey was estimated to be 1.8×10^{21} Mx. This approximates the rate of flux emergence in active regions suggested by Zirin (Section 4). His estimate of two active regions per day corresponds to an average rate of flux emergence of about 8×10^{20} Mx h^{-1}, assuming the somewhat high value of 10^{22} Mx per active region.

Both the number of ephemeral regions observed and the rate of flux emergence deduced therefrom are limited by instrumental and observational conditions. Clearly, both quantities will change as the observations improve. Also, one tends to count only the changes of polarity that are immediately adjacent to each other. The number of reversals observed would probably increase substantially if all reversals on a scale less than a supergranule diameter were counted. With the limited observations available, it is meaningless to speculate about just how probable it is for at least one field reversal to occur in a supergranule cell picked at random. By the same token, we cannot rule out the possibility that it is a rather common feature of the supergranule-network structure.

In a number of cases studied, the bright points in X-ray images of the corona (Chapter II) coincide with small bipolar magnetic regions of the photosphere (Krieger *et al.*, 1971; Vaiana *et al.*, 1973). It is still unknown whether all close lying field reversals lead to enhanced X-ray brightness, but initial studies suggest that all bright X-ray points overlie bipolar magnetic regions.

The location of fine scale features of the solar magnetic fields in quiet solar regions resembles in most respects the locations of the rosettes, and, indeed, the magnetic filaments show a close correspondence with regions of bright K_2 and K_3 emission (Leighton *et al.*, 1962). This correlation, which was first established for active regions (Leighton, 1959; Howard, 1959) appears to be more or less universal in areas where the average fields are less than 300 G (Tsap, 1966). The same correlation is found at line center in Hα, viz., bright Hα emission corresponds to enhanced vertical magnetic field strength. A magnetic field map together with an image of the same solar region made + 2.5 Å from the center of the K line are shown in Figure IV-8. The detailed correlation between the magnetic field and brightness is clearly evident. Some authors, e.g., Zirin (1970), go so far as to say that all regions of vertical field (except sunspots) are bright in Hα center and all regions of horizontal field are dark. Thus, it is stated that all fibrils are manifestations of horizontal fields and that the fibrils lie parallel to the field lines. It is claimed also that no brightness changes exist without associated magnetic field changes, regardless of the scale. This is, of course, an extreme position that cannot be fully backed-up by observations and is probably overstated. However, it does correctly emphasize the close association between Hα or K brightness and the vertical field strength.

The correlation described above does not necessarily imply that the emission has any direct physical dependence on magnetic field strength. What is implied is that in regions of vertical magnetic field there is an excess dissipation of mechanical energy as heat. Thus, vertical magnetic fields will increase the heating which, in turn, increases the emission. We will return to this interesting point in Chapter IX.

Since chromospheric spicules appear to originate in the rosette structure, the spicules occur preferentially in regions of vertical magnetic field. This is consistent with the fact that spicule motions are primarily vertical. As previously noted, the dark Hα fibrils extending outwards from the rosettes appear to be mainly horizontal features. If we assume, with Zirin, that these features follow the horizontal field regions we are forced to conclude that within the chromosphere the magnetic filaments exhibit a rapid divergence of their field lines. This is not incompatible with the results in Table II-2 which suggest that the strongest fields to be found in the upper chromosphere, away from active regions, will be only a few gauss.

There are few, if any, actual measurements of magnetic field strength that can be termed as quiet chromospheric fields. Observations in the Na D lines (Stepanov, 1960) which extend into the low chromosphere, indicate a field gradient of 0.026 to 0.035 G km^{-1}. Such measurements are difficult to interpret, however, unless they clearly resolve the individual magnetic filaments, which is not the case. Severny (1966) has discussed the observations of chromospheric fields measured in Hα, Hβ and K$_3$ in active regions. He discounts the Hβ and K$_3$ results as being unreliable due to blends in Hβ and the extremely low sensitivity in K$_3$. In Hα the small scale profile changes and the low sensitivity limit the usefulness of the observation to low spatial resolution. No results are available for quiet solar regions.

The fact that the chromospheric network is still well defined throughout the chromosphere and the chromosphere-corona transition region together with the well established correlation between the bright patches of the network and the underlying photospheric field clearly suggests that the field lines retain their relative concentration over the cell walls up to at least the base of the corona. Nevertheless the wealth of fibril structure seen in the Hα chromosphere together with the characteristic field strengths given in Table II-2 suggest that considerable divergence of the network fields exists at all levels. All that appears to be required to preserve the basic network structure is to have a relative concentration of vertical field in the network area. It does not matter, so far as

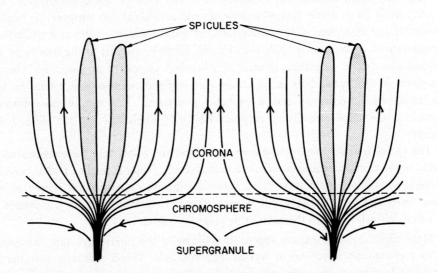

Fig. IV-13.   A schematic illustration of the magnetic field structure over the network and supergranule cells. The field strength decreases with height within the network but retains a relative maximum at the location of the network (after Kopp and Kuperus, 1968).

we now know, whether the concentration results in fields of 10^3 G or in fields of 10 G. Thus, the vertical network fields could diminish by orders of magnitude between the photosphere and upper chromosphere without necessarily destroying the network as such. A schematic illustration of how this might happen is shown in Figure IV-13, which follows a suggestion by Kopp and Kuperus (1968) modified to include a uniform boundary between the chromosphere and corona.

The concentration of solar magnetic fields first into the network then into the network knots is undoubtedly a result of convective currents in the solar atmosphere. Parker (1963) and Pikel'ner (1963) have independently noted that the preponderance of fluid kinetic energy density associated with an outflow velocity of 0.4 km s^{-1} in the photospheric supergranule cell over the energy density in a magnetic field of a few gauss will result in a concentration of field lines at the network overlying the supergranule borders. More detailed computations have been made by Meyer *et al.* (1974) using a mosaic of hexagonal supergranule cells. Their results show field concentrations forming at the vertices of the cells. Their specific application is to the development of sunspots with flux concentrations considerably larger than those found in the magnetic knots of the network. Nevertheless, their results are of interest in that they show the localization of the field into compact areas of high field strength.

6. Coronal Magnetic Fields

Direct observation of magnetic field strengh in the corona have not been made except in prominences. The use of prominence data to infer the general properties of coronal fields is unwarranted because of the singular nature of the prominence phenomena. Nevertheless, the prominence data can and do provide some guide as to the nature of the fields.

Magnetic field strengths observed in prominences are summarized by Tandberg-Hanssen (1974). Observed values of the longitudinal field strength vary with the type of prominence observed. Filaments generally show relatively weak fields of 3 to 8 G. Loop prominences over sunspots, however, show field strengths up to 100 G and other active prominence types show fields up to 200 G.

The available evidence indicates that loop structures and prominence threads follow the magnetic lines of force. Although filaments lie between regions of opposite polarity along a magnetic neutral line in such a manner as to suggest that the field lines run through the filament in its narrow dimension, this is not observed to be the case. The field lines within the filament tend to lie more nearly parallel to the filament axis, suggesting that the filament induces a shear in the magnetic lines of force.

Still a further inference about coronal magnetic fields near active centers can be deduced from motions of material streaming out of coronal clouds (a type of active prominence is sometimes seen suspended above active regions). This streaming material often follows curved trajectories indicative of magnetic lines of force. From an estimate of the centripetal force on the flowing material it is possible to estimate a lower limit to the field strength by setting

$$\frac{B^2}{8\pi} > \frac{n_H m_H v^2}{r}, \qquad\qquad\qquad \text{(IV-4)}$$

where v is the velocity of prominence material in an individual knot and r is the radius of curvature. Reasonable values for the quantities on the right hand side of inequality (IV-4) are: $r \approx 10^5$ km, $v \approx 200$ km s^{-1} and $n_H = 10^{11}$. Thus, we find $B > 0.1$ G.

Various other indirect means have been used to estimate coronal magnetic field intensities. For the quiet Sun, these methods include:

(1) magnetic (viscous) damping of prominence oscillations;
(2) confinement of polar plumes as discrete structures; and
(3) backwards extrapolation of the observed interplanetary fields.

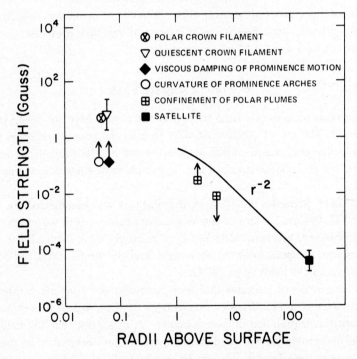

Fig. IV-14.   A plot of extimated field strengths in the corona (after Newkirk, 1967).

The results of these methods together with the information from prominences are summarized in Figure IV-14. For the active Sun, additional methods are available for estimating field strengths. These methods mostly utilize solar radio burst data applicable to the areas of strong fields and frequently yield field strengths of a few hundred gauss in the inner corona. For a review of these methods the reader is referred to Newkirk (1967).

Most of what is currently known (perhaps 'believed' is more appropriate) about coronal field strengths and configuration in the first radius above the solar surface, is obtained by extrapolation of the photospheric fields. A complete set of values for the longitudinal field strength over the solar surface can be used, in principle, to construct the magnetic field pattern in the solar atmosphere under the assumption that the field is a potential field or that the field is force free (all currents flow along field lines). In order to effect such a construction it is necessary to have a complete magnetic map of the photosphere, which involves observations over a complete solar rotation. Thus, it is necessary to assume that the fields are steady for a period of about one month.

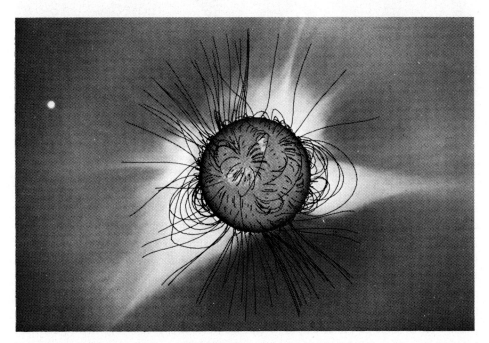

Fig. IV-15. Selected (strong) magnetic field lines computed for the solar corona from the observed photospheric fields for the period of the total solar eclipse in 1966. An image of the corona photographed at the eclipse together with an Hα image of the disk are superposed on the field lines (Altschuler, 1971).

Furthermore, it is assumed that the coronal fields become radial beyond some radial distance R_1. The restriction that the fields are radial beyond R_1 requires that a system of currents are flowing near R_1. The imposition of radial fields beyond $R_1 = 2.5R_\odot$ is consistent with observed coronal structures and solar wind theory.

Figure IV-15 shows a sample of the coronal fields constructed by Newkirk and Altschuler (1970) (see also Altschuler and Newkirk, 1969) in the manner described above

Fig. IV-16.   The same as in Figure IV-15 except that the time period corresponds to the eclipse of
1970.

for $R_1 = 2.5R_\odot$. This particular model was constructed for the epoch of the total solar
eclipse of 1966. An eclipse photograph of the white light, corona and a disk image made
in Hα are superposed on the magnetic field map. Figure IV-16 shows a similar montage of
the magnetic field lines, white light coronal and Hα disk for the eclipse of 1970. In this
figure, only the lines of high magnetic field strength are included. The general
correspondence in Figures IV-15 and IV-16 between the types of features observed, open
field and closed field, and their approximate location suggests that the magnetic model is
reasonable. Because of the assumptions that are necessary, particularly that of constant
field geometry for a period of one month, a detailed correspondence of features cannot
be expected. With this reservation in mind, the general correspondence appears to validate
the method.

One of the characteristics of the computed coronal fields is the tendency for some
field lines to reconnect to the solar surface at a large distance from their point of origin,
often in another active region. This tends to confirm the conclusions from flare data and
the associated blast wave phenomena as well as the evidence from X-ray structures that
widely separated regions are magnetically connected. However, it should be noted that
this distant reconnection of field lines in the computed models is strongly influenced by
the detailed field observations and the fact that the magnetic fluxes of opposite polarity
in a given active region do not usually balance each other in the observations. The

observations are influenced, of course, by their angular resolution and by their level of sensitivity. Both the very weak fields and the very strong fields are selectively discriminated against by most observations and this may selectively discriminate against a particular polarity in a given region.

An additional, though probably less important, bias in the data is deliberately imposed in order to balance the total solar flux to zero. The photospheric observations normally do not give zero net flux and to achieve zero flux a uniform field of given polarity must be added. This added field, even though it is small, changes the net flux balance in each active region and may give some spurious results. The effect is believed, however, to be relatively unimportant.

References

Altschuler, M.D.: 1971, *Sky Telesc.* 41, 146.
Altschuler, M. D. and Newkirk, G.: 1969, *Solar Phys.* 9, 131.
Babcock, H. W.: 1961, *Astrophys. J.* 133, 572.
Babcock, H. W. and Babcock, H. D.: 1955, *Astrophys. J.* 121, 349.
Bumba, V. and Howard, R.: 1965, *Astrophys. J.* 141, 1502.
Dicke, R. H.: 1970, *Astrophys. J.* 159, 1.
Dicke, R. H.: 1973, *Astrophys. J.* 180, 293.
Evans, J. W.: 1966, Proceedings, *Solar Magnetic Fields and High Resolution Spectroscopy*, Rome, 14–16 September, p. 123.
Frazier, E. N.: 1970, *Solar Phys.* 14, 89.
Frazier, E. N.: 1974, private communication.
Gillespie, B., Harvey, J., Livingston, W., and Harvey, K.: 1973, *Astrophys. J.* 182, L11.
Gopaysuk, S. E., Kotov, V. A., Severny, A. B., and Tsap, T. T.: 1973, *Solar Phys.* 31, 307.
Harvey, J. and Hall, D. N. B.: 1974, private communication.
Harvey, J. and Livingston, W.: 1969, *Solar Phys.* 10, 283.
Harvey, K. and Harvey, J.: 1973, *Solar Phys.* 28, 61.
Harvey, K. and Martin, S. F.: 1973, *Solar Phys.* 32, 389.
Howard, R.: 1959, *Astrophys. J.* 130, 193.
Howard, R.: 1972, *Solar Phys.* 25, 5.
Howard, R. and Stenflo, J. O.: 1972, *Solar Phys.* 22, 402.
Howard, R., Bumba, V., and Smith, S. F.: 1967, *Atlas of Solar Magnetic Fields*, Carnegie Institution, Wash., D.C., Publ. 626.
Kopp, R. A. and Kuperus, M.: 1968, *Solar Phys.* 4, 212.
Krieger, A. S., Vaiana, G. S., and Van Speybroeck, L. P.:1971, in R. F. Howard (ed.), 'Solar Magnetic Fields', *IAU Symp.* 43, 397.
Leighton, R. B.: 1959, *Astrophys. J.* 130, 366.
Leighton, R. B.: 1964, *Astrophys. J.* 140, 1547.
Leighton, R. B.: 1969, *Astrophys. J.* 156, 1.
Leighton, R. B., Noyes, R. W., and Simon, G. W.: 1962, *Astrophys. J.* 135, 474.
Livingston, W. and Harvey, J.: 1969, *Solar Phys.* 10, 294.
Livingston, W. and Harvey, J.: 1971, in R. F. Howard (ed.), 'Solar Magnetic Fields', *IAU Symp.* 43, 51.
Meyer, F., Schmidt, H. V., Weiss, N. O., and Wilson, P. R.: 1974, in R. G. Athay (ed.), *Chromospheric Fine Structure*, D. Reidel Publishing Co., Dordrecht, Holland, p. 235.
Newkirk, G., Jr.: 1967, *Ann. Rev. Astron. Astrophys.* 5, 213.
Newkirk. G., Jr. and Altschuler, M. D.: 1970, *Solar Phys.* 13, 131.
Parker, E. N.: 1963, *Astrophys. J.* 138, 552.
Parker, E. N.: 1970, *Astrophys. J.* 160, 383.
Piddington, J. H.: 1974, in R. G. Athay (ed.), *Chromospheric Fine Structure*, D. Reidel Publishing Co., Dordrecht, Holland, p. 269.
Pikel'ner, S. B.: 1963, *Soviet Astron. A. J.* 6, 757.

Severny, A. B.: 1966, *Astron. Zh.* **43**, 465.
Sheeley, N. R., Jr.: 1967, *Solar Phys.* **1**, 171.
Simon, G. W. and Zirker, J. B.: 1974, *Solar Phys.* **35**, 331.
Stenflo, J. O.: 1972, *Solar Phys.* **23**, 307.
Stenflo, J. O.: 1973, *Solar Phys.* **32**, 41.
Stepanov, V. E.: 1960, *Izv. Obs.* **23**, 291.
Tandberg-Hanssen, E.· 1974, *Solar Prominences*, D. Reidel Publishing Co., Dordrecht, Holland.
Tsap, T. T.: 1966. *Izv. Obs.* **35**, 161.
Vaiana, G. S., Davis, J. M., Giaconni, R., Krieger, A. S., Silk, J. K., Timothy, A. F., and Zombeck, M.:
    1973, *Astrophys. J.* **185**, L47.
Vrabec, D.: 1974, in R. G. Athay (ed.), *Chromospheric Fine Structure*, D. Reidel Publishing Co.,
    Dordrecht, Holland, p. 201.
Zirin, H.: 1968, *The Solar Atmosphere*, Ginn and Blaisdell, Waltham.
Zirin, H.: 1970, *Solar Phys.* **14**, 328.
Zirin, H.: 1974, in R. G. Athay (ed.), *Chromospheric Fine Structure*, D. Reidel Publishing Co.,
    Dordrecht, Holland. p. 161.

CHAPTER V

SPECTRAL CHARACTERISTICS

1. Visual and Near Infrared Disk

The chromosphere can be observed against the solar disk in a number of strong Fraunhofer lines and in the weak He I lines at λ 10 830 and λ 5876. In general, the stronger the line and the nearer to line center at which one observes the higher one sees in the chromosphere. According to Figure II-2, for example, the center of the Ca II K line is formed near 1500 km in the chromosphere when observed near disk center. At the location of the K_2 peaks, the absorption coefficient has dropped to approximately 0.005 of its value at line center (cf. Athay and Skumanich, 1968; Linsky and Avrett, 1970) and the region of formation has moved down to about 700 km. In the K_1 region of the profile the radiation arises mainly in the temperature minimum region, and in the line wings the radiation is of photospheric origin. This one line alone, therefore, when observed near disk center, is formed over some 1900 km of depth in the solar atmosphere and carries information on the atmospheric model throughout these layers. Its companion line, Ca II H, is formed only about 100 km lower. Also the infrared triplet lines of Ca II, which share the same upper levels as H and K, are formed still deeper in the atmosphere.

By extending the disk observations to the edge of the solar disk, say, $\mu = 0.2$, the depth of line formation can be moved upwards a factor of five in optical depth. We note from Figure II-2 that this corresponds to a gain in height of 250 to 300 km. Thus, this system of Ca II lines can be used to probe the solar atmosphere through a depth range of approximately 2200 km and provides a powerful means for determining the fluid dynamic and thermodynamic properties of the photosphere, temperature minimum region, low chromosphere and middle chromosphere.

The Balmer-α line of hydrogen (Hα), according to Figure II-2 is formed near 1300 km at line center and at disk center. Higher Balmer lines are formed progressively deeper, and this series of lines provides a second useful set for probing the chromosphere and photosphere in depth. However, because the population of the second quantum level in hydrogen (n_2) varies only slowly with depth and because the relative opacity between successive Balmer lines changes by large factors ($\tau_0(\text{H}\alpha)/\tau_0(\text{H}\beta) \approx 7$), the Balmer lines have proven less useful than the Ca II lines.

Other strong lines of chromospheric origin in the visual spectrum include lines of Fe I Na I, Mg I and Ca I. The cores of the stronger Fe I lines (λ 3720 and λ 3820) are formed near 900 km at disk center (Athay and Lites, 1972; Lites, 1972). Numerous other strong lines of Fe I are present in the spectrum, and these strong lines again provide overlapping coverage of the photosphere and low chromosphere as well as much of the middle chromosphere.

The centers of the Na D lines are formed near 700 km at disk center, and the center of Mg b_1 is formed near 500 km at disk center (Athay and Canfield, 1969). These lines are useful, therefore, only for heights in the low chromosphere and below.

Table V-1 provides a summary of the heights at which several of the stronger visual and near infrared lines reach optical depth unity at line center and at disk center. These heights can be determined, of course, only by model computations (cf. Chapter VIII). They are sensitive to the particular choice of model parameters as well as to the adopted abundance of the element and to atomic parameters. The latter are not accurately known in some cases. Heights given in Table V-1 should be regarded only as illustrative, therefore, and should not be accepted without reservation.

TABLE V-1

Partial list of visual lines that reach optical depth unity at disk center in height range indicated. Above 1000 km the list is complete

Height interval (km)	1600–1400	1400–1200	1200–1000	1000–800	800–600
Lines	K, H	Hα	Ca II: λ 8542	Ca II: λ 8662	Ca II: λ 8498
				Fe I: λ 3720	Na I: D_2
				λ 3535	Fe I: λ 3441,
					λ 3609,
					λ 3816,
					λ 3820,
					λ 3886,
					λ 4064,
					λ 4383,
					λ 4404

600–400	400–200	200–0	0–(–100)
Na I: D_1	Mg I: b_2, b_4	Fe I: λ 4603	Fe I: λ5250
Mg I: b_1	Fe I: λ 5227	λ 5124,	
		λ 5233,	
		λ 6430	
Fe I: λ 4272,			
λ 5270			

It is evident from Table V-1 that the stronger Fraunhofer lines in the visual spectral range provide excellent coverage of the low and middle chromosphere, and one may reasonably ask why any additional data are needed. Also, those not familiar with the problems encountered in the analyses of spectral line data may wonder why it is that these lines were not used long ago to construct reliable models of the low and middle chromosphere. The fact is that these lines have been used successfully to construct model chromospheres only in the last few years. Prior to this the theory of line formation was not sufficiently well developed and computational techniques were not sufficiently advanced.

To illustrate the nature of the line data in relation to the chromosphere, we show in

Figure V-1 a plot of the equivalent black-body temperature, T_R, at the centers of a number of Fraunhofer lines. The abscissa is the reduced equivalent width of the line taken from Moore *et al.* (1966). Central intensities were taken from a variety of scattered sources but mostly from Brault *et al.* (1971). The curve drawn in Figure V-1 is the Harvard-Smithsonian Reference Atmosphere (Gingerich *et al.*, 1971) and has been related

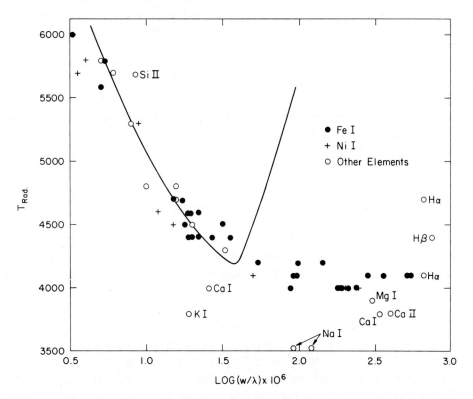

Fig. V-1.   Central brightness temperatures versus the reduced equivalent width of selected Fraunhofer lines. The solid curve is the predicted relationship using the HSRA.

to the equivalent widths of the lines through the classical curves-of-growth under the assumptions of local thermodynamic equilibrium, LTE, a constant, isotropic micro-turbulent velocity, ξ, of 2.4 km s^{-1} and a Voigt parameter, a, of 10^{-3}. Two aspects of the central intensity data are of interest, at this point: Firstly, there is no hint whatever in the central intensities of the strong rise in temperature in the chromosphere. If our assumptions were valid, the central intensities should define a unique curve that follows, more or less, the HSRA. Changes in microturbulence and in the damping parameter, a, will cause some spread in the points about the mean curve but will not grossly distort the curve. Secondly, there is altogether too much scatter in the points to be compatible with the assumption of LTE, or to be explained by changes in ξ and a, and this is true at all depths. These points will be discussed in more detail in Chapter VI and the following. At this point, we note only that the interpretation of the line data is not a simple, straight-forward exercise.

From today's vantage point, we view the strong Fraunhofer lines as perhaps our most valuable set of data for the middle and low chromosphere. The analytical techniques for extracting the fluid dynamic and thermodynamic information from the line data are within reach of current theory and computational capabilities. Perhaps even more importantly, the experience and techniques developed in the analyses of these solar lines will provide the basis for extended exploration of stellar chromospheres.

None of the lines mentioned thus far have any significant contribution in the upper chromosphere. Thus, in order to study the upper chromosphere other data must be used. Similarly, there are no spectral features in the visual disk spectrum that can be attributed to the corona, except for the appearance of prominences in the stronger lines. Attempts have been made to detect forbidden coronal lines at λ 5303 and λ 6374 in the disk spectrum with some reports of success. Bernard Lyot successfully detected the λ 5303 line on the disk using a polarimeter device. However, no quantitatively useful data have been obtained in this way.

2. Visual and Near Infrared Limb

A. CHROMOSPHERIC ECLIPSE DATA

Prior to the development of rather advanced observations from rockets, OSO spacecraft (Orbiting Solar Observatory) and ATM (Apollo Telescope Mount) the primary source of quantitative chromospheric data was solar eclipses. Much of the history and much of the success in solar physics are intimately involved with observations made at eclipse. In the limited space available in the book, we cannot discuss the various aspects of eclipse observing, nor can we even summarize a majority of the data that have been obtained. Instead, we shall review a selected set of data that are particularly useful in later chapters for specifying the chromospheric model. An extensive set of line data from the 1962 eclipse is given by Dunn et al. (1968) and data from the 1952 eclipse are given by Thomas and Athay (1961) and Houtgast (1957).

Beyond the visible photospheric limb the continuum spectrum of the chromosphere and corona are visible at total eclipse. As noted in Chapter I, Section 2, the visible limb occurs near $\tau_c^t = 1$. Thus, by definition, the chromosphere and corona are transparent in the continuum, even at the limb. The scale height for continuum opacity in the photosphere is approximately 60 km, and this scale height persists for some distance beyond the limb. At the base of the chromosphere, which we have arbitrarily set at the temperature minimum and tentatively placed at a height of 200 km beyond the limb, the expected continuum opacity measured tangential to the limb is of the order of 0.04.

A prominent feature of the chromospheric continuum is the Balmer freebound continuum of hydrogen whose head is at λ 3647. The total number of hydrogen atoms in the $n = 2$ level in a radial column above 200 km obtained from Figure II-2 is approximately 2×10^{13} cm^{-2}. When integrated along a column tangential to the limb, the results in Figure II-2 give a column density of $N_2 \approx 1.5 \times 10^{14}$ cm^{-2}. At the head of the Balmer continuum, the absorption coefficient α_{ν_0} is 6.4×100^{-18} Hence, the tangential opacity at the head of the Balmer continuum is approximately 10^{-3}, and we see that this feature in the continuum is optically thin also.

Because the continuum is transparent beyond the limb it is particularly useful for model building. We show in Figure V-2 a plot of the continuum intensity vs height above

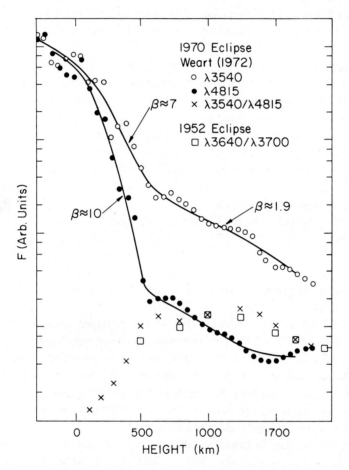

Fig. V-2. The limb profile at two wavelengths in the continuum (Weart, 1972). The shorter wavelength is in the Balmer Continuum. At the bottom of the figure, the ratio of intensities at the two wavelengths is plotted together with the ratio at two similar wavelengths observed at an earlier eclipse.

the limb at two wavelengths: one in the Balmer continuum at λ 3540, and one to the long wavelength side of the Balmer continuum at λ 4815. These data were obtained at the 1970 eclipse by Weart (1972) using photoelectric detectors. The data plotted represent the surface brightness of the chromosphere at the limb. What one observes at eclipse is the integral of the surface brightness over height. Hence, if we designate the surface brightness by F and the observed quantity by E, then E and F are related by

$$E = \int_{h_0}^{h_1} F \, dh. \tag{V-1}$$

To obtain F, therefore, it is necessary to differentiate the data. The plots in Figure V-2 are obtained from direct differencing of the data and the relatively small amount of scatter in the points is indicative of the relatively high level of accuracy in the photoelectric data.

Two features of the data in Figure V-2 are of interest at this point. These are: (1) the strong change in slope that occurs near 600 km, and (2) the strong increase in the ratio of $F(\lambda\, 3540)/F(\lambda\, 4815)$ in the first 600 km. Both of these characteristics are consistent with the interpretation that the emission mechanisms change form at approximately 600 km. As we shall see in Chapter VII, the primary source of continuum emission below 500 km is the H^- ion. This is true even at wavelengths in the Balmer continuum. Above 500 km the dominant emission mechanisms change to electron scattering at $\lambda > 3647$ and to Balmer free-bound emission at $\lambda < 3647$. The Paschen continuum and Raleigh scattering from neutral hydrogen add small amounts.

Values of β indicated in Figure V-2 are the reciprocal scale heights obtained from the approximation

$$F = F_0 e^{-\beta h} \tag{V-2}$$

where h is measured in units of 1000 km. A value $\beta = 10$ corresponds to a scale height of 100 km.

Figure V-3 contains a combined plot of the observed emission in the Balmer continuum at the 1952 and 1962 eclipses. The intensity units* are $erg\, cm^{-2}\, s^{-1}\, dv^{-1}$ *radiated in the total solid angle by a radial slice of chromosphere 1 cm wide.* In the plotted data the observed continuum intensity at $\lambda\, 3700$ has been subtracted from the intensity at $\lambda\, 3640$ leaving only the free-bound contribution. The curve labeled F was obtained by differentiating the smooth curve drawn through the data points. Thus, it represents a derivative of highly smoothed data.

The photographic data in Figure V-3 do not have the accuracy to warrant direct data differentiation as was used for the plots in Figure V-2. However, the derivative of the smoothed E curve again indicates a change in the character of emission beginning near 500 km. Above 500m the free-bound emission intensity decreases approximately exponentially with $\beta = 2.1$, but below 500 km there is relatively little increase in the emission and even a suggestion of a decrease. This is consistent with the data shown in Figure V-2.

Figure V-4 exhibits continuum data at $\lambda\, 4700$ from the 1952 eclipse over an extended height scale reaching well into the corona. It may be seen that within the chromosphere ($h < 3000$ km) the continuum brightness continues to drop relatively quickly, whereas

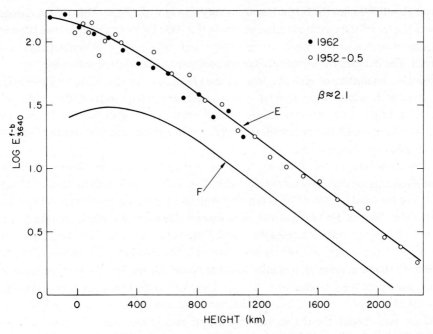

Fig. V-3.   A plot of the free-bound emission in the Balmer Continuum as observed at two eclipses
(see Equation (V-1) for the definition of E and F).

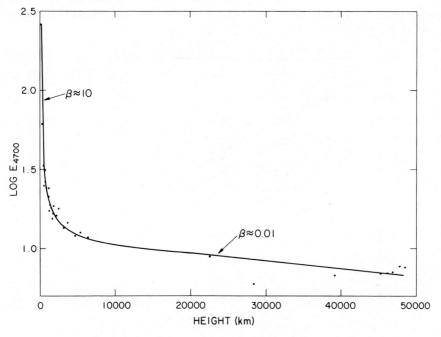

Fig. V-4.   Continuum intensity at λ 4700 as a function of height above the solar limb (1952 data).

above about 5000 km the brightness drops very slowly with height. This second change in the character of the continuum that occurs in the 3000 to 5000m layer is associated with spicular structure in the upper chromosphere and the large rise in temperature in the corona. The dominant emission mechanism continues to be electron scattering.

Relative intensities of emission lines in the chromosphere observed at eclipse change markedly with height. The character of the spectrum becomes increasingly 'harder' with increasing height. In a very coarse way, the emission scale heights increase (β decreases) with increasing ionization potential. However, within a given ion a wide range of emission scale heights may be observed.

Lines of helium in the visual spectrum are optically thin in the chromosphere, and these show a particularly interesting character. Figures V-5 and V-6 show plots of data for λ 7065 of He I and λ 4686 of He II together with the derived curves of surface brightness. Many other lines of He I exhibit the same general character as λ 7065, i.e., the β values are near 1.1 in the upper chromosphere and F shows maxima near 1100 km. The λ 4686 line of He II is the only line of this ion observed in the visual chromospheric spectrum at eclipse. It shows a value of β similar to those found for the He I lines. Also, there is a rather poorly defined maximum in F near 1500 km. Note that β for He I and He II is substantially smaller than β for the Balmer free-bound continuum at heights above 1000 km even though the chromosphere is transparent in each case. This point will be of interest later on in Chapter VII.

Metal lines exhibit a wide range of intensity scale heights in the chromosphere. This occurs for the following two principal reasons: (1) the chromosphere at the limb is opaque in some lines with the result that the surface brightness is relatively constant, and

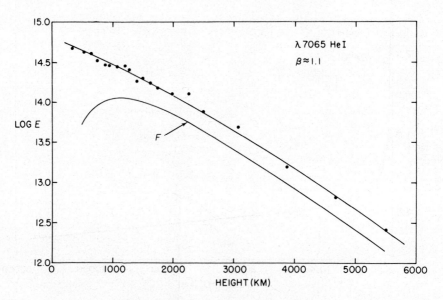

Fig. V-5.   Intensity of the λ 7065 line of He I as a function of height above the limb (1952 data).

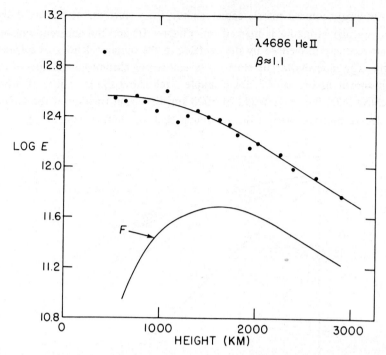

Fig. V-6.   Intensity of the λ 4686 line of He II as a function of height above the limb (1952 data).

(2) excitation conditions change markedly with height. The intensity gradients are influenced additionally by geometrical structures in the chromosphere. This latter effect is particularly evident in the upper chromospheric layers where emission from hydrogen and

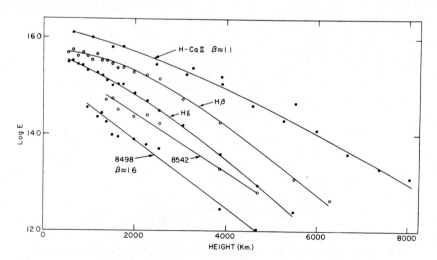

Fig. V-7.   Intensities of strong lines of Ca II and H.

neutral and singly ionized metals comes mainly from spicules. At heights above about
3000 km, spicular structure is marked (cf. Chapter II) and the apparent emission scale
heights are strongly influenced by the decrease in the number density of spicules above
this height. The intensity-height plots for several strong chromospheric lines of Ca II and
hydrogen shown in Figure V-7, for example, each show similar values of β (≈ 1.6) at
heights above 3000 km. At a height of 5000 km, the mean fraction of the limb covered
by spicules is approximately 10%. Assuming that the limb is completely covered by

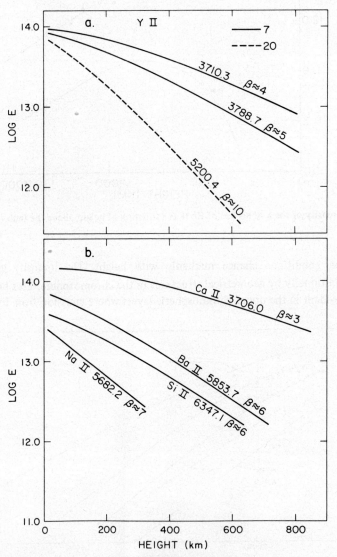

Fig. V-8.   Intensities of selected lines of Fe I (multiplets 21 and 39) and Fe II
(multiplets 27, 38 and 48).

spicules at 3000 km, we find for the effective reciprocal scale height of the spicule area $\beta_a \approx 1.1$. This suggests that possibly all of the decrease in helium brightness and most of the decrease in hydrogen and calcium brightness above 3000 km is due to the decreasing surface area of spicules. This point will be investigated further in Chapter VII.

The wide diversity in emission scale heights for metal lines in the low chromosphere is exhibited in Figures V-8 and V-9. Values of β range from less than 2 to about 10. Much of this variation is due to chromospheric opacity in the metal lines. We noted in Section 1 of

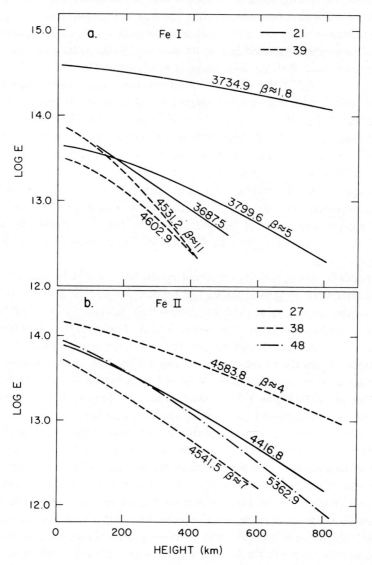

Fig. V-9.   Intensities of lines in multiplets 7 and 20 of Y II and in selected lines of Na II, Si II Ba II and Ca II.

this chapter that many of the strong Fraunhofer lines reach radial optical depth unity in the 600 to 800 km height range. These same lines are expected to have optical thicknesses in the tangential direction that are approximately 150 times larger than in the radial direction. Thus, the low chromosphere at the limb will have high optical thickness in many strong lines and will have appreciable optical thickness in many lines of only moderate strength. Changing excitation conditions also influence the relative emission scale heights.

Chromospheric spectra obtained at eclipse with sufficient spectral resolution to give reliable line profiles have been obtained at a few eclipses in limited spectral regions. Such spectra are obtained either with a slit spectrograph or with a slitless spectrograph in which the thin chromospheric crescent acts as its own slit. In the latter case, the optics are usually designed such that the solar image is highly ellipsoidal and the chromospheric crescent is compressed in the direction of dispersion but not at right angles to it. This technique permits both high spectral resolution and good spatial resolution along the solar limb. In the slit method the observer usually attempts to guide the image so that the slit is just above the lunar limb. This method has the advantage of restricting the height range in the chromosphere over which the line profile is averaged but has the disadvantage of having the position of the slit only crudely known on a fixed height scale.

Chromospheric Doppler widths reported in Chapter III (Figure III-7) from the work of Suemoto (1963) were obtained from analyses of eclipse profile data. The most extensive analysis and report of line profile data for the chromosphere at eclipse are contained in two papers published by Tanaka (1971a, b). Some of his results will be described in the following.

The Fraunhofer absorption lines observed against the solar disk become emission lines in the chromosphere. In general terms, this happens because the line opacity is greater than the continuum opacity and the 'continuum limb' lies inside of the 'line limb'. This process of reversal from absorption to emission in the lines does not occur uniformly in all lines or even within a particular line. In most of the very strong Fraunhofer lines with well developed wings the emission line just beyond the limb shows an absorption dip (self-reversal) near line center. Figures V-10 and V-11 exhibit this effect in two strong lines of Mg I and Ti II. Figure V-11 includes, in addition, lines of Ni I, Ni II and Fe I. An effect noted by several authors (cf. Tanaka, 1971a) is that lines of neutral atoms show a much stronger tendency for self-reversal than lines of ions of the same intensity. This is clearly illustrated in Figure V-11 by the lines of Ni I and Ni II. The Ni II line shows no self-reversal even though it is stronger than the Ni I line, which is strongly reversed. Tanaka accounts for this effect, probably correctly, in terms of the differences in the opacity scale heights. As we noted earlier on this section, the population density of neutral metals tends to decrease more rapidly with height than does the population density for ions. Thus, in a given height interval the lines of ions will tend to show less change in opacity than will the lines of neutral atoms. We will return to this point in Chapter VII.

Reversal from absorption to emission occurs inside the limb for some lines. Figure

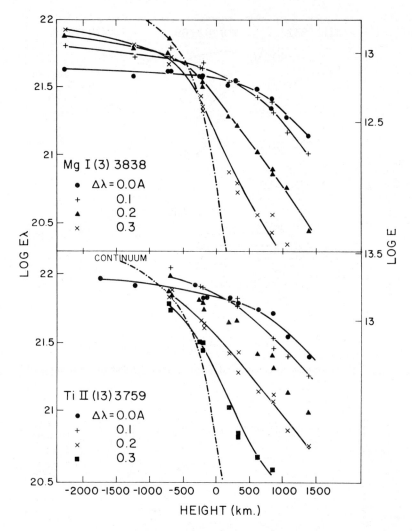

Fig. V-10.   Limb profiles of two lines of Mg I and Ti II at selected wavelengths in the lines. Note the change from absorption ($dE/d\Delta\lambda > 0$) to emission ($dE/d\Delta\lambda < 0$) (after Tanaka, 1971a).

V-12 shows the reversal heights obtained by Tanaka for several lines. The reversal heights are plotted against the reduced equivalent widths observed at disk center. The tendency for rare earth lines to reverse inside the limb has been confirmed by observations outside of eclipse (Canfield, 1969).

Figure V-13 shows the brightness in line center as a function of height for several strong lines. Note the tendency in some lines for the brightness of the chromosphere to increase with height. Similar effects have been noted in other lines by other authors (cf. Thomas and Athay, 1961). The near constancy or outwards increase of the central

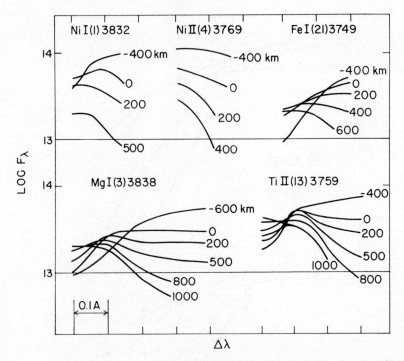

Fig. V-11.   Line profiles of selected metal lines near the solar limb (after Tanaka, 1971).

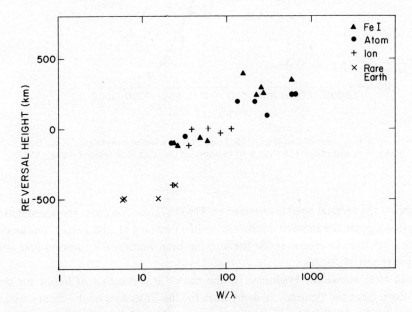

Fig. V-12.   A plot of the distance from the solar limb at which different lines reverse from absorption to emission relative to the continuum (after Tanaka, 1971a).

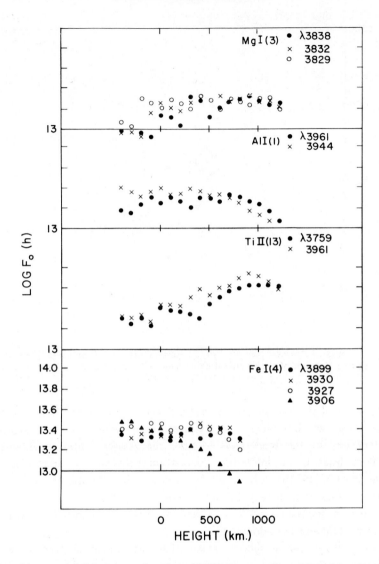

Fig. V-13.   Line center brightness as a function of height for several metal lines (after Tanaka, 1971a).

brightness in the strong lines presents an unambiguous manifestation that the chromosphere, at the limb, is opaque in the centers of these lines out to heights in excess of 1000 km. This high opacity in the lines of neutral metals provides a useful guide as to the nature of the model of the low and middle chromosphere.

An additional effect of particular interest in the chromospheric spectrum is the tendency for the weaker lines to show the greater central brightness. This effect is exhibited in Figure V-14 where the equivalent black body radiation temperature at line center and at a mean chromospheric height below 300 km is plotted against the reduced

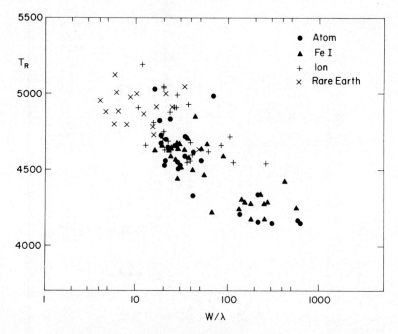

Fig. V-14. Central brightness temperature of several groups of emission lines in the low chromosphere as a function of their reduced equivalent width in the Fraunkofer spectrum near disk center (after Tanaka, 1971a).

equivalent width of the line in absorption at disk center. The rare earth lines show values of T_R near 5000°, which is much above the kinetic temperature in these same layers.

This tendency of the weak lines to have the greater central brightness in the chromosphere leads to an interesting comparison between low dispersion and high dispersion spectra. In the low dispersion spectra the chromospheric spectrum is dominated by lines of hydrogen, helium and the strong lines of Ca II, Fe I, Fe II etc. However, in high dispersion spectra it is the rare earth's and some of the weak lines of heavy elements that appear to dominate.

The half-widths of chromospheric emission lines are influenced strongly by both the chromospheric optical thickness and the microturbulent velocity. The outwards increase in the microturbulence tends to increase the broadening of the lines with increasing height whereas the outwards decrease in optical thickness tends to decrease the broadening. Furthermore, since the rate of decrease of optical thickness is slower for the ions than for the neutral atoms the rate of decrease in line broadening is slower for the lines of ions than for the lines of neutral atoms. These effects are illustrated in Figure V-15 for a number of lines.

B. CHROMOSPHERIC LIMB OUTSIDE OF ECLIPSE

Chromospheric emission lines are readily observed at the solar limb by placing the slit of a spectrograph tangential to a solar image of reasonable size, say, greater than 1 cm. With

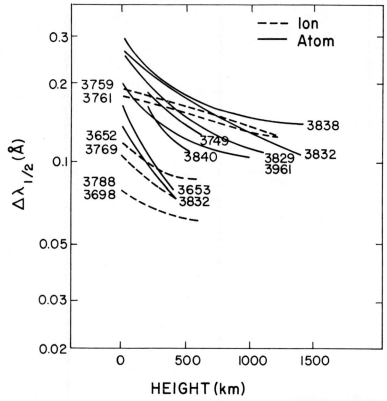

Fig. V-15.   Half-widths of chromospheric emission lines at the limb illustrating the tendency for the
half-widths to decrease more rapidly for neutral metals than for ions (after Tanaka, 1971a).

large solar images radial slits may be used also. The spectrum observed in this way is
contaminated by photospheric light scattered by the optics in the instrument and by the
Earth's atmosphere. Additionally, the seeing excursions of the solar limb are usually large
enough that the photospheric image periodically wanders onto the slit. Thus, one sees
always, with this technique, a mixture of the chromospheric emission spectrum with the
Fraunhofer absorption spectrum.

Because the rare earth lines are relatively weak in the photosphere and relatively bright
at line center in the chromosphere, they are among the easiest lines to observe in emission
at the limb with high dispersion instruments. Other lines that are easily observed include
lines of He I, Ca II, H, O and several lines of singly ionized heavy atoms. Lines of neutral
heavy atoms are also observed, but these are weakened relative to the lines of singly
ionized atoms.

Quantitative studies of the limb spectrum of the low chromosphere observed outside
of eclipse are difficult to carry out because of the contamination problem with
photospheric light. Often the chromospheric line is seen as little more than a 'filling in' of

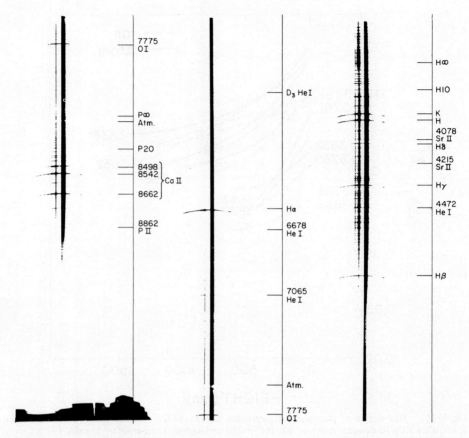

Fig. V-16.   Limb spectra of the chromosphere observed at the total eclipse of 1962, Lae, New Guinea.

the photospheric line. Figure V-16 illustrates spectrograms of the chromosphere at eclipse. Figure V-17 shows two spectrograms taken outside of eclipse. The difference in the character of the spectra observed is striking. Because of the difficulties of analysis associated with the limb spectra taken outside of eclipse, observations of this type have received relatively little attention. The most extensive observations available have been made by A. K. Pierce at Kitt Peak National Observatory. No extensive set of quantitative data has been published.

The stronger, high excitation chromospheric lines, notably those of H, Ca II and He, can be observed in the upper chromosphere and in spicules extending several seconds of arc above the limb. In this case, one of the prime sources of photospheric contamination (seeing excursions of the photosphere onto the spectrographic slit) can be eliminated. By taking additional care to reduce scattered photospheric light within the telescope optics, relatively clean chromospheric spectra can be obtained. Figure V-18 shows images

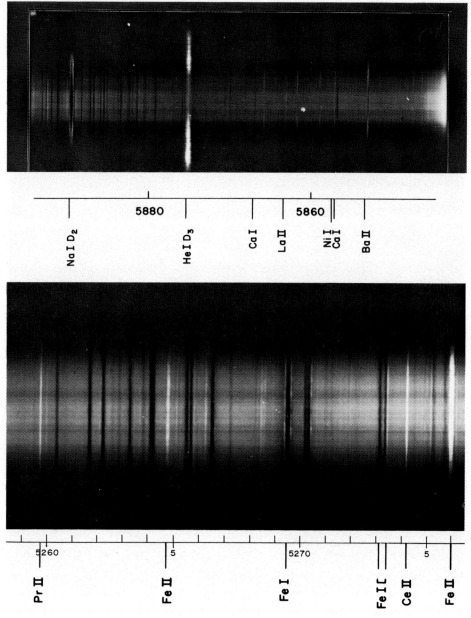

Fig. V-17.   Limb spectra of the chromosphere taken outside or eclipse. Note the broad line of He I
(D₃) as opposed to the very narrow line of La II in part a (Kitt Peak National Observatory).

of the photospheric limb and the chromosphere in Hα and Ca II observed with a radial
slit.

Fig. V-18.   The chromosphere at the limb as observed with a radial slit with both high spatial and spectral resolution. The Hα image in part a and the Ca II K image in part b both show pronounced broadening just above the continuum limb.

White (1963) has been able to show from observations similar to those in Figure V-18 that the D_3 line of He I has maximum surface brightness at a height of approximately 1500 km, which is in general agreement with the eclipse results in Figure V-5.

Fig. V-18b.

Limb observations made with high spatial resolution, as shown in Figure V-19, are capable of giving line profiles for individual spicules. As noted in Chapter II, however, spicules tend to occur in clumps or bushes and it is sometimes not clear whether a given observation resolves individual spicules or only clumps or spicules. Also, the spicule emission is superposed on the scattered photospheric light plus any background or

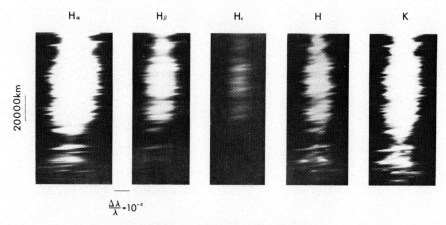

Fig. V-19.   Limb spectra made in the middle and high chromosphere with a tangential slit. The individual features are spicules or clumps of spicules.

foreground emission in the interspicular medium. In some observations such corrections are of moderate importance, but in others the corrections are small.

Beckers (1972) has presented a convenient summary of spicule spectroscopic data, which we repeat in Table V-2. This summary gives only the half widths of the lines and does not indicate the nature of the profile shape. The calcium lines, in particular, have

TABLE V-2
Spectroscopic properties of spicules (from Beckers, 1972)

Line	$W(\times 10^{+4})$ [a]	dW/dh $10^{-4}/1000$ km	I_0 (erg cm^{-2} s^{-1} sr^{-1} cm^{-1}) $\times 10^{12}$				
			2000	4000	6000	8000	10000
Hydrogen							
λ 6563 (H$_\alpha$)	1.6 (6000)	−0.10	45	35	20	12	5
λ 4861 (H$_\beta$)	1.4 (6000)	−0.07	–	40	15	–	–
λ 4340 (H$_\gamma$)	1.6 (4500)	−0.08	17	12	11	–	–
λ 4101 (H$_\delta$)	1.4 (4500)	–	–	2	–	–	–
λ 3970 (H$_\epsilon$)	1.4 (3000)	–	–	–	–	–	–
λ 3889 (H$_\zeta$)	1.2 (2000)	–	3.8	–	–	–	–
Helium							
λ 5875 (D$_3$)	0.8 (5500)	+0.05	–	8	2	0.5	0.2
λ 10830	0.6 (5000)	–	–	–	6	–	–
λ 3889 (Heζ)	0.8 (2000)	–	1.5	–	–	–	–
Calcium							
λ 3933 (K)	1.2 (5000)	−0.10	–	10	5	1.5	–
λ 3968 (H)	1.1 (5000)	−0.10	–	6	3	1.0	–
λ 8498	1.1 (5000)	–	–	–	–	–	–
λ 8542	1.5 (5000)	–	–	–	–	–	–

[a]The numbers in parentheses give the mean height in km to which W refers. W is defined as $\Delta\lambda_2/\lambda$, where $\Delta\lambda_2$ is the total width at half intensity.

shapes that are often strongly non-Gaussian. Relative to the Gaussian shape, the observed profiles are flat topped and sometimes show evidence of weak self-reversal. The Hα line tends also to have a somewhat flat topped profile in many spicules.

Spicule line widths summarized in Table V-2 represent, of course, only the average conditions. Actual profile widths reported vary quite markedly from spicule to spicule and not all authors agree as to the average width. Athay and Bessey (1964), for example, found two classes of spicule profiles for the H line of Ca II. One class of broad profile spicules was observed at low heights and a second class of narrow profile spicules was observed extending from low to relatively large heights. The range of H line reduced widths ($\Delta\lambda/\lambda$) was 2.3×10^{-4} to 0.7×10^{-4} with an average of 1.4×10^{-4}.

The widths of the Ca II lines in spicules are puzzling to understand. A reduced width of 1.1×10^{-4} for a Gaussian profile corresponds to a reduced Doppler width of 0.66×10^{-4}, which, in turn, corresponds to a broadening velocity of 20 km s^{-1}. Other lines, e.g., those of He I, are too narrow for such large velocities. Also, spicules have been observed showing faintly in the oxygen lines at λ 7772 and λ 8446 (Athay, 1961; Nikolskii, 1964). Individual spicule profiles were too weak to measure reliably, but, visually, the spicule profiles were judged to be comparable in width to the background chromospheric line. The chromospheric line gives $W \approx 0.55 \times 10^{-4}$, which is only half the value for the H line. Both the O line reduced widths and the Ca II line reduced widths are too large for thermal motions. There appears to be no possibility for the Ca II broadening to be due to self-absorption.

Interestingly enough, there is no good evidence for systematic changes in the chromospheric spectrum during the sunspot cycle (cf. Athay *et al.*, 1957). Marked temporal changes in the limb spectrum do occur in association with active regions (cf. Thomas and Athay, 1961) but these are relatively rare in terms of the average chromospheric conditions.

C. CORONAL ECLIPSE DATA

Variations in coronal structure, both temporally and spatially, make it somewhat meaningless to give average spectral properties of the corona. Nevertheless, these average spectral properties do provide insight to the basic nature of the corona, and, in any case, they are necessary if one is to discuss an average coronal model. Some of the data presented in the following will represent true average data in the sense that many sets of observations are combined. Other data will represent either the average corona as observed at a particular eclipse or a specific radial slice of the corona at a particular eclipse. The three sets of data are not equivalent, and, therefore, they must be treated with some caution. It is necessary to refer to particular eclipses and to particular points on the solar limb at these eclipses because these are frequently the only data available. In spite of the long history of eclipse observing, efforts to systematically deduce the spectrum of the 'average' corona are relatively rare.

Eclipse observations of the corona still provide one of the primary sources of data for the corona and, historically, have yielded much fruitful information. Figure V-20 shows

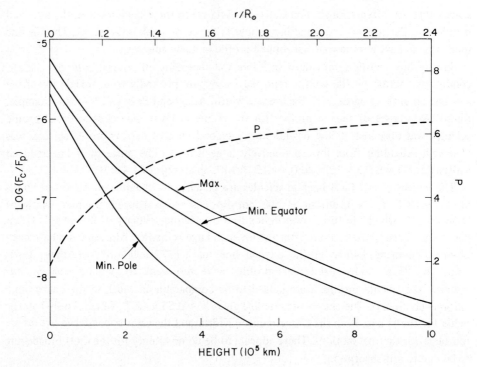

Fig. V-20.   Brightness and polarization of the coronal continuum as a function of distance from the
limb.

plots of the brightness of the K-corona in the visual continuum relative to the average
photospheric disk brightness. Since the 'color' of the coronal continuum is the same as
that of the photosphere, the spectral band pass is irrelevant. Curves are given in Figure
V-20 for the corona at sunspot maximum and at sunspot minimum. Near sunspot
maximum the corona is circularly symmetric to a reasonable degree, but near minimum
the polar corona is markedly reduced in brightness.

   Figure V-20 contains, also, a plot of the percentage polarization of the K-corona as a
function of height above the solar limb. The continuum is linearly polarized with the
electric vector parallel to the solar limb. Efforts to detect a tilt in the plane of
polarization, as might be expected if synchrotron emission were present, have so far
proven fruitless. Both the intensity and polarization of the coronal continuum are
consistent with the supposition that it is formed by electron scattering of photospheric
light (cf. Chapter VII).

   In the inner corona, the average continuum brightness drops by a factor e in about
50 000 km, which corresponds to $\beta = 0.02$. To give some indication as to how much this
gradient may vary, we give in Table V-3 the relative change in continuum brightness
between 3000 and 46 000 km for four points on the solar limb at the 1952 eclipse
(Athay and Roberts, 1955). The intensity given is in units of E rather than F, but this will

TABLE V-3

$$\Delta \log E = \log \frac{E(3000)}{E(46000)} \text{ and equivalent values of } \beta$$

	q-1	q-2	51-M	52-C
$\Delta \log E_C$	0.30	0.26	0.27	0.42
β_C	0.016	0.014	0.015	0.022
$\Delta \log E_{7892}$	0.75	0.90	0.79	0.12
β_{7892}	0.040	0.048	0.043	0.0064
$\Delta \log E_{5303}$	0.17	0.13	0.53	0.38
β_{5303}	0.0090	0.0070	0.029	0.020

not seriously influence the ratios. Three of the regions, two of which (q-1 and q-2) are quiet and one of which is active, have similar values of $\Delta \log E_c$ with an average of 0.28. The equivalent value of β is 0.015. In the fourth region, which is active $\Delta \log E_c = 0.42$ and the equivalent β is 0.023. A plot of the detailed fall off in continuum intensity in one of the regions, q-1, is shown in Figure V-21. It should be noted from this plot that β decreases with increasing height and has a value near 0.013 at 30 000 km.

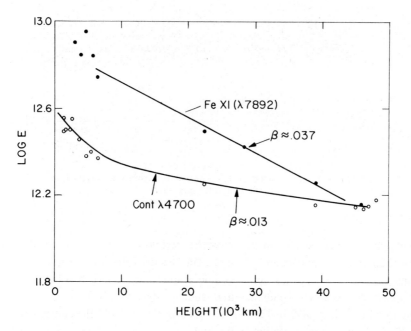

Fig. V-21.   Relative changes with height in region 52-C observed at the 1952 eclipse in the continuum at λ 4700 and in the Fe XI line at λ 7892.

Local changes in continuum brightness in the low corona from point to point along the solar limb are sometimes large. In a particularly bright condensation observed at the

1962 eclipse, whose isophotes are shown in Figure V-22, the maximum brightness at $1.1\,R_\odot$ (70 000 km) is five times greater than the average brightness of the equatorial minimum corona at this same height.

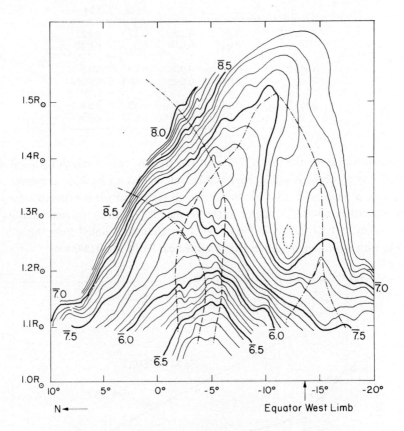

Fig. V-22.   Isophotes of an unusually bright coronal condensation observed at the 1962 eclipse (Saito and Billings, 1964).

Forbidden coronal lines in the visual spectral regions tend to fall off more rapidly with height than does the continuum. However, the specific behavior of the ratios of line to continuum intensities depends upon the line in question, upon height in the atmosphere and upon the particular limb feature studied. Relative changes of intensity with height for two lines, $\lambda\,7892$ of Fe XI and $\lambda\,5303$ of Fe XIV, and the continuum are given in Table V-3. Note that in three of the regions β_{7892} varies from 2.5 to 3.5 times β_c, whereas in the fourth region it is only 0.28 times β_c. The detailed variation of $\lambda\,7892$ relative to the continuum for region q-1 is shown in Figure V-21.

Table V-4 contains values of β for the $\lambda\,10\,747$ line of Fe XIII in the height range 63 000 km ($R = 1.09$) to 140 000 km ($R = 1.2$) obtained at the 1963 eclipse. These

TABLE V-4
Values of β for λ 10 747 of Fe XIII
between 6000 km and 140 000 km at the 1963 eclipse

Position angle	010	040	050	060	110	240	290
β_{10747}	0.020	0.017	0.015	0.014	0.022	0.022	0.018

values of β have been deduced by Eddy (1973) from data published by Eddy et al. (1973).

Gradients in line intensities relative to the continuum intensity are influenced primarily by the coronal temperature, as we shall show in Chapter VII. The height at which the coronal temperature reaches maximum is not known with any degree of precision, but it is certainly above 3000 km in the quiet corona. Thus, the plot in Figure V-21 and the results in Table V-3 for regions q-1 and q-2 reflect the fact that the temperature is increasing outwards. Ionization equilibria for coronal iron ions computed by Jordan show the maximum concentration of Fe XI at 1.2×10^6 K, Fe XIII at 1.7×10^6 K and Fe XIV at 1.8×10^6 K. The fact that β is larger for λ 7892 than for λ 5303 is consistent with an outwards increase of temperature. Values of 'β for λ 10 747 in Table V-4 average 0.018. This is approximately twice the value of β for the continuum in the same height range.   Coronal emission lines in the visual spectrum arise from transitions between different J states in the ground terms of coronal ions. Although these transitions are forbidden in the usual spectroscopic sense, the unusually low density in the corona permits these lines to be moderately strong (cf. Chapter VI). Coronal forbidden lines are excited primarily by a combination of electron collisions and photo-excitations. In the low corona, the electron collisions dominate and the line intensity is proportional to the square of the local density. This accounts, in an approximate way, for the fact that β for the emission lines generally exceeds β for the continuum in the low corona, and for the tendency in some lines for β to have approximately twice the value of β for the continuum.

In the outer corona the situation changes. The coronal density decreases more rapidly with distance from the solar surface than does the photon density in the photospheric radiation, i.e., n_e decreases faster than r^{-2}. As a result, the photo-excitation rate for the visual coronal lines eventually exceeds the electron collision rate. When this happens the line emission will vary linearly with density, and the line and continuum intensities will decrease at approximately the same rate. This effect is illustrated in Figure V-23 where the λ 6374 intensity is plotted relative to the continuum intensity on an arbitrary scale (Shain, 1947).

The results in Figure V-23 indicate that above $r \approx 1.5$ the λ 6374 line is excited by photospheric radiation and that through a distance of some four solar radii the proportionality between the Fe XI concentration and the electron density concentration, n_e, remains relatively constant. Since the total iron concentration should remain at a fixed proportion to the hydrogen concentration, and since the hydrogen concentration is proportional to n_e, we may conclude from Figure V-23 that there is relatively little

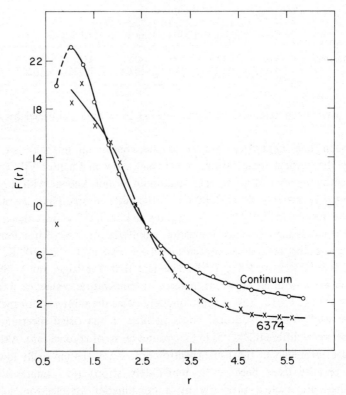

Fig. V-23.   Relative line and continuum intensities in the corona far above the solar limb (after Shain, 1947).

change in the proportionality of Fe XI concentration to the total iron concentration between approximately $r = 1.5$ and $r = 5$. The relative concentration of any one stage of ionization in a given element depends, of course, on the local temperature of the corona (cf. Chapter VI). Thus, the results in Figure V-23 suggest that the coronal temperature does not change markedly over the height range $r \approx 1.5$ to $r \approx 5$. (Note that the maximum height shown in Figure V-21 corresponds to $r = 1.07$.) This interpretation is consistent with current coronal models.

Spectral lines originating in ions of different ionization potential show different spatial distributions of intensity around the solar limb as well as with height in the corona. Figure V-24, for example, shows the relative distributions of intensity with solar latitude for five lines and the continuum as observed at the 1952 eclipse (Athay and Roberts 1955). The lines fall into roughly two classes: (1) Fe X and Fe XI, and (2) Fe XIV, Fe XV and Ni XV. Note that the high excitation group, two of which are observed in region 52-C only, shows stronger equatorial concentration than does the low excitation group. Also, the high excitation group shows a consistent positive correlation with the continuum intensity, but the low excitation group sometimes shows an anticorrelation, i.e., the line intensity drops where the continuum rises. The two groups of lines considered here are members of Class I and Class II lines as tabulated by van de Hulst (1953). Lines in still higher excitation classes tend to show only in active regions. One of these in particular, λ 5694 of Ca XV (Class IV), was observed for the first time at eclipse in region 52-C indicated in Figure V-24.

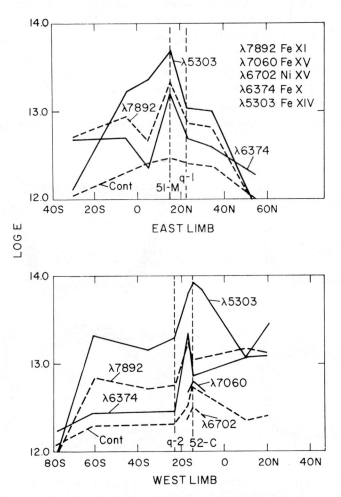

Fig. V-24.   The distribution of intensity with solar latitude for five coronal lines and the coronal continuum as observed at the 1952 eclipse (Athay and Roberts, 1955).

D. CORONAL LIMB DATA OUTSIDE OF ECLIPSE

Coronal continuum observations are made outside of eclipse with the aid of a scanning polarimeter known as the K-coronameter. This device utilizes the fact that the coronal light is strongly plane polarized whereas the sky background near the sun is only weakly polarized. Observations of this type are capable of detecting large scale coronal features, but do not show the fine structure of the corona. They have the distinct advantage over eclipse observations of permitting a more or less continuous monitoring of the gross structure of the inner corona. Present instruments are limited to a spatial resolution of 1.3' (56 000 km or 0.08 $R_\odot$) and are capable of detecting the corona out to $r \approx 2\,R_\odot$

Daily scans of the corona at a fixed distance of $r = 1.125\,R_\odot$ for the month of July for the years 1964, 1965, 1966 and 1967 are plotted in Figure V-25. The observed quantity is PF where P is the degree of polarization ($\approx 35\%$ at $r = 1.125\,R_\odot$, see Figure V-20) and F is measured in units of photospheric brightness times 10^{-8}. Sunspot minimum occurred near 1964 and the corona is seen to be strongly elliptical and of low relative brightness. As activity increases (1966), the northern hemisphere brightens markedly. This is followed by a brightening of the southern hemisphere. Note that the south polar region remains relatively unchanged during this period. Some impression of the day to day changes in structure can be inferred from the spread in the individual intensity plots.

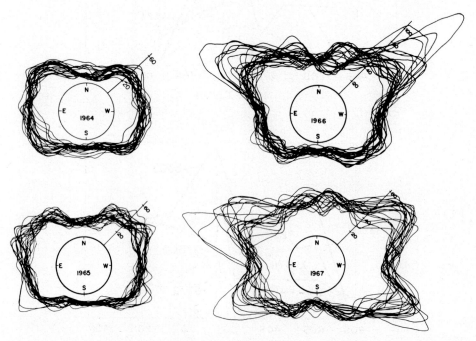

Fig. V-25.   Daily scans of the corona for the months of July for the years 1964, 1965, 1966 and 1967. The contours are isophotes for the product at a fixed distance $N = 1.125\,R_\odot$ from the limb. The tendency for symmetry about the N-S axis is due to the fact that the plots cover a full rotation period of the Sun (Hansen *et al.*, 1969).

Coronal continuum structure has also been observed outside of eclipse using white light coronagraphs carried to high altitude by balloons (Newkirk and Bohlin, 1965), rockets (Koomen *et al.*, 1970; Bohlin *et al.*, 1971) and satellites (Koomen *et al.*, 1971; MacQueen *et al.*, 1974). Such observations have been limited to the outer corona through restrictions on instrument size. However, streamer structure in the corona is readily detected. Photographs of the corona obtained by MacQueen *et al.* (1974) from the ATM-Skylab are shown in Figures II-24 and III-10.

Observations of coronal line emission with coronagraphs have been made routinely for many years at several solar observatories. Such data provide a convenient means of studying the morphology of the corona and of monitoring the level of solar activity. They have been particularly useful for determining the statistical properties of the corona and for the detailed study of individual active centers. The two lines most commonly observed are λ 5303 (Fe XIV) and λ 6374 (Fe XI). Other convenient lines that show up in some active centers are λ 6702 (Ni XV) and λ 5694 (Ca XV). The brightest and most studied line in the visual is λ5303.

Extensive studies of coronal morphology and its relationship to solar activity have been conducted from observations of the visual coronal lines. Since these studies are of interest primarily with regard to the solar activity cycle, they will not be dealt with here. The interested reader is referred to the summary by Billings (1966).

Widths of coronal emission lines have been extensively studied outside of eclipse. They vary from line to line and, to some extent, from point to point on the Sun. Billings (1966) gives as the average broadening temperatures, assuming that only thermal broadening is present, 1.8×10^6 K for λ 6374 (Fe XI), 2.5×10^6 K for λ 5303 (Fe XIV) and 5×10^6 K for λ 5694 (Ca XV). The corresponding mean Doppler velocities and Doppler widths are: 23 km s^{-1}, 0.49 Å (λ 6374); 29 km s^{-1}, 0.51 Å (λ 5303); and 46 km s^{-1}, 0.87 Å(λ 5694).

These are properties of the average profiles only. However, it is clear from these averages that the three ions do not, on the average, occupy the same volume elements. It is suggested rather that the corona usually exhibits a range of temperatures and that each ion selectively occurs in the temperature regions where it is relatively most populous. This conclusion is independently confirmed by the fact that the three lines are identified with different latitudinal distribution on the Sun, i.e., they belong to different classes, as noted in the preceding section.

The λ 5694 line of Ca XV occurs in the coronal spectrum only in association with intense activity. Thus, the line widths quoted above for this line refer only to active regions. Both λ 6374 and λ5303 occur in quiet as well as active regions, but λ 5303 shows a much stronger tendency to be enhanced in active regions than does λ 6374.

3. Infrared, Radio and XUV Continuum Data

A. INFARED AND RADIO DATA

The continuum opacity in the solar atmosphere has a minimum near a wavelength of 1.6 μm. It remains relatively low throughout the visual spectrum then increases in the far UV due to the bound-free absorption of atoms and molecules. On the long wavelength side of 1.6 μm the opacity increases due to free-free absorptions. It follows that the surfaces defined by optical depth unity in the continuum move progressively higher in the solar atmosphere as the wavelength is either increased above 1.6 μm or decreased below about 3000 Å. The increase to the violet is not monotonic because of the nature of the free-bound absorptions, but a general increase occurs nevertheless.

Surfaces of optical depth unity reach the temperature minimum at wavelengths of approximately 200 μm in the far infrared and approximately λ 1700 in the XUV. Broadly speaking, the radiation at mm wavelengths arises in the chromosphere; cm wavelengths arise in the upper chromosphere and low corona and decimeter and meter wavelengths arise in the corona. On the UV side, the continuum radiation remains chromospheric in origin down to X-ray wavelengths. The chromospheric emission includes the strong bound-free Lyman continuum of hydrogen. Other sources of prominent free-bound continua include C I, Si I He I and He II.

It is generally agreed that the free-free continua are formed under conditions that are close to local thermodynamic equilibrium. This fortunate circumstance makes the far infrared and radio continua of unusual interest. At wavelengths below about one meter there is presumably a one to one correspondence between the equivalent black-body temperature of the observed solar intensity and the kinetic temperature at some depths in the solar atmosphere. Thus, a map of the solar brightness temperature at disk center (in the continuum) as a function of wavelength is also a map of kinetic temperature vs depth in the solar atmosphere. All that remains to be done, in principle, is to assign a height of formation to each wavelength. Center-limb data at these wavelengths provide additional information on the temperature model.

In practice, of course, the observations of the solar brightness at disk center are not easy to obtain in the far infrared and radio wavelengths. The infrared observations are handicapped by heavy absorption in the Earth's atmosphere and must be carried out from high altitude. Radio observations suffer from the difficulty of having inadequate spatial resolution and difficulties of calibration of antenna temperatures, i.e., the temperature of the observed radiation. Also, center-limb data are strongly influenced by the inhomogeneous structure of the solar atmosphere. Disk center observations average over inhomogeneous structures of various sorts also, but here the effects are of lesser importance.

Figure V-26 contains a compilation of the observed brightness temperatures at the center of the solar disk at far infrared and mm wavelengths. Figure V-27 extends the plot into meter wavelengths. Unlike the data in Figure V-26, however, the data in Figure V-27 represent the average apparent brightness of a disk of photospheric diameter, i.e., the solar disk is not resolved and the area of the emitting surface is assumed rather than measured. Several features of interest stand out in Figures V-26 and V-27 in spite of the relatively large scattering of observational points about a mean curve. Beginning at shorter wavelengths, the first points of interest are the value of the temperature minimum and its location in wavelength. The observations suggest a minimum temperature between 4000° and 4400 K located near 200 μm. Beyond the temperature minimum, the observed brightness temperature rises gradually to about 8000° at 1 cm where it then begins a relatively rapid rise to near 10^6 K.

At cm and meter wavelengths it is necessary to distinguish the 'quiet Sun' component

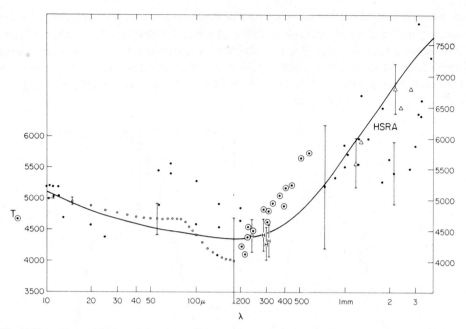

Fig. V-26.   A compilation of the observed central brightness temperature of the Sun as a function of wavelength from 10 μm to 3 mm. The solid line is a prediction based on the HSRA. Observations by different observers are indicated by different points (Mankin *et al.*, 1974).

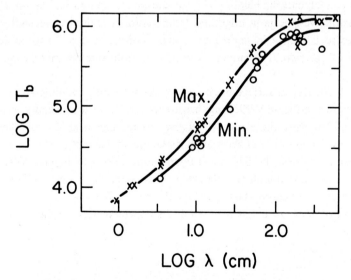

Fig. V-27.   A compilation of central brightness temperatures of the Sun from wavelength of 1 cm to 5 m at epochs near both sunspot maximum and sunspot minimum (after Kundu, 1965).

of radiation from the active component. This distinction is relatively well defined at cm wavelengths but becomes difficult to define at meter wavelengths. Figure V-28 exhibits seven scans across the solar disk with the Nancay interferometer utilizing a fan shaped beam of 3.8′ width at a wavelength of 1.78 m. For the seven days illustrated, the Sun was supposedly free of major activity. Nevertheless, there are clear and pronounced changes in surface brightness from day to day.

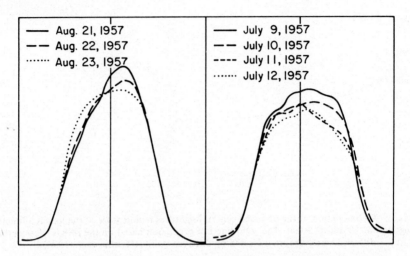

Fig. V-28.   Scans across the solar disk with the fan-beam of the Nancay interferometer at a wavelength of 1.78 m on 7 relatively quiet days (Boischot, 1958).

The two sets of data in Figure V-27 for sunspot minimum and sunspot maximum indicate a distinctive change in brightness temperature at a given wavelength. This could be due either to a change in temperature of the corona or to a change in density or a combination of the two. The curves are displaced from one another by about 0.17 in log λ.

Center-limb changes in surface brightness of the solar disk, as summarized by Kundu (1965), are shown in Figure V-29. Limb brightening occurs as wavelengths between about 8 mm and 1.55 m with the strongest relative brightening near 20 to 30 cm. Recent observations by Kundu (1971) show limb darkening at 1.2 mm and 3.5 mm. If it were not for inhomogeneities, the brightness temperature curves in Figures V-28 and V-29 would lead to pronounced limb brightening at all wavelengths between 200 μm and 1.5 m. Even where the observations show substantial limb-brightening, such as at 21 cm, the amount of brightening is far less than would be expected from a spherically symmetric model.

B. XUV DATA

Continuum observations in the XUV are also of great value for studying the temperature minimum and chromospheric regions. However, the XUV continua are formed by

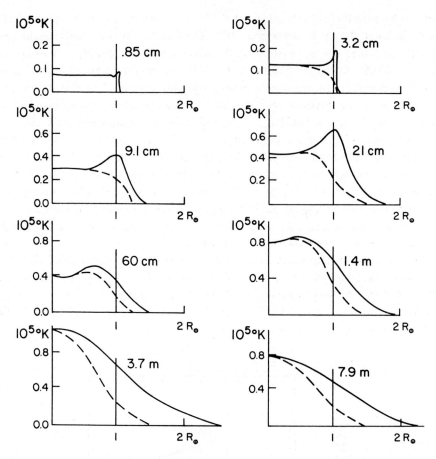

Fig. V-29.   Center-limb changes in brightness temperature at several wavelengths in the radio spectrum (Kundu, 1965).

bound-free transitions, and these are not formed under conditions of local thermodynamic equilibrium (cf. Chapter VI). There is not, therefore, a unique correspondence between the observed radiation temperature and the kinetic temperature at the depth in the solar atmosphere where the observed radiation arises. Indeed, one can relate the observed radiation temperature to the kinetic temperature only after detailed solution of the radiative transfer equations together with the equilibrium equations for the atoms producing the radiation.

   The lack of correspondence between brightness temperature and kinetic temperature in the free-bound continua is directly evident in the observational data. For example, the Lyman continuum of hydrogen is described by Vernazza and Noyes (1972) as having an intensity distribution with wavelength that is mimicked by a combination of two diluted Planck functions. At $\lambda > 700$ Å the spectrum resembles a black body at 8400° diluted in

intensity by a factor of 200, and for $\lambda < 650$ Å it resembles a black body at 20 000°
diluted in intensity by a factor near 10^8. The dilution at short wavelengths can be
partially accounted for by postulating the existence of an optically thin temperature
plateau at 20 000 K, but the dilution at $\lambda > 700$ Å cannot be explained in a similar way.
There is no possibility for the chromosphere to be optically thin in the Lyman continuum
and the dilution in intensity near the head of the continuum can arise only from a
diminution of the source function, i.e., through a loss of radiating efficiency. As we shall
show in Chapter VI, the color temperature of the Lyman continuum is a relatively good
indicator of the kinetic temperature at the depth where the continuum forms.

C I and Si I bound-free continua are formed deeper in the atmosphere than the Lyman
continuum. The C I continuum at λ 1100 from the ground state ($2p^2\ ^3P$) has a color
temperature near 6500° (Vernazza, 1972) but is diluted in intensity from a black body at
6500°. Again, this continuum arises from optically thick layers and is diluted by
departures from local thermodynamic equilibrium.

The Lyman continuum shows limb-darkening near the long wavelength limit but limb
brightening shortward of λ 750. This is illustrated in Figure V-30. Vernazza and Noyes
(1972) attribute the limb-darkening at $\lambda > 750$ Å to the inhomogeneous structure of the
atmosphere. The C I continuum at λ 1100 limb-brightens (Vernazza, 1972), but details of
the limb-brightening have not been published.

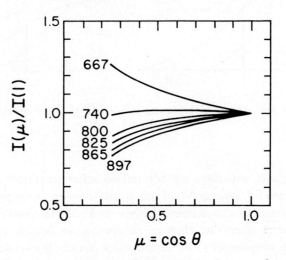

Fig. V-30.   Center-limb changes in Lyman continuum intensity as a function of wavelength
(Vernazza and Noyes, 1972).

Figures V-31, V-32 and V-33 show the solar disk center, continuum spectrum from
λ 4000 to λ 500. In the latter figure, the lines are included as they appear at 3 Å
resolution. Since the lines are narrower than 3 Å, the spectrum in Figure V-33
discriminates against the weaker lines.

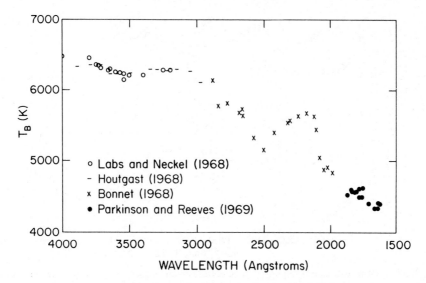

Fig. V-31.   Observed central brightness temperatures for the continuum from λ 4000 to λ 1600
(after Gingerich *et al.*, 1971).

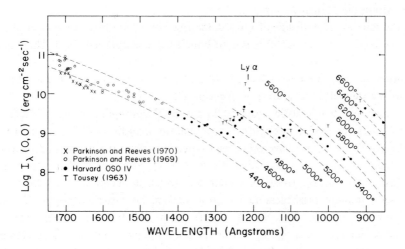

Fig. V-32.   Observed central brightness temperatures for the continuum from λ 1700 to λ 850 (after
Gingerich *et al.*, 1971).

The general increase of the continuum temperature with decreasing λ, for λ < 1300 Å,
is clearly evident in Figure V-33. Prominent features of the continuum are seen near
λ 1100, λ 900 and λ 500. These are, respectively, the free-bound resonance continua of
C I, H and He I. The apparent increase of continuum temperature beginning near λ 1300
(Figure V-32) is due to the broad emission wings of the Lyman-α line. One of the

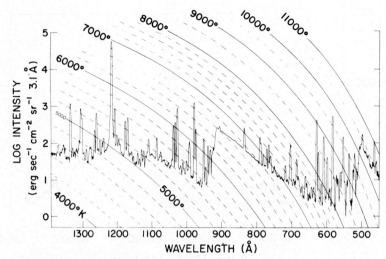

Fig. V-33.   Observed central brightness temperatures from λ 1400 to λ 450 (courtesy R. W. Noyes,
Harvard College Observatory and Smithsonian Astrophysical Observatory).

subordinate free-bound continua of C I shows as an emission feature beginning near
λ 1240 within the Lyman-α wing.

To the long wavelength side of λ 1700, the free-bound continua are in absorption. The
dip in the continuum at λ 2500 is due to bound-free absorption of Mg I. The sharp drop
in intensity near λ 2100 is due to Al I and a second drop near λ 2000 is due to Si I.
Absorption edges of Fe I between λ 1800 and λ 1580 appear to be present but weak.
They are blended with the absorption edges of Si I.

The minimum temperature observed in the EUV continuum is near 4400°. This is
consistent with the minimum brightness temperature of 4400° to 4000° from far infrared
data (Figure V-25). As we have previously noted, however, the free-bound continua are
not formed in LTE. In particular, Cuny (1971) has shown that the brightness
temperatures for the Si I free-bound continua, which dominate near λ 1700 to λ 1600
where the minimum brightness temperature occurs, lie systematically above the true
temperature minimum. This effect results from radiation on either side of the
temperature minimum penetrating the temperature minimum region and forcing the Si I
source functions above the Planck function at these depths. Hence, the observed
minimum brightness temperature in the EUV region represents an upper limit to the
actual value of the temperature minimum.

4. XUV Emission Lines

Chromospheric emission shows in the visual spectrum as faint emission peaks in the Ca II
H and K lines and as enhanced central intensities in nearly all of the strong Fraunhofer
lines with well developed wings. Further to the violet, the resonance lines of Mg II at

$\lambda\,2796$ and $\lambda\,2803$ show strong emission peaks superposed on broad, deep absorption lines. These are analogous to the Ca II emission and are enhanced over the Ca II emission mainly because the Mg abundance in the Sun is approximately ten times that of Ca. Thus, the chromosphere has much higher opacity in the Mg resonance lines than in H and K.

Still further to the violet the continuum intensity decreases relative to the lines until eventually the spectrum shows only emission lines from the chromosphere and corona superposed on a relatively faint continuum. The transition from absorption lines to emission lines is somewhat gradual but occurs mainly in the interval between $\lambda\,1700$ and $\lambda\,1900$. Resonance lines of Si II at $\lambda\,1817$ and $\lambda\,1808$ are strongly in emission. They are surrounded, however, by absorption lines. Below $\lambda\,1700$, the lines appear to be entirely in emission.

Chromospheric emission features extend to wavelengths near $\lambda\,200$, which lies in the resonance continuum of He II. However, lines formed by the permitted transitions of transition region ions and coronal ions begin near $\lambda\,1700$ and extend throughout the shorter wavelengths. Lines of all of the more abundant solar elements are prominent in various stages of ionization. These include especially the lines of H, He, C, N, O, Ne, Mg, Si, S and Fe. At the long wavelength end of the emission spectrum where the continuum is formed in the temperature minimum region and low chromosphere, lines of CO and Fe II are prevalent.

The emission spectrum in the XUV is rich in detail, but has been observed in an exploratory way only. As instrumental techniques in rocket and satellite experiments become more sophisticated and higher spectral resolution is achieved, more and more emission lines appear. The major part of the XUV spectrum has only recently been observed from the ATM-Skylab with sufficient spectral resolution to exhibit the line profiles. These data are not yet available in the literature, and, for the most part, we are still forced to work with only the integrated intensity in spectral bands that are wider than the lines themselves. Wavelengths, intensities and identifications of lines observed between $\lambda\,1400$ and $\lambda\,300$ by OSO-IV are given in Table V-5 (Dupree and Reeves, 1971).

TABLE V-5
OSO-IV quiet Sun spectrum

λ (A)	Intensity $(cm^{-2}\ s^{-1}\ sr^{-1})$ photons	erg	Element	Transition	J_{lower}	J_{upper}	Notes
1393.8	2.07 (13)	2.94 (2)	Si IV	$3s\ ^2S - 3p\ ^2P$	1/2	3/2	
1388.6			Na IX (2)	(694.3×2)			
1381.6	1.12 (13)	1.61 (2)		(690.8×2)			
1371.4			N III (2)	(685.7×2)			
1363.4			Na IX (2)	(681.7×2)			
1356.7	1.06 (13)	1.55 (2)	O I	$2p^4\ ^3p - 2p^3\,3s\ ^5S$	2, 1	2	†
1343.0			N II (2)	(671.5×2)			
1335.5	6.94 (13)	1.03 (3)	C II	$2s^2\,2p\ ^2p - 2s2p^2\ ^2D$	1/2, 3/2	3/2, 5/2	

TABLE V-5 (continued)

λ (Å)	Intensity (cm^{-2} s^{-1} sr^{-1}) photons	erg	Element	Transition	J_{lower}	J_{upper}	Notes
1322.8			S IV (2)	(661.4 × 2)			
1314.7			S IV (2)	(657.3 × 2)			
1309.3	8.53 (12)	1.29 (2)	Si II	$3s^2\,3p\,^2p - 3s3p^2\,^2S$	3/2	1/2	
1305.2	4 72 (13)	7.18 (2)	O I	$2p^4\,^3p - 2p^3\,3s\,^3S$	1, 0	1	†
1302.2	2.61 (13)	3.98 (2)	O I	$2p^4\,^3P - 2p^3\,3s\,^3S$	2	1	
1288.4			O II (2)	(644.2 × 2)			
1277.0				(638.5 × 2)			
1264.8	1.04 (13)	1.64 (2)	Si II	$3p\,^2p - 3d\,^2D$	3/2	3/2, 5/2	
1259.4			O V (2)	(629.7 × 2)			†
1250.6			Mg X (2)	(625.3 × 2)			
1242.8	2.79 (12)	4.47 (1)	N V	$2s\,^2S - 2p\,^2P$	1/2	1/2	*, †
1238.8	3.31 (12)	5.31 (1)	N V	$2s\,^2S - 2p\,^2P$	1/2	3/2	*
1232.6			O II (2)	(616.3 × 2)			
1219.6			Mg X (2)	(609.8 × 2)			
1215.7	3.23 (15)	5.27 (4)	H Ly-α	$1s\,^2S - 2p\,^2P$	1/2	1/2, 3/2	
1206.5	3.27 (13)	5.38 (2)	Si III	$3s^2\,^1S - 3s3p\,^1P$	0	1	*
1199.9	9.08 (12)	1.50 (2)	N I	$2p^3\,^4S - 2p^2\,3s\,^4P$	3/2	1/2, 3/2, 5/2	*, †
1194.1	3.04 (12)	5.04 (1)	Si II	$3s^2\,3p\,^2p - 3s3p^2\,^2P$	1/2, 3/2	1/2, 3/2	*, †
1190.4	1.45 (12)	2.41 (1)	Si II	$3s^2\,3p\,^2p - 3s3p^2\,^2P$	1/2	3/2	*, †
1182.8			He I (2)	(591.4 × 2)			*
1175.7	1.47 (13)	2.48 (2)	C III	$2s2p\,^3p - 2p^2\,^3P$	0, 1, 2	0, 1, 2	*
1168.6			He I (2)	(584.3 × 2)			
1160.8			O II (2)	(580.4 × 2)			
1157.7	2.51 (12)	4.30 (1)	C II				†
1152.2	1.90 (12)	3.28 (1)	O I	$2p^4\,^1D - 2p^3\,3s\,^1D$	2	2	
1148.2	1.70 (12)	2.94 (1)					
1144.6			Ne V (2)	(572.3 × 2)			
1139.7	2.52 (12)	4.39 (1)	{ Al XI (2) / Ne VI (2)	(568.5 × 2)			†
1134.6	1.65 (12)	2.88 (1)	N I	$2s^2\,2p^3\,^4S - 2s2p^4\,^4P$	3/2	1/2, 3/2, 5/2	
1128.3	2.81 (12)	4.94 (1)	Si IV	$3p\,^2P - 3d\,^2D$	3/2	3/2, 5/2	
1125.6			Ne VI (2)	(562.8 × 2)			
1122.5	2.74 (12)	4.85 (1)	Si IV	$3p\,^2P - 3d\,^2D$	1/2	3/2	
1117.2			Ne VI (2)	(558.6 × 2)			
1108.8			Q IV (2)	(554.4 × 2)			
1100.0			Al XI (2)	(550.0 × 2)			*
1085.1	4.35 (12)	7.97 (1)	N II	$2s^2\,2p^2\,^3P - 2s2p^3\,^3D$	0, 1, 2	1, 2, 3	*, †
1077.1	1.00 (12)	1.85 (1)	S III	$3s^2\,3p^2\,^1D - 3s3p^3\,^1D$	2		*
1074.2			He I (2)	(537.1 × 2)			*, †
1067.1	4.85 (11)	9.04					*
1062.7	4.58 (11)	8.57	S IV	$3s^2\,3p\,^2P - 3s3p^2\,^2D$	1/2	1/2	*
1051.6			O III	(525.8 × 2)			*
1044.9	1.54 (11)	2.92					
1042.2			Si XII (2)	(521.1 × 2)			*
1037.6	1.13 (13)	2.16 (2)	O VI	$2s\,^2S - 2p\,^2P$	1/2	1/2	*, †
1031.9	1.38 (13)	2.66 (2)	O VI	$2s\,^2S - 2p\,^2P$	1/2	3/2	*
1025.7	3.48 (13)	6.73 (2)	H Ly-β	$1s\,^2S - 3p\,^2P$	1/2	3/2	*
1021.0	4.38 (11)	8.54					*
1015.8			O III (2)	(507.9 × 2)			*
1010.2	9.07 (11)	1.78 (1)	C II	$2s2p^2\,^4P - 2p^3\,^4S$	1/2, 3/2, 5/2	3/2	*
998.6			Si XII (2)	(499.3 × 2)			*
991.0	3.58 (12)	7.17 (1)	N III	$2s^2\,2p\,^2P - 2s2p^2\,^2D$	1/2, 3/2	3/2, 5/2	*
977.0	4.42 (13)	9.00 (2)	C III	$2s^2\,^1S - 2s2p\,^1P$	0	1	*

Table V-5 (continued)

λ (Å)	Intensity (cm^{-2} s^{-1} sr^{-1}) photons	erg	Element	Transition	J_{lower}	J_{upper}	Notes
972.5	7.99 (12)	1.63 (2)	H Ly-γ	$1s\ ^2S - 4p\ ^2P$	1/2	3/2	*
958.9	1.22 (11)	2.53					*, †
949.7	3.96 (12)	8.27 (1)	H Ly-δ	$1s\ ^2S - 5p\ ^2P$	1/2	3/2	
944.5	6.35 (11)	1.34 (1)	S VI	$3s\ ^2S - 3p\ ^2P$	1/2	1/2	
937.8	2.48 (12)	5.24 (1)	H Ly-ϵ	$1s\ ^2S - 6p\ ^2P$	1/2	3/2	
933.4	7.92 (11)	1.69 (1)	S VI	$3s\ ^2S - 3p\ ^2P$	1/2	3/2	
930.7	1.93 (12)	4.12 (1)	H Ly-ζ	$1s\ ^2S - 7p\ ^2P$	1/2	3/2	†
923.1	2.10 (12)	4.51 (1)	N IV	$2s2p\ ^3P - 2p^2\ ^3P$	0, 1, 2	0, 1, 2	
904.1	1.95 (12)	4.28 (1)	C II	$2s^2\ 2p\ ^2P - 2s2p^2\ ^2P$	1/2, 3/2	1/2, 3/2	*
859.6	5.29 (11)	1.22 (1)					*, †
834.2	4.32 (12)	1.03 (2)	{ O II	$2s^2\ 2p^3\ ^4S - 2s2p^4\ ^4P$	3/2	1/2, 3/2, 5/2	*
			O III	$2s^2\ 2p^2\ ^3P - 2s2p^3\ ^3D$	0, 1, 2	1, 2, 3	*
790.2	2.68 (12)	6.73 (1)	O IV	$2s^2\ 2p\ ^2P - 2s2p^2\ ^2D$	3/2	3/2, 5/2	*
787.7	1.86 (12)	4.70 (1)	O IV	$2s^2\ 2p\ ^2P - 2s2p^2\ ^2D$	1/2	3/2	*, †
780.3	7.19 (11)	1.83 (1)	Ne VIII	$2s\ ^2S - 2p\ ^2P$	1/2	1/2	*
776.0	1.75 (11)	4.48	N II	$2s^2\ 2p^2\ ^1D - 2s2p^3\ ^1D$	2	2	*
770.4	1.25 (12)	3.22 (1)	Ne VIII	$2s\ ^2S - 2p\ ^2P$	1/2	3/2	*
764.6	1.83 (12)	4.75 (1)	{ N IV	$2s^2\ ^1S - 2s2p\ ^1P$	0	1	
			N III	$2s^2\ 2p\ ^2P - 2s2p^2\ ^2S$	1/2, 3/2	1/2	*
760.4	7.09 (11)	1.85 (1)	O V	$2s2p\ ^3P - 2p^2\ ^3P$	0, 1, 2	0, 1, 2	*
750.0	2.44 (11)	6.47					*, †
736.2	4.91 (10)	1.32	Mg IX (2)	(368.1 x 2)			
730.0	5.09 (10)	1.39					*
718.5	4.40 (11)	1.22 (1)	Q II	$2s^2\ 2p^3\ ^2D - 2s2p^4\ ^2D$	3/2, 5/2	3/2, 5/2	*
712.7	8.33 (10)	2.32	S VI	$3p\ ^2P - 3d\ ^2D$	3/2	3/2, 5/2	*
703.4	2.38 (12)	6.73 (1)	O III	$2s^2\ 2p^2\ ^3P - 2s2p^3\ ^3P$	0, 1, 2	0, 1, 2	*
694.3	1.58 (11)	4.51	Na IX	$2s\ ^2S - 2p\ ^2P$	1/2	1/2	*
690.8	1.26 (11)	3.62					*
685.7	9.43 (11)	2.73 (1)	N III	$2s^2\ 2p\ ^2P - 2s2p^2\ ^2P$	1/2, 3/2	1/2, 3/2	*
681.7	2.66 (11)	7.76	Na IX	$2s\ ^2S - 2p\ ^2P$	1/2	3/2	*
671.5	7.47 (10)	2.21	N II	$2p^2\ ^3P - 2p3s\ ^3P$	0, 1, 2	0, 1, 2	*, †
661.4	3.37 (11)	1.01 (1)	S IV	$3p\ ^2P - 3d\ ^2D$	3/2	5/2	
657.3	2.00 (11)	6.03	S IV	$3p\ ^2P - 3d\ ^2D$	1/2	3/2	
650.3	9.40 (10)	2.87					†
644.1	1.52 (11)	4.68	O II	$2s^2\ 2p^3\ ^2P - 2s2p^4\ ^2S$	1/2, 3/2	1/2	
638.5	1.50 (11)	4.66					†
629.7	9.56 (12)	3.02 (2)	O V	$2s^2\ ^1S - 2s2p\ ^1P$	0	1	
625.3	1.32 (12)	4.19 (1)	Mg X	$2s\ ^2S - 2p\ ^2P$	1/2	3/2	†
616.6	1.65 (11)	5.31	O II	$2p^3\ ^2D - 2p^2\ 3s\ ^2P$	3/2, 5/2	1/2, 3/2	
609.8	3.10 (12)	1.01 (2)	Mg X	$2s\ ^2S - 2p\ ^2P$	1/2	1/2	†
599.6	8.28 (11)	2.74 (1)	O III	$2s^2\ 2p^2\ ^1D - 2s2p^3\ ^1D$	2	2	
592.4	1.16 (11)	3.89					†
584.3	9.57 (12)	3.25 (2)	He I	$1s^2\ ^1S - 1s2p\ ^1P$	0	1	
580.4	2.40 (11)	8.23	O II	$2s^2\ 2p^3\ ^2P - 2s2p^4\ ^2P$	1/2, 3/2	1/2, 3/2	†
572.3	2.37 (11)	8.23	Ne V	$2s^2\ 2p^2\ ^3P - 2s2p^3\ ^3D$	2	2, 3	†
568.5	1.74 (11)	6.08	{ Al XI	$2s\ ^2S - 2p\ ^2P$	1/2	1/2	
			Ne V	$2s^2\ 2p^2\ ^3P - 2s2p^3\ ^1D$	0, 1	1, 2	
562.8	2.80 (11)	9.89	Ne VI	$2s^2\ 2p\ ^2P - 2s2p^2\ ^2D$	1/2	3/2, 5/2	†
558.6	2.78 (11)	9.90	Ne VI	$2s^2\ 2p\ ^2P - 2s2p^2\ ^2D$	1/2	3/2	†
554.4	2.69 (12)	9.63 (1)	O IV	$2s^2\ 2p\ ^2P - 2s2p^2\ ^2P$	1/2, 3/2	1/2, 3/2	
550.0	1.59 (11)	5.74	Al XI	$2s\ ^2S - 2p\ ^2P$	1/2	3/2	
542.8	1.58 (11)	5.80	Ne IV	$2s^2\ 2p^3\ ^4S - 2s2p^4\ ^4P$	3/2	1/2, 3/2, 5/2	

Table V-5 (continued)

λ (A)	Intensity (cm^{-2} s^{-1} sr^{-1}) photons	erg	Element	Transition	J_{lower}	J_{upper}	Notes
537.0	8.72 (11)	3.23 (1)	He I	$1s^2\ ^1S - 1s3p\ ^1P$	0	1	†
525.8	2.27 (11)	8.56	O III	$2s^2 2p^2\ ^1D - 2s2p^3\ ^1P$	2	1	
521.1	4.94 (11)	1.88 (1)	Si XII	$2s\ ^2S - 2p\ ^2P$	1/2	1/2	†
515.6	1.39 (11)	5.36	He I	$1s^2\ ^1S - 1s5p\ ^1P$	0	1	†
507.9	5.70 (11)	2.23 (1)	O III	$2s^2 2p^2\ ^3P - 2s2p^3\ ^3S$	0, 1, 2	1	
499.3	6.58 (11)	2.62 (1)	Si XII	$2s\ ^2S - 2p\ ^2P$	1/2	3/2	*
489.5	1.11 (11)	4.50	Ne III	$2s^2 2p^4\ ^3P - 2s2p^5\ ^3P$	0, 1, 2	0, 1, 2	*
482.1	1.23 (11)	5.06	Ne V	$2s^2 2p^2\ ^3P - 2s2p^3\ ^3P$	0, 1, 2	C, 1, 2	*
475.9	2.58 (10)	1.07					*
469.8	1.10 (11)	4.66	Ne IV	$2s^2 2p^3\ ^2D - 2s2p^4\ ^2D$	3/2, 5/2	3/2, 5/2	*
465.2	9.03 (11)	3.86 (1)	Ne VII	$2s^2\ ^1S - 2s2p\ ^1P$	0	1	*, †
460.0	1.60 (11)	6.93					†
447.0	7.91 (10)	3.52					†, #
436.1	4.32 (11)	1.97 (1) {	Mg VIII	$2s^2 2p\ ^2P - 2s2p^2\ ^2D$	3/2	3/2, 5/2	#
			Ne VI	$2s^2 2p\ ^2P - 2s2p^2\ ^2S$	3/2	1/2	
430.5	2.51 (11)	1.16 (1)	Mg VIII	$2s^2 2p\ ^2P - 2s2p^2\ ^2D$	1/2	3/2	#
417.0	3.76 (11)	1.79 (1)	Fe XV	$3s^2\ ^1S - 3s3p\ ^3P$	0	1	†, #
411.0	1.73 (11)	8.38					†, #
401.7	6.02 (11)	2.98 (1)	Ne VI	$2s^2 2p\ ^2P - 2s2p^2\ ^2P$	1/2, 3/2	1/2, 3/2	†, #
368.1	1.49 (12)	8.06 (2)	Mg IX	$2s^2\ ^1S - 2s2p\ ^1P$	0	1	#
361.7	1.84 (11)	1.01 (1)	Fe XVI	$3s\ ^2S - 3p\ ^2P$	1/2	1/2	#
303.8	3.49 (13)	2.28 (3)	He II	$1s\ ^2S - 2p\ ^2P$	1/2,	1/2, 3/2	#

Notes to Table V-5

* Background continuum subtracted from total line intensity; See Table V-2.
† Notes arranged by wavelength (Å). BR refers to identifications given by Burton and Ridgeley (1970).

1356.7	Bl. C I (Mult. 42), 1355.84; C I (Mult. 43), 1354.29 nearby; BR.
1305.2	Bl. Si II (Mult. 3), 1304.37.
1259.4	Bl. Si II (Mult. 4), 1260.42.
1242.8	? Bl. Fe XII, 1242.02, $3p^3\ ^4S_{3/2} - 3p^3\ ^2P_{3/2}$, BR.
1199.9	Bl. O III (2), 599.6 × 2; S III (Mult. 1), 1200.97, BR.
1194.1	Bl. S III (Mult. 1), 1194.02/40, BR.
1190.4	Bl. S III (Mult. 1), 1190.17, BR.
1157.7	Unresolved C I (Mult. 19), 1156; C I (Mult. 16), 1158; BR.
1139.7	Unresolved C I (Mult. 23), 1139.0, BR.
1085.1	Bl. Ne IV (2), 542.8 × 2.
1074.2	Bl. S IV (Mult. 1), 1073.52/2.99.
1037.6	Bl. C II (Mult. 2), 1037.02/6.34.
1021.0	? S III (Mult. 2), 1021.32/10
958.9	? He II (Mult. 18), 958.72/67
930.7	Bl. Ne VII (2), 465.2 × 2.
859.6	? C II (Mult. 4), 858.6/1
787.7	Bl. S V (Mult. 1), 786.48.
750.0	? Mg IX, 751.6, $2s2p\ ^1P_1 - 2p^2\ ^1D_2$
671.5	? Bl. O II (Mult. 12), 673.22.
650.3	? S VI, 650.43/648.63, $3d\ ^2D_{5/2} - 4p\ ^2P_{1,3/2}$
638.5	? Al X, 638.8, $2s^2\ ^1S_0 - 2s2p_1\ ^3P_1$, observed off-limb in OSO-VI spectra (Munro, 1970).
625.3	? Bl. O IV (Mult. 6), 625.
609.8	? Bl. O IV (Mult. 2), 609.83/8.40.

Notes to Table V-5 (continued)

592.4	? In part He I (Mult. 1), 591.41, observed off-limb in OSO-VI spectra (Munro, 1970).
580.4	? Bl. Si XI, 580.85, $2s^2 \ ^1S_0 - 2s2p \ ^3P_1$.
572.3	? Bl. Ca X, 574.01, $3s \ ^2S_{1/2} - 3p \ ^2P_{1/2}$.
562.8	Bl. Ne VII, 562, $2s2p \ ^3P_{2,1} - 2s2p^2 \ ^3P_{2,1,0}$.
558.6	Bl. Ne VII, 559.95/8.61, $2s2p \ ^3P_{1,0} - 2s2p^2 \ ^3P_{2,1}$; ? Ca X, 557.74, $3s \ ^2S_{1/2} - 3p \ ^2P_{3/2}$.
537.0	Bl. O II (Mult. 2 and 7), 538; C III (Mult. 5), 538.
521.1	Bl. Ne IV (Mult. 8), 521.82/74; He I (Mult. 4), 522.21.
515.6	Bl. O II (Mult. 17), 515.64/50.
465.2	? Bl. Ca IX, 466.23, $3s^2 \ ^1S_0 - 3s3p \ ^1P_1$.
460.0	Observed in off-limb spectra from OSO-VI (Munro, 1970).
447.0	Broad blended feature.
417.0	? Bl. Ne V (Mult. 4), 416.20.
411.0	? Na VIII, 411.15, $2s^2 \ ^1S_0 - 2s2p \ ^1P_1$ (Munro, 1970).
401.7	? Bl. Mg VI, 400, $2s^2 2p^3 \ ^4S_{3/2} - 2s2p^4 \ ^4P_{1,3,5/2}$

Extremely weak line due to low detector sensitivity in this wavelength region; the background noise was removed and a correction for non-linearity in the detection system was applied. The intensities should be treated with caution.

In the case of the stronger lines, such data give a reliable indication of the line strength and its variation over the solar disk. For weaker lines, however, there is no clear way of determining in low resolution data whether the observed emission represents a single line or a blend of lines. Line profiles are available in current literature for a few selected chromospheric lines.

Samples of spectra observed from ATM are shown in Figure V-34. These spectra have sufficient resolution to give good profiles for the broader lines, but, for the weaker lines they do not show the finer details of the profiles. A few profiles observed with somewhat higher spectral resolution from rockets and high altitude balloons are available also. Examples of profiles observed from ATM as well as some observed from rockets and balloons are shown in Figures V-35 through V-40.

Many of the stronger chromospheric lines are flat-topped or self-reversed. This is clearly evident in all of the wider profiles shown in Figures V-35 through V-40. In the case of the O I lines part of the central absorption core is due to oxygen in the Earth's atmosphere. The Lyman-α line of hydrogen shows a definite terrestrial absorption core, but this has been eliminated from the profile in Figure V-38.

The tendency of strong chromospheric lines to exhibit self-reversal is readily understood in terms of the behavior of the line source function to be demonstrated in the next chapter. It is a natural consequence of the escape of photons from the external layers of the chromosphere and does not require the introduction of any special mechanism or geometry.

Table V-6 summarizes the values obtained for the full width at half maximum for some of the profiles shown in Figures V-35 through V-38. There appears to be a definite tendency for the profiles to broaden from center to limb as is observed for all lines in the visual. There is little tendency, however, for any systematic change in half width from the limb to +4″.

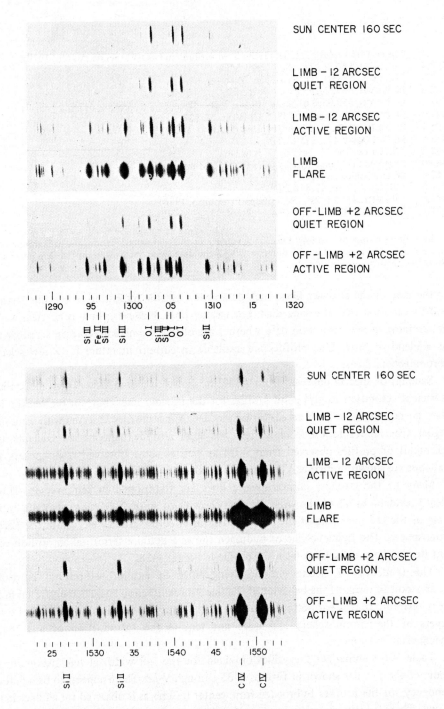

Fig. V-34.   Samples of spectra obtained by ATM-Skylab near λ 1300 and λ 1540. The samples show both active and quiet regions and both disk and limb regions (Naval Research Laboratory).

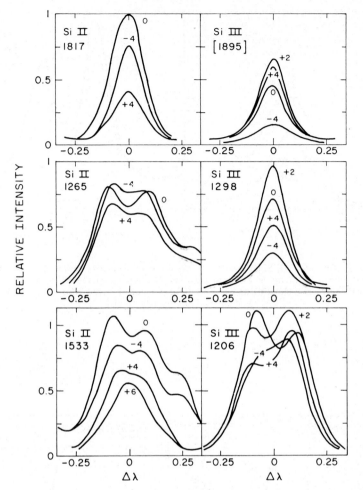

Fig. V-35.   Profiles of Si II and Si III lines observed near the solar limb by ATM-Skylab (Naval
Research Laboratory).

The Si III line at λ 1895 is from a forbidden transition and has a low f-value. In view
of this, it is highly probable that the chromosphere is optically thin in this line. The
λ 1298 line of Si III is a subordinate line whose lower level is the metastable upper level
of the λ 1895 transition lying 6.5 eV above the ground level. Thus, it, too, will have much
lower optical thickness in the chromosphere than the strong ground transition λ 1206.
This accounts, qualitatively, for the striking differences between the observed profiles for
λ 1206 and the remaining two lines of Si III.

The Si II lines in Figure V-35 are all from ground state transitions. However, the *gf*
values for these lines are 4.3, 0.53 and 0.023 for λ 1265, λ 1533 and λ 1817, respectively.
Thus, the optical thickness of the chromosphere in the λ 1812 line is smaller by a factor

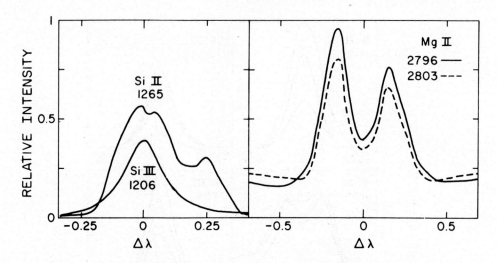

Fig. V-36.   Profiles of Si II, Si III and Mg II lines observed near disk center. The silicon lines were observed in a rocket flight by E. C. Bruner, Jr. and E. Chipman (Laboratory for Atmospheric and Space Physics, Univ. of Colorado) and the Mg II lines were observed from a high altitude balloon (Lemaire and Skumanich, 1973).

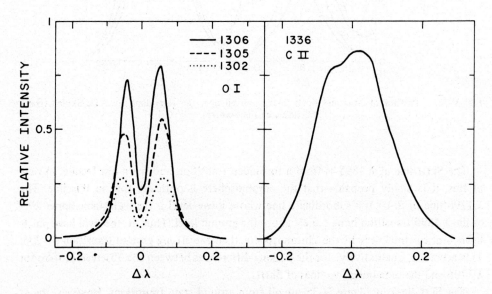

Fig. V-37.   Profiles of O I lines observed by E. C. Bruner, Jr. and E. Chipman (Laboratory for Atmospheric and Space Physics, Univ. of Colorado) and a C II line (Berger *et al.*, 1970) observed on the solar disk.

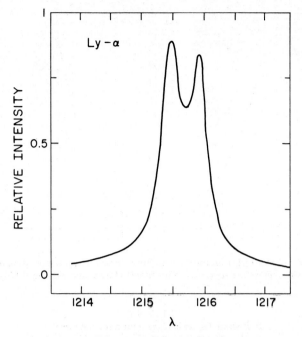

Fig. V-38. The Lyman-α profile observed on the disk by E. C. Bruner, Jr, and E. Chipman (Laboratory for Atmospheric and Space Physics, Univ. of Colorado).

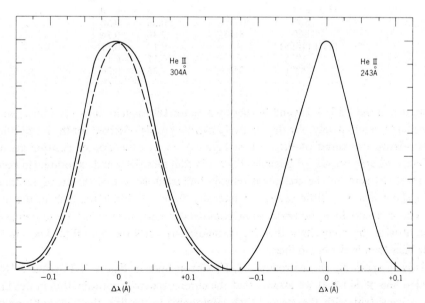

Fig. V-39. Profiles of two He II lines. The dashed profile for λ 304 was observed near disk center whereas the solid profiles were observed in the eastern quarter of the disk (Feldman and Behring, 1974).

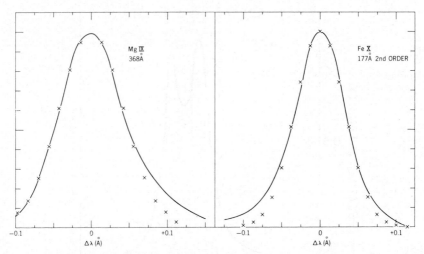

Fig. V-40.   Profiles of two high transition region lines formed near a temperature of 1×10^6 K. The crosses represent pure Gaussian profiles (Feldman and Behring, 1974).

TABLE V-6
Full width (Å) at half maximum for the profiles
shown in Figures V-35 and V-37

Line	Disk center	Limb	+4″
C II 1336	0.24		
O I 1306	0.15		
Si II 1817		0.18	0.18
1533		0.36	0.30
1265	0.30	0.36	0.36
Si III [1895]		0.17	0.17
1298		0.17	0.17
1206	0.20	0.36	0.36

22 than it is in the λ 1533 line and smaller by a factor 180 than in the λ 1265 line. Again, this accounts, qualitatively, for the relatively narrow profile observed for the λ 1817 line.

The relative intensities of the lines in Figure V-35 as a function of position are not completely reliable because of possible changes in film sensitivity and exposure. However, the general character of the changes is reliably indicated. Note that the broad, saturated profiles show relatively little change in intensity from −4″ inside the limb to the limb itself. The narrower lines, however, show considerable limb brightening. These trends are consistent with the supposition that the chromosphere has lower optical thickness in the narrow lines than in the broad lines.

Relatively speaking, the narrowest of the silicon lines in Figure V-35 is the λ 1895 forbidden line of Si III. If we assume that the chromosphere is optically thin in this line, which is consistent with the marked limb brightening in the line, the half width of the line can be used to give a measure of the broadening velocity. For a Gaussian profile, the full width at half maximum is 1.66 times the Doppler width. Thus, the Doppler width of the λ 1895 line is near 0.1 Å, which corresponds to a broadening velocity of 16 km s⁻¹. A slightly higher velocity would be obtained from the λ 1817 line, on the same assumption.

Jordan's (1969) computations of ionization equilibrium for silicon indicate that Si III has maximum concentration near 35 000°, in the solar atmosphere. This may be an overestimate of the temperature for maximum Si III concentration since Jordan's computations ignore the influence of radiative excitations followed by ionization from the excited level. The strong, broad character of the λ 1206 line clearly suggests that radiative excitation may be of importance. Nevertheless, it seems clear that the Si III lines are formed in the high chromosphere, probably in the temperature plateau near 20 000° indicated in Figure I-1. The thermal velocity of hydrogen atoms at 20 000° is 12.9 km s⁻¹, but silicon has a thermal velocity of only 2.4 km s⁻¹ at this temperature. Thus, the Si III, λ 1895, profile is consistent with the idea that the non-thermal broadening motions in the upper chromosphere are comparable to the thermal velocity of hydrogen atoms as suggested by the results in Figure III-7.

The Si II lines are formed considerably deeper in the chromosphere than are the Si III lines. Thus, both the thermal velocity and non-thermal velocity should be lower for the Si II lines than for the Si III lines. It appears, therefore, that even the λ 1817 line has considerable saturation. Also, the lines of C II, O I and Mg II shown in Figures V-36 and V-37 are broadened by saturation effects.

Spectral scans at resolution that does not show the line profiles are shown in Figure V-33. Representative spectroheliograms showing brightness changes over the solar disk are shown in Figure V-41. Notice that some of the lines show strong limb-brightening whereas others show a flat distribution across the dolar disk. Notice also that the contrast between the quiet disk and active regions varies considerably from line to line.

Objective grating spectroheliograms made by the Naval Research Laboratory (Tousey et al., 1973; Brueckner and Bartol, 1974) have sufficient spatial resolution to show some of the finer details of the solar disk in the XUV. The λ 304 spectroheliogram shown in Figure II-9 is an example of this type of observation. Network structure is visible throughout the chromosphere and transition region. No evidence of network structure has yet been observed in lines of coronal origin (see Figure V-41). The X-ray images of the Sun shown in Figures II-25 and I-5 show a wealth of detail, but there is no evidence of a pronounced network structure.

Although the XUV data have been limited in spatial and spectral resolution they have nevertheless provided the basis for nearly all that is known about the temperature and density model of the upper chromosphere, the transition region and lower corona. They represent, therefore, an unusually valuable addition to our means of studying the outer solar atmosphere. The improvement of these data to yield useful line profiles with good spatial resolution will represent another major step ahead. Only then will we be able to discuss the chromosphere and corona with the same level of understanding and confidence as is now possible for the solar photosphere.

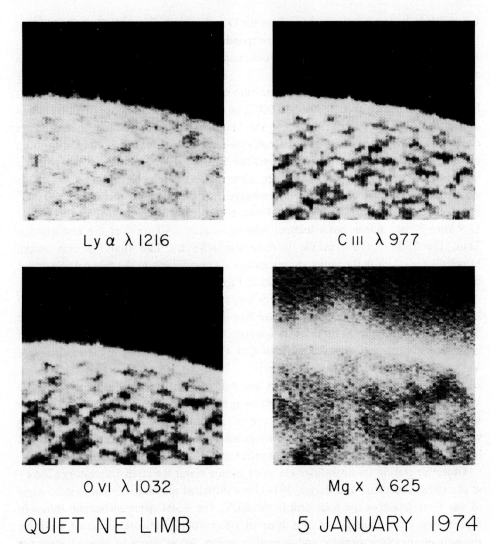

Ly α λ 1216   C III λ 977

O VI λ 1032   Mg X λ 625

QUIET N E LIMB   5 JANUARY 1974

Fig. V-41.   Spectroheliograms near the solar limb in four lines of differing excitation potential. Note the relative absence of limb-brightening in Lyman-α as opposed to strong limb-brightening in Mg X. The irregular limb seen in the Lyman-α, C III and O VI images is apparently due to spicule structure (Skylab photographs, courtesy Center for Astrophysics, Harvard College and Smithsonian Astrophysical Observatories).

References

Athay, R. G.: 1961, *Astrophys. J.* **134**, 755.
Athay, R. G. and Bessey, R. J.: 1964, *Astrophys. J.* **140**, 1174.
Athay, R. G. and Canfield, R. C.: 1969, *Astrophys. J.* **156**, 695.
Athay, R. G. and Lites, B. W.: 1972, *Atrophys. J.* **176**, 809.
Athay, R. G. and Roberts, W. O.: 1955, *Astrophys. J.* **121**, 231.
Athay, R. G. and Skumanich, A.: 1968, *Solar Phys.* **3**, 181.

Athay, R. G., Menzel, D. H., and Orrall, F. Q.: 1957, *Smithsonian Contr. to Astrophys.* **2**, 35.
Beckers, J. W.: 1972, *Ann. Rev. Astron. Astrophys.* **10**, 73.
Berger, R. A., Bruner, E. C., Jr., and Stevens, R. J.: 1970, *Solar Phys.* **12**, 370.
Billings, D. E.: 1966, *A Guide to the Solar Corona*, Academic, New York.
Bohlin, J. D., Koomen, M. J., and Tousey, R.: 1971, *Solar Phys.* **21**, 408.
Boischot, A.: 1958, *Ann. Astrophys.* **21**, 273.
Brault, J. W., Slaughter, C. D., Pierce, A. K., and Aikens, R. S.: 1971, *Solar Phys.* **18**, 366.
Brueckner, G. E. and Bartol, J.-D. F.: 1974, *Solar Phys.* **38**, 133.
Burton, W. M. and Ridgeley, A.: 1970, *Solar Phys.* **14**, 3.
Canfield, R. C.: 1969, *Astrophys. J.* **157**, 425.
Cuny, Y.: 1971, *Solar Phys.* **16**, 293.
Dunn, R. B., Evans, J. W. Jefferies, J. R., Orall, F. Q., White, O. R., and Zirker, J. B.: 1968, *Astrophys. J. Suppl.* **15**, 275.
Dupree, A. K. and Reeves, E. M.: 1971, *Astrophys. J.* **165**, 599.
Eddy, J. A.: 1973, unpublished.
Eddy, J. A., Lee, R. H., and Emerson, J. P.: 1973, *Solar Phys.* **30**, 351.
Feldman, U. and Behring, W. E.: 1974, *Astrophys. J.* **189**, L45.
Gingerich, O., Noyes, R. W., Kalkofen, W., and Cuny, Y.: 1971, *Solar Phys.* **18**, 347.
Hansen, R. T., Garcia, C. J., and Hansen, S. F.: 1969, *Solar Phys.* **7**, 417.
Henze, W., Jr.: 1969, *Solar Phys.* **9**, 56.
Houtgast, J.: 1957, *Rech. Astron. Obs. Utrecht.* **13** (No. 3).
Hulst, H. C., Van de: 1953, in G. Kuiper (ed.), *The Sun*, Univ. of Chicago, Chicago.
Jordan, C.: 1969, *Monthly Notices Roy. Astron. Soc.* **142**, 501.
Koomen, M. J., Purcell, J. D., and Tousey, R.: 1970, *Nature* **226**, 1135.
Koomen, M. J., Brueckner, G. E., and Tousey, R.: 1971, *Bull. Am. Astron. Soc.* **3**, 440.
Kundu, M. R.: 1965, *Solar Radio Astronomy,* Interscience, New York.
Kundu, M. R.: 1971, *Solar Phys.* **21**, 130.
Lemaire, P. and Skumanich, A.: 1973, *Astron. Astrophys.* **22**, 61.
Linsky, J. L. and Avrett, E. H.: 1970, *Publ. Astron. Soc. Pacific* **82**, 169.
Lites, B. W.: 1972, Thesis Univ. of Colorado (National Center for Atmospheric Research Cooperative Thesis No. 28, Boulder).
MacQueen, R. M., Eddy, J. A., Gosling, J. T., Hildner, E., Munro, R. H., Newkirk, G. W., Jr., Poland, A. I., and Ross, C. L.: 1974, *Astrophys. J.* **187**, L85.
Mankin, G. W., Eddy, J. A., Lee, R. H., and MacQueen, R. M.: 1974, *Proceedings Soc. Photo-optical Engineers* **44**, 133.
Moore, C. E., Minnaert, M. G. J., and Houtgast, J.: 1966, *The Solar Spectrum 2935 Å to 8770 Å*, Monograph 61. Dept. of Commerce, National Bureau of Standards, Wash., D.C.
Newkirk, G. A., Jr. and Bohlin, J. D.: 1965, *Ann. Astrophys.* **28**, 234.
Nikolskii, G. M.: 1964, *Geomag. and Aeronomy* **4**, 209 (English Edition 4, 163).
Saito, H. and Billings, D. E.: 1964, *Astrophys. J.* **140**, 760.
Shain, G. A.: 1947, *Izv. Krymsk. Astrofiz. Obs.* **1**, 102.
Suemoto, Z.: 1963, *Pub. Astron. Soc. Japan* **15**, 531.
Tanaka, K.: 1971a, *Publ. Astron. Soc. Japan* **23**, 185.
Tanaka, K.: 1971b, *Publ. Astron. Soc. Japan* **23**, 217.
Thomas, R. N. and Athay, R. G.: 1961, *Physics of the Solar Chromosphere*, Interscience, New York.
Tousey, R., Bartol, J.-D., Bohlin, J. D., Breuckner, G. E., Purcell, J. D., Sherrer, V. E., Sheeley, N. R., Schumacher, R. J., and van Hoosier, M. E.: 1973, *Solar Phys.* **33**, 365.
Vernazza, J. E.: 1972, Thesis Harvard Univ.Vernazza, J. E. and Noyes, R. W.: 1972, *Solar Phys.* **22**, 358.
Weart, S.: 1972, unpublished.
White, O. R.: 1963, *Astrophys. J.* **138**, 1316.

CHAPTER VI

ANALYTICAL METHODS FOR SPECTROSCOPIC DATA

1. Scattering Continuum

A. ELECTRON SCATTERING

A variety of evidence indicates that the continuum radiation from the corona and the upper chromosphere arises from electron scattering of photospheric radiation. This evidence includes the following points: (1) the coronal continuum has approximately the same color (spectral distribution) as the photospheric spectrum including line blocking. (2) The coronal continuum is partially plane polarized with the electric vector parallel to the limb. (3) The continuum intensity in the corona decreases with height at the same rate as would be expected if the emissivity were proportional to the density. (4) The Fraunhofer absorption lines are missing in the coronal continuum.

It is immediately evident from the low surface brightness of the corona that the corona is transparent in the continuum. To a reasonable approximation, the coronal surface brightness, F_c, due to scattering of photospheric radiation is given by

$$F_c = \tau_c F_p,\tag{VI-1}$$

where τ_c is the optical thickness of the corona and F_p is the photospheric surface brightness. Since F_c/F_p is observed to be less than 10^{-5}, it follows that $\tau_c < 10^{-5}$. Thus, multiple scatterings are unimportant and we may treat the corona as a single scattering medium.

Equation (VI-1) is not sufficiently accurate for a detailed analysis of the coronal continuum and does not include the polarization properties of the radiation. Hence, it is necessary to consider the problem in more detail. The problem of coronal scattering has been discussed at length by a number of authors including van de Hulst (1950) and Billings (1966). Since the problem is one of geometry, the derivation of the scattering equations is not very instructive and will not be repeated here.

Van de Hulst (1950) finds for the tangentially and radially polarized components of the coronal intensity in erg cm^{-2} s^{-1}:

$$F_{t\nu}(r) = C_\nu \int_r^\infty n_e(\rho) A(\rho) \frac{\rho \, d\rho}{(\rho^2 - r^2)^{\frac{1}{2}}}\tag{VI-2}$$

and

$$F_{r\nu}(r) = C_\nu \int_r^\infty n_e(\rho) \left\{ \left(1 - \frac{r^2}{2}\right) A_\nu(\rho) + \frac{r^2}{\rho^2} B_\nu(\rho) \right\} \frac{\rho \, d\rho}{(\rho^2 - r^2)^{\frac{1}{2}}},\tag{VI-3}$$

where the constant $C_\nu = (8\pi/3)R_\odot \sigma F_{p\nu} = 4.33 \times 10^{-13} F_{p\nu}$ cm^3, $F_{p\nu}$ is the *average* surface brightness of the Sun's disk at frequency ν and per unit solid angle σ is the Thompson scattering cross section and $A_\lambda(\rho)$ and $B_\lambda(\rho)$ are tabulated functions depending on the limb darkening of the disk at frequency ν (van de Hulst, 1950). The observed coronal brightness is

$$F_\nu(r) = F_{t\nu}(r) + F_{r\nu}(r), \tag{VI-4}$$

and the degree of polarization is

$$P_\nu(r) = \frac{F_{t\nu}(r) - F_{r\nu}(r)}{F_\nu(r)}. \tag{VI-5}$$

Most observations of the coronal continuum brightness are made in broadband and only averages over wavelengths of $F_t(r)$ and $F_r(r)$ are obtained. The functions C_ν, $A_\nu(r)$ and $B_\nu(r)$ are readily averaged.

Analysis of the continuum data gives directly $n_e(\rho)$, or, equivalently $n_e(r)$, provided a geometry is imposed on the distribution of n_e. Most analyses are made assuming that the distribution of $n_e(r)$ is spherically symmetric. Values of $n_e(r)$ obtained in this way are very valuable for understanding the nature of the upper chromosphere and coronal model.

B. RAYLEIGH SCATTERING

In the low and middle chromosphere ·Rayleigh scattering by netural hydrogen may add significantly to the continuum brightness relative to electron scattering. The ratio of intensity due to Rayleigh scattering to that due to electron scattering is given by (Unsold, 1955) as

$$\frac{n_H}{n_e}\left(\frac{1026}{\lambda}\right)^4, \qquad \lambda \text{ in Å.}$$

At $\lambda 5000$ this ratio is $1.8 \times 10^{-3} n_H/n_e$. Hence, the contribution of Rayleigh scattering is important only when n_H exceeds n_e by over a factor of 10^2.

2. Bound-Free Continua: Effectively Thin Case

A. H⁻

Since we have defined the limb of the Sun by the condition that the continuum optical thickness be unity in the tangential direction, it follows that in the temperature minimum region and above the tangential optical thickness is less than unity. The surface brightness may therefore be written as (see Figure I-2):

$$F_\nu(h) = \int_{-\infty}^{\infty} j_\nu(h)\, dy, \tag{VI-6}$$

where $j_\nu(h)$ is the emissivity. Throughout the temperature minimum region and the chromosphere we shall assume that $j_\nu(h)$ may be written in the form

$$j_\nu(h) = j_\nu(0) \sum_i a_i e^{-\beta_i h}. \tag{VI-7}$$

We then use Equation (I-4) to obtain

$$j_\nu(y) = j_\nu(0) \sum_i a_i e^{-\beta_i y^2/2R_\odot} e^{-\beta_i h_0}. \tag{VI-8}$$

and from Equation (VI-6) we find

$$F_\nu(h) = j_\nu(0) (2\pi R_\odot)^{1/2} \sum_i a_i \beta_i^{-1/2} e^{-\beta_i h_0}. \tag{VI-9}$$

In the particular case where Equation (VI-7) is sufficiently accurate with a single exponential term only, we note that Equation (VI-9) becomes simply

$$F_\nu(h) = \left(\frac{2\pi R_\odot}{\beta}\right)^{1/2} j_\nu(h). \tag{VI-10}$$

For the H^- ion, the absorption coefficient is usually tabulated as a cross section for absorption per neutral hydrogen atom and per unit electron pressure and is computed under the assumption of local thermodynamic equilibrium. This latter assumption is not valid at small optical depths, so we cannot use the tabulated results without appropriate modification.

We may write, with complete generality,

$$j_\nu \equiv \kappa_\nu S_\nu, \tag{VI-11}$$

where S_ν is the source function. In order to introduce a departure from local thermodynamic equilibrium in such a way as to relate S_ν to the Planck function B_ν, we express the population density of the H^- ion in terms of the Saha equation with a departure coefficient b_{H^-}. Thus, we write

$$n_{H^-} = \left(\frac{h^2}{2\pi mk}\right)^{3/2} \frac{n_e n_H T^{-3/2}}{4} b_{H^-} e^{\chi_{H^-}/kT}. \tag{VI-12}$$

With this introduction of b_{H^-}, the bound-free H^- source function for the total solid angle becomes

$$S_\nu = \frac{8\pi h \nu^3}{c^2} \frac{1}{b_{H^-} e^{h\nu/kT} - 1}. \tag{VI-13}$$

The bound-free absorption coefficient $\kappa_\nu(bf)$ may be written

$$\kappa_\nu(bf) = \alpha_\nu(bf) n_H n_e T, \tag{VI-14}$$

where $\alpha_\nu(bf)$ is the tabulated bound-free cross-section (Geltman, 1962) and is a function of T as well as a function of ν. For T in the range $5000°$ to $10\,000°$ and for λ near 5000 Å, $\alpha_\nu(bf)$ varies approximately as $T^{-3.6}$.

Equations (VI-11), (VI-12) and (VI-13) yield

$$j_\nu(bf) = \frac{8\pi h\nu^3}{c^2} \frac{kn_H n_e T\alpha_\nu(bf)}{(b_{H^-}e^{h\nu/kT} - 1)}. \tag{VI-15}$$

The departure coefficient b_{H^-} has been computed by Gebbie and Thomas (1970) for conditions somewhat approximating those in the solar atmosphere. Representative values are given in Table VI-1. However, the reader should note that b_{H^-} is dependent upon the model atmosphere through n_H, n_e and T, and that the results in Table VI-1 are illustrative only.

In the visual and violet spectral regions bound-free absorption predominates over free-free absorption for H^-. Also, stimulated emissions are relative unimportant in these spectral regions. Hence, we see that the H^- emission is reduced approximately by the amount $b_{H^-}^{-1}$ from its LTE value. The free-free absorption of H^- is discussed in Section 3.

TABLE VI-1
Representative values of b_{H^-}

τ_s	10^{-2}	10^{-3}	10^{-4}	10^{-5}
n_H	1.7×10^{16}	3.8×10^{15}	4.7×10^{14}	7.0×10^{13}
n_e	2.1×10^{12}	6.4×10^{11}	4.6×10^{11}	4.7×10^{11}
T	4720	4755	5025	5495
b_{H^-}	1.0088	1.0383	1.295	1.854

It is evident from the form of Equation (VI-15) that the H^- bound-free emission is a strong function of temperature since both n_H and n_e are strong functions of temperature. Analysis of the continuum data in regions where H^- emission is important will therefore provide a sensitive indication of the temperature as well as of the density. Because the H^- emission is proportional to the neutral hydrogen density n_H and because hydrogen is predominantly neutral in the low chromosphere and photosphere, the H^- emission must decay rapidly with height. It is evident from Figure V-2 that the H^- emission, in fact, is of little importance at heights above 500 km.

B. BALMER AND PASCHEN BOUND-FREE CONTINUA

Estimates of the chromospheric opacity in the Balmer lines of hydrogen based both on model calculations and on empirical evidence from eclipse data yield values for the population density of hydrogen atoms in the $n = 2$ level of excitation lying between 10^6 cm^{-3} and 10^5 cm^{-3} in the low chromosphere (cf. Thomas and Athay, 1961). Also, the scale height, H, is of the order of 10^8 cm. From Equation (I-7), the effect path length in the horizontal direction is $(2\pi RH)^{1/2} = 6.6 \times 10^9$ cm. Thus, a cm^2 column tangentially through the low chromosphere contains between 6.6×10^{15} and 6.6×10^{14} atoms in the $n = 2$ level. The absorption cross section at the head of the Balmer continuum is

1.4×10^{-17} cm^2, and we see that the optical thickness of the chromosphere is considerably less than unity. We shall consider the Balmer continuum (also the Paschen) therefore as being optically thin in the chromosphere. It is only necessary in this case to consider the emission coefficients.

Emission coefficients in the bound-free continua of hydrogen are well known and are given by

$$j_\nu = 2.15 \times 10^{-32} g_{fb} n^{-3} z^4 n_e n_p T^{-3/2} e^{-h(\nu - \nu_0)/kT} \tag{VI-16}$$

where g_{fb} is the Gaunt factor (Menzel and Pekeris, 1935), z is the net charge and n is the principal quantum number of the level.

Near the head of the Balmer continuum we take $g_{fb} = 0.88$, which gives

$$j_{\nu B}(\lambda \approx 3600) = 2.34 \times 10^{-33} n_e n_p T^{-3/2}. \tag{VI-17}$$

For the Paschen continuum, we take $g = 0.95, 0.97, 0.99$ and 1.0 at $\lambda = 6000, 5000, 4000$ and 3600, respectively. This gives

$$j_{\nu p}(\lambda = 6000) = 7.6 \times 10^{-34} n_e n_p T^{-3/2} e^{-6500/T}$$
$$j_{\nu p}(\lambda = 5000) = 7.7 \times 10^{-34} n_e n_p T^{-3/2} e^{-11100/T}$$
$$j_{\nu p}(\lambda = 4000) = 7.9 \times 10^{-34} n_e n_p T^{-3/2} e^{-18500/T} \tag{VI-18}$$
$$j_{\nu p}(\lambda = 3600) = 8.0 \times 10^{-34} n_e n_p T^{-3/2} e^{-22300/T}.$$

Although it appears superficially that the bound-free continuum intensity is a slow function of temperature near the series limit, this is not the case. Both n_e and n_p are strong functions of temperature since they are products of the ionization of hydrogen.

Analyses of Balmer continuum data in the chromosphere observed at eclipse provide, in conjunction with the electron scattering and H$^-$ continua, very convenient and very sensitive determinations of density and temperature. We note from Figure V-3 that the Balmer continuum flux is well determined between heights of about 500 km and 2500 km.

It is of interest at this point to note that the ratio of Balmer continuum emissivity to H$^-$ emissivity at λ 3640 is given approximately by

$$\frac{j_{\nu B}}{j_{\nu H^-}} (\lambda\, 3640) = \frac{2.34 \times 10^{-33} n_e n_p T^{-3/2}}{8.6 \times 10^{-42} n_e n_H T} b_{H^-} e^{39600/T}$$

$$= 2.7 \times 10^6 \frac{n_p b_{H^-} e^{39600/T}}{n_H T^{5/2}} \tag{VI-19}$$

In thermodynamic equilibrium the Saha equation for hydrogen ionization gives

$$n_H = \left(\frac{h^2}{2\pi mk}\right)^{3/2} \frac{n_e n_p}{2} T^{-3/2} e^{157000/T}, \tag{VI-20}$$

where we assume that all of the hydrogen is in the ground state. This equation gives

$$\frac{n_p}{n_H} = \frac{0.48 \times 10^{16} T^{3/2}}{n_e e^{157000/T}}, \tag{VI-21}$$

and from Equation (VI-19) we find

$$\frac{j_B}{j_{H^-}} (\lambda\, 3640) = 1.3 \times 10^{22} b_{H^-} n_e^{-1} T^{-1} e^{-117000/T}. \tag{VI-22}$$

Since n_e is not a strong function of height in the low chromosphere the right hand side of Equation (VI-22) is primarily a function of temperature. As a mean value for $n_e T$ in the low chromosphere, we take 10^{15}, and we set $b_{H^-} = 1$. Equation (VI-22) then gives ratios of j_B/j_{H^-} of 0.026, 9 and 450 at $4000°$, $5000°$ and $6000°$, respectively. Thus, within the temperature range expected between the temperature minimum and the low chromosphere, the ratio j_B/j_{H^-} at $\lambda\, 3640$ will change by orders of magnitude. The assumptions we have made are not quite accurate, of course, but this simple calculation serves to illustrate the radically different behavior between the H^- continuum evidenced in Figure V-2 and the Balmer continuum shown in Figure V-3. It appears that the H^- continuum intensity decreases by over two orders of magnitude between 0 and 500 km, whereas the Balmer continuum intensity may even increase in this same height range.

C. HELIUM BOUND-FREE CONTINUA

The chromosphere is optically thick in the bound-free resonance continuum of He I at $\lambda\, 504$. However, optical depth unity is reached in the middle chromosphere where the temperature is about $7000°$, and the observed intensity is much too high for radiation from the middle chromosphere. It is far more likely, therefore, that the He I bound-free continuum is formed in the high chromosphere where the temperature is higher and where He I is appreciably ionized. There is insufficient helium in the Sun for the high chromosphere to have optical thickness unity in the helium continua, so, again, we may treat the optically thin case.

The absorption coefficient in the He I continuum is (Allen, 1963)

$$\alpha_\nu = 7.6 \times 10^{-18} \left(\frac{\nu_0}{\nu}\right)^2. \tag{VI-23}$$

Hence, the emission coefficient is given by

$$j_\nu = n_1^* \alpha_\nu B_\nu = 7.6 \times 10^{-18} n_1^* \left(\frac{\nu_0}{\nu}\right)^2 \frac{8\pi h \nu^3}{c^2} e^{-h\nu/kT}, \tag{VI-24}$$

where n_1^* is the ground state population of He I in thermodynamic equilibrium. We write for n_1^*

$$n_1^* = \left(\frac{h^2}{2\pi mk}\right)^{3/2} \frac{n_e n_{II} T^{-3/2}}{4} e^{h\nu_0/kT}, \tag{VI-25}$$

so that j_ν becomes

$$j_\nu = 7.6 \times 10^{-18} \left(\frac{h^2}{2\pi mk}\right)^{3/2} \frac{2\pi hc}{\lambda_0^3} \left(\frac{\lambda_0}{\lambda}\right) n_e n_{II} T^{-3/2}$$

$$= 3.05 \times 10^{-32} \left(\frac{\lambda_0}{\lambda}\right) n_e n_{II} T^{-3/2} e^{-h(\nu-\nu_0)/kT}. \tag{VI-26}$$

The quantity n_{II} denotes the density of He II ions, and j_ν is in units of erg s^{-1} dν^{-1}. For He II, the bound-free emissivity is given directly by Equation (VI-16) as

$$j_\nu = 2.75 \times 10^{-31} n_e n_\alpha T^{-3/2} e^{-h(\nu-\nu_0)/kT}, \tag{VI-27}$$

where n_α is the density of alpha particles and where we have set $g = 0.8$. The He II resonance continuum occurs at λ 228, and, like the He I continuum, it is optically thin.

3. Free-Free Continua

Free-free absorption due to H$^-$ is the principal source of continuum opacity from wavelengths of a few microns to several hundred microns. There is no evidence that the emissivity and absorptivity of H$^-$ in the free-free transitions departs from local thermodynamic equilibrium. Thus, we may write

$$j_\nu = \kappa_\nu B_\nu. \tag{VI-28}$$

John (1966) gives κ_ν as

$$\kappa_\nu = \alpha_\lambda n_H kT n_e \tag{VI-29}$$

with

$$\alpha_\lambda = 5 \times 10^{-27} \left(\frac{6300}{T}\right)^{0.85} \lambda^2, \tag{VI-30}$$

where λ is in μm. Equations (VI-29) and (VI-30) give

$$\kappa_\lambda = 1.17 \times 10^{-39} n_H n_e T^{0.15} \lambda^2 \ (\mu\text{m}). \tag{VI-31}$$

In the radio wavelengths absorptions due to free-free transitions of electrons in the fields of positive ions dominates over H$^-$ absorptions. The absorption cross section in this case is (cf. Smerd and Westfold, 1949)

$$K_{i\lambda} = 1.98 \times 10^{-23} Z_i^2 g n_e n_i T^{-3/2} \lambda^2, \tag{VI-32}$$

where λ is in cm and where, for radio waves

$$g = 1.27(3.38 + \log T - \tfrac{1}{3} \log n_e). \tag{VI-33}$$

In the chromosphere $Z_i^2 n_i$ may be replaced by n_p, but in the transition region and in the corona where helium is doubly ionized the effective value of the sum of $Z_i^2 n_i$ is given by

$$\sum_i Z_i^2 n_i = n_p \left(1 + 4\frac{n\alpha}{np}\right) = 1.4 n_p. \tag{VI-34}$$

It is of interest to compare the relative values of κ_ν given by equations (VI-31) and (VI-32) for the chromosphere. We find

$$\frac{K(n_H - n_e)}{K(n_p - n_e)} = 0.6 \times 10^{-8} \frac{n_H T^{1.65}}{n_p g}. \tag{VI-35}$$

For $T = 6000°$ and $n_e = 10^{11}$, which are typical of the middle chromosphere, Equation (VI-33) gives $g = 4.4$ and the right hand side of Equation (VI-35) is equal to 2.5×10^{-3} n_H/n_p. Hence, as soon as hydrogen is ionized to where $n_H/n_p \lesssim 400$ the free-free opacity due to electron-proton interactions will dominate.

At mm and cm wavelengths the solar plasma is non-dispersive. However, at meter wavelengths the index of refraction is considerably less than unity and the corona is a dispersive medium.

The index of refraction in a plasma is given by

$$n^2 = 1 - \omega_p^2/\omega^2, \tag{VI-36}$$

where the plasma frequency ω_p is given by

$$\omega_p^2 = \frac{4\pi e^2 n_e}{m} \tag{VI-37}$$

and ω is given by

$$\omega = 2\pi c/\lambda. \tag{VI-38}$$

Hence, we find

$$n^2 = 1 - \frac{e^2}{\pi c^2 m} n_e \lambda^2$$

$$= 1 - 8.9 \times 10^{-14} n_e \lambda^2 \text{ (cm)}. \tag{VI-39}$$

Under coronal conditions n begins to depart from unity at the surface where $\tau_\lambda = 1$ at $\lambda \approx 10^2$ cm. Note from Equation (VI-32) that $K_\lambda \propto n_e n_i \lambda^2$, or, since $n_e \approx n_i$, $K_\lambda \propto n_e^2 \lambda^2$. As λ is decreased from λ_1 to λ_2, therefore, the surface of $\tau_\lambda = 1$ will move to a depth where n_e has increased by the approximate amount

$$\frac{n_{e2}}{n_{e1}} = \frac{\lambda_1}{\lambda_2} \tag{VI-40}$$

On the other hand, the change in $n_e\lambda^2$ is given by

$$\frac{n_{e2}\lambda_2^2}{n_{e1}\lambda_1^2} = \frac{\lambda_2}{\lambda_1}, \tag{VI-41}$$

and we see that the second term on the right hand side of Equation (VI-39) is decreased by a factor λ_2/λ_1. The change to non-dispersive propagation therefore takes place rather quickly at wavelengths below about 10^2 cm.

4. Total Line Intensities: Effectively Thin Case

A. TWO-LEVEL ATOM

Most of the data currently available for lines in the XUV give only the total intensities (equivalent width) of the lines, either as an average over the solar disk or as an average of some portion of the disk. Similarly, coronal and chromospheric lines observed at eclipse or with coronagraphs usually do not have sufficient spectral resolution to give accurate line profiles so we are again left with only the integrated line intensity.

The usual procedure for determining the equivalent width of a line starts with a computation of the monochromatic intensity, which is then integrated over the line profile. For emission lines, such a procedure is unnecessarily complex, and we may compute the total energy in the line without having to first determine the line profile.

To illustrate the computation of line intensities, we begin with the simplest cases and proceed to more difficult cases. Although it is not necessary to do so, we will formulate even the simplest cases in terms of quantities that are useful for the more difficult cases as well.

The first problem we consider is that of a two-level atom in a plane parallel atmosphere. We assume that excitation of a line may occur either through absorption of photons or by collisions with electrons. De-excitation occurs by the inverse processes of spontaneous emission and superelastic collisions. For algebraic convenience, we will ignore stimulated emissions. These can be included without much added complexity, but they are of relatively minor consequence in the solar case.

The quantity I_ν defined in Chapter I is the specific intensity per unit solid angle. Two moments of I_ν are convenient to introduce. These are

$$J_\nu = \int I_\nu \frac{d\omega}{4\pi} \tag{VI-42}$$

and

$$H_\nu = \int I_\nu \mu \frac{d\omega}{4\pi}, \tag{VI-43}$$

which define, respectively, the average intensity and the net flux. It is convenient, also, to express the opacity in the line as

$$K_\nu = \phi_\nu K_0,$$ (VI-44)

where K_0 is the opacity at line center and ϕ_ν is the shape of the line absorption coefficient normalized to unity at line center. A second normalization of ϕ_ν defined by

$$\int \Phi_\nu \, d\nu = 1$$ (VI-45)

normalizes Φ_ν to unit area under the profile. For the cases of interest here

$$\Phi_\nu = \frac{1}{\pi^{1/2} \Delta \nu_D} \phi_\nu.$$ (VI-46)

Throughout this book we will assume that ϕ_ν gives the shape of the emission coefficient as well as the absorption coefficient. This is equivalent to the assumption of complete redistribution or of completely non-coherent scattering. We assume, in other words, that photon absorption and re-emission are two distinct events and are uncorrelated in frequency. For a more complete discussion the reader is referred to Hummer (1962).

An added quantity of interest at this point is the 'escape coefficient', ρ, which we define as

$$\rho = 1 - \frac{\int J_\nu \Phi_\nu \, d\nu}{S_\nu}.$$ (VI-47)

The significance of this quantity will become apparent in a moment. Other authors have frequently referred to ρ as the net radiative bracket.

When we observe a stellar spectrum that is relatively steady in time, all that we really know is that a steady state has been reached. In the case of our hypothetical two-level atom, the steady state condition can be expressed by the equation

$$(C_{12} + B_{12} \int J_\nu \Phi_\nu \, d\nu)n_1 = (C_{21} + A_{21})n_2,$$ (VI-48)

where C_{12} and C_{21} are the electron transition rates between levels 1 and 2 due to electron collisions, B_{12} is the Einstein absorption coefficient and A_{21} is the Einstein spontaneous transition probability. B_{12} and A_{21} are related by

$$\frac{B_{12}}{A_{21}} = \left(\frac{2h\nu^3}{c^2}\right)^{-1} \frac{\omega_2}{\omega_1},$$ (VI-49)

and $\tilde{\omega}_2$ and $\tilde{\omega}_1$ are the statistical weights of levels 2 and 1. C_{12} and C_{21} are related by

$$\frac{C_{12}}{C_{21}} = \frac{\tilde{\omega}_2}{\tilde{\omega}_1} e^{-h\nu/kT}.$$ (VI-50)

Note that

$$\frac{C_{12}}{C_{21}} \frac{A_{21}}{B_{12}} = \frac{2h\nu^3}{c^2} e^{-h\nu/kT} = B_\nu,$$ (VI-51)

where, as before, B_ν is the Planck function (without stimulated emission, i.e., the Wein function). We also define

$$\frac{C_{21}}{A_{21}} = \epsilon,$$

(VI-52)

so that

$$\frac{C_{12}}{B_{12}} = \epsilon B_\nu.$$

(VI-53)

Using the above definitions, we rewrite Equation (VI-48) as

$$\frac{n_2}{n_1} = \frac{B_{12} \int J_\nu \Phi_\nu \, d\nu + C_{12}}{A_{21} + C_{21}}$$

$$= \frac{B_{12}}{A_{21}} \left(\frac{J_\nu \Phi_\nu \, d\nu + \epsilon B_\nu}{1 + \epsilon} \right),$$

(VI-53)

or

$$\frac{2h\nu^3}{c^2} \frac{\omega_1}{\omega_2} \frac{n_2}{n_1} = \frac{\int J_\nu \Phi_\nu \, d\nu + \epsilon B_\nu}{1 + \epsilon}.$$

(VI-54)

The emissivity of the two-level atom is

$$j_\nu = \frac{h\nu}{4\pi} A_{21} n_2 \Phi_\nu$$

(VI-55)

and the absorbtivity is

$$K_\nu = \frac{h\nu B_{12}}{4\pi} n_1 \Phi_\nu.$$

(VI-56)

Since we are dealing with emission lines, we ignore continuum, absorption and emission. Thus,

$$S = \frac{j_\nu}{K_\nu} = \frac{A_{21}}{B_{12}} \frac{n_2}{n_1} = \frac{2h\nu^3}{c^2} \frac{\tilde{\omega}_1}{\tilde{\omega}_2} \frac{n_2}{n_1},$$

(VI-57)

and from Equation (VI-54) we see that

$$S = \frac{\int J_\nu \Phi_\nu \, d\nu + \epsilon B}{1 + \epsilon},$$

(VI-58)

where we drop the ν subscript on B. Both S and B are regarded as being independent of ν over the small range in ν within the line profile.

Note from Equations (VI-47) and (VI-58) that

$$S = \frac{\epsilon B}{\rho + \epsilon},$$

(VI-59)

and

$$\rho S = \epsilon B - \epsilon S.$$

(VI-60)

Also, from Equations (VI-47) and (VI-57)

$$\rho = 1 - \frac{B_{12} n_1 \int J_\nu \Phi_\nu \, d\nu}{A_{21} n_2}.$$

(VI-61)

We are now in a position to derive convenient expressions for the net flux in an emission line. By operating on Equation (I-16) with the operator $\int d\omega / 4\pi$, we obtain

$$\frac{dH_\nu}{d\tau_\nu} = J_\nu - S.$$

(VI-62)

We rewrite this as

$$\frac{dH_\nu}{d\tau_0} = \phi_\nu J_\nu - \phi_\nu S$$

(VI-63)

and integrate our frequency to obtain

$$\int \frac{dH_\nu}{d\tau_0} \, d\nu = \pi^{1/2} \Delta\nu_D (\int J_\nu \Phi_\nu \, d\nu - S)$$

$$= \pi^{1/2} \Delta\nu_D \rho S.$$

(VI-64)

We further rewrite this as

$$dH = \pi^{1/2} \Delta\nu_D \rho S \, d\tau_0,$$

(VI-65)

where

$$dH = \int dH_\nu \, d\nu.$$

We now integrate Equation (VI-65) over depth and use Equation (VI-60) to replace ρS. This yields

$$\int_0^{\tau_0} dH = \pi^{1/2} \int \Delta\nu_D \epsilon (B - S) \, d\tau_0.$$

(VI-66)

In situations where $\epsilon \ll 1$, Equation (VI-58) requires that $S \ll B$ unless $\int J_\nu \Phi_\nu \, d\nu = B$. This latter condition is fulfilled deep in an atmosphere where the line photons are effectively trapped and cannot escape to outer space. However, if the photons can escape and if $\epsilon \ll 1$, there is no mechanism to prevent the quantity $\int J_\nu \Phi_\nu d\nu$ from falling much below B, and we will find $S \ll B$. This point will be clearly demonstrated in the next section of this chapter. Here, we wish only to note that in the external layers where the photons are not trapped we may approximate $\epsilon(B-S)$ by ϵB in cases where ϵ is small. Since photon escape is just as effective in reducing S below B regardless of where in the line the escape occurs, the escape of photons will continue to be of importance even when $\tau_0 \gg 1$ and escape occurs primarily in the line wings.

We define the 'effectively thin' condition as the condition that $S \ll B$ which implies that photon escape is dominant in determining the magnitude of S. Equation (VI-66) then gives

$$\int_0^{\tau_0} dH = \pi^{1/2} \int_0^{\tau_0} \Delta\nu_D^{1/2} \, \epsilon B \, d\tau_0. \tag{VI-67}$$

It follows from Equation (VI-53) and from the definition of $d\tau_0$ that

$$\int_0^h dH = \frac{h\nu}{4\pi} \int_0^h C_{12} n_1 \, dh. \tag{VI-68}$$

Hence the net flux in the line arising between the physical heights 0 and h is just given by the total number of collisional excitations in the line. Note that the effectively thin case under discussion may still be valid for $\tau_0 \gg 1$ when $\epsilon \ll 1$ and is much less restrictive than the 'optically thin' condition, which requires $\tau_0 \ll 1$.

As an estimate of ϵ let us assume a collisional de-excitation cross section of πa_0^2 and a mean electron velocity of 5×10^7 cm s^{-1} ($T \approx 10\,000°$). The downward collision rate is then given approximately by

$$C_{21} = 5 \times 10^7 \, \pi a_0^2 n_e = 4.4 \times 10^{-9} n_e.$$

In the chromosphere $n_e \approx 2 \times 10^{11}$, so $C_{21} \approx 10^3$. Thus, any line formed in the chromosphere for which $A_{21} \gg 10^3$ will satisfy the condition $\epsilon \ll 1$. Similarly, in the low corona $n_e \approx 3 \times 10^8$ and the mean electron velocity is about 5000 km s^{-1}. This gives $C_{21} \approx 10$ and any line for which $A_{21} \gg 10$ will again satisfy the condition $\epsilon \ll 1$. These conditions for $\epsilon \ll 1$ can be seen to include many 'forbidden' lines as well as the permitted lines.

It is of interest at this point to note that in the two-level approximation the line flux depends only on the product $n_1 C_{12}$. So-called, forbidden lines have the property that A_{21} is small, but they do not have the property that C_{12} is small. In electron collisions, the two electrons may exchange places and the normal selection rules include both electrons. The exchange cross sections are usually an appreciable fraction of the total cross section, and, as a result, the collisional excitation rates for forbidden transitions are of the same order of magnitude as for permitted transitions. Equation (VI-68) predicts, therefore, that forbidden lines will have comparable intensities to permitted lines. The low transition probability for the forbidden lines means only that n_2 will increase to a large number. It is a characteristic of the coronal spectrum that the forbidden and permitted lines do, in fact, have comparable intensities.

At densities higher than those found in the corona, e.g., the chromosphere, the line intensities increase for both permitted and forbidden lines. However, the forbidden lines are effectively thin only when τ_0 is relatively small, and the effectively thin approximation fails much quicker for forbidden lines than for permitted lines. As a result, the permitted lines grow in intensity much more than do the forbidden lines. An added effect arises from the break down of the two-level approximation in the forbidden lines. The large build up of the n_2 population in these lines enhances the liklihood that the $n = 2$ level will interact with other excited levels. Thus, an enhanced collision rate may not lead to a corresponding increase in the n_2 population. Instead, the electrons excited to the $n = 2$ level may be removed to other levels rather than return to the $n = 1$ level. These effects account for the predominance of permitted lines over forbidden lines in the chromospheric spectrum.

Not all spectral lines can be adequately represented by the simple two-level approximation we have used. In fact, the majority of lines formed in stellar atmospheres cannot be treated so simply. However, for the majority of the stronger lines formed in the upper chromosphere and corona the approximation is reasonably good.

It must be remembered, of course, that the net flux H, as we have defined it, is not the observed flux. The observed flux includes H plus any scattered radiation that may be present. If the contribution of scattering is unimportant, Equation (VI-48) reduces to

$$n_1 C_{12} = A_{21} n_2 \qquad\qquad (VI-69)$$

and Equation (VI-68) becomes

$$\int_0^h dH = \frac{h\nu}{4\pi} \int A_{21} n_2 \, dh, \qquad\qquad (VI-70)$$

i.e., H is given simply by the total number of emissions and is now equal to the observed flux, F.

The usual approximation made in the low corona is to use Equation (VI-68) with $H = F$, i.e.,

$$\int_0^h dF = \frac{h\nu}{4\pi} \int_0^h C_{12} n_1 \, dh. \qquad\qquad (VI-71)$$

In application, this equation is transformed in a variety of ways. One usually expresses n_1 as

$$n_1 = \frac{n_1}{n_i} \frac{n_i}{n_t} \frac{n_t}{n_H} \frac{n_H}{n_e} n_e \qquad\qquad (VI-72)$$

where n_i is the total density of ions of which level 1 is the ground state, n_t is the total density of ions of all stages of ionizations for the particular atomic specie and n_H is the total hydrogen density (neutral plus ionized). We shall write

$$\frac{n_1}{n_i}\frac{n_i}{n_t} = g(T, n_e),$$

$$\frac{n_t}{n_H} = A$$

and

$$n_H = \chi n_e.$$

The function $g(T, n_e)$ is yet to be specified, A is the abundance of the element in question and χ gives the ratio of n_H to n_e including electrons from all sources. We assume χ to be constant. Also, we write C_{12} as

$$C_{12} = Q_0 f(T) n_e.$$

With these definitions, Equation (VI-71) becomes

$$\int_0^h dF = \frac{h\nu}{4\pi} A Q_0 \chi \int_0^h G(T, n_e) n_e^2 \, dh. \tag{VI-73}$$

For many ions $G(T, n_e)$ is a slow function of n_e and exhibits a sharp maximum as a function of T. Thus, for these cases it is reasonable to write Equation (VI-73) as

$$F = \frac{h\nu}{4\pi} A Q_0 \chi \langle G(T, n_e) \rangle \int_{h(T_1)}^{h(T_2)} n_e^2 \, dh \tag{VI-74}$$

where the average of $G(T, n_e)$ is taken between the temperature limits T_1 and T_2. In practice it is found that $T_2/T_1 \approx 2$, and $\langle G(T, n_e) \rangle \approx G\langle (T_2 + T_1)/2, n_e \rangle$. The integral in Equation (VI-74) is the so-called 'emission measure' introduced by Pottasch (1964). Using the equation in this form, one normally interprets F as being the emission from a layer of atmosphere where $T_1 \leqslant T \leqslant T_2$. It is implicit in this interpretation that T is a monotonic function of n_e.

A variation of Equation (VI-73) of particular use in the transition region is obtained by the substitutions

$$n_e^2 = (n_e T)^2 T^{-2}$$

and

$$dh = \left(\frac{dT}{dh}\right)^{-1} dT.$$

One then assumes that $n_e T$ and dT/dh are approximately constant in the layers between T_1 and T_2. In this case Equation (VI-73) becomes

$$F = \frac{h\nu}{4\pi} A Q_0 \chi (n_e T)^2 \left(\frac{dT}{dh}\right)^{-1} \int_{T_1}^{T_2} G(T,n_e) T^{-2} \, dT. \qquad \text{(VI-75)}$$

This latter form carries the same implication that T is a monotonic function of n_e. The justification for assuming $n_e T = \text{const}$ is that the transition region is thin relative to the pressure scale height. Such an assumption is well supported by current models of the transition region. The assumption $dT/dh = \text{const}$ is much more serious. Within the transition region dT/dh is proportional to $T^{-5/2}$ (see Chapter VII), and if $T_2/T_1 = 2$, $(dT/dh)_1/(dT/dh)_2 \approx 5.6$. In spite of this difficulty, Equation (VI-75) is useful for estimating the average value of dT/dh between T_1 and T_2.

As we shall later show, there is good reason to believe that the conductive energy flux

$$F_c = kT^{5/2} \frac{dT}{dh}$$

is constant, or nearly so, through much of the transition. If we assume that both F_c and $n_e T$ are constant, Equation (IV-75) becomes

$$F = \frac{h\nu}{4\pi} \frac{A Q_0 k}{F_c} (n_e T)^2 \int_{T_1}^{T_2} G(T,n_e) T^{-1/2} \, dT. \qquad \text{(VI-76)}$$

Equations (VI-73), (VI-75) and (VI-76) appear to be quite different in character. In practice, however, they give very similar results. The function $G(T,n_e)$ varies sufficiently rapidly with T that it controls the value of the integral.

Jefferies et al. (1972) have called attention to the fact that F can be expressed in terms of a distribution function $\mu(T,n_e)$ defined by

$$dN_e(T,n_e) = N_e \mu(T,n_e) \, dT \, dn_e, \qquad \text{(VI-77)}$$

where N_e is the total columnar density of n_e in the line of sight. The normalization of $\mu(T,n_e)$ is such that

$$\int_0^\infty \int_0^\infty \mu(T,n_e) \, dT \, dn_e \equiv 1.$$

With this introduction of $\mu(T,n_e)$, Equation (VI-75) becomes

$$F = \frac{h\nu}{4\pi} A Q_0 \chi N_e \int_0^\infty \int_0^\infty G(T,n_e) \mu(T,n_e) \, dn_e \, dT. \qquad \text{(VI-78)}$$

Given enough data of reliable quality it is possible, in principle, to determine the function $\mu(T,n_e)$. However, in practice this has not proven practicable because of insufficient data and Jefferies et al. (1972) revert to a form

$$F = \frac{h\nu}{4\pi} A Q_0 \chi N_e \int_0^\infty G(T,n_e)\phi(T)\,dT \qquad\qquad\qquad \text{(VI-79)}$$

where

$$\phi(T) = \int_0^\infty \mu(T,n_e)\,dn_e. \qquad\qquad\qquad\qquad\qquad \text{(VI-80)}$$

Equation (VI-79) is of the same form as Equation (VI-75). However, there is the important distinction that $\phi(T)$ is related directly to dT/dh only if there is a monotonic relationship between T and n_e. The function $\phi(T)\,dT$ simply gives the fraction of n_e that lies between T and $T + dT$ regardless of the atmospheric geometry. Any information about geometry (including the determination of dT/dh) requires the imposition of additional constraints that are derived from observational data.

B. THREE-LEVEL ATOM

When more energy levels are added to the model atom, the photon energy in one spectral line interacts with the photon energy in other lines of the same atom. This can, and often does, lead to a more complex situation than is adequately described by the simple two-level case. As a convenient and useful illustration of the effects of adding more energy levels, we consider the case of three levels, numbered 1, 2, and 3 in order of increasing excitation energy.

Equation (VI-65) is still valid for the multi-level case, but the expression for S must be modified to include the interaction terms. For the three-level case, we write

$$R_{11}n_1 = R_{21}n_2 + R_{31}n_3 \qquad\qquad\qquad\qquad \text{(VI-81)}$$

and

$$R_{22}n_2 = R_{12}n_1 + R_{32}n_3. \qquad\qquad\qquad\qquad \text{(VI-82)}$$

R_{ij} is defined by

$$R_{ij} = C_{ij} + B_{ij} \int J_\nu \Phi_\nu d\nu. \qquad\qquad\qquad\qquad \text{(VI-83)}$$

Similarly,

$$R_{jj} = \sum_{i \neq j} R_{ij}, \qquad\qquad\qquad\qquad\qquad \text{(VI-84)}$$

Solution of Equations (VI-81) and (VI-82) yields

$$\frac{n_2}{n_1} = \frac{R_{12}R_{33} + R_{13}R_{32}}{R_{21}R_{33} + R_{31}R_{23}}, \qquad\qquad\qquad \text{(VI-85)}$$

$$\frac{n_3}{n_1} = \frac{R_{13}R_{22} + R_{12}R_{23}}{R_{31}R_{22} + R_{21}R_{32}}, \qquad\qquad\qquad \text{(VI-86)}$$

and

$$\frac{n_3}{n_2} = \frac{R_{23}R_{11} + R_{21}R_{13}}{R_{32}R_{11} + R_{12}R_{31}}.$$ (VI-87)

We next divide the equation for n_j/n_i by $A_{ji}R_{kk}$, where $k \neq i, j$, and rewrite the equations in the form

$$\frac{n_j}{n_i} = \left(\frac{2h\nu^3}{c^2}\right)^{-1} \frac{\tilde{\omega}_j}{\tilde{\omega}_i} \frac{\int J_\nu \Phi_\nu \, d\nu + \epsilon_{ji}^* B}{1 + \epsilon_{ji}^\dagger}.$$ (VI-88)

The quantities ϵ_{ji}^* and ϵ_{ji} are given by

$$\epsilon_{ji}^* = \epsilon_{ji} + \frac{R_{ik}R_{kj}e^{h\nu/kT} \dfrac{\tilde{\omega}_i}{\tilde{\omega}_j}}{A_{ji}R_{kk}},$$ (VI-89)

and

$$\epsilon_{ji}^\dagger = \epsilon_{ji} + \frac{R_{ki}R_{jk}}{A_{ji}R_{kk}}.$$ (VI-90)

Note that ϵ_{ji}^* and ϵ_{ji} each contain terms involving J_ν for at least one of the other transitions.

The source functions for the three transitions may now be written in the general form

$$S_{ji} = \frac{\int J_\nu \Phi_\nu \, d\nu + \epsilon_{ji}^* B}{1 + \epsilon_{ji}^\dagger}.$$ (VI-91)

Hence, Equation (VI-65) gives

$$dH_{ji} = \pi^{1/2} \Delta\nu_D (\epsilon_{ji}^* B - \epsilon_{ji}^\dagger S_{ji}).$$ (VI-92)

In each of the above equations, B has the implicit indices ji as does J_ν, Φ_ν and ν.

It is of interest to apply Equations (VI-89), (VI-90) and (VI-92) to problems of particular interest. One such problem arises in the case of coronal ions in the He I isoelectronic sequence. Transitions in these ions from 2^3P to 1^1S and from 2^3S to 1^1S are known, in the literature, as the intercombination and forbidden lines, respectively. We will designate 2^3P as level 3, 2^3S as level 2 and 1^1S as level 1. Transitions between 2^3S or 2^3P and excited singlet levels will be ignored.

Transition probabilities in the forbidden and intercombination lines increase markedly as Z increases. For the coronal ions, these transitions are moderately strong. The 2^3P to 2^3S transition is permitted and is strong in all cases. In applying the general equations for the three-level case to these particular transitions, we denote $A_{ij} = B_{ij} \int J_\nu \Phi_\nu \, d\nu$ and we set $A_{32} \gg C_{32}$, $C_{12} \gg A_{12}$, $C_{13} \gg A_{13}$, $A_{21} \gg C_{21}$ and $A_{31} \gg C_{31}$. The second and third of these conditions are equivalent to setting $J_\nu = 0$ for the two transitions, which, in turn, is equivalent to setting $\rho_{ji} = 1$. For $\rho_{ji} = 1$, $\rho_{ji} \gg \epsilon_{ji}^{\dagger}$ and $\rho_{ji} S_{ji} = \epsilon_{ji}^* B$. Hence, we must have

$$dH_{ji} = \pi^{1/2} \Delta \nu_D \epsilon_{ji}^* B \, d\tau_0 \tag{VI-93}$$

to the same approximation.

The intercombination and forbidden lines have close to the same wavelength and we may equate the values of $\pi^{1/2} \Delta \nu_D B$ for the two lines with reasonable accuracy. In this case, the ratio of intensities in the two lines is given by

$$\frac{dH_{31}(I)}{dH_{21}(F)} = \frac{(C_{13}R_{22} + R_{12}R_{23})}{(C_{12}R_{33} + R_{13}R_{32})} \frac{d\tau_{31}}{d\tau_{21}}, \tag{VI-94}$$

where we have made use of the identity

$$\epsilon_{ji} A_{ji} e^{-h\nu/kT} \frac{\omega_j}{\omega_i} \equiv C_{ji} e^{-h\nu/kT} \frac{\omega_j}{\omega_i} \equiv C_{ij}, \tag{VI-95}$$

Since the intercombination and forbidden lines have the same ground state and close to the same wavelengths, we may set

$$\frac{d\tau_{31}}{d\tau_{21}} = \frac{A_{31}}{A_{21}}. \tag{VI-96}$$

With this simplification plus the inequalities between C_{ji} and A_{ji} stated earlier, Equation (VI-94) becomes

$$\frac{dH_{31}(I)}{dH_{21}(F)} = \left[\frac{C_{13}(A_{21} + C_{23} + A_{23}) + C_{12}(C_{23} + A_{23})}{C_{12}(A_{32} + A_{31}) + A_{32}C_{13}} \right] \frac{A_{31}}{A_{21}}$$

$$= \frac{A_{21} + \gamma(A_{23} + C_{23})}{A_{21}(\gamma\alpha - 1)} \tag{VI-97}$$

with

$$\gamma = \frac{C_{12} + C_{13}}{C_{13}} \tag{VI-98}$$

and

$$\alpha = \frac{A_{31} + A_{32}}{A_{31}} \tag{VI-99}$$

Both γ and α are independent of electron density. Thus, the intensity ratio of the lines, as given by the final form of Equation (VI-97), depends upon density only through the term C_{23}. Under coronal conditions A_{23} and C_{23} are often competitive with C_{23} dominating in regions of high density and A_{23} dominating in regions of low density. The line ratios for the He I isoelectronic sequence are useful, therefore, as indicators of density in the high density regions.

Note that γ is of the order of 2 and α is of the order of A_{32}/A_{31}. Thus, $\gamma\alpha - 1$ is of the order of A_{32}/A_{31}, and the right side of Equation (VI-98) is of the order of $(A_{31}/A_{21})\langle(A_{21} + A_{23} + C_{23})/A_{32}\rangle$. If the two lines were each approximated by the two-level case, the ratio of line intensities would be just A_{31}/A_{21}. Hence, the approximate factor $(A_{21} + A_{23} + C_{23})/A_{32}$ may be interpreted as a correction factor for the influence of the interactions on the intercombination line. Because A_{32} is strongly permitted, this correction factor is less than unity and the flux in the intercombination line is reduced. The reduction results, of course, from the escape of electrons from level 3 into level 2 via the permitted 3-2 transition. In effect, a large population can accumulate in level 2 but not in level 3.

A second problem of similar nature occurs in the case of Lyman-α (Lα) and Lyman-β (Lβ) in the chromosphere. In an optically thin gas with negligible incident radiation, the Lα to Lβ flux ratio is correctly given by Equation (VI-97). For this case, however, $C_{12} \gg C_{13}, A_{31} \approx A_{32}$ and $A_{21} \gg \gamma(A_{23} + C_{23})$. Hence,

$$\frac{dH(\beta)}{dH(\alpha)} \approx \frac{C_{13}}{2C_{12}}, \tag{VI-100}$$

i.e., approximately half the electrons excited to $n = 3$ cascade 3-2-1.

If we now imagine that the optical thickness of the atmosphere increases, the Lα photons will begin to entrap resulting in an increase in J_ν in Lα. On the other hand, so long as the atmosphere has low opacity in Hα the Lβ photons cannot be entrapped. At every second scattering, on the average, an Lβ photon will degenerate to an Hα and Lα pair. Since the Hα photon readily escapes the atmosphere, the process is not balanced by its inverse.

Under the latter set of conditions Equation (VI-97) is no longer valid. However, the assumptions on which this equation is based are incorrect only in the statement that $C_{12} \gg A_{12}$. If we reverse this inequality, we need only replace C_{12} with A_{12} in Equation (VI-98).

During the initial buildup of Lα photons due to entrapment, A_{21} will remain larger than $\gamma(A_{23} + C_{23})$. The flux ratio, in this case, is given by

$$\frac{dH(\beta)}{dH(\alpha)} = \frac{1}{\gamma\alpha} \approx \frac{C_{13}}{2A_{12}}. \tag{VI-101}$$

Thus, the Lα flux increases in direct proportion to the increase in J_ν in Lα whereas the Lβ flux remains near its optically thin value.

Even if we view an optically thick medium, the regions from which the observed Lα and Lβ fluxes arise will be optically thin in Hα and we may expect Equation (VI-101) to be approximately correct. In the case of the solar chromosphere, the observed flux ratio is Lβ/L$\alpha \approx 0.01$. The collision rate ratio, by comparison, is $C_{13}/C_{12} \approx 0.06$ at a temperature of 2×10^4 K.

5. Profiles of Lines and Bound-Free Continua: Effectively Thick Case

A. A FORM OF THE LINE TRANSFER EQUATION FOR NON-COHERENT SCATTERING

Radiation in bound-bound and bound-free transitions in stellar atmospheres is not formed in local thermodynamic equilibrium as a general rule. By definition, a stellar atmosphere is that part of a star from which photons escape relatively easily. Those photons that escape no longer interact with the atmosphere, and the escape process, therefore, influences the photon equilibrium. Whenever a bound state is involved so that the photon distribution is restricted to a narrow band of frequencies within which the opacity varies markedly, the effect of escape on the equilibrium distribution may be large.

In the preceding section of this chapter, we have discussed specific cases of extreme departures from local thermodynamic equilibrium. There we discussed several cases in which the mean radiation intensity, J_ν, was negligibly small. For the two-level case, this results in the condition expressed by Equation (VI-69). This is to be contrasted with the case of thermodynamic equilibrium (or local thermodynamic equilibrium) for which the conditions

$$n_1^* C_{12} = n_2^* C_{21} \qquad\qquad\qquad\qquad\qquad \text{(VI-103)}$$

and

$$n_1^* B_{12} \int J_\nu \Phi_\nu \, d\nu = n_2^* A_{21} + n_2^* B_{21} \qquad\qquad\qquad \text{(VI-104)}$$

must hold in detail. The ratio of n_1^*/n_2^* to n_1/n_2 as given by Equations (VI-103) and (VI-69) is

$$\frac{n_1^*}{n_2^*} \frac{n_2}{n_1} = \frac{C_{21}}{C_{12}} \frac{C_{12}}{A_{21}} = \epsilon \qquad\qquad\qquad\qquad \text{(VI-105)}$$

and we note from Equation (VI-58) that this is just the result

$$\frac{S}{B} = \epsilon. \qquad\qquad\qquad\qquad\qquad\qquad \text{(VI-106)}$$

Since the conditions leading to Equations (VI-105) and (VI-106) completely ignore J_ν, these equations evidently reflect the extreme departures from local thermodynamic equilibrium that may occur. By ignoring J_ν, we have set the escape coefficient ρ equal to unity, and, hence, we have considered the case where eventual photon escape is a certainty. It is a characteristic of stellar atmospheres that ϵ is small. Even in the solar photosphere, for example, C_{21} is of the order

$$C_{21} \approx \pi a_0^2 v_e n_e = 8.8 \times 10^{-17} \times 4 \times 10^7 \times 6 \times 10^{13}$$

$$= 2.1 \times 10^5 \ \text{s}^{-1}.$$

A line with a transition probability of $10^8 \ \text{s}^{-1}$ therefore has an ϵ of approximately 2×10^{-3}. In the chromosphere and corona, ϵ is smaller still. When ϵ is small and ρ is allowed to approach unity, Equation (VI-59) reduces to Equation (VI-106). This is the case for which photon escape is completely dominant.

We note from Equation (VI-59) that for $\epsilon \ll 1$ we find

$$\frac{S}{B} = \frac{\epsilon}{\rho}. \tag{VI-107}$$

Thus, even when $\rho \ll 1$, S may be appreciably less than B. For example, if $\epsilon = 10^{-4}$, S will be less than B whenever $\rho > 10^{-4}$. To have $\rho = 10^{-4}$ it is necessary to allow only one photon to escape for every 10^4 transitions, and it is clear, therefore, that photon escape may be important even when the optical depths are quite large.

Two effects may be present to drive S closer to B than is given by the extreme result of Equation (VI-106). One has just been mentioned, viz., an increase in optical thickness will increase the local value of J_ν and, hence, will decrease ρ. The second arises from multilevel (interlocking) effects, which may raise the effective values of ϵ^* and $\epsilon^\dagger$ defined by Equation (VI-91). Since the basic reason for a difference between S and B is the escape of photons from the atmosphere, the interlocking effects in multilevel atomic models cannot eliminate the problem. The interlocking effects may modify the ratio of S/B, but they may decrease it as well as to increase it.

We have defined the effectively thin condition for the two-level atom in terms of the escape of line photons. If a photon eventually escapes the atmosphere before it is reconverted to thermal energy, we say that the atmosphere is effectively thin. As we shall show for the two-level case, this is synonymous with the statements

$$\rho > \epsilon \quad \text{and} \quad S < B.$$

For the two-level case, there is no problem with this definition of effectively thin. However, for the multilevel case photon escape is complicated by the fact that the photons are now shared by the system of lines and the probability of escape will be greater in some lines than in others. Thus, the statements that photons escape and $S \neq B$ are not synonymous, and we must choose between them for the definition of effective thinness. The customary choice, which we adopt, defines effectively thin as the condition $S \neq B$. The alternative condition of photon escape, where 'escape' includes escape through an interlocked line, is referred to as photon degradation.

The effectively thick case is defined by the condition that the optical thickness is sufficiently large that $S = B$ at some depth. This implies that at some wavelengths the radiation temperature is equal to the kinetic temperature. There are neither lines nor bound-free continua for which the corona is effectively thick. The chromosphere, however, is effectively thick in some of the strongest transitions in the more abundant elements.

In order to discuss the effectively thick case, it is necessary to couple the radiative transfer equations with the steady state equations for the atom and to solve the resultant set of equations simultaneously. We refer to this coupled set of equations as the 'kinetic equilibrium' equations. The coupling occurs through J_ν or, equivalently, through ρ.

Solutions are performed numerically on large computers. The solutions give $S(\tau_0)$, which is sufficient to determine both line profiles and equivalent widths. As input, it is necessary to specify $T, n_e,\ n_H$ and ϕ_v as functions of height. However, n_e and h_H can be determined self-consistently with the solution of the kinetic equilibrium equations for hydrogen and the hydrostatic equilibrium equations.

A complete discussion of the established methods for solving the kinetic equilibrium equations or a comprehensive review of the results obtained from such solutions is beyond the scope of this book. Indeed, such a discussion requires a full text. The reader is referred for this purpose to texts by Athay (1972a), Mihalas (1970), and Jefferies (1968).

Our discussion here of the kinetic equilibrium approach to the effectively thick problem will be restricted to a solution of the two-level problem for a few idealized cases and to a brief review of a few multilevel solutions for a model solar atmosphere. These examples will be sufficient to indicate the general nature of the results, but will not be exhaustive in any way.

Throughout the remainder of this section and the following Sections 5b, c and d we shall treat the line source function as being formed by non-coherent scattering. This renders the line source function frequency independent and allows the emission coefficient to be equated to the absorption coefficient at each frequency. The total source function, which is a weighted average of the line and continuum source functions, is still frequency dependent, but frequency dependence from this source does not particularly complicate the transfer problem. Cases where the line source function itself becomes frequency dependent through the mechanism of coherent scattering will be discussed in Section 5e together with the case of the bound-free continua.

Equation (VI-91) may be rewritten in the form

$$\int J_v \Phi_v \, dv = S_L + \epsilon^\dagger S_L - \epsilon^* B, \tag{VI-108}$$

where the subscripts j and i have been dropped for convenience and where S_L now denotes the line source function. The mean intensity J_v is related to S_L via the lambda transform

$$J_v = \Lambda(\tau_v) S_v \tag{VI-109}$$

where S_v is the weighted average of the line and continuum source function given by

$$S = \frac{\phi_v}{\phi_v + r_0} S_L + \frac{r_0}{\phi_v + r_0} B \tag{VI-110}$$

and where

$$\Lambda(\tau_v) S_v = \tfrac{1}{2} \int_{\tau_v}^{\infty} S_v E_1(t - \tau_v) \, dt + \tfrac{1}{2} \int_0^{\tau_v} S_v E_1(\tau_v - t) \, dt. \tag{VI-111}$$

The quantity r_0 is defined by

$$r_0 = \frac{d\tau_c}{d\tau_0},$$ (VI-112)

where $d\tau_c$ and $d\tau_0$ are, respectively, the continuum and line center elements of optical depth. The exponential integrals are defined by

$$E_n(x) = \int_1^\infty \frac{e^{-xt}}{t^n} dt.$$ (VI-113)

Integrals of order n are related to those of order $n - 1$ by the recursion relation

$$E_n(x) = \frac{e^{-x} - xE_{n-1}(x)}{n - 1}$$ (VI-114)

and the derivative relation (cf. Kourganoff, 1963)

$$\frac{d}{dx} E_n(x) = - E_{n-1}(x).$$ (VI-115)

Equation (VI-109) is a form of the equation of radiative transfer and, when combined with Equation (VI-108), provides a coupling between the transfer equation and the statistical equilibrium equation. It is understood, of course, that S_L, B, ϵ^* and $\epsilon^\dagger$ are each functions of τ. Also, ϵ^* and $\epsilon^\dagger$ depend explicitly on the values of J_ν for transitions other than the one in question. Thus, Equation (VI-108) represents a set of simultaneous equations involving all of the transitions that interlock with each other. In practice, this set of equations must be solved self-consistently. A number of techniques for doing this have been developed (cf. Mihalas, 1970, Athay, 1972a). Here, we shall treat only one particular method for solving Equations (VI-108) and (VI-109) for a given transition.

The reader should note that Equation (VI-108) ignores the influence of stimulated emissions. The inclusion of stimulated emissions introduces an added term in $\epsilon^\dagger$ containing $\int J_\nu \Phi_\nu \, d\nu$ for the transition being treated. This, in turn, introduces additional non-linearities in the problem. However, in most cases, the stimulated emission terms can be handled without adding substantially to the difficulty of the numerical problem. We omit them here in order to simplify the algebraic formulation.

Equation (VI-110) can be rewritten in the form

$$S_\nu = \left(1 - \frac{r_0}{\phi_\nu + r_0}\right) S_L + \left(\frac{r_0}{\phi_\nu + r_0}\right) B,$$ (VI-116)

which combines with Equations (VI-108) and (VI-109) to give

$$S_L + \epsilon^\dagger S_L - \int \Phi_\nu \Lambda(\tau_\nu) \left(1 - \frac{r_0}{\phi_\nu + r_0}\right) S_L \, d\nu$$

$$= \epsilon^* B + \int \Phi_\nu \Lambda(\tau_\nu) \left(\frac{r_0}{\phi_\nu + r_0}\right) B \, d\nu.$$ (VI-117)

This equation can be solved numerically by dividing frequency and depth spaces into discrete sets of points and performing the indicated operations numerically.

The lambda transform of a vector A, of order N, can be represented by a matrix of order N. Let us assume, for example, that A is a step function with values that are constant between adjacent depth points. Specifically, we take A to have the value A_i between the depth points τ_b and τ_a defined by

$$\tau_b = (\tau_i \tau_{i+1})^{1/2} \qquad \text{(VI-118)}$$

and

$$\tau_a = (\tau_{i-1} \tau_i)^{1/2}. \qquad \text{(VI-119)}$$

For optical depths $\tau_j < \tau_a$, the contribution of A_i to J_j is given by

$$J_j(A_i) = \tfrac{1}{2} \int_{\tau_a}^{\tau_b} A_i E_1(t - \tau_j)\, dt$$

$$= \frac{A_i}{2} \left[E_2(\tau_a - \tau_j) - E_2(\tau_b - \tau_j) \right]. \qquad \text{(VI-120)}$$

Similarly, for $\tau_a < \tau_j < \tau_b$, i.e., $\tau_j = \tau_i$,

$$J_j(A_j) = \tfrac{1}{2} \int_{\tau_j}^{\tau_b} A_j E_1(t - \tau_i)\, dt + \tfrac{1}{2} \int_{\tau_a}^{\tau_j} A_j E_1(\tau_j - t)\, dt$$

$$= A_j - \frac{A_j}{2} \left[E_2(\tau_j - \tau_a) + E_2(\tau_b - \tau_j) \right] \qquad \text{(VI-121)}$$

and for $\tau_j > \tau_b$

$$J_j(A_i) = \tfrac{1}{2} \int_{\tau_a}^{\tau_b} A_i E_1(\tau_j - t)\, dt$$

$$= \frac{A_i}{2} \left[E_2(\tau_j - \tau_b) - E_2(\tau_j - \tau_a) \right]. \qquad \text{(VI-122)}$$

It is clear from the form of Equations (VI-120), (VI-121) and (VI-122) that we may write J_ν in the form

$$J_\nu = S_\nu - S_\nu^* + \mathbf{E}_\nu S\nu, \qquad \text{(VI-123)}$$

where the vector S_ν^* has components

$$S_{\nu j}^* = \frac{S_{\nu j}}{2} \left[E_2(\tau_{\nu j} - \tau_{\nu a}) + E_2(\tau_{\nu b} - \tau_{\nu j}) \right] \qquad \text{(VI-124)}$$

and where the matrix E_v is of the form

$$E_v = \frac{1}{2} \begin{pmatrix} 0 & E_2(\tau_{va} - \tau_{vj}) - E_2(\tau_{vb} - \tau_{vj}) \\ E_2(\tau_{vj} - \tau_{vb}) - E_2(\tau_{vj} - \tau_{va}) & 0 \end{pmatrix}, \qquad \text{(VI-125)}$$

where j is the row index and i is the column index.

The integral of J_v over frequency, from Equation (VI-123), gives

$$\int J_v \Phi_v \, dv = \int S_v \Phi_v \, dv - \int S_v^* \Phi_v \, dv + \int E_v \Phi_v S_v \, dv, \qquad \text{(VI-126)}$$

which may be rewritten as

$$\int J_v \Phi_v \, dv = S_L - \delta S_L + \delta B - (P_j - P_j^c)S_L - P_j^c B + (X - Y)\mathcal{S}_L + YB, \qquad \text{(VI-127)}$$

where

$$\delta = \int \frac{\Phi_v r_0}{\phi_v + r_0} \, dv, \qquad \text{(VI-128)}$$

$$P_j = \frac{1}{2} \int \Phi_v [E_2(\tau_{vj} - \tau_{va}) + E_2(\tau_{vb} - \tau_{vj})] \, dv, \qquad \text{(VI-129)}$$

$$P_j^c = \frac{1}{2} \int \frac{r_0 \Phi_v}{\phi_v + r_0} [E_2(\tau_{vj} - \tau_{va}) + E_2(\tau_{vb} - \tau_{vj})] \, dv, \qquad \text{(VI-130)}$$

$$X_{ij} = \int \Phi_{vi} E_{vij} \, dv \qquad \text{(VI-131)}$$

and

$$Y_{ij} = \int \left(\frac{r_0 \Phi_v}{\phi_v + r_0} \right)_i E_{vij} \, dv. \qquad \text{(VI-132)}$$

It follows from Equations (VI-108) and (VI-127) that

$$(\epsilon^\dagger + \delta + P_j - P_j^c - X + Y)S_L = (\epsilon^* + \delta - P_j^c + Y)B, \qquad \text{(VI-133)}$$

or that

$$S_L = (\epsilon^\dagger + \delta + P_j - P_j^c - X + Y)^{-1} (\epsilon^* + \delta - P_j^c + Y)B. \qquad \text{(VI-134)}$$

The quantities ϵ, δ, P_j and P_j^c are represented here as diagonal matrices.

Equation (VI-134) represents the formal matrix solution for S_L. Alternatively, Equation (VI-133) may be solved by iteration, which, in many cases tested, proves to be faster than the matrix inversion. To solve Equation (VI-133) by iteration, rewrite it in the form

$$(-X + Y)S_{LX-1} = (\epsilon^* + \delta - P_j^c + Y)B - (\epsilon^\dagger + \delta + P_j - P_j^c)S_{LX}, \qquad \text{(VI-135)}$$

where S_{LX-1} is the previous iterate and S_{LX} is the current iterate. The matrices X and Y and the vectors P_j and P_j^c need be computed only once for a given set of tau. Hence, the iterations proceed rapidly and consume but little computing time. With the equations in this form, it is convenient to set $S_{L0} = 0$ as a starting solution.

In order to understand the nature of the transfer problem, it is helpful to consider the meaning of the various terms in Equation (VI-134) in somewhat more detail. For a multilevel atom, ϵ^* and $\epsilon^\dagger$ are complicated functions involving interlocking effects. In the particular case of a two-level atom, however,

$$\epsilon^* = \epsilon^\dagger = \epsilon = \frac{C_{21}}{A_{21}}$$

and ϵ has a simple interpretation. Thus, the probability that a photon absorption is followed by a collisional de-excitation is given by

$$\frac{C_{21}}{A_{21} + C_{21}} = \frac{\epsilon}{1 + \epsilon}.$$

Similarly, the probability that a photon absorption in the line is followed by a photon emission, which, in turn, is followed by a continuum capture is given by

$$\frac{A_{21}}{A_{21} + C_{21}} \int \Phi_y \frac{d\tau_c}{d\tau_y + d\tau_c} \, dy = \frac{1}{1 + \epsilon} \int \Phi_y \frac{r_0}{\phi_y + r_0} \, dy = \frac{\delta}{1 + \epsilon}.$$

The sum of the two probabilities is $(\epsilon + \delta)/(1 + \epsilon)$ and represents the probability that a line photon will be destroyed following its next absorption. Collisional de-excitation destroys the photon by converting it to thermal energy and continuum absorption destroys the line photon by removing it from the wavelength of the line.

To understand the meaning of the E_2 terms, consider the probability that a photon travels from depth i to depth j. For a photon at frequency ν and depth i traveling in direction μ, the probability for going from i to j is

$$e^{-(\tau_{\nu i} - \tau_{\nu j})/\mu}.$$

The probability for going in a direction between μ and $\mu + d\mu$ that is in the correct hemisphere is $d\mu/2$, and the average over μ is

$$\tfrac{1}{2} \int_0^1 e^{-(\tau_{\nu i} - \tau_{\nu j})/\mu} \, d\mu = \tfrac{1}{2} \int_1^\infty \frac{e^{-(\tau_{\nu i} - \tau_{\nu j})x} \, dx}{x^2} = \tfrac{1}{2} E_2 (\tau_{\nu i} - \tau_{\nu j}).$$

A further averaging over the line profile yields

$$\tfrac{1}{2} \int \Phi_\nu E_2 (\tau_{\nu i} - \tau_{\nu j}) \, d\nu,$$

which has the same form as the elements in the X matrix. Thus, the terms in P_j represent the probability for photons at τ_j to escape the interval τ_a to τ_b, which is centered on τ_j. Similarly, the off-diagonal elements represent the probabilities for photons to leave some other interval, say, τ_a' to τ_b', which is centered on τ_i and to subsequently arrive at τ_j. It can be seen, therefore, that P_j represents the loss of photons from the interval τ_a to τ_b to the rest of the atmosphere, whereas the X matrix represents the net gain of photons in the interval τ_a to τ_b from the rest of the atmosphere.

The quantities P_j^c and Y have similar meaning except that they represent the combined probabilities for photons to travel from i to j and to undergo continuum absorption.

Rather than pursue a discussion of Equation (VI-133) and its numerical solutions, we shall, instead, turn to a discussion of a much simpler equation whose solution bears remarkable similarities to the solutions of Equation (VI-133) and which has the advantage of having analytic solutions. Before writing this alternative equation, we define the escape probability P_e, and a mean scattering depth, N, to be used in the equation.

The probability, P_e, for a photon to escape from optical depth τ_j to the surface of the atmosphere is given by

$$P_e = \tfrac{1}{2} \int \Phi_v E_2(\tau_{jv})\, dv. \tag{VI-136}$$

It follows from the nature of the integrals involved in the definitions of P_e and P_j that most of the contributions to both P_e and P_j come from frequencies where the argument of E_2 is of the order of unity or less. It further follows that, to a reasonably good approximation, $P_j \approx 2P_e$.

Note that the escape probability, P_e, is defined quite differently from the escape coefficient, ρ. Thus, P_e is equivalent to the probability that an emission event *will not be followed by* an absorption event, whereas ρ is equivalent to the probability that an emission event *is not the result of* an absorption event. In particular, ρ depends upon both n_2 and n_1 whereas P_e depends only upon n_1. This distinction is particularly important near the surface of the atmosphere where ρ approaches a value $\epsilon^{1/2}$ as contrasted with $P_e \approx \tfrac{1}{2}$.

It follows from Equation (VI-59) that S_L saturates to B at a depth where $\rho \approx \epsilon$. If we set $r_0 = 0$ and $\epsilon^* = \epsilon^\dagger = \epsilon$ in Equation (VI-134), which conditions are analogous to those underlying Equation (VI-59), we find

$$S_L = (\epsilon + P_j - X)^{-1} \epsilon B.$$

At large optical depths, the terms in the X matrix are small compared to P_j, and, again to a reasonable accuracy, we may set $X = 0$. Thus, for large optical depths,

$$S_L \approx \frac{\epsilon B}{\epsilon + P_j} \approx \frac{\epsilon B}{\epsilon + 2P_e}.$$

This latter form suggests that S_L saturates to B near $P_e = \epsilon/2$ or near a depth where $P_e \approx \rho/2$.

We next define a mean scattering depth N to be used in place of τ. Since P_e is the probability for a single scattering event to lead to a photon escape, P_e^{-1} represents the number of scattering events required before an escape occurs. Since it is P_j that enters Equation (VI-134) rather than P_e, and since $P_j \approx 2P_e$, we define N by the relation

$$N = \frac{1}{2P_e}. \tag{VI-137}$$

Approximate expressions for $N(\tau_0)$ can be derived by replacing $E_2(\tau_\nu)$ with a step function of value zero for $\tau_\nu > x$ and unity for $\tau_\nu \leqslant x$. Thus, we find

$$P_e = \int_{y_x}^{\infty} \Phi_y \, dy, \tag{VI-138}$$

where y_x is the frequency at which $\tau_y = x$. For Φ_y of Gaussian form, P_e is just the error function and, for $y_x \gtrsim 1.5$, is given by

$$P_e = \frac{e^{-y_x^2}}{2\pi^{\frac{1}{2}} y_x}.$$

Since y_x is defined by

$$x = \tau_0 e^{-y_x^2},$$

we find

$$P_e = \frac{x}{2\pi^{\frac{1}{2}} \tau_0 \left[\ln \dfrac{\tau_0}{x} \right]^{1/2}}, \qquad \tau_0 \gtrsim 10.$$

The corresponding value of N for the Guassian profile is

$$N_G = \frac{\pi^{\frac{1}{2}} \tau_0 \left[\ln \dfrac{\tau_0}{x} \right]^{1/2}}{x}, \qquad \tau_0 \gtrsim 10.$$

For a dispersion profile of the form

$$\Phi_y = \frac{a}{\pi y^2}, \tag{VI-139}$$

a similar set of steps leads to

$$N_D = \left(\frac{\pi^{3/2} \tau_0}{4ax} \right)^{1/2}, \qquad \tau_0 \gtrsim \frac{10}{a}.$$

This latter condition on τ_0 applies to the case of a Voigt profile where the dispersion wings begin near $\tau_0 = 1/a$.

Figure VI-1 shows plots of $\log N$ vs $\log \tau_0$ for both Gaussian and Voigt profiles. The solid curves are drawn from numerical integration of Equation (VI-137), and the plotted circles and crosses are the empirically fitted relations

$$N_G = 1 + 4\tau_0 \left[\ln(\tau_0 + 1)\right]^{1/2} \tag{VI-140}$$

and

$$N_D = 2 \left(\frac{\tau_0}{a}\right)^{1/2} \tag{VI-141}$$

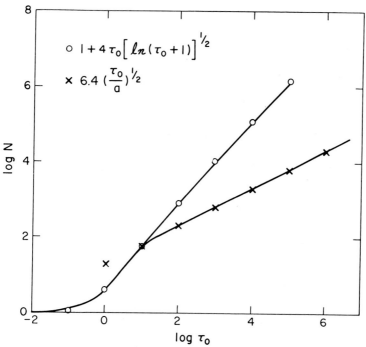

Fig. VI-1. The number of scattering steps, N, required for escape from optical depth τ_0 for a Doppler profile (O) and for a Voigt profile (X) with $a = 10^{-2}$.

Note that the empirically fitted curves are relatively accurate and that $N_G = \epsilon^{-1}$ near $\tau_0 = 10^{-1} \epsilon^{-1}$ and $N_D = \epsilon^{-1}$ near $\tau_0 = 0.25a/\epsilon^2$.

The value of τ_0 at which S_L saturates to B is commonly referred to as the *thermalization depth*, τ_{th}, and the preceding values of τ_0 provide a reasonably accurate measure of τ_{th}. The length τ_{th} gives the optical thickness of the boundary layer in which the photons are partially decoupled from the thermal energy of the atmospheric gases. This decoupling results from the escape of photons through the boundary.

In order to further clarify the nature of Equation (VI-133), we note, that each of the terms in the equation are in the nature of photon sources and sinks. For the two-level case, for example, ϵB and δB are the source terms arising from collisions and continuum emissions. These terms represent creation of photons from thermal energy. In a like manner, ϵS_L and δS_L represent photon sinks due to collisional and de-excitation and continuum absorption. Photon energy, in these cases, is reconverted to thermal energy. As we have seen, the terms in P_j, P_j^c, $\mathbf{X}$ and $\mathbf{Y}$ represent photon sinks and sources arising from the transport of photons from one point to another within the atmosphere. If there were no net sinks or sources due to transport within the atmosphere, the main effect of such transport would be represented by the gradient of S and the excape of photons from the surface. Thus, we should be able to write, quite generally, that

$$\Sigma \text{ sink rates} - \Sigma \text{ source rates} = \frac{dS}{dN}, \qquad (\text{VI-142})$$

provided the sink and source rates are given in appropriate units.

We have thus far identified three sink terms $(P_e + \epsilon + \delta)$ S_L and two source terms $(\epsilon + \delta)B$. Hummer (1964) has shown that, for the case $\Phi_y \neq \Phi_y(\tau_0)$, photon escape by a single step is indeed one of the primary photon sink terms. This results from the fact that photons travel large distances primarily by diffusing in frequency into the line wings where $\tau_v < 1$. Thus, they undergo many scatterings near line center, during which they travel only a short distance, then, by chance, they are emitted far enough from line center to escape the atmosphere. Obviously, this is not the exclusive means of photon transport and the neglect of all other processes of photon transport may lead to appreciable errors. Nevertheless, we shall proceed by assuming that

$$\frac{dS}{dN} = (p_e + \epsilon + \delta)S_L - (\epsilon + \delta)B, \qquad (\text{VI-143})$$

which we rewrite, using Equation (VI-137), as

$$\frac{dS}{dN} = \left(\frac{1}{2N} + \epsilon + \delta \right) S_L - (\epsilon + \delta)B. \qquad (\text{VI-144})$$

It has been demonstrated by the author (Athay, 1972b) that an equation of this form does, in fact, mimic quite closely the exact solution to Equation (VI-133). The factor ½ in the term involving S_L/N is important. Without this factor, the formal parallellism between Equations (VI-133) and (VI-144) is lost. It cannot be claimed that the arguments leading to the factor ½ in Equation (VI-137) are in any way rigorous. Also, the supposed derivation of the factor ½ in the author's (1972b) original discussion is unacceptable. A more satisfactory, but still nonrigorous, derivation has been given by Delache (1974). Thus, Equation (VI=144) should be regarded, at this time, as simply an empirical whose solution is known to mimic the solution to Equation (VI-133). We shall refer to Equation (VI-144), in the following, as the pseudo transfer equation. For convenience, we omit the continuum sink and source terms. To include these, it is sufficient to replace ϵ by $\epsilon + \delta$ in the subsequent discussion.

B. SOLUTION OF THE PSEUDO LINE TRANSFER EQUATION

The formal solution of Equation (VI-144) with $\delta = 0$ is

$$S = e^u \left[C \int_b^N \epsilon B e^{-u} \, dx \right],$$ (VI-145)

where

$$u = \frac{1}{2} \int_a^x \left(\frac{1}{\omega} + 2\epsilon \right) d\omega.$$ (VI-146)

Consider the case of $\epsilon = $ const. Equation (VI-146) then gives

$$u = \ln N^{1/2} + \epsilon N$$ (VI-147)

and

$$e^u = N^{1/2} e^{\epsilon N},$$ (VI-148)

where the constant a is chosen to be zero. Substitution of Equation (VI-148) into Equation (VI-145) yields

$$S = N^{1/2} e^{\epsilon N} \left[C - \epsilon \int_b^N B \frac{e^{-\epsilon x}}{x^{1/2}} \, dx \right],$$

or, with

$$\lambda = \epsilon N = t^2$$ (VI-149)

this latter equation for S becomes

$$S = N^{1/2} e^\lambda \left[C - 2\epsilon^{1/2} \int_b^{\lambda^{1/2}} B e^{-t^2} \, dt \right].$$ (VI-150)

As a boundary condition we require $S = B$ as N tends to infinity. This requires, in turn, that we choose $C = 0$ and $b = \infty$. Hence, we find

$$S = 2\lambda^{1/2} e^\lambda \int_{\lambda^{1/2}}^\infty B e^{-t^2} \, dt.$$ (VI-151)

For large t (large N) the depth variation of e^{-t^2} will completely dominate any normal variation of B and we may consider B as relatively constant. The integral in Equation (VI-151) is then the error function, which integrates for large t to give

$$\int_{\lambda^{1/2}}^\infty e^{-t^2} \, dt = \frac{1}{2} \frac{e^{-\lambda}}{\lambda^{1/2}}, \qquad \lambda \gg 1.$$ (VI-152)

Hence, we find $S = B$ as required.

Let us consider now the case of B = const. For $\lambda \ll 1$, i.e., for $P_e \gg \epsilon$, the lower limit on the integral in Equation (VI-151) can be taken as zero, and we find

$$\frac{S}{B} = \pi^{1/2}\lambda^{1/2}e^{\lambda} \approx \pi^{1/2}\lambda^{1/2}. \tag{VI-152}$$

At the surface $\tau_0 = 0$, $N = 1$ and $\lambda = \epsilon$. Hence the solution at the surface is

$$S/B = \pi^{1/2}\epsilon^{1/2}, \qquad \tau_0 = 0. \tag{VI-153}$$

We note from Equation (VI-140) that in the Doppler core of a line N varies as $\tau_0(\ln \tau_0)^{1/2}$. Thus, for some distance beyond $\tau_0 = 1$ we expect S/B to increase as $\tau_0^{1/2}(\ln \tau_0)^{1/4}$; or essentially as $\tau_0^{1/2}$. According to Equations (VI-140) and (VI-141) $N(\text{Gaussian}) = N(\text{wing})$ at $\tau_0 \ln \tau_0 \approx 1/a$. Near this value of τ_0, S/B will slow its increase from being proportional to $\tau_0^{1/2}$ to being proportional to $\tau_0^{1/4}$, provided, of course, that λ is still small.

Finally, we note from the form of Equation (VI-151) that the saturation of S to B occurs at a characteristic depth near $\lambda = 1$, i.e., when $2P_e \approx \epsilon$. This is the same as the saturation, or thermalization, depth discussed earlier.

Figure VI-2 illustrates the variation of S/B for the cases $\epsilon = 10^{-6}$ and $a = 0$ and $a = 10^{-2}$. Exact numerical solutions are given as well as the approximate solutions given by Equation VI-151.

It can be seen from Figure VI-2 that the solution for Equation (VI-144) is strikingly

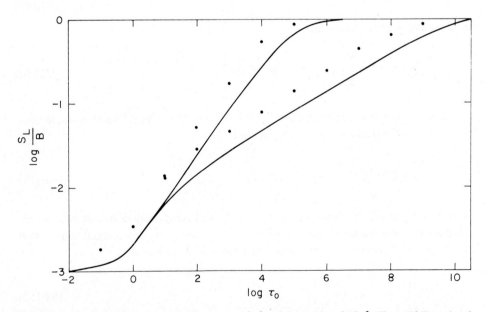

Fig. VI-2.   A plot of Equation (VI-161) for $\epsilon = 10^{-6}$ and for $a = 0$ and 10^{-2}. The solid lines give the exact numerical solutions to Equation (VI-133) for the corresponding cases. The lower set of curves is for the Voigt case ($a = 10^{-2}$).

similar to the solution for Equation (VI-133). The former solution gives values of S that are too high by a factor of $\pi^{\frac{1}{2}}$ at $\tau_0 < 1$ and by a factor of 2 for $\tau_0 \gg 1$. Similarly, it gives a thermalization depth that is too small by a factor of 4. Both solutions have a slope given by $S/B \propto \tau_0^{\frac{1}{2}} (\ln \tau_0)^{\frac{1}{4}}$ throughout most of the range $1 \lesssim \tau_0 \leqslant \tau_{th}$ or $1 \leqslant \tau_0 \leqslant 1/a$ depending upon whether $1/a$ is greater than or less than τ_{th}. In the range $1/a \lesssim \tau_0 \lesssim \tau_{th}$, $S/B \propto \tau_0^{\frac{1}{4}}$. For both Gaussian and Voigt profiles, the τ_0 dependence is correctly expressed by the solution (VI-152) giving $S \propto N^{\frac{1}{2}}$.

The drop in the slope of S/B from essentially $\tau_0^{\frac{1}{2}}$ to $\tau_0^{\frac{1}{4}}$ at depths where $\tau_0 > 1/a$ is a result of the dispersion wings added to the Voigt profile. The larger value of Φ_y in the wings results in a larger probability that photons will be emitted at values of y large enough for τ_0 to be less than unity. Hence, the stronger wings on Φ_y facilitate the escape of photons and decrease N at large depths. For $\tau_0 > 1/a$, photon escape occurs primarily in the dispersion wings.

The correspondence between the solutions for Equations (VI-133) and (VI-144) becomes almost exact if ϵ is replaced by $\epsilon/4$ in Equation (VI-144), or, equivalently, if N is replaced by $N/4$. However, we can find no logical basis for such a substitution, and we note, again that Equation (VI-144) is a completely heuristic analog of Equation (VI-133) and not a derived analog.

One purpose in introducing Equation (VI-144) is to utilize it to demonstrate the effect of a chromospheric contribution to B, which we do in the following section.

C. THE INFLUENCE OF A CHROMOSPHERE

A star that has a chromosphere will have a different visual spectrum from a star which is otherwise identical but which has no chromosphere. However, the differences in the visual spectra of the two stars may be quite subtle depending upon the precise nature of the chromosphere that is present. The influence of the chromosphere on the spectral lines arises in several ways. Three of the more important influences arise from: (a) the chromospheric temperature rise, (b) the relative increase in electron density associated with the temperature rise, and (c) the increase in Doppler width of the line absorption coefficient in the chromosphere.

An increase in temperature produces an increase in the Planck function, which we may characterize by setting

$$B = B_0(1 + Ae^{-c\tau_c}). \tag{VI-154}$$

For $\tau_c \gg c^{-1}$, this equation gives $B \approx B_0$, and for $\tau_c \ll c^{-1}$ it gives $B = B_0(1 + A)$. Thus A is a measure of the chromospheric rise in B, and c is a measure of the continuum optical depth where the rise occurs. We may rewrite Equation (VI-154) as

$$B = B_0(1 + Ae^{-\gamma N}), \tag{VI-155}$$

where, as before, N is the scattering depth and γ characterizes the reciprocal scattering depth at which the rise in B takes place. Both γ and N, of course, differ from line to line. Equations (VI-151) and (VI-155) yield

$$S = 2\lambda^{1/2}e^{\lambda}B_0 \left[\int_{\lambda^{1/2}}^{\infty} e^{-t^2} dt + A \int_{\lambda^{1/2}}^{\infty} e^{-\gamma N - t^2} dt \right],$$ (VI-156)

or with

$$t_1^2 = t^2 + \gamma N = t^2(1 + \gamma/\epsilon) = \lambda_1,$$ (VI-157)

$$S = 2\lambda^{1/2}e^{\lambda}B_0 \left[\int_{\lambda^{1/2}}^{\infty} e^{-t^2} dt + \frac{A}{(1 + \gamma/\epsilon)^{1/2}} \int_{\lambda_1}^{\infty} e^{-t_1^2} dt_1 \right].$$ (VI-158)

Consider the case where $\lambda_1 \ll 1$ and $\lambda \ll 1$. Equation (VI-158) then gives

$$\frac{S}{B_0} = \pi^{1/2}\lambda^{1/2}e^{\lambda} \left[1 + \frac{A}{(1 + \gamma/\epsilon)^{1/2}} \right],$$ (VI-159)

and we see that the chromospheric rise in B increases S/B_0 by a factor

$$1 + \frac{A}{(1 + \gamma/\epsilon)^{1/2}}$$

at depths where

$$\epsilon N(1 + \gamma/\epsilon) = \epsilon N + \gamma N \ll 1.$$

For cases where $\epsilon N + \gamma N \gg 1$, there is little or no influence from the chromospheric term. Also, if $\gamma \ll \epsilon$, S/B_0 is increased by a factor $(1 + A)$, i.e., S is fully coupled to the chromosphere through the region of temperature rise.

If we denote N_1 as the value of N at the base of the chromosphere, Equation (VI-159) predicts a chromospheric rise in S/B_0 by the approximate factor $\epsilon^{1/2}N_1^{1/2}(1 + A)$. For a Gaussian profile, this is equivalent to the factor

$$2\epsilon^{1/2}\tau_{01}^{1/2}(\ln \tau_{01})^{1/4}(1 + A),$$

and for a Voigt profile with $\tau_{01} > 1/a$ it is the equivalent of the factor

$$\epsilon^{1/2}(\tau_{01}/a)^{1/4}(1 + A).$$

At this point we again note that in all of the preceding discussion of Equation (VI-144) ϵ could be replaced by $\epsilon + \delta$. It is convenient in the following discussion to make this replacement.

The increase of S in the chromosphere by the above factors may result in a local maximum in S. In order to estimate the value of this maximum, we note from Equation (VI-158) that S is represented by a linear superposition of two terms, one of which represents the chromosphere. In this sense, the chromosphere acts essentially as a separate layer of finite optical thickness. When the chromosphere is effectively thin, i.e., $\tau_{01} \ll \tau_{th}$, all of the photons generated within the chromosphere escape without reconversion to thermal energy. Under such circumstances, the mean number of scatterings experienced by a photon is given by the ratio of the total number of emissions per unit time to the total number of created photons per unit time in a radial column through the chromosphere. The former quantity is given by

$$\int_0^\infty d\nu \int_0^{\tau_{01}} 4\pi\Phi_\nu S_\nu h\nu \, d\tau_0$$

and the latter is given by

$$\int_0^\infty d\nu \int_0^{\tau_{01}} 4\pi\Phi_\nu \frac{\epsilon + \delta}{1 + \epsilon} Bh\nu \, d\tau_0.$$

Since Φ_ν is rapidly varying in ν and S_L and B are each independent of ν, we may cancel the factors $4\pi h\nu$ and write

$$N_1 = \frac{\int_0^{\tau_{01}} (S_L - \delta S_L + \delta B) \, d\tau_0}{\int_0^{\tau_{01}} \frac{\epsilon + \delta}{1 + \epsilon} B \, d\tau_0}. \tag{VI-160}$$

If we take ϵ, δ and B to be constant and make use of the fact that most of the contribution to the integral

$$\int_0^{\tau_0} S_L \, d\tau_0$$

comes near $S_L = S_L(\text{max})$, we find

$$N_1 \approx \frac{S_L(\text{max})(1 - \delta) + \delta B_0(1 + A)}{(\epsilon + \delta)B_0(1 + A)}$$

or

$$S_L(\text{max}) \approx (\epsilon + \delta)B_0 N_1(1 + A) \tag{VI-161}$$

for $\delta \ll 1$ and $N_1 \gg 1$. For a Gaussian profile,

$$\frac{S_L(\text{max})}{B_0} \approx 4(\epsilon + \delta)\tau_{01}(\ln \tau_{01})^{1/2}(1 + A), \tag{VI-162}$$

and for a Voigt profile with $N_1 \gg 1/a$

$$\frac{S_L(\text{max})}{B_0} \approx 2(\epsilon + \delta)\left(\frac{\tau_{01}}{a}\right)^{1/2}(1 + A). \tag{VI-163}$$

Exact solutions of Equation (VI-133) for Gaussian profiles show that the scaling laws for $S_L(\max)/B$ are indeed of the form (VI-162) but with a numerical coefficient nearer to unity. Thus, we again find good agreement between these approximate solutions and the exact solutions if N is reduced by a factor ¼.

Exact solutions of Equation (VI-133) for $A = 10$ and various combinations of ϵ, r_0 and a are shown in Figure (VI-3). Note that for these cases $\tau_{01} = \tau_{c1}r_0^{-1} = 10^{-4}r_0^{-1}$. Thus, the dependence of the solutions upon r_0 reflects the dependence upon τ_{01}. The value $A = 10$ corresponds to a temperature rise from 4300° to 6000° at λ 4000.

The maxima in the ratio S/B shown in Figure VI-3 are typical of the source function behavior for most strong chromospheric lines. As previously noted, this, more often than

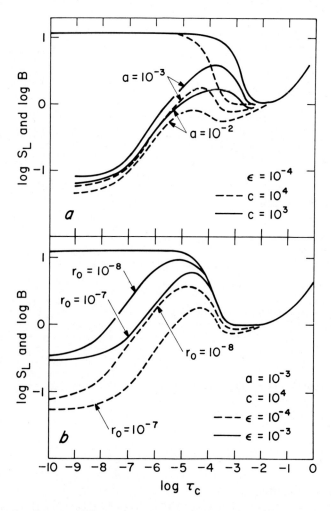

Fig. VI-3.   Exact solutions of Equation (IV-133) for two values of each of the parameters C, ϵ, a, and r_0. The ratio r_c/r_0 gives the value of τ_0 in the line at which the chromospheric rise in B occurs. B increases in this case from a minimum value of unity to a maximum value of 11.

not, results in S itself exhibiting a maximum at some value of $\tau_0 > 1$. A consequence of this is that the emergent line profile has a central reversal, i.e., the intensity first increases away from line center and passes through a maximum before falling to the continuum level.

The profiles shown in Figures V-35 through V-38 contain a variety of cases where the central reversal is clearly evident. A number of factors may tend to obscure the central reversal. Interlocking effects, for example, may remove the maximum in S, and even when S has a maximum, gradients in the broadening velocity in ϕ_ν may suppress the maxima in the observed profile. These effects will receive further discussion in subsequent chapters.

The increase in electron density associated with the increase in temperature in the chromosphere will produce an increase in ϵ for lines for which collisional destructions are important. This increase in ϵ increases the coupling of S to the chromosphere in a manner analogous to an increase in τ_0.

An increase in Doppler width with height in the chromosphere inhibits the escape of line photons and increases the value of N at the base of the chromosphere. This, in turn, decreases γ and increases the chromospheric coupling between S and B. The effects of increased electron density and/or increased Doppler width in the chromosphere can be of comparable importance to the increase in temperature.

The illustrations in Figure VI-3 exhibit the effects of a temperature rise with ϵ and $\Delta\lambda_D$ held constant. In Figure VI-4 we show the effect of increasing the Doppler width according to the relation

$$\Delta\lambda_D = \Delta\lambda_D(0)[1 + 2e^{-(c\tau_0)^{1/2}}]. \tag{VI-164}$$

Both ϵ and T are held constant, i.e., there is no chromospheric rise in B. Values of c are indicated in the figure. The profile of Φ_y for this illustration has a Voigt form with $a = 10^{-3}$. Transfer in the Gaussian core dominates for $\tau_c \ll 10^{-3}$ and transfer in the damping wings dominates for $\tau_c \gg 10^{-3}$.

Chromospheric influences on spectral lines in the visual spectrum can exhibit themselves in a variety of ways including greater residual intensity, emission reversals, and greater line widths. Most strong solar lines exhibit the first and third of these effects and only a few, notably H and K of Ca II, exhibit the emission reversals. Even in the case of H and K, the chromosphere is effectively thin. Emission reversals occur when S has a local maximum, as in the examples shown in Figure VI-3. When the maximum in S is weak or nonexistent, the line simply has higher residual intensity.

The reader should note that the chromospheric influence on lines is enormously weaker than it would be if the lines were formed in local thermodynamic equilibrium. Conversely, if one infers chromospheric structure assuming that the lines are formed in local thermodynamic equilibrium the magnitude of the chromospheric temperature rise will be grossly under-estimated. This is clearly illustrated by a comparison of computed and observed profiles for the λ 3734 line of Fe I shown in Figure VI-5. Profiles computed under the assumptions of local thermodynamic equilibrium (LTE) are shown in Part a of

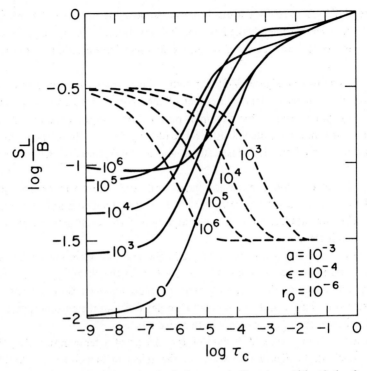

Fig. VI-4.   The effect on S_L of an outwards increase in Doppler width of the form given by
Equation (VI-164). Values of C are indicated on the dashed curves, which show the relative increase in
$\delta\lambda_0$ on a linear scale. The corresponding curves for S_L/B are labeled with the values of C also. No
chromospheric rise in B is included in this illustration.

the figure. The cores of the lines are of chromospheric origin and are totally
misrepresented by the LTE assumption. Profiles computed for kinetic equilibrium (KE)
in which the coupled steady-state and radiative transfer equations are solved simultaneously are shown in Part b of the figure. These profiles give good agreement with the
observed profiles in the line cores. They were computed using a model iron atom
containing 15 energy levels (Athay and Lites, 1972; Lites, 1972).

Emission reversals in the H and K lines of Ca II are a well recognized chromospheric
phenomenon. As is the case for the Fe I line illustrated in Figure VI-5, the Ca II lines
would exhibit enormously strong chromospheric emission cores if they were formed in
LTE. Without the chromosphere the emission reversals would disappear and the entire
line core would be darker than is observed. Similarly, the core of the λ 3734 line shown
in Figure VI-5 would be much darker if the chromosphere were absent. The same is true
of most of the stronger solar Fraunhofer lines.

The influence of a corona on the spectral lines is similar qualitatively to the influence
of a chromosphere. Quantitative'y, however, the two effects are quite different. The
coronal ions do not exist in any appreciable quantity in the chromosphere and
photosphere and the corona canrot be regarded as an extension of an optically thick

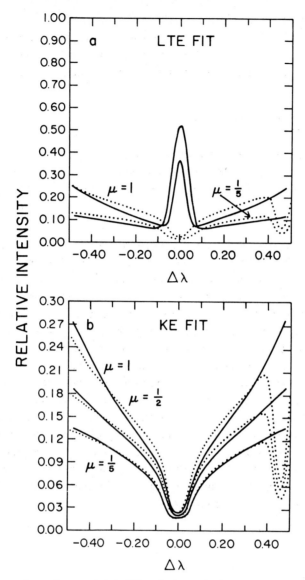

Fig. VI-5.   A comparison of computed (solid) and observed (dots) profiles for the λ 3734 line of
Fe I in the solar spectrum using a model atmosphere closely approximating the HSRA. Note the
change in the vertical scale in the kinetic equilibrium fit. The strong emission core found in the local
thermodynamic equilibrium case is typical of what could be expected in many strong lines if LTE
were valid (after Lites, 1972).

atmosphere. Instead, the corona must be treated as an optically thin shell. For this case,
we may ignore J and write the source function simply as

$$S = \epsilon B = \epsilon A B_0.$$ (VI-165)

A temperature rise from $6000°$ to 1.6×10^6 K gives $A \approx 10^{4.2}$ at λ 4000, $A \approx 10^{6.5}$ at λ 2000 and $A \approx 10^{11.4}$ at λ 1000. To a crude approximation, $\epsilon \approx 10^{-7}$ for coronal permitted lines, and we see that ϵA exceeds unity only for $\lambda < 2000$. The intensity of emission from the corona, assuming that it is optically thin, is

$$I_\nu = \int_0^{\tau_0} S \, d\tau = \int_0^{\tau_0} \epsilon A B_0 \, d\tau, \tag{VI-166}$$

which is the formal equivalent of Equation (VI-67).

In the case of coronal forbidden lines ϵ is of the order of unity, but τ_0 is very small. In fact the product $\epsilon \tau_0$ is approximately the same for permitted and forbidden lines.

D. MULTILEVEL EFFECTS

Although most coronal lines are satisfactorily treated with a simple two level approximation, this is not true of chromospheric lines. The chromospheric effects illustrated in Figure VI-3 are for the simple two-level case in which the only photon sources are thermal, i.e., collisional excitation. Formally, the photon source term for thermal processes is $P_d B$, where P_d is the photon destruction probability, and J and S respond readily to changes in B. In multilevel problems this is not necessarily true. The main source of photons for a particular spectral line may, in fact, come from interlocking with other lines or from recombination from the continuum.

A specific example of the type just mentioned, viz., when the main source of line photons is supplied by photoionizations followed by recombinations, may produce a line that is not at all sensitive to the chromospheric temperature rise even though the line is very strong. If, as is often the case, the photoionizing radiation is formed at depths below the chromospheric temperature rise, the photon source term is largely independent of the local temperature and the line will show little or no response to the chromospheric temperature rise. Figure VI-6 illustrates this case. For purposes of this illustration $\epsilon^* B$ and $\epsilon^\dagger$ in Equation (VI-91) are replaced by

$$\epsilon^* B = \epsilon B + \eta B^*, \tag{VI-167}$$

and

$$\epsilon^\dagger = \epsilon + \eta, \tag{VI-168}$$

where ηB^* is independent of the local temperature. Relative values of ϵB and ηB^* and of ϵ and η are indicated in the figure. Note that when $\eta B^* > \epsilon B$, S follows neither the drop in B in the temperature minimum region nor the rise in B in the chromosphere, i.e., lines of this type give very little indication of the temperature structure in the external layers.

An in-depth discussion of multilevel effects in line transfer problems is beyond the scope of this book. The interested reader is referred for this purpose to references cited earlier. In the following, we provide only a brief commentary on some of the more clearly established results of multilevel studies.

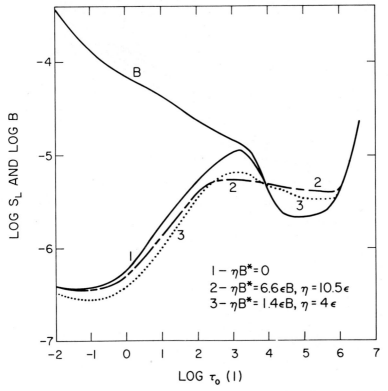

Fig. VI-6. An illustration of one effect on S_L of photoionization from the upper level of the line transition (see text for explanation).

The simplest multilevel case to discuss is that involving two lines in a close doublet with one energy level in common. For convenience we label the common level as level 1 and the two remaining levels as levels 2 and 3. Transitions between levels 2 and 3 are assumed to be optically forbidden but to have relatively large collisional exchange rates.

If the common level is the lower level, as in the case of Ca II H and K lines and the Na D lines, the photon interaction between the two lines occurs via the 2-3 collisions and their inverse. Doublets of this type are likely to have approximate source function equality at a given geometrical depth in the atmosphere. The exact degree of equality in the source functions depends upon the relative opacities in the two lines and the ratio C_{23}/A_{21}. In the domain where transfer in the Gaussian core is dominant the solution with $C_{23} = 0$ gives

$$\frac{S_{21}}{S_{31}} = \left(\frac{f_{12}}{f_{13}}\right)^{1/2} = \left(\frac{\widetilde{\omega}_2 f_{21}}{\widetilde{\omega}_3 f_{31}}\right)^{1/2} = \left(\frac{\widetilde{\omega}_2}{\widetilde{\omega}_3}\right)^{1/2}, \qquad \text{(VI-169)}$$

since the emission oscillator strengths are the same for doublets of this type. As C_{23} is increased, S_{21} and S_{31} approach each other but do not become exactly equal until C_{23}/A_{21} approaches unity. Figure VI-7a illustrates this effect for a case where

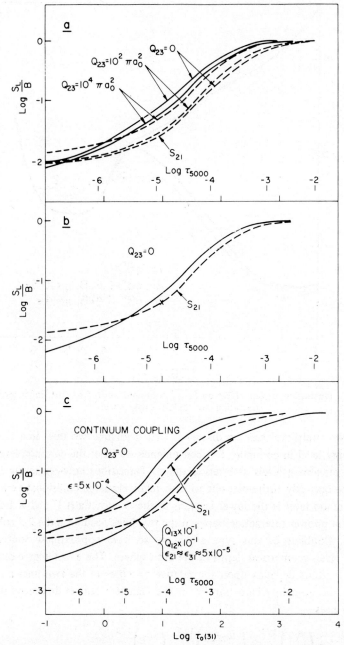

Fig. VI-7.   Source function coupling for lines of a doublet with common lower level and upper levels of the same excitation energy but with statistical weights whose ratio is 10:1. Case a illustrates the effect of electron collisions between the two excited levels; Case b illustrates the effect of having the rate of photon destruction by continuum absorption exceed the rate of collisional deexcitations $(\delta \gg \epsilon)$; and Case c illustrates the effect of photoionizations and recombinations for an atom mimicking neutral sodium.

$\widetilde{\omega}_2/\widetilde{\omega}_3 = 10^{-1}$ and $A_{21} = A_{31} = 10^8 \text{ s}^{-1}$ and where the main sources and sinks are collisional, i.e., $\epsilon \gg \delta$.

Actually, of course, collisional exchange between levels 2 and 3 is not the only way of coupling levels 2 and 3 together. Transitions from 2 to 3 via photoionization and recombination or via photoexcitation from level 2 to a fourth level and spontaneous decay from 4 to 3 may be more important than the direct collisional coupling. Figure VI-7c illustrates the continuum coupling for a model atom in the solar atmosphere having excitation and ionization potentials similar to Na I. Unlike the Na D lines, however, the ratio $\widetilde{\omega}_2/\widetilde{\omega}_3$ has been retained at a tenth to provide analogy with the results in Part a of the same figure. The continuum coupling is seen, for this case, to provide the same coupling as $Q_{23} \approx 10^4 \pi a_0^2$, and provides relatively close equality of the source functions.

When the main sources and sinks of photons result from continuum absorption and emission, i.e., $\delta \gg \epsilon$ there is a further source of coupling between the levels 2 and 3. This is illustrated in Figure VI-7b for the same atom as in Part a of the figure and with $Q_{23} = 0$. Note that coupling of this type is relatively strong.

By choosing $\widetilde{\omega}_2/\widetilde{\omega}_3 = 10^{-1}$ we have exaggerated the expected differences between S_{21} and S_{31} for doublets of this type. More typically, $\widetilde{\omega}_2/\widetilde{\omega}_3$ is of the order of ½. Thus, it seems safe to conclude that in close doublets with a common lower level the two source functions are closely equal, say, within 10 or 20%, but that they are not exactly equal. The same results can be generalized to more complex multiplets.

Multiplets in which the upper level is the common level, such as the Mg b lines, are always partially coupled regardless of the rates between the lower levels. Also, even a modest collision rate between the lower levels is sufficient to produce relatively close equality of the source functions. The common upper level provides an automatic sharing of photons between the different lines, and the collision rates between the lower levels competes only with the *upward* excitation rates. Since the upward rates are usually a few orders of magnitude smaller than the downward rates, strong coupling requires only modest collision rates.

Figure VI-8 illustrates a doublet with common upper level and $\widetilde{\omega}_2/\widetilde{\omega}_3 = 10^{-1}$. Solutions are shown for $Q_{23} = 0$ and $10\pi a_0^2$. The solutions with $Q_{23} = 10\pi a_0^2$ already show close equality of the source functions. It seems safe to conclude, therefore, that multiplets with a common upper level will exhibit close equality between their source functions.

The reader should note that for both the common upper level case and the common lower level case there is no general tendency for the source functions to be closer to B than in the case of a single line. Thus, source function equality in no way implies LTE.

Figure VI-9 shows 12 examples of energy level configurations involving from 1 to 4 spectral lines. In each example one line is labeled with a circle. The indicated line has, in each case, the same wavelength, the same ϵ and approximately the same r_0. The interlocking lines each have the same transition probability and ϵ as the comparison line. Source functions computed for the 12 comparison lines using a model solar atmosphere are shown in Figure VI-10.

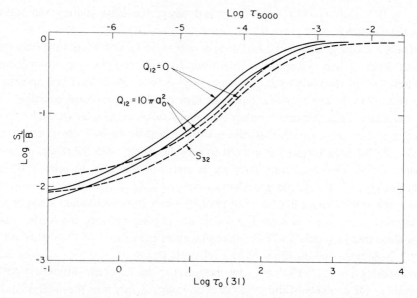

Fig. VI-8.   Similar to Figure VI-7a, but for a case in which the doublet has a common upper level.

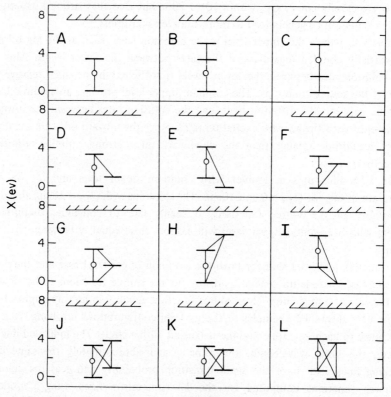

Fig. VI-9.   Idealized atomic configurations selected for comparison of source functions as shown in Figure VI-10.

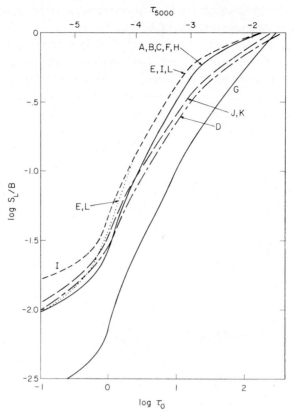

Fig. VI-10. Comparison of the source functions computed for the lines indicated in Figure VI-9.

Although the results in Figure VI-10 show significant differences for most of the cases, the only strongly different result occurs for case G. The unique feature of this case, which we refer to as the Lyman-β case, is that the transition crosses an intermediate level connected to both the upper and lower states of the comparison line by strongly permitted transitions. The double photon escape routes in the 1-2 and 2-3 transitions depresses the 3-1 source function. Note also that the largest source function in the group is for case I, which is the Balmer-α transition in the same three-line loop of transitions.

Apart from the Lyman-β case, there is a large measure of similarity in the behavior of the source functions for the different cases. Much of this similarity, however, results from the fact that λ, ϵ and r_0 are kept constant for the different cases. Within a given atom, of course, each of these parameters will differ from line to line. An example of the three line loop computed for a case where the ϵ's are the same for each of the three lines but where the r_0's and λ's are different is shown in Figure VI-11. A solution for S_{31} with $A_{21} = 0$ is shown for comparison. The crosses on the curves mark the locations of $\tau_0 = 1$ in each of the lines. Note that, in this case, the three source functions have markedly different behavior with the Lyman-β source function falling markedly below either the Lyman-α or the Balmer-α source functions throughout most of the line forming layer.

The closed loop of transitions illustrated in Figure VI-II illustrates one of the extreme examples of source function differences. The remaining cases illustrated in Figure VI-9 generally show considerably better agreement between the individual source functions for the different transitions within a given atomic model than do the cases G, H and I. (For a more detailed discussion of multilevel effects see Athay (1972a).)

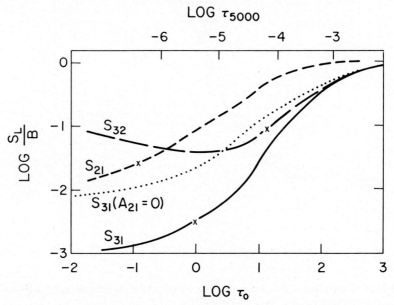

Fig. VI-II.   Source functions computed for a three-level atom with realistic values of r_0 and λ for the three lines. The dotted line shows the behavior of the 31 source function when level 2 is metastable.

A further effect that requires special note in multilevel problems arises from optical pumping. Since there is a strong tendency for $S \approx J$ in many spectral lines, and since J for the continuum exceeds B in the violet end of the spectrum, lines in the violet end of the spectrum may show $S > B$ throughout part or all of the atmosphere. Furthermore, if such lines interlock with lines in the red-green spectral range they may even drive $S > B$ for these latter lines in special cases. Certain of these effects are of importance in discussions of the free-bound continua formed in the temperature minimum region (Chapter VIII) and in attempts to explain the appearance of emission lines observed on the disk in the wings of the H and K lines of Ca II as·well as the reversal of rare Earth lines from absorption to emission inside the solar limb (Canfield 1971a, b).

E. LINES FORMED BY COHERENT SCATTERING AND BOUND-FREE CONTINUA

The case of bound-free continua requires, only a modest revision of the formulation leading up to Equation (VI-121). The source function, in this case, is frequency dependent. However, the frequency dependence is explicitly known, and it is not difficult to take it into account. We note from Equation (VI-57) that the source function may be written in the form

$$S = \frac{2h\nu^3}{c^2} \frac{\tilde{\omega}_1}{\tilde{\omega}_2} \left(\frac{n_2}{n_1}\right)^* \left(\frac{n_1}{n_2}\right)^* \left(\frac{n_2}{n_1}\right),$$

(VI-170)

where stimulated emissions are ignored. This may be rewritten as

$$S = B \left(\frac{n_1}{n_c}\right)^* \left(\frac{n_c}{n_1}\right)$$

(VI-171)

for the particular case where level 2 is the continuum.

The product $(n_1/n_c)^* (n_c/n_1)$ is a pure number independent of frequency. Thus, the frequency dependence of the bound-free source function is the same as the frequency dependence of B. Since B may be written as

$$B = B_1 \left(\frac{\nu}{\nu_1}\right)^3 e^{-h(\nu - \nu_1)/kT},$$

(VI-172)

where B_1 is the value of B at the head of the bound-free continuum, the bound-free source function may be expressed in terms of its value, S_1, at the head of the continuum and a known frequency dependence. It is only S_1 that is modified by the transfer effects, and the problem is therefore entirely analogous to the line transfer problem. For details of the bound-free transfer problem, the reader should consult Mihalas (1970) or Athay (1972a).

It was argued for many years that the effects of Doppler redistribution are sufficient to render the scattering of line photons as effectively non-coherent in so far as the determination of S is concerned. In order for this to be strictly true, a photon absorbed near $\Delta\lambda_2$ should have as much chance of being emitted at a given distance from line center, say $\Delta\lambda_1$, as a photon absorbed near $\Delta\lambda_1$. If the primary reason that the absorption near $\Delta\lambda_2$ occurs is that the absorption coefficient of the absorbing atom is Doppler shifted, the probability for reemission near $\Delta\lambda_1$ is independent of $\Delta\lambda_2$ and the reemission will be non-coherent. On the other hand, if $\Delta\lambda_2$ is large and if the absorption at $\Delta\lambda_2$ occurs because of the intrinsic width of the energy states in the atom, and if the absorbing electron does not undergo a secondary transition before reemission occurs, than the reemission will occur preferentially near $\Delta\lambda_2$, i.e., the scattering will be more nearly coherent in wavelength.

In the case of a Voigt absorption profile, the photons absorbed in the damping wing are absorbed there because of the intrinsic widths of the energy levels and not because of a large Doppler shift. Thus, unless the absorbing atom, in its excited state, is perturbed by a collision with another particle, the reemission will also occur in the line wings. This means that photons absorbed near line center are preferentially reemitted near line center and photons absorbed in the damping wings are preferentially reemitted in the damping wings. The effect of this is to restrict the escape probability for photons near line center to a value near that given by a pure Gaussian profile. Thus, the source function near line center behaves much like the source function for the Gaussian profile. In the line wings, on the other hand, the photons are not redistributed back to the Doppler core and the escape probability for these photons is relatively high. This has the effect of reducing the source function markedly in the line wings.

The effects of coherent scattering on chromospheric lines have been investigated in a few cases only. Vernazza *et al.* (1973) were the first to note that the wings of the Lyman-α line (Figure V-38) were strongly reduced in intensity by the effects of coherent scattering. They found a good empirical fit to the observed profile when the scattering in the wings was assumed to be 93% coherent. In a more detailed study of Lyman-α, Milkey and Mihalas (1973) treated the coherency problem explicitly and obtained good agreement with observations. These same authors (1974) have subsequently shown that the resonance lines of Mg II also are strongly influenced by coherent scattering. Again, the effect is to depress the line wings in the region of the minima (Figure V-36) just outside the line core.

Examples of the Mg II source functions computed by Milkey and Mihalas (1974) for three wavelengths in the K line (λ 2796) are shown in Figure VI-12. The source function,

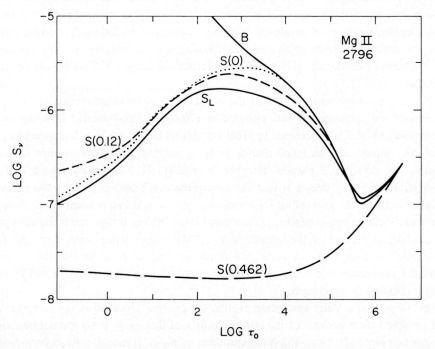

Fig. VI-12.   Source functions for the Mg II resonance line at λ 2796 as a function of wavelength in the line. The frequency dependence arises from coherency effects in the line wings (Milkey and Mihalas, 1974).

S_L, computed for the purely non-coherent case is shown for comparison. Note that in the region of temperature minimum S_L lies definitely above B whereas the source function $S(0.462)$, where the minimum in the K line occurs, lies definitely below B. Prior to these computations it was thought that the brightness temperature at the K_1 minimum in the Mg II profiles was an upper limit to the temperature minimum in the Sun. It is now clear, however, that the minimum brightness temperature in K_1 is well below the actual temperature minimum.

The source function near line center, $S(0)$, lies above S_L, particularly for $\tau_0 > 10^2$. This arises from the inhibition of photon escape near line center with the consequence that the effective absorption profile is Gaussian. The computation for S_L was made with a Voigt profile for ϕ_ν.

It is not clear, at this point, how extensive the effects of coherent scattering are. In both of the cases cited, the line in question arises from energy levels that are relatively isolated from other energy levels. This minimizes the effects of collisional perturbations and interlocking transitions and, thus, maximizes the effects of coherency. It is to be expected that the effects of coherent scattering will be diminished in more complex energy level structures, such as are encountered in heavy atoms and in subordinate transitions in both heavy and light atoms. On the other hand, it seems likely that the resonance lines of many other relatively simple atoms and ions, such as He I, He II, C I, C II, O I and Si II, to name a few, will be noticeably influenced by coherent scattering. More work is needed before definitive conclusions can be drawn.

References

Allen, C. W.: 1963, *Astrophysical Quantities*, 2nd Edit., Athlone, London.
Athay, R. G.: 1972a, *Radiation Transport in Spectral Lines*, D. Reidel Publishing Co., Dordrecht, Holland.
Athay, R. G.: 1972b, *Astrophys. J.* **176**, 659.
Athay, R. G. and Lites, B. W.: 1972, *Astrophys. J.* **176**, 809.
Billings, D. E.: 1966, *A Guide to the Solar Corona*, Academic, New York.
Canfield, R. C.: 1971a, *Astron. Astrophys.* **10**, 54.
Canfield, R. C.: 1971b, *Astron. Astrophys.* **10**, 64.
Delache, P.: 1974, *Astrophys. J.* **193**, 475.
Gebbie, K. B. and Thomas, R. N.: 1970, *Astrophys. J.* **161**, 229.
Geltman, S.: 1962, *Astrophys. J.* **136**, 935.
Hulst, H. C., van de: 1950, *Bul. Astron. Inst. Neth.* **11**, No. 410.
Hummer, D. G.: 1962, *Monthly Notices Roy. Astron. Soc.* **125**, 21.
Hummer, D. G.: 1964, *Astrophys. J.* **140**, 276.
Jefferies, J. T.: 1968, *Spectral Line Formation*, Blaisdell, Waltham.
Jefferies, J. T., Orrall, F. Q., and Zirker, J. B.: 1972, *Solar Phys.* **22**, 307.
John, T. L.: 1966, *Monthly Notices Roy. Astron. Soc.* **131**, 315.
Kourganoff, V.: 1963, *Basic Methods in Transfer Problems*, Dover, New York. (Orig. publ. 1952, Oxford.)
Lites, B. W.: 1972, Thesis Univ. of Colorado, (National Center for Atmospheric Res. Cooperative Thesis No. 28.)
Menzel, D. H. and Pekeris, C. L.: 1935, *Monthly Notices Roy. Astron. Soc.* **96**, 77.
Mihalas, D.: 1970, *Stellar Atmospheres*, Freeman, San Francisco.
Milkey, R. W. and Mihalas, D.: 1973, *Astrophys. J.* **185**, 709.
Milkey, R. W. and Mihalas, D.: 1974, *Astrophys. J.* **192**, 769.
Pottasch, S. R.: 1964, *Space Sci. Rev.* **3**, 816.

Smerd, S. F. and Westfold, K. C.: 1949, *Phil. Mag.* **40**, 121.
Thomas, R. N. and Athay, R. G.: 1961, *Physics of the Solar Chromosphere,* Interscience, New York.
Unsold, A.: 1955, *Physik Sternatmosphären*, 2nd Edition, Springer, Berlin.
Vernazza, J. E., Avrett, E. H. and Loeser, R.: 1973, *Astrophys. J.* **184**, 605.

CHAPTER VII

EMPIRICAL CHROMOSPHERIC AND CORONAL MODELS

1. Summary Models

In a certain sense, the construction of spherically symmetric models of the solar atmosphere runs counter to the discussion in Chapters II, III and IV. This may seem particularly true for the upper chromosphere and the corona where the geometric structure of the atmosphere seems clearly to depart very far from spherical symmetry. On the other hand, the 'mean' properties of even highly structured atmospheres are of importance for a number of purposes. The mean properties serve as a reference to which a given structure may be compared and for the elaboration of models that contain suitable geometric structures.

Modeling of the solar atmosphere has hardly progressed beyond the stage of spherically symmetric models. The few notable attempts at two- or three-component models that have been made, by and large, have been based on a few specific observational effects. No attempt has yet been made to synthesize a large body of data with a multicomponent atmosphere.

Part of the reason for the relative lack of interest in multicomponent solar models prior to this time is simply that the spherically symmetric models have not been successfully developed to the point where they have achieved a semblance of universality, i.e., there has not been sufficient agreement on the characteristics of the 'best' spherically symmetric model. This difficulty is now beginning to disappear, however, and 'reference' model atmospheres have acquired an established role. The first such model to attain recognition as a genuine reference atmosphere was the *Bilderberg Continuum Atmosphere* published by Gingerich and de Jager (1968) as an outgrowth of a working conference involving some thirty selected participants. The BCA covered the height range (as used here) from about −400 km to about 1900 km, and was based solely on disk continuum data. It made no serious attempt to satisfy any of the spectral lines in the solar disk spectrum or to satisfy either line or continuum data obtained at eclipse. The BCA (Figure VII-1) is characterized by a broad, flat temperature minimum of 4600 K extending from $\tau_5 \approx 0.006$ to 0.0006 (−40 to 100 km), followed by a chromosphere with a relatively gradual temperature rise out to about 1500 km where $T \approx 6500$ K and a relatively steep rise to about 1900 km where $T \approx 9500$ K.

The failure of the BCA to incorporate either Fraunhofer data or eclipse data was the result of a deliberate choice of the conference and was made out of frustration more than out of logic. It was recognized at the conference that the model to emerge as the BCA would be in sharp discord with certain models based on eclipse data and with certain models based on Fraunhofer data. The conference delegates were unable to resolve these difficulties and elected to restrict the data on which the model was based in order to arrive at a well defined model.

Several models of the chromosphere based on eclipse data were published prior to the BCA. Models taken from Thomas and Athay (1961) and Hiei (1963) are shown in Figure VII-1 along with the BCA. The eclipse models referenced are characterized by considerably higher temperatures and electron densities than are given by the BCA and were derived so as to reproduce the observed visual and Balmer continuum intensities in the chromosphere. Differences between the Thomas-Athay model and the Hiei model reflect mainly differences in the continuum data that they used. The BCA predicts intensities in the visual continuum and in the Balmer continuum in the chromosphere at the limb that are too low by large factors.

A further failure of the BCA occurs in the magnitude of the gas pressure in the upper chromosphere. The runs of gas pressure with height in the BCA and in the Thomas-Athay

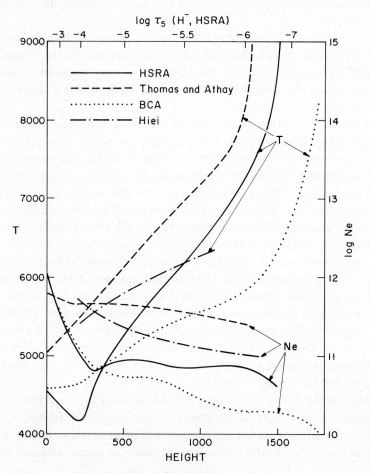

Fig. VII-1.   Comparison of the BCA to HSRA and other model atmospheres.

model are shown in Figure VII-2. Also shown in this Figure is an arrow indicating the most commonly accepted value of the gas pressure in the corona. Note that in the BCA model the gas pressure falls to a value nearly an order of magnitude lower than the coronal pressure, which, of course, is physically unacceptable. This failure of the BCA results from extending the cooler regions of the chromosphere ($T < 7000°$) over too large a height range, i.e., the temperature gradient is too low.

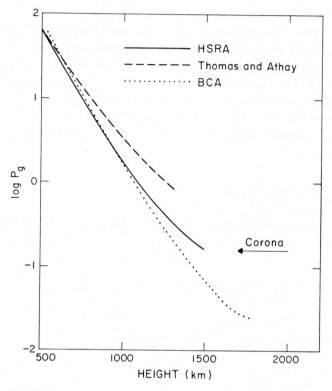

Fig. VII-2.   Gas pressure vs height in three reference atmospheres.

An additional serious shortcoming of the BCA was the use of LTE for computing the ionization equilibrium of hydrogen and metals. This assumption fails significantly in the upper photosphere and fails badly in the chromosphere. Since the ionization equilibrium is incorrectly computed, the electron densities and opacities in the model are not consistent with the assumed temperatures.

Models of the chromosphere based upon data from the 1962 eclipse appeared shortly after the BCA (Athay, 1969; Henze, 1969) and resulted in electron densities in close agreement with those obtained earlier by Hiei from independent data. Athay (1969) solved the Lyman continuum transfer problem to obtain the ionization of hydrogen self-consistently with the adopted temperature model. However, the model inadvertently used hydrogen densities that were too high by a factor of about two and was therefore somewhat inconsistent with other data.

A feature of interest in this latter model was that it relied solely on data for the Balmer continuum (see Chapter V, Section 2A). These data give the quantity $n_e n_p T^{-3/2}$ as a function of height. The hydrostatic equilibrium equation gives the relationship between $T(h)$ and $n_H(h)$ and the kinetic equilibrium equations together with known solar abundances of elements give the additionally needed relationships between $n_e(h)$, $n_p(h)$ and $n_H(h)$.

A second feature of importance in the Athay model was the explicit adoption of a lower limit to the chromospheric gas pressure, as imposed by currently accepted coronal models, and the requirement that the gas pressure decrease monotonically outwards. This limit to the gas pressure limits the vertical extent of the 'low temperature' regime in the atmosphere to about 1500 km if hydrostatic equilibrium is assumed. The vertical extent can be increased somewhat by adding a contribution from 'turbulent pressure' to the normal thermal pressure. Thus, following an original proposed by McCrea (1929), we write the pressure as

$$p_g = \sum_j n_j k T_j = \sum_j \left(n_j kT + \frac{n_j}{3} m_j \xi^2 \right), \tag{VII-1}$$

where ξ is the microturbulent velocity. For the case where most of the gas pressure results from hydrogen, helium and electrons, as is true in the chromosphere, Equation (1) becomes

$$p_g = (n_e + n_H + n_{He})kT + \tfrac{1}{3}(n_e m_e + n_H m_H + n_{He} m_{He})\xi^2, \tag{VII-2}$$

or

$$P_g = (1 + x + i)n_H kT + \tfrac{1}{3}(1 + 4x)n_H m_H \xi^2, \tag{VII-3}$$

where

$$x = \frac{n_{He}}{n_H}, \tag{VII-4}$$

and

$$i = \frac{n_e}{n_H} \tag{VII-5}$$

and where the mass of the electrons is ignored relative to the mass of the hydrogen atom.

The density scale height in an isothermal hydrostatic atmosphere is given by

$$H_{st} = \frac{kT}{\mu m_H g}, \tag{VII-6}$$

where, for the case being considered, the mean molecular weight, μ, is given by

$$\mu m_H = \frac{n_e m_e + n_H m_H + n_{He} m_{He}}{n_e + n_H + n_{He}}$$

$$= m_H \frac{1 + 4x}{1 + x + i} . \qquad \text{(VII-7)}$$

Equations (VII-6) and (VII-7) can be rewritten as

$$H_{st} = \frac{(1 + x + i)kT}{m_H g(1 + 4x)} = \frac{p_g(st)}{n_H m_H g(1 + 4x)} . \qquad \text{(VII-8)}$$

The addition of a constant microturbulence changes H to

$$H = \frac{p_g}{n_H m_H g(1 + 4x)} = \frac{kT}{\mu m_H g} + \frac{1}{3} \frac{\xi^2}{g} . \qquad \text{(VII-9)}$$

We rewrite this as

$$H = \frac{kT}{\mu m_H g} \left(1 + \frac{1}{3} \frac{\xi^2}{g} \frac{\mu m_H g}{kT} \right)$$

$$= H_{st} \left(1 + \frac{1}{3} \frac{\xi^2}{g H_{st}} \right)$$

$$= H_{st} \left(1 + \frac{H_{turb}}{H_{st}} \right) \qquad \text{(VII-10)}$$

with

$$H_{turb} = \frac{1}{3} \frac{\xi^2}{g} . \qquad \text{(VII-11)}$$

H_{turb} increases from 11 km at $\xi = 3$ km s^{-1} to 121 km at $\xi = 10$ km s^{-1}. Since H_{st} is of the order of 100 km, H_{turb} becomes significant for $\xi \gtrsim 3$ km s^{-1}. We note from Figure III-7 that velocities of this order are reached near 500 km and, hence, that we must expect some extension of the chromosphere beyond the hydrostatic models. However, we note also that the microturbulence is not of sufficient amplitude to increase H by a factor of order two, say.

The BCA gave way, as a reference atmosphere, to the *Harvard Smithsonian Reference Atmosphere* (HSRA) (Table VII-1) published by Gingerich *et al.* in 1971. This model incorporated new continuum data in the submilimeter and XUV regions and included, in particular, the Lyman continuum of hydrogen. The model recognized the need to incorporate electron density data from eclipse observations and to limit the gas pressure. Additional improvements included more realistic computations of the ionization equilibrium of metals as well as for hydrogen.

TABLE VII-1
The Harvard Smithsonian Reference Atmosphere[a]

Height	τ_s	T	Gas pressure	Electron pressure	N_e	N_H
−340	1	6390	1.31 + 5	5.64 + 1	6.4 + 13	1.4 + 17
−277	0.316	5650	8.31 + 4	1.12 + 1	1.4	9.8 + 16
−202	0.1	5160	4.56 + 4	3.95	5.5 + 12	5.9
−128	0.0316	4840	2.41 + 4	1.78	2.7	3.3
−57	0.01	4660	1.27 + 4	0.90	1.4	1.8
+12	0.00316	4520	6.65 + 3	0.47	7.5 + 11	9.8 + 15
+80	1.0 − 3	4380	3.46 + 3	0.24	4.0	5.3
148	3.16 − 4	4250	1.77 + 3	0.12	2.0	2.8
217	1.0 − 4	4170	8.68 + 2	0.061	1.1 + 11	1.4
314	3.16 − 5	4660	3.38 + 2	0.040	3.4 + 10	4.9 + 14
418	1.58 − 5	5040	1.35 + 2	0.052	7.5 + 10	1.8
500	1.0 − 5	5300	67.9	0.068	9.3 + 10	8.3 + 13
607	6.31 − 6	5590	29.2	0.068	8.8	3.5
680	5.01 − 6	5750	16.6	0.065	8.6	1.9
780	3.98 − 6	5950	8.23	0.063	7.7	9.1 + 12
990	3.16 − 6	6180	3.79	0.060	7.0	4.0
1000	2.51 − 6	6440	1.81	0.062	7.0	1.8
1090	2.00 − 6	6720	1.00	0.069	7.4	9.3 + 11
1170	1.58 − 6	6970	0.65	0.073	7.6	5.5
1280	1.00 − 6	7360	0.38	0.078	7.7	2.7
1400	3.98 − 7	7820	0.23	0.068	6.3	1.4
1450	2.00 − 7	8090	0.19	0.059	5.3	1.1
1500	5.01 − 8	8510	0.16	0.050	4.3	8.6 + 10
1510	1.58 − 8	8810	0.15	0.048	3.9	7.9

[a]The height scale of the HSRA has been corrected by −340 km to correspond to the definition of $h = 0$ used in this text. 3.16 − 4 means 3.16×10^{-4}.

Figures VII-1 and VII-2 illustrate the HSRA in comparison to the BCA, the Thomas-Athay model and the Tanaka-Hiei (1972) model. A further comparison of HSRA to the Athay model and other models is shown in Figure VII-3a and 3b. The models by Tanaka and Hiei (1972), Athay (1969) and Henze (1969) are based on limb data whereas the model by Vernazza et al. (1973) is basically a revision of the HSRA. This latter model includes the effect of microturbulence on the scale height and thus allows for a somewhat more extensive low temperature region.

The most apparent discrepancy between the HSRA and models based on eclipse data as illustrated in Figure VII-3 is in the electron densities with the HSRA densities being too low relative to the eclipse densities by about a factor of two.

Models of the upper chromosphere in the temperature regime from 10 000 K to about 100 000 K have not begun to approach the status where a reference model would have meaning. A two-component model by Athay and Menzel (1956) contained layers some 2000 to 3000 km thick at temperatures of 18 000 to 19 000 K. Thomas and Athay (1961) revised this earlier model by elevating the temperature in the same layers to 50 000 K. These early models were based on hydrogen and helium line intensities observed at eclipse. A model by Morton and Widing (1961) based on the observed

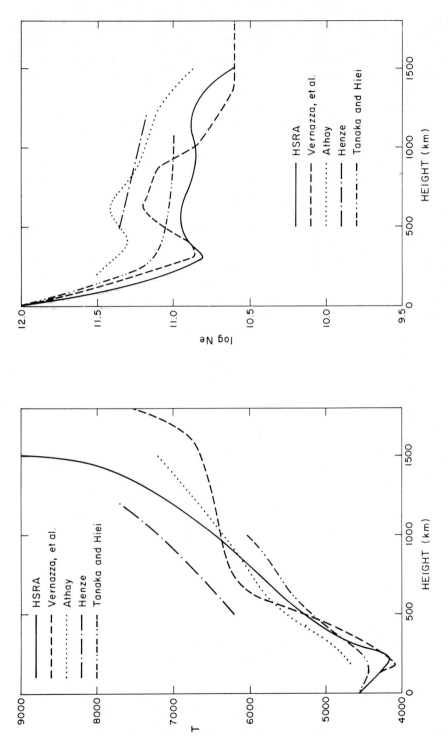

Fig. VII-3. Comparison of HSRA to other recent chromospheric models; (a) temperature models, (b) density models.

Lyman-α profile for the disk chromosphere suggested the existence of a layer of appreciable thickness with temperatures in the range 70 000 K to 90 000 K. A later model by Athay (1965) based on the XUV fluxes in the lines and continua of H, He I and He II pointed to a model in which T_e rose steadily from a value of about 25 000 K near a height of 1000 km to about 50 000 K at heights 3000 to 4000 km. Still more recently, models by Chipman (1971) and Vernazza et al. (1973) based on XUV line profile data give, respectively, a temperature plateau at about 18 000° and width of about 60 km, and a temperature plateau at about 20 000 K and width of about 170 km. In each of these latter two cases the temperature plateaus lie at heights near 2000 km.

The model of Vernazza et al. is shown in Figure VII-3 for $h \leqslant 2000$ km. We will adopt it as a reference model of the upper chromosphere for the subsequent discussion in this chapter.

Models of the chromosphere-corona transition region are derived from XUV line fluxes following methods outlined in Chapter VI. Spherically symmetric models of this region leave little room for ambiguity in the model and lead to models in which the temperature rises extremely rapidly from the upper chromospheric plateau to temperatures of several hundred thousand degrees. Details of the model, an example of which is shown in Figure VII-4, will be discussed in a subsequent section of this chapter.

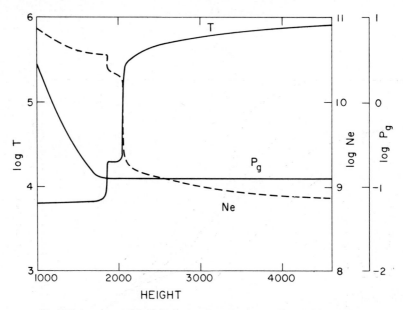

Fig. VII-4.   A model of the upper chromosphere and transition region.

Reference models of the corona are usually confined to the isothermal case, i.e., a 'coronal temperature' is given without reference to the gradients in temperature. Electron densities in the corona are readily obtained from continuum data. Both the density and gas pressure in the corona decrease outwards from the base of the corona with a scale height in the inner corona of about $0.1 R_{\odot}$, or about 70 000 km. The exact values are variable from point to point on the Sun and with the epoch in the solar cycle. The average density gradients correspond to an isothermal, hydrostatic atmosphere at a temperature of about 1.5×10^6 K.

We turn now to a more detailed consideration of the model beginning somewhat below the temperature minimum region and extending to the corona.

2. Conditions at the Optical Limb

Since observations at the solar limb are made with reference to geometrical height whereas those on the disk are made with reference to optical depth, it is necessary in solar studies to relate the two depth scales. We have arbitrarily set $h = 0$ at the point in the solar atmosphere where the *tangential* optical thickness at λ 5000 is unity. Other workers may prefer an alternate scale with $h = 0$ at a *radial* optical depth at λ 5000 of unity. As noted in Chapter 1, $\tau_5^t = 1$ corresponds to $\tau_5 = 3.7 \times 10^{-3}$ and the geometrical distance between $\tau_5 = 1$ and $\tau_5^t = 1$ is 340 km.

Aside from the arbitrariness in the choice of $h = 0$ there is a real question as to how the depths $\tau_5 = 1$ and $\tau_5^t = 1$ are defined observationally. The Eddington-Barbier relation defines $\tau_5 = 1$ as simply the 'mean depth' to which one sees at λ 5000 in the continuum. Thus, all properties of the atmosphere, such as temperature, density, velocity, etc., that are inferred from observations at λ 5000 are assigned to a depth of $\tau_5 = 1$. As was noted in Chapter I, however, the Eddington-Barbier relation is only an approximation and is not exact.

The operational definition of $\tau^t = 1$ has difficulties of comparable nature. For an atmosphere in which the temperature (source function) is constant with height, the specific intensity observed at the limb is given by

$$I = \int_0^{\tau^t} S\, e^{-t} dt$$

$$= S(1 - e^{-\tau^t}). \qquad \text{(VII-12)}$$

Differentiating twice with respect to height, we find

$$\frac{d^2 I}{dh^2} = S\, e^{-\tau^t} \left[\frac{d^2 \tau^t}{dh^2} - \left(\frac{d\tau^t}{dh} \right)^2 \right]. \qquad \text{(VII-13)}$$

Thus, at the point of maximum slope in I as a function of h

$$\frac{d^2 \tau^t}{dh^2} = \left(\frac{d\tau^t}{dh} \right)^2. \qquad \text{(VII-14)}$$

If we let τ^t vary exponentially with height, Equation (VII-14) is satisfied for $\tau^t = 1$. This provides an operational definition of $\tau^t = 1$, viz., the location of the maximum in dl/dh.

The two assumptions we have made in the preceding definition of $\tau^t = 1$ ($S = \text{const}$ and τ^t varying exponentially) are not too serious. Near the limb the HSRA shows the Planck function at λ 5000 varying by a factor of e in about 340 km. The scale height for τ_5^t, by comparison, is only about 60 km. Thus, the assumption that S is constant relative to $e^{-\tau^t}$ is quite good. Also, τ^t does vary exponentially with height to a relatively good approximation. It appears, therefore, that the operational definition of $\tau^t = 1$ is no more uncertain perhaps than the definition $\tau = 1$ using the Eddington-Barbier relation.

The condition $\tau^t = 1$ together with the condition that τ varies exponentially with height allows us to make a reliable estimate of the density of hydrogen atoms at $h = 0$. From Equation (I-7), the dual condition $\tau^t = 1$ and $h = 0$ requires that

$$(2\pi RH)^{\frac{1}{2}} K_0 = 1. \tag{VII-15}$$

For K_0 we write

$$K_0 = \alpha_{H^-} n_H n_e kT, \tag{VII-16}$$

where α_{H^-} is the absorption coefficient of the H^- ion per hydrogen atom and per unit electron pressure. The value of T at the limb is near $4550°$ and we may assume that hydrogen, helium, carbon, nitrogen and oxygen are all very weakly ionized because of their high ionization potentials. Hence, the electrons will come from the more abundant heavy elements, Mg, Si and Fe. These three metals have relative abundances ($n_H = 12$) of 7.4, 7.5 and 7.3. Their sum gives an abundance of 7.5×10^{-5} relative to hydrogen, and since these elements will be mostly singly ionized at 4500 K we may set $n_e = 7.5 \times 10^{-5}$ n_H. The value of α_{H^-} at 4550 K is 1.2×10^{-25} and the value of H is 60 km. Equations (VII-15) and (VII-16) give (with $n_e = 7.5 \times 10^{-5}$ n_H)

$$n_H = \frac{115}{(2\pi RH)^{\frac{1}{4}}(kT\alpha_{H^-})^{\frac{1}{2}}}$$

$$= 1.0 \times 10^{16} \text{ cm}^{-3}.$$

This value is within 10% of the value given by the HSRA at a height of -340 km, which corresponds to $h = 0$ on the scale we are using. The combined conditions $T = 4550$ K, $n_H = 1.0 \times 10^{16}$ and $n_e = 7.5 \times 10^{11}$ serve as useful lower boundary conditions for chromospheric models.

3. The Temperature Minimum Region

A. RADIATION LOSSES

Two quantities of particular interest in the temperature minimum region are the location of the minimum in optical depth and the difference δT between the temperature that occurs there and the temperature that would occur in pure radiative equilibrium. The product $\delta T \tau_{\min}$ provides a measure of the departure from radiative equilibrium in this region of the atmosphere. To demonstrate this, we consider a step increase in the Planck function defined by $\delta B =$ const, $\tau \leqslant \tau_{min}$, and $\delta B = 0$, $\tau > \tau_{min}$. The increase in outgoing flux at $\tau = 0$ due to such a layer is given by

$$2\pi\delta H(0) = \int_0^{2\pi} \mu\delta I_\nu \, d\omega \qquad\qquad (VII\text{-}17)$$

The increase in I_ν is given by

$$\delta I_\nu(\mu) = \int_0^{\tau_{min}} \delta B \, e^{-t/\mu} \frac{dt}{\mu} = \delta B(1 - e^{-\tau_{min}/\mu}) = \frac{\delta B \tau_{min}}{\mu}. \qquad (VII\text{-}18)$$

Whence, we find

$$2\pi\delta H(0) = 2\pi\delta B \tau_{min}. \qquad\qquad (VII\text{-}19)$$

There is an equal flux loss *inward* at $\tau = \tau_{min}$ so the total flux loss in both hemispheres is

$$2\pi\delta H(0) + 2\pi\delta H(\pi_{min}) = 4\pi\delta H(0) = 4\pi\delta B(\tau_{min}). \qquad (VII\text{-}20)$$

Since δB is integrated over frequency, we have

$$B = \frac{\sigma T^4}{\pi} \qquad\qquad (VII\text{-}21)$$

and

$$\delta B = \frac{4\sigma T_{min}^3}{\pi} \delta T. \qquad\qquad (VII\text{-}22)$$

Finally, then, we obtain for the additional energy loss due to δB

$$4\pi\delta H = 16\sigma \, T_{min}^3 \, \delta T \tau_{min}. \qquad\qquad (VII\text{-}23)$$

If τ_{min} is small enough, it may well be the case that S departs from the LTE condition. We should then write

$$S = \frac{B}{b} = \frac{\sigma T^4}{\pi b} \qquad\qquad (VII\text{-}24)$$

and

$$\delta S = \frac{4\sigma T_{min}^3 \, \delta T}{\pi b} - \frac{\sigma T^4}{\pi b^2} \frac{db}{dT}. \qquad\qquad (VII\text{-}25)$$

According to Praderie and Thomas (1972) b is proportional T^3 in the solar case when τ_{min} is very small. Hence, we have $1/b \, db/dT = 3\delta T/T$, and Equation (VII-25) becomes

$$\delta S = \frac{\sigma T_{\min}^3 \, \delta T}{\pi b}, \qquad \tau_{\min} \ll 1. \tag{VII-26}$$

Equations (VII-20) and (VII-26) then yield the energy loss as

$$4\pi \delta H(0) = \frac{4\sigma}{b} \, T_{\min}^3 \, \delta T \, \tau_{\min},$$

$$\tau_{\min} \ll 1, \tag{VII-27}$$

which is reduced below the loss given by Equation (VII-23) by a factor $(4b)^{-1}$. Equation (VII-27) is valid only when b departs appreciably from unity. Values of b computed for H^- for the solar case by Gebbie and Thomas (1970) range from 1.04 at $\tau_5 = 10^{-3}$ to 1.3 at $\tau_5 = 10^{-4}$. It seems questionable, therefore, whether Equation (VII-27) or (VII-23) is more appropriate for discussing the solar temperature minimum which lies near $\tau_5 = 10^{-4}$.

In either case, the results demonstrate that the additional energy loss is proportional to $\delta T \, \tau_{\min}$. It is of interest to compare the added energy loss given by Equation (VII-23) to the net solar flux $\pi H_\odot = \sigma T_{\rm eff}^4$. The ratio of these quantities is

$$\frac{4\pi \delta H(0)}{\pi H_\odot} = \frac{16 T_{\min}^3 \, \delta T \tau_{\min}}{T_{\rm eff}^4}. \tag{VII-28}$$

For $T_{\min} = 4200$ K and $T_{\rm eff} = 5785$ K, Equation (VII-28) gives

$$\frac{4\pi \delta H}{\pi H_0} = 1.06 \times 10^{-3} \delta T \tau_{\min}. \tag{VII-29}$$

Estimates of the net energy loss from the upper chromosphere and corona by radiation place the value (cf. Chapter IX) at approximately 6×10^5 erg cm^{-2} s^{-1}, which is equal to $10^{-5} \, \pi H_0$. According to Equation (VII-29) this same loss would be experienced from the temperature minimum region for $\delta T \, \tau_{\min} = 0.01$. The best current estimates place $\tau_{\min}$ at 10^{-4}. Thus, we would require $\Delta T = 100$ K in order for the net radiative loss from the temperature minimum region to be comparable to the loss from the upper chromosphere and corona.

The determination of a difference of only 100 K between the actual temperature in the temperature minimum region and the radiative equilibrium temperature at that same depth presents a dual challenge to solar physicists. We must determine both the actual temperature and the radiative equilibrium temperature to an accuracy of better than 50 K, i.e., to an accuracy of better than about 1%. Neither half of the problem has been satisfactorily solved, but the realization of that objective is not hopelessly beyond the bounds of current capabilities.

The realization of an important need to specify the temperature minimum accurately together with the realization that the techniques for doing so are perhaps within reach has spurred much recent interest in this problem.

B. EMPIRICAL DETERMINATION OF T_{min}

Efforts to determine the minimum temperature in the solar atmosphere have, until recent years, been thwarted by the difficulty either of obtaining appropriate data or of interpreting such data as existed. When viewed against the disk, the temperature minimum region can be observed in the continuum in the far IR near 200 to 300 μm, in the XUV continuum near λ 1600Å and in strong Fraunhofer lines.

The far IR and XUV regions have been observed only in recent years. They have furnished, by now, however, some of the most important sources of information concerning the temperature structure of the high photosphere and low chromosphere as well as the temperature minimum region itself. Fraunhofer line data have been readily available for many years, but adequate techniques for interpreting these data are only now emerging.

The temperature minimum region can be observed at eclipse also. However, the effective path length of integration at the limb for a *constant* emissivity in a layer of thickness L is $2(2RL)^{1/2}$ (Equation (I-4)), and, in effect, limb observations in the H^- continuum lose a factor of $2(2R/L)^{1/2}$ in height discrimination due to the line-of-sight integration. Since the thickness of the minimum region is at most 100 to 200 km, $(2R/L)^{1/2}$ is small, and the limb observations have an appreciable disadvantage with respect to disk center observations. An additional difficulty with observations at the limb is that lack of spatial resolution in the data translates directly into lack of height resolution. Thus, a $1''$ resolution at the limb integrates over $1''$ in height, which is several scale heights, in addition to the line of sight integration. This difficulty can be overcome to some extent at eclipse by using the motion of the Moon's limb to provide the height definition. Unfortunately, the Moon's limb is not smooth, and seeing effects in the Earth's atmosphere blend the emission from adjacent regions of the lunar limb, which have different projected heights on the solar image. This effect produces a loss of resolution that may easily amount to an appreciable fraction of a second of arc, i.e., one or two scale heights.

For practical purposes, little information has been gained about the temperature minimum region from limb data other than to place upper limits on the height at which it occurs. This aspect of the limb data will be discussed in connection with the discussion of the low chromosphere model. Thus, we turn now to a discussion of the disk data.

Equation (IV-31) gives the far IR opacity due to free-free transitions of electrons in the field of H^- ions as

$$K_\lambda = 1.17 \times 10^{-39} \ n_H n_e T^{0.15} \lambda^2, \qquad (\lambda \text{ in } \mu\text{m}).$$

At a visual wavelength of λ 5000, we may write

$$K_5 = \alpha_5(\tau) n_H n_e kT,$$

where $\alpha_5(T)$ is the usual tabulated absorption coefficient for H^- at λ 5000. Thus the ratio K_λ/K_5 is given by

$$\frac{K_\lambda}{K_5} = \frac{8.5 \times 10^{-24}\lambda^2\,(\mu m)}{\alpha_5(T)T^{0.85}}.$$

(VII-30)

For the moment, we adopt $T_{min} = 4200$ K and $\alpha_5(T_{min}) = 1.8 \times 10^{-25}$ and write Equation (VII-30) as

$$\frac{K_\lambda(min)}{K_5} = 0.04\,\lambda^2(\mu m).$$

(VII-31)

This gives $K_\lambda = 1$ at $K_5 = 6.4 \times 10^{-4}$ for $\lambda = 200$ μm and at $K_5 = 2.8 \times 10^{-4}$ for $\lambda = 300$ μm.

Inspection of Figure (V-26) shows that the minimum observed radiation temperature probably lies between 200 and 300 μm and has a value of, say, 43000 K ± 200 K. On this basis, then, we expect T_{min} to be near 4300 K and to occur between $\tau_5 \approx 6 \times 10^{-4}$ and 3×10^{-4}. The most recent far IR data shown in Figure (V-26) viz., those of Mankin *et al.* (1974) favor the larger value of τ_5 for the location of T_{min}.

Continuum data in the XUV shown in Figure (V-32) show a minimum radiation temperature near λ 1650 of approximately 4400 K. Interpretations of the XUV data are complicated by two factors: geometry and a breakdown of the LTE hypothesis. The difficulty with geometry arises from the fact that for $\lambda = 1650$ Å and T near 4200 K the Planck function changes approximately by a factor of ten for a 10% change in T_{min}. Thus, any inhomogeneities in temperature may seriously influence the deduced 'homogeneous' temperature. For example, 10% of the surface covered by regions where $\Delta T/T_{min} = 0.1$ would approximately double the surface brightness.This doubling of the apparent brightness would result in a 3% (120 K) increase in the inferred temperature.

The second difficulty in the XUV, i.e., the failure of LTE, arises because the main sources of opacity are due to the free-bound continua of Si I (Cuny 1971) and Fe I (Gingerich *et al.*, 1971; Athay and Lites, 1972). These continua have source functions that depart appreciably from the Planck function, and, in the region of T_{min}, are generally greater than the Planck function (cf. Cuny, 1971; Athay and Lites, 1972). Thus, the observed radiation temperature in the region of T_{min} lies above the kinetic temperature and the observed T_{min} of 4400 K serves as an upper limit to the true T_{min}. The difference between $T_{min}(obs)$ and $T_{min}(true)$ arising from this source could easily amount to 200 K but is not likely to be as large at 400 K. The former value results for $S/B \approx 3$ and the latter for $S/B \approx 10$.

Values of the observed T_{min} in the XUV obtained by Sandlin and Widing (1967) are higher than those obtained by Parkinson and Reeves (1969) by about 250 K. The magnitude of the effect upon the observed T_{min} arising from inhomogeneities and from departures from LTE are not known with any precision. No reliable estimate has been made of the range of variation in T_{min} between supergranule cells and the network, for example. Also, the free-bound continuum source function is difficult to evaluate in complex atoms because of the need to treat several lines simultaneously with the continua in order to obtain proper opacities. Assuming that the opacity near λ 1650 is due mainly to Si I Cuny (1971) found that the observed intensity in the range $\lambda\lambda$ 1680 to 1520 is virtually independent of T_{min} and is compatible with $T_{min} < 4400°$ at $\tau_5 = 10^{-4}$. However, the Fe I opacity is non-negligible in this region of the spectrum, and the Fe I problem has not been properly solved (cf. Athay and Lites, 1972; Lites, 1972). It seems safe to conclude at this time that the XUV data, even if those of Sandlin and Widing are used, are not incompatible with a value of T_{min} near 4200 K located near $\tau_5 = 10^{-4}$.

A strong additional source of information on T_{min} comes from Fraunhofer line data. Analyses of line cores, many of which are formed at heights spanning the minimum region, has generally not yielded precise information about the temperature minimum region. For the strongly permitted lines, the cores tend to form in the thermalization layer where S and B are only weakly coupled and where, as a result, the observed intensity is only weakly dependent upon T_{min}. An exception occurs in the case of forbidden lines such as λ 4571 of Mg I and λ 5166.3 and λ 4427.3 of Fe I. These lines are strong enough that their centers are formed near T_{min} and, because they are forbidden, they are formed under conditions close to LTE (cf. Athay and Lites, 1972). A study of the λ 4572 line by White *et al.* (1972) gives a radiation temperature of 4190 K at $\mu = 0.2$. Since the line shows limb-darkening out to at least this value of μ, this represents an upper limit to T_{min}. The Fe I lines mentioned have not been carefully studied.

Extremely strong lines, such as the H and K lines of Ca II and the corresponding lines of Mg II, have wings that form in layers spanning the temperature minimum. In such strong lines, the thermalization layers may lie entirely above the temperature minimum region, in which case, we should find $S = B$ in this region. Solutions for the Ca II (cf. Linsky and Avrett, 1070) H and K lines and for the Mg II (cf. Athay and Skumanich, 1968) resonance lines, assuming frequency independent source functions, indicated that this was indeed the case for these lines. However, a recent treatment by Milkey and Mihalas (1974) of frequency dependence in the line source function due to the effects of coherent scattering shows that, in the case of the Mg II resonance lines, the source function at the K_1 region of the profile is not thermalized at the temperature minimum. This result is exhibited by the curve labeled $S(0.462)$ in Figure VI-12. It can be seen from this curve that, at T_{min}, the source function lies below B and is decreasing outwards. Thus, at the wavelength of K_1 in the Mg II line, the observed value of T_{rad} will lie below T_{min}. Similar computations have not yet been carried out for the Ca II lines. However, the results for Mg II suggest that similar effects will be found for Ca II.

The observed values of T_{rad}(min) at K_1 and at $\mu = 1$ are near $4200°$ in both the Mg II and Ca II lines. The results of Milkey and Mihalas (1974), therefore, place a lower limit on T_{min} of $4200°$. One of the observed properties of both the Mg II and Ca II lines is that the K_1 minima exhibit limb darkening. This would not happen if S were equal to B and if the Doppler width of Φ_ν were reasonably constant. The source function, $S(0.462)$, shown in Figure VI-12 decreases outward and leads quite naturally to limb darkenings as a consequence of the departure of S from B. A different mechanism for producing T_{rad}(min) $\neq T_{min}$ has been proposed by Athay and Skumanich (1967). In their suggestion, S is equal to B, but $I_\nu(\min)$ is not equal to $S_\nu(\min)$. The inequality between $I_\nu(\min)$ and $S_\nu(\min)$ is accomplished by invoking an outwards increase in the microturbulent Doppler width starting at a depth just below $T_{\min}$. This increase in Doppler width renders the region of $T_{\min}$ optically thin, i.e. $\Delta\tau \ll 1$ and leads to values of $T_{\mathrm{rad}}(\min)$ that lie above $T_{\min}$. The increase in $T_{\mathrm{rad}}(\min)$ is greatest at $\mu = 1$ and K_1 therefore limb darkens. In this interpretation, $T_{\min}$ necessarily lies below $4200°$. However, there is no direct observational evidence that the microturbulent Doppler width does, in fact, increase with height at depths below $T_{\min}$, and the results of Milkey and Mihalas (1974) render the mechanism proposed by Athay and Skumanich (1967) as unnecessary. It seems clear, therefore, that $T_{\min}$ does not lie below $4200°$.

In spite of the uncertainties as to the value and location of $T_{\min}$, most of the information seems consistent with a value $4300° \pm 100$ K and with a location $\log\tau_5 = -3.5 \pm 0.5$. In terms of the accuracy specified in Section 3A of this chapter, neither $T_{\min}$ nor $\tau_{\min}$ are determined accurately enough. Thus, the problem of empirically determining $T_{\min}$ and $\tau_{\min}$ continues to be of strong interest in solar physics.

The preceding discussion of $T_{\min}$ reflects a considerable change since the BCA was introduced. In that model $T_{\min}$ was placed at 4600 K at an optical depth of $\tau_5 = 6 \times 10^{-4}$. This, in turn, represented a strong revision in the location of $T_{\min}$, which in the Thomas-Athay (1961) model, and in most models contemporary with it, was located near $\tau_5 = 100^{-2}$.

The other half of the temperature minimum problem centers around the radiative equilibrium model of the upper photosphere. This problem will be dealt with in Chapter IX.

4. The Low Chromosphere

At heights above the temperature minimum the temperature rises rapidly. This is evidenced by the plots of far IR and XUV continuum in Figures V-26 and V-32, which show a relatively rapid rise in T_{rad} towards longer (IR) and shorter (XUV) wavelengths from the wavelengths of $T_{\min}$. Perhaps an even more dramatic evidence of the rise in T is given by the Balmer continuum intensity observed at eclipse (Figures V-2 and V-3).

Throughout the remainder of this chapter we will define the low, middle and high chromosphere in terms of temperature intervals. We adopt this convention in order to relate the chromospheric divisions to the spectral properties of the divisions rather than the geometric heights, which vary from model to model. The low chromosphere is defined as the region between the temperature minimum and a temperature of 6000 K. This latter limit is somewhat arbitrary, but, as will be shown, it marks the approximate upper limit of the region in which H^- emission dominates the visual continuum of the chromosphere.

At a height of 500 km the quantity $n_e n_p \, T^{-3/2}$, inferred from the Balmer continuum intensity (Figure V-3) has a mean value of 1.5×10^{17} cm^{-6}. Since n_p cannot exceed n_e, and since T is of the order of 5500 K,, n_e must be approximately equal to or greater than

2×10^{11} cm^{-3}. Let us now compare this with the hydrogen density. In Section 2 of this chapter we estimated n_H at $h = 0$ as 1×10^{16}. An approximate value of n_H at 500 km can be obtained by computing the hydrostatic equilibrium value for a mean temperature of 4500 K. At this temperature the mean molecular weight is 1.3 and the hydrostatic density scale height is 110 km. Thus, in 500 km, n_H drops by a factor 10^2 from 1×10^{16} to 1×10^{14} cm^{-3}. This value will be somewhat in error because of the approximate value of T we have used but it should not contain a large error.

From the above estimates of n_e and n_H, we note that, at 500 km, n_H/n_e is approximately 500 as compared to a ratio of 1.3×10^4 for the case where hydrogen ionization does not contribute to n_e, i.e., at $h = 0$ and in the vicinity of T_{min}. It is evident, therefore, that the relative low value of n_H/n_e at 500 km reflects a large increase in the degree of ionization of hydrogen. This, in turn, must reflect a relatively large increase in T.

Further evidence of the increase in T and n_e/n_H below 500 km is given by the ratios F_{3540}/F_{4815} and F_{3640}/F_{3700} shown in Figure V-2. The ratio F_{3540}/F_{4815} is on an arbitrary scale, but it shows an increase of a factor 10 between $h = 0$ and 500 km. The ratio F_{3640}/F_{3700}, in Figure V-2, is on an absolute scale and has a value of about 10 at 500 km.

Continuum emission at these wavelengths and these heights arises from a combination of mechanisms, including H$^-$, Balmer free-bound, Paschen free-bound, Rayleigh scattering and Thompson scattering. It is readily shown that for the conditions in the chromosphere below 600 km the H$^-$ and Balmer free-bound continua are strongly dominant (cf. Thomas and Athay, 1961, Chapter VI). For this case, we may write the ratio F_{3640}/F_{3700} as

$$\frac{F_{3640}}{F_{3700}} = \left(\frac{\beta_{3640}}{\beta_{3700}} \right)^{1/2} \left[\frac{2.34 \times 10^{-33} n_e n_p T^{-3}}{4\pi n_H n_e k T \alpha_{3700}(T) B_{3700}} {}^2 + 1 \right] \qquad \text{(VII-32)}$$

(cf. Equations (VI-17) and (VI-11)). We have replaced S_λ in equation (VI-11) with B_λ, which is sufficiently accurate for the present purposes. From Figure (V-2), we estimate $\beta_{3640}/\beta_{3700}$ as 0.7. Equation (VII-32), for a given value of F_{3640}/F_{3700}, then gives a relationship between n_{hH}/n_p and T. The relationship is shown in Table VII-2 for $F_{3640}/F_{3700} = 10$.

TABLE VII-2

Values of N_H/N_P as a function of T from Equation (VII-32). $F_{3640}/F_{3700} = 10$

T	4200	5000	6300
N_H/N_P	5400	1300	430

For the type of models indicated in Figure VII-3, the results in Table VII-2 lead to an estimate of $n_H/n_p \approx 1000$. Earlier in this section, we estimated $n_e = 2 \times 10^{11}$ cm^{-3} for

$n_e = n_p$, and we estimated the hydrostatic equilibrium value of n_H at 1×10^{14} cm^{-3}. These earlier results give $n_H/n_p \approx 500$, which compares favorably with the results from Equation (VII-32). It seems clear from the continuum data, therefore, that, at $h = 500$ km, $n_p \approx n_e \approx 10^{-3}$ n_H. For this to be true, most of the electrons must be contributed by hydrogen rather than the metals.

In the temperature minimum region the value of $n_{/np}$ given by the Saha Equation (Equation (VI-20)) for $T = 4200°$ and $n_e = 10^{-4}$ $n_H \approx 10^{11}$ is 3×10^6. Thus, for $n_H = 10^{15}$. which is the approximate value at T_{min}, $n_p \approx 3 \times 10^8$, and we note that n_p evidently *increases* by about three orders of magnitude between the temperature minimum and a height of 500 km. This is consistent with the inference from Figure V-3 that the Balmer continuum emissivity $(n_e n_p T^{-3/2})$ increases outwards in the lowest levels of the chromosphere.

Since hydrogen is still only about 0.1% ionized at 500 km, a continued rapid increase in T above 500 km would lead to a continued upward increase in n_p. The data shown in Figure V-3 clearly show that this is not the case. There is, however, a continued increase in n_p/n_H as evidenced by the relatively large scale height (~450 km) for the Balmer continuum emission. It may be inferred that the temperature continues to increase outwards above 500 km but that the temperature gradient is less steep than it is below 500 km. Most recent chromospheric models, including the HSRA, reflect this trend.

Continuum data at wavelengths longer than the Balmer series limit provide by themselves, a valuable guide to the model of the low chromosphere independently of the Balmer continuum data. It was noted in connection with the continuum data shown in Figure V-2 that the character of the continuum emission shows a sharp break near 600 km. Below 600 km, the intensity falls rapidly with height and near 600 km the rate of decrease drops markedly. This change in character of the continuum emission is associated with a change from predominantly H$^-$ emission in the low chromosphere to predominantly Paschen free-bound emission and electron scattering in the middle chromosphere. Near the height where $\beta^{1/2}$ F is the same for the two segments of the curve for λ 4815 in Figure V-2, the two mechanisms will contribute equally to the observed emission. This equality occurs near a height of 700 km.

We note from Equations (VI-11), (VI-18) and (VI-2) that the ratio of H$^-$ emissivity to electron scattering plus Paschen emissivity at λ 4815 is given by

$$\frac{\epsilon H^-}{\epsilon_e} = \frac{4\pi n_H n_e kT\alpha_\lambda(T)S_\lambda}{2\pi\sigma_T n_e I_p + 7.7 \times 10^{-34} n_e n_p T^{-3/2} e^{-12200/T}} \qquad \text{(VII-33)}$$

Thus, the two emissivities are equal at

$$n_H = \frac{\sigma_T I_p + 1.23 \times 10^{-34} n_p T^{-3/2} e^{-12200/T}}{2kT\alpha_\lambda(T)S_\lambda} \qquad \text{(VII-34)}$$

For the present, we set $S_\lambda = B_\lambda$. To obtain n_p, we set $n_p = n_e$ and evaluate n_e from the relation $n_e = 2.07 \times 10^{16} T^{3/4} (j_{\nu B})^{1/2}$ given by Equation (VI-17). The value of $j_{\nu B}$

inferred from the Balmer continuum data at 700 km in Figure V-3 is 1.5×10^{-8}. Thus, $n_e = 3.1 \times 18^8 T^{3/4}$. Values of n_H vs T obtained from Equation (VII-34) are given in Table VII-3. The Paschen term in the numerator of Equation (VII-34) contributes

TABLE VII-3
Relationship between N_H and T near 700 km from Equation (VII-34)

T	4200		5040	6300
N_H	3.0×10^{13}		2.0×10^{13}	1.3×10^{13}

approximately 30% of the total at 6320° and 15% at 4200°. Since the indicated relation between n_H and T is not a strong function of T, the change from H⁻ emission to electron scattering near 700 km serves mainly to confirm that the value of n_H at this height is near 2×10^{13} cm⁻³, as predicted by hydrostatic equilibrium models. Figure VII-5 shows the relationship between n_H and T given at 700 km by various models as compared to the results from Table VII-3 (solid line). The HSRA and the Vernazza model are in

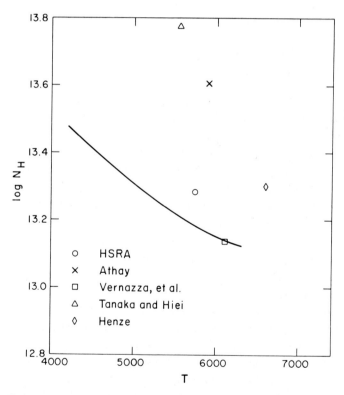

Fig. VII-5.   The relationship between N_H and T at a height of 700 km. The solid line shows the results given in Table VII-3.

good agreement with the results from Table VII-3.   It should be noted, however, that it is not strictly legitimate to set $S_\lambda = B_\lambda$ in Equation (VII-34). If, instead, S_λ is set equal to B_λ/b_{H^-}, the values of n_H in Table VII-3 should be increased by about 0.7 b_{H^-}. From Table V-2, we see that $b_{H^-} \approx 1.8$ at $\tau_5 = 10^{-5}$ and we know that b_{H^-} approaches an upper limit of 2 at smaller values of τ_5. Thus, the close agreement between the Vernazza et al. model and the results in Table VII-3 are fortuitous, and the solid curve in Figure VII-5 should properly be displaced upwards by an amount 0.1 to 0.15 in the logarithm. As was noted earlier n_H is too high in the Athay model by a factor of about 2.

Below 700 km where H^- emission is dominant in the visual continuum, hydrogen is mostly unionized and the value of n_H is quite insensitive to the temperature. If the ratio n_H/n_e and T were constant with height, the product $n_H n_e$, and hence the H^- emissivity, would decrease with height with a scale height equal to half that of hydrogen. Thus, we would expect $H = 65$ km or $\beta \approx 15.4$. The value of β indicated in Figure V-2, however, is 10, which corresponds to $H = 100$ km. In a height range of 500 km, this difference between the isothermal scale height and the actual scale height represents a difference in the ratio $\epsilon(500)/\epsilon(0)$ in the two cases amounting to a factor of 15. Thus, we again have an indication in the H^- emission that the ratio n_H/n_e is some 15 times larger at 500 km than at $h = 0$.

The low chromosphere portions of both the BCA and the HSRA are based heavily on continuum intensities in the far IR. The HSRA incorporates, additionally, the XUV disk continuum and the Balmer continuum observed at eclipse. One of the most obvious failures of the BCA was its overestimates of n_H/n_e in the low chromosphere. This difficulty was partially, but not completely, overcome by the HSRA ($n_e = 9 \times 10^{10}$ at 500 km).

The failure of the HSRA to yield sufficiently large values of n_e to reproduce the Balmer continuum data at the limb indicates that the far IR and XUV disk continuum intensities adopted for the model are somewhat inconsistent with the limb data. This apparent discrepancy could represent a fault in the data or it could represent a real difference between the mean disk model and the mean limb model. Such a difference should be expected if the inhomogeneities known to be present in the chromosphere produce a significant effect on the mean model. If, for example, the chromosphere is divided into two types of features, one of which has an optical thickness in the far IR that is appreciably different from the other, then the observed disk surface will be roughened and the inferred 'mean' disk model clearly may differ from the mean limb model. On the other hand, if all of the features are optically thin the mean disk and mean limb models should still agree.

It is too early to ascertain whether the failure of the HSRA to fully represent eclipse continuum data is a fault of the data or is a symptom of inhomogeneous structure. However, the evolution of reference disk models is clearly towards better and better agreement with the limb model. Also, the latest far IR measurements of Mankin et al. (1974) shown in Figure V-26 indicate a more rapid rise in temperature above the temperature minimum (greater λ) than is given by the HSRA. This more rapid rise in temperature in the low chromosphere indicated by the recent far IR data is in the direction needed to bring the disk and limb models into agreement and, quantitatively, appears to be more than sufficient in magnitude.

It seems clear from the preceding discussion that between the temperature minimum region and a height of 500 km the temperature rises by approximately 1000 K and exceeds 5000 K at 500 km. The height at which a temperature of 6000 K is reached, which we chose as an arbitrary boundary for the low chromosphere, is not definite. In the models in Figure VII-3a it ranges from less than 500 to 1000 km. However, three of the models show a range of only 650 to 800 km, and it seems safe to conclude that the height of the $6000°$ surface and the height where the character of the continuum emission in the visual changes from predominantly H^- to electron scattering plus Paschen free-bound approximately coincide.

Analyses of Fraunhofer line data lead to models that clearly support a model of the type indicated by the continuum data. Because of the complexity of the line transfer problems (see Chapter VI) and uncertainties in atomic collision cross sections, photoionization cross sections and, in some cases, f-values, the analyses of line data have not led to a precisely defined model for the low chromosphere. Models based upon line data necessarily involve additional aspects of the model, notably, the macroscopic velocity field. Both the residual intensities in the lines and the line shapes depend upon the magnitude and gradient of the so-called micro- and macroturbulence (see Figure VI-3).

One of the primary effects of the temperature structure of the low chromosphere on the lines is an indirect effect arising from the dependence of n_e upon T. An increase in T, for example, produces an increased ionization of hydrogen and, hence, an increased n_e. The increased n_e increases the collision rates thereby increasing the destruction probability, P_d, which, in turn, increases the coupling between S and B.

Analyses of the Na D and Mg b lines (Athay and Canfield, 1969), XUV lines of O I (Athay and Canfield, 1970) and Fe I lines (Lites, 1972) show a preference for a model having electron densities near those of the limb models as opposed to the lower density HSRA. On the other hand, this is only a preference. The lines could perhaps be suitably explained with the HSRA by adopting a more rapid increase of microturbulent velocity with height in the low chromosphere or by adopting larger collision cross sections. Attempts to fit the above mentioned lines with models of this type have not been fully exploited, but sufficient attempts have been made to justify the conclusion that the models based on limb data lead to an explanation of the profiles with more readily acceptable values of microturbulent velocities and collision cross sections.

Linsky and Avrett (1970) have successfully fitted the Ca II profiles with a model that is close to the HSRA. They did not investigate models of the type derived from limb continuum data. Other authors (cf. Athay and Skumanich, 1968) have shown that the Ca II lines can be fitted with the limb models. In each of these cases, however, the effects of coherent scattering were ignored and the results are now suspect.

A more complete discussion of the models based upon Fraunhofer line data is given in

Paragraph 7 of this Chapter following the discussion of the middle and high chromosphere, and in Chapter VIII.

5. The Middle Chromosphere

A. GENERAL COMMENTS

We define the middle chromosphere as the region through which the temperature rises from 6000 to 10 000 K. The Vernazza *et al.* model shown in Figure VII-4 gives the upper boundary of this layer near 1800 km and the preceding discussion of the low chromosphere gives the lower boundary near 700 km. Again, however, we emphasize that the actual heights are quite uncertain.

Within the approximate height range of the middle chromosphere the spectrum at the limb shows: (1) a predominance of electron scattering and Paschen free-bound emission in the visual continuum (Figure V-2), (2) a steady decrease of Balmer continuum flux (Figure V-3), and (3) a rise to maximum in the He I and He II line fluxes. In the disk spectrum, the middle chromosphere gives rise to: (1) the Lyman continuum, (2) to the cores of Hα and the H and K lines plus a few other strong lines of Ca II and Fe I (see Table I-3), (3) to many emission lines in the XUV, and (4) to the radio spectrum between approximately 1 mm and 1 cm.

The most useful data for deriving models of the middle chromosphere have been limb data for the Balmer free-bound and the electron scattering continua and disk data for the Lyman continuum. The Balmer, Paschen and electron scattering continua are optically thin whereas the Lyman continuum is optically thick.

Rayleigh scattering by neutral hydrogen also contributes to the continuum in the visual spectrum. In Paragraph 1B of Chapter VI, the ratio of intensities due to Rayleigh and electron scattering is given by

$$\frac{n_H}{n_e}\left(\frac{1026}{\lambda}\right)^4 = 0.0021\frac{n_H}{n_e} \quad \text{and} \quad 0.0059\frac{n_H}{n_e}$$

at $\lambda\,4815$ and $\lambda\,3700$, respectively. As noted in the preceding section on the low chromosphere, n_H is approximately 2×10^{13} at the lower boundary of the middle chromosphere and n_e is approximately 1.4×10^{11}. Hence, n_H/n_e is approximately 140 and the intensity due to Rayleigh scattering is approximately 30% of the electron scattering intensity at $\lambda\,4815$ and approximately 80% at $\lambda\,3700$. We expect the ratio of n_H/n_e to decrease with height as the temperature increases so that the contribution due to Rayleigh scattering becomes progressively less important. In a careful analysis of the chromospheric continuum, Rayleigh scattering should, of course, be included.

B. VISUAL CONTINUUM

It was noted in Chapter V in the discussion of Figure V-2, that the ratios of surface brightnesses F_{3540}/F_{3700} and F_{3640}/F_{4815} were approximately constant between heights of 500 to 1500 km and that they decreased slowly between 1500 and 1800 km. The absolute value of the ratio F_{3640}/F_{3700} is approximately 10. Since the ratio is essentially constant with height, we may equate the surface brightness ratio to the emissivity ratio.

From the equations for the appropriate emissivities given in Chapter VI, we write

$$\frac{j_\nu(\text{Balmer})}{j_\nu(\text{Paschen}) + j_\nu(\text{Thompson}) + j_\nu(\text{Rayleigh})} + 1 = 10,$$

or

$$\frac{2.34 \times 10^{-33} n_e n_p T^{-3/2}}{8.00 \times 10^{-34} n_e n_p T^{-3/2} e^{-21340/T} + 7.31 \times 10^{-29} n_e + 4.31 \times 10^{-31} n_H} = 9$$

(VII-35)

Setting $n_e = n_p$ and using the modified Boltzmann-Saha equation for n_H, we obtain

$$1 = 3.1 e^{-21340/T} + \frac{2.8 \times 10^5 T^{3\frac{1}{2}}}{n_e} + 7.0 \times 10^{-13} b_1 e^{157\,300/T} \qquad \text{(VII-36)}$$

To satisfy the observational data, therefore, the combination of terms on the right hand side of Equation (VI-36) must be relatively constant.

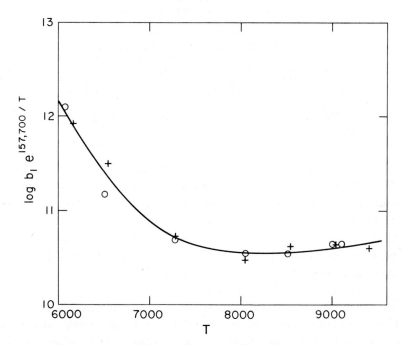

Fig. VII-6.   A comparison of computed, +, (Noyes and Kalkofen, 1970) and empirical, 0, (Vernazza and Noyes, 1972) values of b_1 as a function of T.

The quantity b_1 should be obtained for radiative transfer effects in the Lyman lines and continuum and in the early Balmer lines. Instead of following that procedure here, however, we shall adopt solutions due to other workers. Noyes and Kalkofen (1970) have published computed values of b_1 for a model very similar to the HSRA and Vernazza and Noyes (1972) have determined empirical values of b_1 from the observed Lyman continuum data on the solar disk. Values of the quantity $b_1 e^{157\,300/T}$ obtained by Noyes and Kalkofen (1970) and Vernazza and Noyes (1972) are plotted in Figure VII-6. The solid curve gives the approximate average of these results. Note that the results appear to be quite unique even though b_1 is determined by entirely different procedures in the two cases.

The uniqueness of the results in Figure VII-6 arises from the fact that in the chromospheric models used by both Noyes and Kalkofen and Vernazza and Noyes a given value of T is associated with nearly the same values of n_e and optical depth in the Lyman lines and continuum. It is this tendency for uniqueness between T, optical depth and n_e that results in the uniqueness between b_1 and T. Thus, the results in Figure VII-6 are probably not applicable to other stars.

Using the curve in Figure VII-6 for $b_1 e^{157\,300/T}$, we obtain the relative values for the three terms on the right hand side of Equation (VII-36) given in Table VII-4. The last column in the table gives the values of n_e necessary to satisfy Equation (VII-36).

TABLE VII-4
Relative values of terms in Equation (VII-36)

T	Paschen	Electron scattering	Rayleigh scattering	n_e
6000	0.09	$1.3 \times 10^{11}\, n_e^{-1}$	1.0	–
6500	0.12	$1.5 \times 10^{11}\, n_e^{-1}$	0.18	2.1×10^{11}
7000	0.15	$1.6 \times 10^{11}\, n_e^{-1}$	0.06	2.0×10^{11}
8000	0.22	$2.0 \times 10^{11}\, n_e^{-1}$	0.02	2.5×10^{11}
9000	0.29	$2.4 \times 10^{11}\, n_e^{-1}$	0.03	3.5×10^{11}
10 000	0.37	$2.8 \times 10^{11}\, n_e^{-1}$	0.04	4.7×10^{11}

The apparent increase in n_e with increasing T in Table VII-4 should not be taken literally. It could result from either of two causes: (a) a too severe restriction on the data, or (b) a too severe restriction on the model atmosphere. In the former case, we have forced the left hand side of Equation (VII-35) to be exactly 9, which is certainly too restrictive. In the latter case, we have required that the atmosphere be spherically symmetric, which we know is not true. The inhomogeneities in the atmosphere will have temperature and densities that differ from the average and, as a result, they will emit in different proportions in the various continua.

It is clear from the Balmer continuum data that n_e decreases with height and that T does not increase rapidly with height in the middle chromosphere, otherwise the Balmer continuum intensity would continue to increase with height. If we adopt the Vernazza model (Figure VII-4) as a reference, we see that $T^{-3/2}$ decreases by about a factor of 2 in the height interval 650 to 1850 km. In this same interval, the Balmer continuum intensity drops by a factor 12.4, and it follows that n_e drops by a factor 2.5 (assuming $n_e = n_p$).

In terms of a spherically symmetric model, the results in Table VII-4 suggest that the gradient in T should be relatively small in the middle chromosphere, i.e., T should not rise to values in excess of $8000°$ too quickly. On the other hand, the results could be interpreted to mean that the chromosphere, in fact, has significant departure from spherical symmetry in these height layers.

C. LYMAN CONTINUUM

We turn now to a brief discussion of the Lyman continuum disk data in order to gain further insight to the mid-chromosphere model. Analyses of Lyman continuum data require computation of the Lyman continuum source function given by

$$S_\lambda = \frac{B_\lambda}{b_1}, \qquad (VII-37)$$

when stimulated emissions are ignored. To proceed one usually assumes T (hence, B) as a function of height then solves the combined radiative transfer, statistical equilibrium and hydrostatic equilibrium equations to obtain b_1, n_H, n_e and τ_0 (τ_0 is the optical depth at the head of the Lyman continuum) self consistently as functions of height. This provides $S_\lambda(\tau_0)$ from which the Lyman continuum intensity is readily computed. The correct $T(h)$ distribution is obtained by trial and error computations to synthesize the observed intensities as a function of λ and μ. An added restriction on the models is provided by the coronal gas pressure of 0.16 dyn cm^{-2}, which acts as a lower limit to the chromospheric pressure. This latter restriction severely limits the range of chromospheric models that can be considered.

Figure VII-7 shows a model obtained from the Lyman continuum data (Figures V-26 and V-27) by Noyes and Kalkofen following the above procedure. This model is the basis of and is essentially identical to the HSRA in the height interval shown. It should be noted that the Lyman continuum data on the disk determines only the distribution $S_\lambda(\tau_0)$ and contains no direct information on geometrical height. The relationship between τ_0 and h comes entirely via the hydrostatic equilibrium assumption and must be regarded as somewhat arbitrary. If, as was suggested in Part 1 of this chapter, the atmospheric scale height is extended by macroscopic motion, the height scale may change quite markedly. The gas pressure at a height of 500 km in the HSRA is 68 dyn cm^{-2}. It drops to 0.16 dyn cm^{-2} at 1500 km. Thus, in 1000 km the pressure drops by a factor 420 corresponding to six scale heights. An increase in the scale height by 50% would produce the same pressure drop in 1500 km and would extend the middle chromosphere by an additional 500 km.

Representative optical depths in the Lyman continuum are indicated on the curves in Figure VII-7. Note that $\tau_0 = 1$ occurs near $8300°$, which agrees with the observed color temperature of the continuum at disk center. The source function at the head of the continuum *decreases* slowly outwards due to a rapid outwards increase in b_1. At the depth $\tau_0 = 1$, $b_1 \approx 200$. At shorter wavelengths where b_λ is a stronger function of T, S_λ

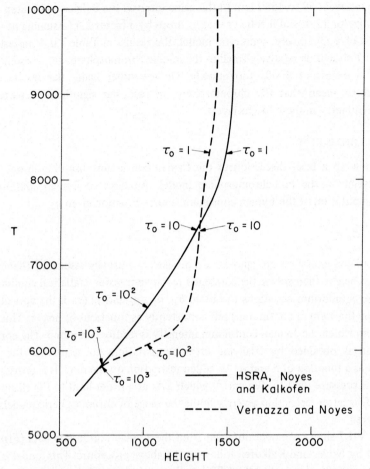

Fig. VII-7.   Two models of the middle chromosphere. The model given by Vernazza and Noyes is
the spherically symmetric component of a two-component model. Discrete cool features are assumed
to overlie the spherically symmetric component, and this modifies the interpretation of the
center-limb behavior of the Lyman continuum data. Values of τ_0 indicate the optical depth at the
head of the Lyman continuum.

increases slowly outwards. These effects produce limb-darkening at the head of the
continuum and limb-brightening at shorter wavelengths.

Noyes and Kalkofen (1970) were not able to reproduce all of the observed
characteristics of the Lyman continuum with the spherically symmetric model they
proposed. Vernazza and Noyes (1972) revised the model to allow for departures from
spherical symmetry and for the existence of a narrow temperature plateau in the upper
chromosphere at $T \approx 22\,000$ K.

Their model of the middle chromosphere consists of a spherically symmetric,
hydrostatic equilibrium component on top of which are relatively cool dense absorbing

features identified by Vernazza and Noyes with spicules, but which are more properly identified as fibrils. The spherically symmetric component is shown in Figure VII-7. Since the overlying fibrils are relatively cool and dense, they have a lower source function than the spherically symmetric component and they modify the predicted limb-darkening and limb-brightening. This, in turn, requires that the temperature distribution in the spherically symmetric component be modified in order that the combined effects of the two components reproduce the observed center-limb properties. The model does improve the synthesis with the observed continuum data. It is artificial, of course, in that the fibrils reside entirely above the spherically symmetric component rather than penetrate through it. Nevertheless, the model does demonstrate the need to consider departures from spherical symmetry in refined analyses. This conclusion is entirely consistent with the Balmer continuum data discussed earlier and with the wealth of chromospheric structure observed in strong Fraunhofer lines.

Strong additional evidence of the influence of non-spherical geometry comes from radio data. The notable lack of limb-brightening at mm wavelengths (cf. Figure V-29) stands in contradiction to any spherically symmetric model of the middle chromosphere in which T increases monotonically outwards. In order to suppress the expected limb-brightening for such models it is necessary to roughen the model in somewhat the same manner as Vernazza and Noyes have done. Thus, the Vernazza-Noyes model represents one of several models of similar nature.

The 22 000° plateau proposed by Vernazza and Noyes falls outside the range of the middle chromosphere and is discussed in the following section on the high chromosphere. The plateau is optically thin in the Lyman continuum, but, because of its higher temperature, it produces a distinctly observable increase in intensity at shorter wavelengths in the Lyman continuum.

D. SPECTRAL LINES ON THE DISK

We turn, next, to a brief consideration of disk data for chromospheric lines. Lines of interest include the stronger Fraunhofer lines and XUV emission lines. Again, we restrict the remarks here to only brief comments and defer some of the discussion to Chapter VIII.

Spectral lines formed in the middle chromosphere exhibit the prominent network structure described in Chapter II. It is clear from the nature of the network structure and the smaller scale structure of rosettes, fibrils, etc., that the lines cannot be treated in detail without recourse to non-spherical models. However, nearly all analyses of line data up to this time have assumed plane-parallel models. This is partly a reflection of the fact that detailed line profile data with sufficient spatial resolution to resolve individual chromospheric features has not been generally available and partly a reflection of the difficulty of solving the necessary radiative transfer equations in other than plane-parallel geometries.

Analyses of line data have the added complication of introducing the macroscopic velocity field as a fundamental parameter in the model atmosphere. The simple division of the macroscopic velocity into a 'microturbulence' and 'macroturbulence' is probably grossly inadequate for a proper treatment of the lines. An acceptable alternative has not yet appeared, however, so analyses still proceed based upon the artificial division into microturbulence and macroturbulence. In fact, the motions are probably of a wave nature with some more or less continuous distribution in wavelength and velocity as suggested by the oscillatory phenomena.

From still another standpoint, the line analyses often depend upon atomic parameters, such as collisional excitation and photoionization cross sections, that are not well known. These uncertainties result in uncertainties in the model atmosphere.

Model chromospheres constructed from line data are of the same general nature as the models shown in Figure VII-3a. The temperature minimum in such line models is near the same temperature and occurs at nearly the same depth as in continuum models. Also, the line models prefer an initial rapid increase in temperature in the low chromosphere followed by a more gradual increase in the middle chromosphere. The details of such models, however, are contingent upon the particular choices of microturbulence and macroturbulence velocities and, in particular, upon the velocity gradients that are chosen.

The model proposed by Vernazza *et al.* (1972) and shown in Figure VII-3a is a modification of the Vernazza-Noyes model shown in Figure VII-7. Modifications were made in order to better represent some of the line data as well as the Lyman continuum and free-bound continua of C I. The lines used include: Lyman-α, Lyman-β, Hα, Hβ, Pα and Ca II, H, K and the IR triplet. Other lines that have been treated in some detail include the Na D and Mg b lines, strong lines of Fe I, resonance lines of Mg II and XUV lines of O I and C II. These analyses, including those due to Vernazza *et al.*, are discussed in Chapter VIII. We note here only that the line analyses bring important additional information to the model atmosphere without requiring major revisions in the model.

E. HELIUM LINES

No discussion of the middle chromosphere is complete without attempting to account for the lines of He I and He II. Helium lines observed at the limb rise to maximum intensity some 1000 to 1500 km above the limb (see Figures V-5 and V-6). On the other hand, lines of several neutral metal atoms, including Fe, Mg, Ca, Al and Na, persist to well above 1500 km. Na has an ionization potential of only 5.14 eV whereas the λ 4686 line of He II observed at eclipse has an upper excitation potential of 50.8 eV. It is difficult to see how the He II line could be excited at temperatures low enough to leave sufficient abundance of Na I to permit observation of the D lines. Conversely, it is difficult to see why there should be any appreciable Na I emission at temperatures high enough to excite the λ 4686 line of He II. The same arguments apply to other neutral metals as well.

Two alternatives for explaining the helium emission require consideration. These are: (a) excitation of helium by XUV radiation from the corona and chromosphere-corona transition region, and (b) excitation of helium in high temperature regimes in a non-spherical chromosphere. Before considering these two alternatives it is instructive to consider some of the properties of the helium emission in more detail.

The relative intensities among the different subordinate lines of He I remain constant with height above 1100 km, i.e., the lines each exhibit the same scale height. This suggests that the chromosphere is optically thin in the subordinate lines, which is consistent with all current chromospheric models. In addition, it suggests that the mechanism of excitation is relatively constant with height also.

Since the chromosphere is optically thin in the helium lines, the observed line intensities may be converted directly into the populations, n_n, of the emitting levels by equating j_ν to $h_\nu A n_n$ in Equation (VI-10). We write the modified Saha Equation for He I in the form

$$n_j = 1.04 \times 10^{-16} \widetilde{\omega}_j n_e n_{II} T^{-3/2} b_j 10^{5040 x_j/T}, \tag{VII-38}$$

which may be rewritten as

$$\log \frac{10^{16} n_j}{1.04 \widetilde{\omega}_j} = \log n_e n_{II} T^{-3/2} + \frac{5040 x_j}{T} \log b_j. \tag{VII-39}$$

n_{II} is the population density of ionized helium. Thus, if we plot the quantity on the left hand side of Equation (VII-39) vs x_j, the intercept at $x_j = 0$ will be $\log n_e n_{II} T^{-3/2}$ and the slope of the line connecting a given point in the plot (given spectral line) to the intercept will be $(5040/T) \log b_j$. This latter quantity may be rewritten as $5040/T_{ion}$, where T_{ion} is the ionization temperature defined by $T_{ion} = T/\log b_j$. Such a plot is exhibited in Figure VII-8 taken from the work of Athay and Menzel (1956). The plot was constructed for a

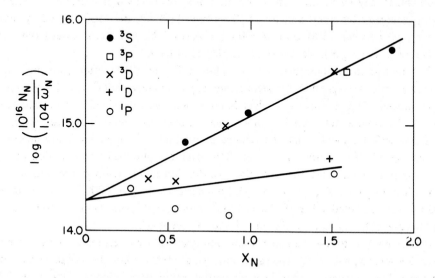

Fig. VII-8. Relative excitation of the He I singlets and triplets. The slopes of the lines are inversely proportional to the excitation temperature (after Athay and Menzel, 1956).

height of 1500 km and shows an intercept corresponding to $n_e n_{II} T^{-3/2} = 1.7 \times 10^{14}$. Since this analysis was completed, Henze (1969) has revised the absolute and relative intensity calibrations in the 1952 data. His corrected intensities reduce the 1952 intensities by a factor of 2.5 at λ 4000. The correction decreases to unity at λ 5500 and longer. For the plot in Figure VII-8, the lines near the left hand side of the plot lie near λ 4000 whereas those near the right hand side lie to the red of λ 5500. Thus, the plot should be corrected to reduce the intercept and steepen the slopes of the lines drawn through the points.

The corrected intercept in Figure VII-8 is estimated at $n_e n_{II} T^{-3/2} = 6.8 \times 10^{13}$. By comparison, the Balmer continuum data for the same height give $n_e n_p T^{-3/2} = 2.0 \times 10^{16}$. Hence, the ratio n_p/n_{II} is approximately 300 at a height of 1500 km if it is assumed that the hydrogen and helium emissions arise in a spherically symmetric atmosphere. The total hydrogen density (neutral plus ionized) at 1500 km is 2×10^{11} cm^{-3} in the Vernazza et $al.$ model, 8×10^{10} cm^{-3} in the HSRA and 1.1×10^{11} cm^{-3} in the Athay model (density reduced by a factor of two). From the above value of $n_e n_p T^{-3/2}$, we obtain $n_e = 1.3 \times 10^{11}$ for $n_e = n_p$ and $T = 8500°$ and n_e 1.1×10^{11} for $T = 7000°$. It seems reasonable to assume therefore that hydrogen is at least 50% ionized at 1500 km. Thus, the ratio $n_p/n_{II} = 300$ implies $n_H/n_{II} \approx 600$. The relative helium abundance in the Sun is near 10% and we conclude that about 98% of the helium at 1500 km is He I.

The ratio n_p/n_{II} determined in the above way decreases with height with a scale height of 1000 km for heights above 1100 km. Below 1100 km, however, n_{II} decreases rapidly and the ratio n_p/n_{II} therefore increases rapidly with decreasing height. Also, the ratio n_H/n_p increases rapidly with decreasing height reaching a value of about 25 at a height of 1100 km. It may be inferred, therefore, that in a spherically symmetric model the ratio n_H/n_{II} decreases by about a factor of 20 between 1100 km and 1500 km and by another factor of 3 between 1500 km and 2000 km (part of this last factor arises from further ionization of hydrogen). We will return to this point later.

The ionization temperatures for the plot in Figure VII-8, after correction of the intensity scale as suggested by Henze, are approximately 5000° for the triplets and approximately 10 000° for the singlets. Similar values were obtained by Hirayama (1971) from 1962 eclipse data. Studies of the helium excitation equilibrium (cf. Athay and Johnson, 1960) indicate that the relative populations of triplet levels are determined mainly by photoexcitation from the 2^3S level by photospheric radiation. Such a mechanism predicts that the excitation and/or ionization temperature will be given approximately by $B_\lambda(T_{ex}) = \frac{1}{2}B_\lambda(T_{rad})$. For an average T_{rad} of 6000° and an average λ of 5000 Å, the predicted value of T_{ex} is 5250°, which is sufficiently close to the ionization temperature determined from Figure VII-8 to be acceptable.

It is of interest to use the triplet excitation temperature to estimate the population of the 2^3S level at $x_j = 4.75$ eV. The value obtained at 1500 km is $n_2 3_S \approx 500$. We concluded from the above discussions that the total neutral helium population at 1500 km is 10% of the hydrogen population, or about 1×10^{10} cm^{-3}. The modified Boltzmann equation gives

$$\frac{n_1 1_S}{n_2 3_S} = \frac{\tilde{\omega}_1 1_S}{\tilde{\omega}_2 3_S}\frac{b_1 1_S}{b_2 3_S} e^{h\nu/kT_{ex}}$$

$$= \frac{\tilde{\omega}_1 1_S}{\tilde{\omega}_2 3_S} e^{h\nu/kT_{ex}}$$

and with $n_1 1_S = 10^{10}$ cm^{-3} and $n_2 3_S = 500$ cm^{-3} we find $T_{ex} = 12\ 700^8$. Similar estimates for He II at 1500 km yield $n_4 = 2 \times 10^{-3}$ cm^{-3}, $n_1 = 2 \times 10^8$ and $T_{ex} = 21\ 000°$. It is clear, therefore, that helium is excited either by a high temperature radiation field or in high temperature regions of the chromosphere that are much hotter than any of the spherically symmetric models allow.

There is no doubt that some of the helium excitation in the chromosphere is due to illumination by XUV radiation. On the other hand, there is no doubt that the chromosphere at heights of 1000 to 2000 km is highly structured and that spherically symmetric models are bound to fail. It is certainly possible, therefore, that both high temperature excitation and XUV illumination play important roles in exciting helium. However, for the time being, we will continue to discuss the illumination mechanisms assuming that it alone is responsible for the helium excitation.

Two questions must be asked of the XUV radiation and its influence on the chromosphere: (1) is there enough radiation to excite helium to the observed extent?; and (2) can the radiation penetrate far enough? The arguments by Athay and Menzel (1956) later repeated by Thomas and Athay (1961) that one XUV photon is required for each photon emitted in a singlet or triplet subordinate line of helium is not valid. Some of the subordinate line photons, perhaps nearly all of them, are produced by scattering of photospheric radiation. All the coronal photons do is provide a sufficient population in the $n = 2$ levels of helium. The same argument applies to the λ 4686 line of He II.

The He I and He II ions clearly will be excited thermally at some level in the upper chromosphere or in the chromosphere-corona transition region. Back radiation in λ 304 could then, presumably, excite He I at lower levels in the chromosphere.

Observations of the photon fluxes at XUV and X wavelengths by Hinteregger *et al.* (1965) yield photon fluxes at Earth of 3.8×10^9 cm^{-2} s^{-1} in λ 304 and 10.8×10^9 cm^{-2} s^{-1} shortward of λ 228 where He II ionization occurs. Thus, the coronal photon flux seems, at first glance, to be sufficient to produce the He II excitation. However, if He II is only a minor constituent in the chromosphere where the helium excitation occurs most of the incident coronal radiation will be absorbed by hydrogen and He I. The relative absorption coefficients at λ ⩽ 228 have the values:

$$\alpha_{0,H}\left(\frac{\lambda}{912}\right)^3 : \alpha_{0,\,He\,I}\left(\frac{n_{He\,I}}{n_H}\right)\left(\frac{\lambda}{504}\right)^2 : \alpha_{0,\,He\,II}\left(\frac{n_{He\,II}}{n_H}\right)\left(\frac{\lambda}{228}\right)^3$$

for H : He I : He II. The values of α_0 are: $\alpha_{0,H} = 6.3 \times 10^{-18}$ cm^{-2}, $\alpha_{0,HeI} = 7.6 \times$

10^{-18} cm^{-2} and $\alpha_{0,\text{He\,II}} = 1.8 \times 10^{-18}$ cm^{-2}. For λ 228, $n_{\text{He\,I}}/n_{\text{H}} = 0.1$ and $n_{\text{He\,II}}/n_{\text{H}} = 0.0017$, which are the values estimated at 1500 km, we find H : He I : He II = 0.39 : 0.60 : 0.01. For λ 114 the values are H : He I : He II = 2.4 : 0.75 : 0.01. If the He I to He II relative densities are correct, therefore, only about 1% of the incident coronal photons for $\lambda < 228$ will be absorbed by He II. The Balmer-α transition of He II at λ 1640 has a photon flux at Earth of approximately 2 x 10^9 photons cm^{-2} s^{-1} (Athay, 1965). This is about one fifth of the coronal flux below λ 228 and is clearly too large to be entirely due to fluorescence with coronal radiation. We cannot conclude from these arguments that the He I and He II emissions in the middle and upper chromosphere are not excited by XUV photons from the corona, but it does seem clear that such photons are not the sole source of He II excitation in the solar atmosphere.

The next question is whether the coronal radiation excites the He I flux. At $\lambda < 505$ the coronal photon flux, exclusive of λ 304, is about 30 x 10^9 cm^{-2} s^{-1}. The He I, λ584, photon flux from the chromosphere is 1.6 x 10^9 cm^{-2}, s^{-1}, so the coronal flux exceeds the He I, λ 584, flux by a factor of about 20. Furthermore, as we have seen a large fraction of the photons will be absorbed by He I. This raises the question as to whether the coronal photons can, in fact, penetrate sufficiently deep in the chromosphere to excite He I. At 1100 km, where the He I lines reach maximum intensity the optical depth at the head of the Lyman continuum is about 10^2 (Figure VII-7). At λ 504 the Lyman continuum optical depth at 1100 km is approximately 17 and the He I optical depth is about 12. Thus, the total optical depth at λ 504 is approximately 30 at a height of 1100 km. The optical depth drops to unity at this height near λ 150 and is due mainly to He I. Coronal radiation below λ 150 amounts to about 3 x 10^9 photons cm^{-2} s^{-1}. Only about one-quarter of the recombinations to He I will produce a λ 584 photon, but ample additional photons are available at wavelengths just above λ 150 where the optical depth is still not much in excess of unity. It appears, therefore, that coronal XUV radiation may possibly account for the He I excitation. Again, however, the He II emission apparently cannot be explained entirely by this process.

Added evidence that coronal radiation is not the sole source of excitation for chromospheric helium comes from the observed association of the λ 584 and λ 304 emission (Tousey, 1967) and λ 10 830 absorption (Zirin and Howard, 1966; Giovanelli *et al.*, 1972) with the chromospheric network. The absence of any pronounced network structure in coronal emission lines rules out the possibility that the enhanced helium excitation in the network could arise from enhanced coronal radiation.

An important clue, to the nature of the helium excitation, not yet fully resolved, is provided by the triplet to singlet intensity ratio. The different ionization temperatures for singlets and triplets results from different b_j coefficients for levels with equal values of x_j. It may be inferred from Figure VII-8 that the ratio b_{3^3}/b_{3^1} is approximately 10:1.

Athay and Johnson (1960) have shown that the overpopulation of the triplets relative to the singlets arises from a depopulation of the singlets to the ground state, 1^1S, via the escape of photons in the resonance lines, most notably the $2^1p - 1^1S$ line at λ 584. They find that the observed triplet:singlet ratio requires that the escape coefficient ρ_{21} in the

λ 584 line lie within the range 10^{-2} to 10^{-3}. These are relatively large values for the escape coefficient and they suggest that the optical depth in the λ 584 line is not too large. In a spherically symmetric model, this will not be the case unless helium is predominantly ionized, and on this basis Athay and Johnson concluded that the helium lines were formed in regions of high temperature in an inhomogeneous chromosphere. However, a satisfactory evaluation of the escape coefficients in helium has not yet been accomplished. The problem is a very difficult one and has defied a number of attempted solutions.

Many sources of evidence suggest that the middle chromosphere cannot be adequately described by a single component model. Departures from spherical symmetry probably become even more pronounced at greater heights where spicules are clearly observed. We will continue to discuss the mean model in terms of a spherically symmetric atmosphere, but we will comment specifically on spicule models in the next section.

6. The High Chromosphere

A. MEAN SPHERICALLY SYMMETRIC MODEL

The upper chromosphere is defined as the region between $T = 10^4$ K and $T = 10^5$ K. In all of the models of the middle chromosphere discussed in the preceding section, the gas pressure at the top of the middle chromosphere has already decreased to within a factor of two or three of the coronal value. On the other hand, the particle number density is still of the order of 10^{11}, which is some 150 to 200 times larger than the particle number density in the corona. Similarly, the temperature is still some 150–200 times lower than it is in the corona. Models of the upper chromosphere and the transition region are limited, therefore, by the constraint that $n_e T$ is approximately constant since hydrogen is predominantly ionized and n_e is proportional to the total particle density. This constraint requires that the temperature rise rapidly, i.e., that there are no regions in which the density drops appreciably without a corresponding increase in temperature. This does not mean that 'temperature plateaus' cannot exist in the upper chromosphere, but it does mean that any such plateau must extend less than a density scale height.

At temperatures above 10 000° in the solar atmosphere the density scale height is expanded by the reduced mean molecular weight resulting from ionization of hydrogen. The isothermal scale height is given by

$$H = \frac{kT}{\mu m_H g},$$

where μ is the mean molecular weight. For fully ionized hydrogen and helium and 10% helium, by number, $\mu = \Sigma m / \Sigma n \approx 14/23 = 0.61$. Thus, at $T = 20\ 000°$, $H \approx 500$ km.

The temperature plateau at 20 000° in the Vernazza *et al.* model shown in Figure VII-4 is only 150 km thick and the pressure has been assumed constant across the plateau. Earlier models of upper chromosphere temperature plateaus by Athay (1965),

Thomas and Athay (1961), and Athay and Menzel (1956) gave much broader plateaus at higher pressures. No serious attempt was made in these earlier models to insure pressure continuity with adjacent layers of the atmosphere and the models are incompatible with XUV data, particularly that for the Lyman continuum.

The existence of a temperature plateau near $22\,000°$ is suggested by Lyman continuum data for $\lambda < 750$ Å. At these wavelengths the continuum shows a color temperature of $22\,000° \pm 4000°$ (Vernazza and Noyes, 1972), as would be expected from the optically thin layer whose kinetic temperature is at this value. The intensity (or brightness temperature) of the Lyman continuum fixes the optical thickness of the plateau at about 0.05 at the head of the Lyman continuum and limits the geometrical thickness to about 200 km.

XUV line data also suggest the existence of a similar temperature plateau. Analyses of C II lines by Chipman (1971) and of Lyman-α and Lyman-β by Vernazza (1972) each yield independent evidence for such a plateau. Each of these emission lines is self-reversed. The self-reversal is sensitive to both the temperature and geometrical thickness of the plateau and it does not appear to be possible to explain the observed line profiles without a plateau similar to that required for the Lyman continuum (see Chapter VIII).

As yet, no conclusive evidence has been found in support of more than one temperature plateau in the upper chromosphere. Perhaps such evidence will appear when more detailed observations and analysis of XUV emission lines have been completed. Until there is some evidence to the contrary, it seems best to conclude that the temperature rises steeply from a value somewhat above $22\,000°$ to 10^5 °. It will be shown in the following section of this chapter that the steep rise in temperature continues above 10^5 °.

From the discussion thus far it appears that the most acceptable spherically symmetric model is one in which the temperature rises rather steeply from 10^4 ° to 10^5 ° broken only by a narrow plateau of about 200 km thickness near a temperature of $22\,000°$. The exact temperature gradients and their locations in height are not well known. The discussion of the low chromosphere model indicates that the 10^4 ° level should be reached near 1500 to 1800 km, and it may be inferred from this that the 10^5 level lies near 2000 km, perhaps lower.

In connection with the above conclusion, it is of interest to examine the limb data in the height interval around 2000 km. If, indeed, the chromosphere were spherically symmetric at these heights and the temperature rose rapidly from $10^4°$ to $10^5°$ at nearly constant pressure the emission spectrum at the limb should undergo dramatic changes. We would expect, for example, that the Balmer continuum intensity, which is proportional to the quantity $n_e m_p T^{-3/2}$, would decrease over this interval in T by a factor $10^{-3.5}$. Similarly the intensities of neutral and singly ionized metal lines should decrease sharply. Instead, both the Balmer line and continuum intensity and the metal line intensities (see Figures V-3, V-5 and V-7) decrease smoothly with height with more or less constant scale heights similar to those in the low chromosphere.

Obviously, there is something radically wrong with the model. What is wrong, of course, is that the model simply cannot be approximated, at the limb, by a spherically symmetric layer. This does not mean, however, that the spherically symmetric model is inapplicable to observations near disk center as well. It was shown in Chapter II that when even 0.1 to 1% of the surface is covered with spicule-like objects the effects at the limb are dramatic. It is entirely possible, and probable as well, that nearly all of the emission seen at the limb in lines and continua of neutral and singly ionized atoms above heights of 1500 km arises in spicules covering less than 1% of the solar surface. These same spicules would have a quite negligible effect on much of the emission observed near disk center.

The plausibility of this suggestion is demonstrated by the plot in Figure VII-9 showing the relative rates of decrease of spicule numbers (taken from Chapter II) and the emission in the Hβ and D$_3$ lines. The spicule number distribution decreases between 3000 and

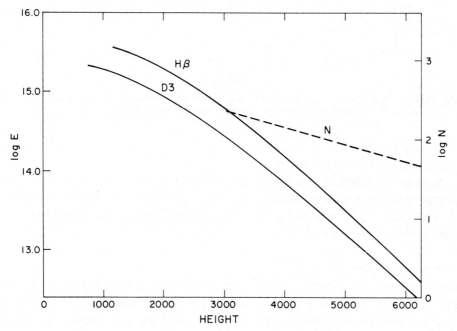

Fig. VII-9. A comparison of the gradient in spicule numbers, N, to the intensities in the Hβ and D$_3$ lines as observed at eclipse.

6000 km with a characteristic scale height of approximately 2200 km. By comparison, the D$_3$ and Hβ line emissions have scale heights of about 700 km. Intensity gradients in other strong lines of hydrogen and helium are shown in Figure VII-10 for comparison. Thus, the spicule number distribution falls off more slowly with height than does any of the line intensities. It follows that if the spicules are effectively thin and have outwardly decreasing material densities they can account for the observed line emission. Different

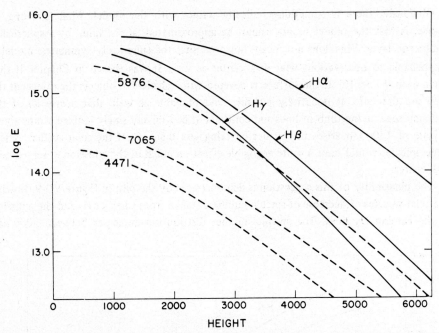

Fig. VII-10.   Relative height gradients and intensities of strong lines of hydrogen and He I (dashed
lines). The emission above a height of about 2500 km is predominantly from spicules.

scale heights observed for different lines could easily arise from slowly changing
excitation conditions within the spicules.

Because the effects of spicules, and possibly other chromospheric structures, are so
evident in the comparison of limb and disk models of the upper chromosphere we digress,
at this point, from considering just the spherically symmetric atmosphere to a
consideration of spicules models.

B. SPICULE MODELS

Direct estimates of the electron density in spicules has been made by Makita (1972) using
the intensity of continuum emission near $\lambda\,6900$ from isolated spicules observed at
eclipse. The value obtained averages $1.8 \times 10^{11}\,\mathrm{cm}^{-3}$ for four spicules. No height
information is given in Makita's results so the value of n_e obtained represents an average
density.

Estimates of electron density may be made also from the intensities of Balmer lines
observed at eclipse together with spicule counts. This point is discussed further in the
following section (C).

The different scale heights for spicule numbers and line intensities shown in Figure
VII-9 imply that n_e is decreasing with height in spicules at a rather significant rate. In the
range 3000 to 6000 km, the ratio $N/F_{H\beta}$ increases by a factor 20. Assuming T to be
constant, we could interpret this as a decrease in n_e by a factor $20^{1/2} = 4.5$ between

3000 and 6000 km. Above 6000 km, the rate of decrease of N increases appreciably. There are no reliable observations of the rate of decrease of Balmer line intensities above 6000 km for the average chromosphere. However, a number of observers have measured the brightness of individual spicules at the limb of the Sun outside of eclipse. Most of these data pertain to $H\alpha$, $H\beta$, H and K of Ca II and D_3 of He I. Beckers (1972) has summarized these data in an excellent review article. Between heights of 4000 km and 10 000 km, Krat and Krat (1971) report a decrease in the central intensities of the $H\alpha$ and D_3 lines by a factor of 7 and 40, respectively. The half width of the D_3 line increases by about 20% whereas the half width of $H\alpha$ decreases by about 40% over this height range. The combined changes in central intensity times the half widths are therefore inferred to be about a factor of 10 in $H\alpha$ and about a factor of 30 in D_3. By analyzing these and similar data, Beckers (1972) finds values of n_e decreasing from 1.5×10^{11} cm^{-3} at 4000 km to 3.4×10^{10} cm^{-3} at 10 000 km.

The overall rate of decrease of electron density found by Beckers is considerably less than is implied by the data in Figure VII-9. However, Beckers results are consistent with the increased rate of decrease of N with height above 6000 km.

In most of the spicule models it is assumed that the electron density is representative of the total hydrogen density, i.e., hydrogen is assumed to be largely ionized. If this is indeed the case, then the gas pressure in spicules is of the same order as the gas pressure in the surrounding atmosphere. As we have noted earlier, the product of $n_e T$ in the upper chromosphere and low corona is close to the value 6×10^{14}. For an electron density of 1×10^{11} cm^{-3}, this value of $n_e T$ is obtained for $T = 6000°$. Although the spicule temperature may be somewhat higher than this, it is certainly no lower and we see that the spicule gas pressure is of the order of the ambient pressure only so long as hydrogen is heavily ionized. It would not be too objectionable, perhaps, to have the spicule gas pressure somewhat above the ambient pressure, but it would be unreasonable to suppose that it were an order of magnitude higher.

Since it seems clear that hydrogen is strongly ionized, the continuity equation can be used to infer reasonable limits on the rate of density decrease. Observations of spicule velocities show no evidence of a systematic change with height. Also, spicule diameters seem to be more or less constant with height. This suggests that, in order to preserve continuity, the spicule density should be relatively constant. It might be supposed that material flows through the spicule, i.e., that the material velocity exceeds the spicule velocity, and hence that the material velocity could increase upwards. However, spicule Doppler shifts are consistent with their growth rates and this clearly requires that the material velocity is close to the apparent velocity. Some diffusion of material through the spicule borders may occur, of course, but it seems unlikely that this would occur to a marked extent without some evident increase in spicule diameter. It seems, therefore, that the continuity requirement limits the rate of decrease of electron density to something less than a factor of ten, say, over the length of the spicule. The preceding discussion of density changes based on line intensities is consistent with this conclusion.

Spicule temperatures are difficult to determine to any degree of precision using existing techniques. Values reported in the literature have ranged from 60 000° (Athay, 1958) to near 6000° (Vernazza and Noyes, 1972; Kawabata and Sofue, 1972). The higher values were obtained from early observations giving widths of hydrogen and helium lines and are clearly incompatible with more recent data.

The two main arguments for high spicule temperatures come from line widths and from the relative strength of He I emission. Many spicules exhibit relatively wide spectral lines with inferred broadening velocities of the order of 16 km s^{-1}. However, the widths of lines of different atomic species are inconsistent with the assumption that the lines are broadened by a simple combination of thermal and microturbulent motions. The lines of Ca II, in particular, are too wide in many spicules for such an interpretation. Average values of the reduced half width, w, defined as the total width at half intensity divided by the wavelength, as summarized by Beckers (1972) for a height of 6000 km are: hydrogen 1.3×10^{-4}, He I 0.7×10^{-4} and Ca II 1.1×10^{-4}. The reduced width of the Ca II lines should be less than the reduced width for the helium lines if the lines are formed in the same regions and if the broadening is a combination of microturbulent and thermal motions.

Several difficulties occur in the measurement and interpretation of spicule line profiles. The observations have been made outside of eclipse and always include a superposition of spectra from scattered disk and chromosphere light in addition to the spectrum of the spicule itself. In some cases the corrections are of small amplitude and can be readily accomplished. However, at low heights and at times other than those of excellent seeing the corrections can be large. It is not always clear exactly what the source of the scattered light is (photospheric or chromospheric) and thus what the spectrum of the scattered light is. Additional problems arise from the difficulty of isolating individual spicules and of knowing whether a given feature is a single spicule or a cluster of spicules. If it is the latter, the line may be spuriously broadened by different line-of-sight motions of individual spicules, i.e., the different spicules may have different Doppler shifts. Some authors (Athay and Bessey, 1964; Krat and Krat, 1971) distinguish between two classes of profiles in the Ca II lines with line widths differing by about a factor of three. Other authors (cf. Zirker, 1962) have noted the same duality in widths and have attributed the narrow class to single spicules and the broader class to clusters of spicules. This explanation, however, is not consistent with the overall Doppler shifts. As noted by Athay and Bessey (1964) the wide features show the same distribution in line shifts as do the narrow features. This would not be expected if the wide features themselves were a composite of differentially shifted features. Also, the broad features occur too frequently to be regarded as random statistical clustering and, in addition, a random statistical clustering should lead to a single peaked distribution in widths rather than a double peaked distribution.

Spicule line profiles are often asymmetric, sometimes reversed and frequently flat topped relative to a Gaussian profile. The mean reduced widths of 1.3×10^{-4}, 0.7×10^{-4} and 1.1×10^{-4} quoted for hydrogen, He I and Ca II lines correspond to mean broadening velocities for equivalent Gaussian profiles of 23, 13, and 20 km s^{-1}, respectively. The Ca II velocities are too large to be thermal and the Ca II profiles are usually more non-Gaussian than are the hydrogen and He I profiles.

Spicules are known to occur in regions of relatively strong magnetic fields. It seems highly unlikely that the ionized matter in spicules can move in the magnetic field with velocities of 25 km s^{-1} without inducing some type of ionic motion in directions normal to the bulk velocity. Spiraling of the ions around lines of force or a channeling of the ions along field lines in a helical field would broaden the line profiles. In this connection it should be noted that a number of authors (e.g. Pasachoff *et al.*, 1968) have observed line profiles that are tilted with respect to the line of dispersion, i.e., the red and blue wings lie at different apparent geometrical positions. Evidence that this is a real effect and not just the blending of line profiles of two or more nearby spicules comes from the persistence of the effect over a large height range and its occurrence at heights where only a few spicules are observed. The tilt of the profile is interpreted by some observers (e.g. Pasachoff *et al.*, 1968) as being due to a bodily rotation of the spicule as a whole. Certainly, such motion should not be surprising for some fraction of the spicules, and it suggests the possibility that bulk motions of related nature may substantially influence the line profiles in many spicules.

It would be of interest to compare profiles of lines of ions other than Ca II to see if the lines from positively charged ions differ systematically from lines of neutral elements. Unfortunately, it has not been possible to make such a comparison because of the difficulty of observing fainter emissions against the noise background of scattered light present in all observations made outside of eclipse. Spicules have been observed in a number of lines arising from positive ions, but the spicule emission is faint and reliable profiles have not been attained.

Since the Ca II lines show such strong evidence of being broadened and distorted by motions evidently related to the bulk motion of the spicule, it does not seem appropriate to assume that the hydrogen and He I lines are free of such effects. Thus, interpretation of these profiles as a result of a simple thermal and microturbulent component undoubtedly leads to unrealistic results, and one should not place much weight on spicule temperatures determined in this way. The mean reduced widths for hydrogen and helium of 1.3×10^{-4} and 0.7×10^{-4} would result from a temperature of 30 000° and a microturbulent velocity of 6 km s^{-1} if these were the sole sources of broadening. Other evidence indicates that this temperature is much too high and hence that other sources of broadening must be present.

A second general method for obtaining spicule temperatures comes from the use of the observed line intensities in spicules. Beckers (1972) proposes a value of $T = 16\,000°$, which supposedly accounts for all of the known line intensities. Also, this temperature is high enough that hydrogen is strongly ionized and there is no difficulty with the gas pressure in the spicule.

Still a third method of obtaining spicule temperatures comes from limb-darkening data for the Lyman continuum and for mm wavelengths. These data show a relatively flat disk intensity, whereas the spherically symmetric models predict limb-brightening. A number of authors, beginning with Coates (1958), have explained the absence of limb-brightening at mm wavelengths as being indicative of a relatively cool, opaque atmospheric structure that is seen preferentially near the limb. For this latter condition to be fulfilled, the cool components must have a significant vertical extent, and it has been natural to associate this component with spicules.

Analyses of limb darkening data yield information on the fraction of the solar surface covered by the cool components, and, in some cases, information on the temperature structure. Thomas and Athay (1961), for example, found from 8.6 mm data a temperature for the cold components of $T_c = 6500°$ and a fractional area of $a_c = 0.4$. Vernazza and Noyes (1972), using 1.2 and 3.5 mm data, find $T_c \approx 5090°$ and $a_c = 0.05$ to 0.1, and Kawabata and Sofue (1972), using 8.6 mm data, find $T_c < 6000°$ and $0.05 \leqslant a_c < 0.1$.

Vernazza and Noyes (1972) find that the same values of T_c and n_c that are needed to explain the 1.2 and 3.5 mm data predict the correct limb-darkening behavior in the Lyman continuum. Also, Withbroe (1970) finds that spectral lines formed in the chromosphere-corona transition region and that lie within the Lyman continuum show a marked reduction in limb-brightening as compared to lines lying outside the Lyman continuum. He interprets this, again, as an obscuration of the transition region at the limb by relatively cold, opaque features. Values of a_c inferred from this effect range from 0.05 to 0.2 for different lines. The temperature of the cold component is unspecified in Withbroe's model, but it must be low enough that the features are opaque in the Lyman continuum.

In each of the cases involving limb-darkening data, there is no direct evidence that the cold features are to be identified with spicules. In fact, the cold features studied in this way most probably represent the most abundant type of structure present at the altitude where the obscuration takes place. The Lyman continuum and mm spectrum for $\lambda \leqslant 8.6$ mm form below 2000 km. Also, the transition region is near 2000 km for the temperature range 5×10^4 to 3×10^5 K. Thus, all of the data used thus far in analyses of this type give the properties of the cold components below heights of about 3000 km. It follows that the features whose properties are inferred are very probably not spicules but are other features, such as the low lying fibrils of the network structure (see Chapter II).

Simon and Zirin (1969) have noted that the failure of limb-brightening at radio wavelengths to be as pronounced as is expected from spherically symmetric models extends to wavelengths of at least 75 cm. Again, they attribute the reduced limb brightening as evidence of cool elements in an inhomogeneous atmosphere. At 75 cm, the radiation arises in the upper transition region but at heights where spicules are still numerous. Thus, radiation at these wavelengths could perhaps provide useful information on spicule temperature, but no detailed analyses have been carried out for this purpose.

Although the limb-darkening data at mm and Lyman-continuum wavelength suggest cold components whose temperature is less than 6000°, there is no reason, on this basis, to reject Beckers' conclusion from line intensities that the spicule temperature is near

16 000°. Thus, we shall consider the analysis of line intensity data in somewhat greater detail.

C. SPECTRAL LINE INTENSITIES

Spicules show bright emission in the hydrogen Balmer lines, the Ca II H and K lines and the stronger singlets and triplets of He I. The excitation of these lines is influenced both by the thermodynamic conditions within the spicules and by the radiation incident on the spicules from the chromosphere and corona. Since spicules extend above the chromosphere proper and penetrate into the corona, they are bathed in the radiation from both the chromosphere and the corona.

If the spicules were optically thin at all wavelengths, it would be a relatively simple matter to compute the equilibrium conditions within the spicules and, hence, to compute their line emission. However, spicules are optically thick in the Lyman lines of hydrogen and in the resonance lines of He I and He II. Furthermore, they have non-negligible opacity in the Hα line and in H and K lines. The radiation transfer problem, therefore, is nontrivial. It is further complicated by the fact that spicules are cylindrical in form and special techniques are required for solving transfer problems in cylindrical geometries.

A proper solution of the radiation transfer problem for a cylindrical spicule model has been done only in the case of the K line using a model two-level atom. Avery and House (1969) treat the problem of a spicule represented by a cylinder 800 km in diameter and 10 000 km in length standing vertically above the chromosphere. Within the cylinder, temperature and density are allowed to vary in the axial direction but not in the radial direction. Calcium is assumed to be entirely singly ionized so that the only external radiation of consequence is the chromospheric emission in the K line. This is properly included in the transfer problem.

Avery and House (1969) use a Monte Carlo technique to compute the K line profile emergent from the spicule. Computed profiles are synthesized to profiles reported by Zirker (1962) for four spicule heights by adjusting the model parameters. The resultant model is characterized by an outwards increase in temperature from 6000° near the base of the spicule to 15 000° near the top of the spicule. Electron density decreases outwards from 8×10^{11} cm^{-3} at the spicule base to 4×10^{10} cm^{-3} at the top. In order to produce the observed widths of the profiles a random microturbulence of 20 km s^{-1}, uniformly in depth, is needed. In a later model (Avery, 1970) the spicule is rotated about the cylindrical axis with uniform angular velocity. Because of the rotation, less microturbulence is needed and a uniform value of 10 km s^{-1} is adopted arbitrarily. The spicule model is further relaxed by allowing T and n_e to vary radially as well as axially, but in the radial direction the condition $n_e T = $ const is imposed. Also, the spicule is assumed to be optically thin in the K line at all wavelengths and at all spicule heights. This latter assumption necessarily leads to different spicule models than found by Avery and House (1969), which was not optically thin.

Avery (1970) finds good agreement between computed and observed profiles, but his revised model is 10 000 to 15 000° hotter and 5 to 10 times more dense than the stationary model. Rotational velocities, at the periphery, are 36 km s^{-1} at a height of 3400 km decreasing to 22 km s at a height of 7300 km. Although the electron densities required in this model are too high to be compatible with continuum data and Balmer line data, the model is interesting in that it shows the direction in which rotation and optical thinness modify the model.

The Ca II equilibrium in a plane parallel slab of finite thickness has been treated by Giovanelli (1967a). Avery et al. (1969) have shown that the radiative transfer solution for a long cylinder is very similar to that of a plane parallel slab of the same total thickness. Hence, Giovanelli's results can be applied to spicules without introducing large errors through differences in geometry. Giovanelli (1967b) arrives at a spicule model in which $T_e \approx 20\,000°$ and $n_e = 3 \times 10^{10}$ cm^{-3} at a height of 6000 km. He uses the same data as used by Avery and House (Zirker, 1962). Using the same computations for Ca II but somewhat modified data, Beckers arrives at a model in which T increases from 9000° at 2000 km to 16 000° at 8000 km. Electron density decreases over this height interval from 1.6×10^{11} cm^{-3} to 4.3×10^{10} cm^{-3}.

Excitation and ionization of hydrogen in slab geometries illuminated by coronal and chromospheric radiation have been computed by Giovanelli (1967a) and Poland et al. (1971). These authors find that the populations of the n_2 and n_3 levels in hydrogen are nearly independent of T in the range 6000° to 20 000° and are closely proportional to n_e^2 for $10^{10} \leqslant n_e \leqslant 10^{12}$. Thus, the hydrogen lines are useful indicators of the mean value of n_e but are of little value for determining T.

Values of n_3 for hydrogen given by Zirker (1962) decrease from 69 cm^{-3} at 6000 km to 10 at 8600 km. Dunn (1960) gives a value of 9.5 cm^{-3} at 10 000 km. These estimates are based on observations of individual spicules at the limb. Similar estimates may be made using eclipse data. The average brightness of the chromosphere in Hα, as observed at eclipse, is of the order of 1.3×10^6 erg cm^{-2} s^{-1} at a height of 6000 km. From spicule counts at the limb, the deduced number of spicules in a 12° arc in equatorial regions at a height of 6000 km above the limb is 80. Assuming a mean spicule diameter of 800 km, we find the average line-of-sight intersection of spicule material to be $80\pi \times 4^2 \times 10^{14}/1.5 \times 10^{10} = 2.7 \times 10^7$ cm. (The denominator gives the linear length of 12° at the solar limb.) Hence, the average volume emissivity of spicules in Hα (assuming that spicules are optically thin) is $1.3 \times 10^6/2.7 \times 10^7 = 4.8 \times 10^{-2}$ erg cm^{-3} s^{-1}. By equating this emissivity to $h\nu A n_3$, we find $n_3 = 380$ cm^{-3}. This is a factor of 5.5 larger than the value given by Zirker. From the excitation tables given by Giovanelli (1967a), we find that $n_3 = 380$ cm^{-3} corresponds to $n_e \approx 8 \times 10^{10}$ cm^{-3}. The lower values of n_3 given by Zirker (1962) and Dunn (1960) lead to considerably lower values of n_e. The value of $n_e = |8 \times 10^{10}$ cm^{-3} is a little lower than the value obtained by Makita (1972) from continuum data, but it is in good agreement with the value adopted by Beckers (1972) on the basis of Ca II, Hα and He I line intensities. Evidently the n_3 values given by Zirker (1962) are too low.

A further point of interest in connection with the hydrogen excitation and ionization is the ratio n_H/n_e. Both Giovanelli (1967a) and Poland et al. (1971) find that for spicule size features and electron densities of 10^{11} cm^{-3}, or less, n_H/n_e is of order 1 to 2 for $T \geqslant 6000°$. Computations have not been carried out for lower values of T. However, at 6000° and for $n_H < 10^{11}$ cm^{-3} much of the ionization is due to the external radiation and it does not seem that reducing the temperature would change the ionization by a large factor. There seems to be no difficulty, therefore, with excessive gas pressure in a low temperature spicule model and n_e near 10^{11} cm^{-3}.

Excitation of helium lines in the chromosphere was discussed in Section 5 of this chapter. The discussion there applies equally as well to the spicule problem. Clearly, the low temperature spicule model ($T < 6000°$) would require coronal XUV as the sole source of helium excitation. The more moderate model with T near 15 000° provides some thermal excitation as well as the fluorescent excitation. A satisfactory solution to the problem of helium excitation in spicules has not been achieved and the questions raised by the helium lines cannot be resolved without a detailed solution of the radiative transfer problems, including the relevant incident radiation.

Some authors have found it necessary to introduce radial gradients in temperature across a spicule. Models of this type have usually been divided into core and sheath components, i.e., into two temperature regimes. In most models of this type, the core is assumed to be cold and the sheath hot, although in some cases, the inverse has been proposed. The primary reason for proposing a sheath-core model stems from the belief that the helium lines are thermally excited, and, hence, that a hot component is necessary. This may, in fact, be the case, but until such time as the helium problem is solved satisfactorily an isothermal spicule model seems to be all that is justified.

If it turns out that the cold features inferred from limb-darkening data at mm wavelengths and in the Lyman-continuum are indeed spicules rather than some basically different type of structure, it will probably be necessary to return to multicomponent spicule models. For the time being, however, we adopt the interpretation that the cold obscuring features at mm and Lyman-continuum wavelengths are low-lying fibrils that are distinct from spicules. For the spicules, themselves, we adopt a mean temperature of about 15 000° and a mean electron density of 1×10^{11} cm^{-3}. At this temperature, hydrogen is predominantly ionized so that the total hydrogen density is near 1×10^{11} cm^{-3} also. This choice of parameters gives a gas pressure that is a factor of 2.5 above the ambient pressure. It would be difficult to reduce the electron density below 1×10^{11} cm^{-3} by any appreciable amount without conflicting with the hydrogen line and the electron scattering continuum data. Pressure equality with the ambient medium could be achieved by reducing the temperature to 6000°. Poland $et\ al.$ (1971) find that even at this low temperature features, such as spicules, that are optically thin in the Lyman continuum are still heavily ionized so that the electron density remains close to the total hydrogen density. Thus, the gas pressure drops in proportion to the temperature for a fixed value of n_e. The primary objection to the 6000° spicule model is that it requires all of the helium excitation to be by XUV radiation from the corona. The objection to this, in turn, is that the network regions where spicules occur are markedly bright in the helium lines but not in coronal lines. Also, the Ca II emission and the widths of line profiles in spicules favors the higher temperature model. Nevertheless, we regard the adopted spicule model as tentative and in need of further study.

7. The Chromosphere-Corona Transition Region

The chromosphere-corona transition region, hereafter referred to simply as the transition region, is observed at cm wavelengths in the radio spectrum and at XUV wavelengths in the spectral lines of multiply ionized elements. Radio data have been available for some 20 yr, but they have not led to successful models of the transition region, for reasons to be discussed shortly. The availability of XUV data from rocket and satellite experiments has sparked renewed interest in the transition region.

A given stage of ionization for a particular ion exists in high relative abundance through only a limited range of temperature, whereas the transition region spans the temperature range from approximately 5×10^4 ° to 1×10^6 °. For heavy elements, the corona is characterized by the presence of ions that are ionized some 10 to 15 times. These same elements are only 2 or 3 times ionized in the upper chromosphere. The several intervening ionization stages occur in the transition region. Line intensities observed for the transition region ions yield direct information on the structure of the transition region. Because the extent of the transition region in temperature is much broader than the domain in which any one ion is populous, the spectral lines provide a sensitive probe of the transition region. They reveal a model of remarkable characteristics and one that poses many challenging problems to theorists.

It is readily shown that the solar corona is effectively thin ($\tau_0 < \tau_{th}$) in all spectral lines in the XUV and visual spectrum, and, indeed, that it is optically thin ($\tau_0 < 1$) in most of the lines. Consider, for example, an abundant element such as oxygen. Assume that all of the oxygen atoms are in one ion stage, that all ions of that stage are in the ground state and that the relative abundance of oxygen is 10^{-3} that of hydrogen. The hydrogen density at the base of the corona is about 3×10^8 cm^{-3} and the coronal density scale height is 10^{10} cm. Thus, there are 3×10^{18} hydrogen atoms and 3×10^{15} oxygen atoms in a vertical column of 1 cm^2 cross section. The absorption coefficient at the center of a spectral line is given by $\alpha_0 = \pi \epsilon^2 f \lambda / mc \sqrt{\pi} v$. For $f = 1$, $\lambda = 1000$ Å, and $v = (2kT/m)^{1/2}$, where $T = 1.8 \times 10^{6}$°, we find $\alpha_0 = 3.4 \times 10^{-14}$ cm^2. Thus, the optical thickness of the corona, in this highly idealized case, is $3 \times 10^{15} \times 3.4 \times 10^{-14} = 10^2$. This is small compared to the thermalization length in the corona, which, for a strongly permitted line, may exceed 10^6 (see Chapter VI).

The transition region is more dense than the corona but also much thinner in geometrical extent. As we shall soon see, the lower, denser layers of the transition region are also the thinnest layers and, as a result, the optical thickness of the transition region in spectral lines is even less than the optical thickness of the corona.

In the effectively thin case, the flux in a spectral line is given by Equation (VI-74), which we repeat here in the form

$$2\pi H = \tfrac{1}{2} h\nu AQ_0 x < G(\tau_1 n_e)> \int_{h(T_1)}^{h(T_2)} n_e^2 \, dh. \tag{VII-40}$$

The quantity $2\pi H$ corresponds to the flux emitted *outwards* at the Sun. This will give rise to an observed flux at Earth of $2\pi H(R_{AU}/R_\odot)^{-2} = 2\pi H/4.62 \times 10^4$. Inward radiation is absorbed in the chromosphere or photosphere by continuum absorption and is reradiated at different wavelengths.

Computation of the function $G(T, n_e)$ requires a proper knowledge of atomic parameters. The function may be obtained from the relative ion concentrations defined by the set of equations

$$R_m \sum_n n_m = \sum_n R_{nm} n_n, \tag{VII-41}$$

where n_m is the density of ions of stage m. Included in $R_{m+1,m}$ are the various rates of recombination such as two-body radiative recombination, dielectronic recombination and three-body collisional recombination. The latter is usually negligible in the transition region and in the corona. Included in $R_{m,m+1}$ are direct collisional ionizations as well as auto-ionizations following collisional excitation to auto-ionizing states. Radiative excitations and ionizations are negligible in the transition region and in the corona owing to the relatively weak radiation field.

The need for including di-electronic recombinations was first noted by Burgess (1964) and Burgess and Seaton (1964). Inclusion of these rates strongly modifies the equilibrium in certain ion stages. Goldberg *et al.* (1965) have noted the importance of auto-ionization following collisional excitation. This effect compensates partially for the effect of di-electronic recombination, but is generally less effective. Thus, the combined effects of di-electronic recombination and auto-ionization result in a marked change in the ionization from the simple case of radiative recombination and direct collisional ionization, which are all that were normally considered prior to 1964.

Jordan (1969, 1970) has computed n_m/n_{el} for a number of elements for conditions appropriate to the solar corona and the transition region. She considers two cases: one for which

$$\alpha_m = \alpha_{rm} + \alpha_{dm} \tag{VII-42}$$

and a second for which

$$\alpha_m = \alpha_{rm} + \alpha_{dm}(\text{eff}) + \alpha_{bm}, \tag{VII-43}$$

where α_{dm} is the di-electronic recombination coefficient, $\alpha_{dm}(\text{eff})$ is the same coefficient corrected for a density dependent term, α_{rm} is the radiative recombination coefficient summed over low-lying energy levels and α_{bm} is the radiative transition rate from all levels that are close enough to the continuum to be in equilibrium with it, and which lie below the first ionization limit, i.e., levels that are to be treated as part of the continuum (Wilson, 1967). The first case is applicable at densities below $n_e \approx 10^9$ cm^{-3} and is appropriate for the corona. The second case is applicable at densities greater than

$n_e \approx 10^9$ cm^{-3} and is appropriate for the transition region. However, the multiplicative factors for correcting α_{dm} to α_{dm}(eff) are not well known. Burgess (1965b) has computed density dependent correction terms, D, defined by α_{dm}(eff)$= D\alpha_{dm}$, for Ca II. Jordan uses these same correction terms for all ions. Expressions used by Jordan for α_{rm}, α_{dm} and α_{bm} are:

$$a_{rm} = 1.3 \times 10^9 n_e(m + 1)^2 x^{\frac{1}{2}}{}_m T^{-1}, \qquad T \geqslant 6 \times 10^5, \qquad (\text{VII-44})$$

Burgess and Seaton (1964);

$$\alpha_{rm} = 0.97 \times 10^{-12} n_e x_m n_0 g T^{-1/2}, \qquad T < 6 \times 10^5, \qquad (\text{VII-45})$$

Elwert (1952) (here x_m is the ionization potential in eV, n_0 is the ground state total quantum number, $g = 4$ for Fe and Ni and $g = 5$ for lighter ions);

$$\alpha_{dm} = 3.0 \times 10^{-3} n_e T^{-3/2} B(z) \sum_j f(j, i) A(x) e^{-E/kT}, \qquad (\text{VII-46})$$

where

$$A(x) = z^{\frac{1}{2}}(1 + 0.105\, x + 0.015 x^2)^{-1}, \qquad x > 0.05,$$

$$x = (z + 1)\epsilon_{ij},$$

$$B(z) = z^{\frac{1}{2}}(z + 1)^{5/2}(z^2 + 13.4)^{-\frac{1}{2}}, \qquad z \leq 20,$$

$$\bar{E}/kT = 1.58 \times 10^5 (z + 1)^2 \epsilon_{ij} T^{-1} a^{-1},$$

$$a = 1 + 0.015 z^3 (z + 1)^{-2}, \qquad E/kT \lesssim 5$$

$$\epsilon_{ij} = \nu_i^{-2} - \bar{\nu}_j^{-2},$$

and where ν_i and ν_j are the effective principal quantum numbers of states i and j, z is the charge on the recombining ion, $f(j,i)$ is the oscillator strength of the recombining ions, L is the initial state of the recombining ion and the sum in Equation (VII-46) is over all j such that $\epsilon_{ij} > 0$ (Burgess, 1965a);

$$\alpha_{bm} = 1.2 \times 10^{-6} n_e z^4 T^{-3/2} n_t^{-1} e^{\phi t/kT},$$

$$n_t = 1.26 \times 10^2 z^{14/17} n_e^{2/17} \left(\frac{kT}{z^2 E_H} \right)^{1/17} \exp \left(\frac{4 z^2 E_H}{17 n_t^3 kT} \right),$$

$$\phi_t^7 = 6 \times 10^{-28} \frac{x_m}{kT} n_e^2,$$

and E_H is the ionization energy of hydrogen (Wilson, 1967).

The collisional ionization rate used by Jordon (1969) is

$$q_{mH} = C_{m+1} + q_{auto,\, m+1}, \qquad (\text{VII-47})$$

where

$$C_{m+1} = 2.0 \times 10^{-8} n_e T^{1/2} \sum_j \xi_i(n,l) x_i^{-2}(n,l) 10^{-5040 x_i/T} \qquad \text{(VII-48)}$$

(Seaton, 1964) and

$$q_{auto,m+1} = 1.7 \times 10^{-3} n_e \sum_j f_{ij} x_{ij}^{-1} T^{-\frac{1}{2}} 10^{-5040 x_{ij}/T_{\bar{g}}} \qquad \text{(VII-49)}$$

$$\bar{\bar{g}} = \int_0^\infty \bar{g} \, e^{-E'/kT} \, d(E'/kT), \qquad \text{(VII-50)}$$

$$E' = E - E_{ij},$$

(Seaton, 1964), and where, as before, x_{ij} is the excitation energy in volts and E_{ij} is the excitation energy in erg, $\bar{g}$ is a tabulated gaunt factor of order unity, and $\xi(n,l)$ is the number of electrons with quantum numbers (n,l). The sum in Equation (VII-48) is taken over the different combinations of (n,l) and the sum in Equation (VII-49) is over all j for which autoionization occurs preferentially over radiative decay.

Representative results from Jordan's computations for silicon and iron are shown in Figure VII-11. Results obtained earlier by House (1964) in which α_d, α_b and q_{auto} were not included are shown for comparison. Note that di-electronic recombination extends somewhat the range of temperature for which a given ion is populace and shifts the temperature of maximum concentration to a higher value. The results plotted in Figure VII-11 are for α_{tot} given by Equation (VII-42), i.e., they contain an estimate for the reduction in α_{dm} due to density dependence. For the model adopted by Jordan, this reduces α_{dm} for temperatures less than about 5×10^5 ° and, at these temperatures, the results lie intermediate to House's results and the results using Equation (VII-44). For example, the Si VI ion reaches maximum concentration at 3×10^5 ° for $\alpha_d = 0$ (House's results) at 3.6×15^5 ° for the density decreased α_d and at 4.5×10^5 ° for the full α_d. (Note that the curves in Figure VII-11 are labeled by the number of electrons removed so that the curves labeled 5 correspond to Si VI, etc.) .

Computations of the ionization equilibrium, such as are illustrated in Figure VII-11, give the function $g(T,n_e)$ defined in Chapter VI. The function $f(T_e)$ is usually taken to be of the form given by Equation (VII-48) with the factor n_e omitted and $\epsilon_1(\tau)$ is given by the Boltzmann equation. Thus the function $G(T,n_e)$ in Equation (VI-40). given by the product of $f(T)$ and $g(T,n_e)$ is readily integrated over temperature.

Pottasch (1964) noted that $G(T,n_e)$ was a sharply peaked function of T for a given ion and removed an average value of $G(T,n_e)$ from the integral as in Equation (VII-40). He then interpreted the remaining integral

$$\int_{h_1}^{h_2} n_e^2 \, dh$$

as giving the 'emission measure' for the ion between the limits h_1 and h_2 where T had the limits T_1 and T_2 and the mean value T_0. This leaves as unknown on the right hand side of Equation (VII-40): the two factors A_{el} and $\int_{h_1}^{h} n_e^2 \, dh$. By using observed fluxes in lines

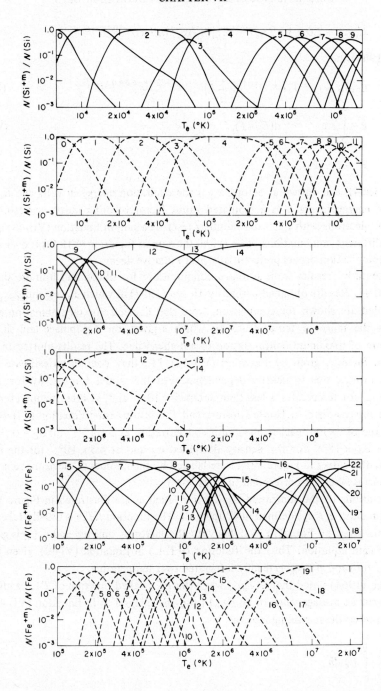

Fig. VII-11.   Concentration of ionic species of Si and Fe as functions of temperature (after Jordan, 1969). The dashed curves are results obtained by House (1964) neglecting dielectronic recombinations and auto-ionizations.

of several ions of a given element, therefore, Pottasch was able to construct a curve relating $A_{el} \int_{h_1}^{h_2} n_e^2 \, dh$ to T_0. He next argued that for a given T_0 there can be only one value of $\int_{h_1}^{h_2} n_e^2 \, dh$ since the characteristic limits h_1 and h_2 have approximately the same ratio for each ion. Hence, by adjusting the curves of $A_{el} \int_{h_1}^{h_2} n_e^2 \, dh$ vs T_0 for each element he was able to obtain both the abundances of the elements relative to a standard element (Si) and a well defined curve of $A_{Si} \int_{h_1}^{h_2} n_e^2 \, dh$ vs T_0. By using radio brightness temperatures, Pottasch obtains the value A_{Si}, and by adopting a particular density model (a composite model from different authors) he finally obtains T as a function of height.

One of the striking results of Pottasch's work was the steepness of the transition region. The temperature was found to increase from 5×10^4 ° to 5×10^5 ° in only 460 km and to 1×10^6 ° in an additional 3200 km. By comparison the hydrostatic, isothermal scale heights at 5×10^4 ° and 5×10^5 ° are 2500 km and 25 000 km, respectively. It may be inferred from this that the thickness of the transition region is much less than the pressure scale height, or, in other words, that the gas pressure is nearly constant within the transition region.

Since the gas pressure appears to be nearly constant within the transition region, it is useful to introduce this condition directly in the equation for the line flux. Also, since this eliminates n_e from the integral it is of further value to introduce a change of variable from h to T. Thus, Equation (VII-40) may be replaced by Equation (VI-75), where it is assumed that dT/dh is constant between T_2 and T_1. Athay (1966) used the equation in this form, and, following a procedure parallel to that adopted by Pottasch, obtained a curve giving $A_{Si} P_0^2 (dT/dh)^{-1}$ vs T_0 and the abundances of elements relative to silicon. Since A_{Si} and P_0 are presumably constant throughout the transition region, the analysis gives the relationship between $(dT/dh)^{-1}$ and T.

Athay (1966) found from an analysis of the type just mentioned that between approximately 10^5 ° and 10^6 ° T and dT/dh were related to close approximation by the condition $T^{5/2} \, dT/dh = $ const. In the absence of a magnetic field, or in a vertical magnetic field, the energy flux carried by thermal conduction, F_c, is proportional to $T^{5/2} \, dT/dh$. We denote this latter quantity by F_0.

Elwert and Raju (1972), Kopp (1972), and Dupree (1972) have noted that since F_0 seems to be constant this condition should be utilized in Equation (VII-40) rather than the condition $dT/dh = $ const. Thus, Equation (VII-40) is replaced by Equation (VI-76). The curve fitting procedure followed by Pottasch now yields $A_{Si} P_0^2 F_0^{-1}$. Elwert and Raju (1972) have analyzed XUV data using Equation (VI-76) with Jordan's (1969) density dependent ionization equilibria and with relative abundances of elements as given by Pottasch (1967) for the corona. For T in the approximate range 10^5 ° to 10^6 °, and for $P_0 = 6 \times 10^{14}$ cm ^{3}, they obtain (Figure VII-12)

$$A_{Si}^{-1} F_0 = 6.7 \times 10^{15} \, \text{cm}^{-1}. \tag{VII-51}$$

It should be noted that outside the range 10^5 ° to 10^6 ° the assumption that F_0 is constant is quite bad. Earlier results by Athay (1966) and Dupree and Goldberg (1967) suggest that for T between 2×10^4 ° and 10^5 ° a form $T^{-n} \, dT/dh = $ const, where n is of

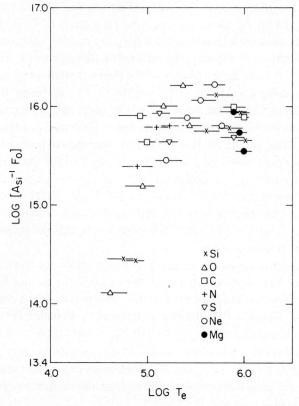

Fig. VII-12.   Values of $A_{Si}^{-1} F_0$ obtained by Elwert and Raju (1972) for the transition region under the assumption $F_0/A_{Si}P_0^2$ is constant. A specific value of $P_0 = 6 \times 10^{14}$ has been used.

order 4, is a better approximation. No analysis has been carried out using an appropriate form for $T^{-n} \, dT/dh$ for $T < 10^5 \,°$.

Dupree (1972) has carried out an analysis similar to that of Elwert and Raju (1972). Her analysis differs from that of Elwert and Raju in the following important respects. She uses: (1) line fluxes observed at the center of the solar disk rather than averaged over the disk; (2) density-dependent ionization equilibria ($\alpha_m = \alpha_{rm} + \alpha_{dm}$, $C_{m+1} = q_{m+1} + q_{auto,m+1}$) (Allen and Dupree, 1969); (3) relative abundances that are nearer to photospheric values than those obtained by Pottasch (1967); and (4) corrections to line intensities for certain ions, particularly Li-like ions, for which part of the observed radiation is of coronal origin. This latter correction arises from the fact that the Li-like ions have a high temperature 'tail' on the ionization curve, as is evidenced by curve 13 (Si XII) in Figure VII-11. Dupree's results for $A_{Si} = 3.5 \times 10^{-5}$ are shown in Figure VII-13. Between the temperature limits $1.6 \times 10^5 \,°$ and $1.6 \times 10^6 \,°$, Dupree finds $P_0^2 F_0^{-1} \approx 4 \times 10^{17}$, and for $P_0 = 6 \times 10^{14}$ (the value used by Elwert and Raju) and $A_{Si} = 3.5 \times 10^{-5}$ (Dupree's adopted value) this corresponds to

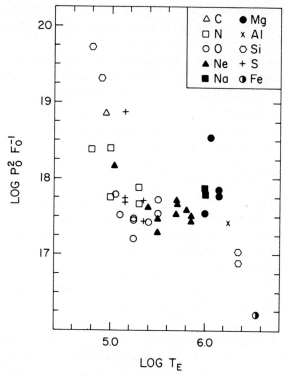

Fig. VII-13. Values of $P_0^2 F_0^{-1}$ obtained by Dupree (1972) for the transition region under the assumption $F_0/A_{Si} P_0^2$ is constant. A specific value of $A_{Si} = 3.5 \times 10^{-5}$ has been used.

$$A_{Si}^{-1} F_0 \approx 2.5 \times 10^{16}. \tag{VII-52}$$

This value is a factor of 3.7 above the value found by Elwert and Raju (1972). The difference evidently results from a combination of the four factors listed above. It may well be that the structure of the transition region is a function of the solar cycle, in which case an additional source of differences in F_0 is present. Dupree used data observed in 1967, near sunspot maximum, whereas Elwert and Raju used data from several sources obtained at various epochs in the sunspot cycle. The analysis by Dupree is deficient in the temperature regimes $T < 10^{5\,\circ}$ and $T > 1.6 \times 10^{6\,\circ}$ for the same reasons as were discussed in connection with the Elwert-Raju analysis.

Munro and Withbroe (1972) have studied the transition region within a coronal hole. From the intensity of coronal radiation from the hole, they conclude that the coronal density is decreased by a factor of three from the normal coronal density. Assuming that this same reduction in density occurs in the transition region, we find that the quantity P_0^2 is reduced by about an order of magnitude in the transition region. However, Munro and Withbroe (1972) note that the transition region ions show little, if any, decrease in intensity in the coronal hole. Hence, they conclude that the quantity $P_0^2 (dT/dh)^{-1}$ (see

Equation (VI-75)) has the same value in the hole as in normal regions of the corona. Since P_0^2 is reduced by an order of magnitude in the hole, it follows that (dT/dh) must also be reduced by a similar factor.

Kopp (1972) has considered a non-planar model of the transition region in which he argues that the mean value of F_0 is markedly reduced from the values given by either Equation (VII-51) or Equation (VII-52). His analysis is rather involved but can be reduced, in essence, to a simple form. Basically, he argues that the transition region can be divided into two components of fractional areas a_1 and a_2. He divides the quantity $P_0^2 F_0^{-1}$ obtained from the line data into two components

$$\langle P_0^2 F_0^{-1} \rangle = a_1 P_1^2 F_1^{-1} + a_2 P_2^2 F_2^{-1}. \tag{VII-53}$$

He then sets $a_2 P_2^2 F_2^{-1} \ll a_1 P_1^2 F_1^{-1}$ and $a_1 \ll a_2$. These latter two conditions require $P_2^2 F_2^{-1} \ll P_1^2 F_1^{-1}$, which is presumably due to the condition $P_2 \ll P_1$. Kopp then equates the mean value $\langle F_0 \rangle_1$ of F_0 due to region 1 to $a_1 F_1$, i.e., $\langle F_0 \rangle_1$ is the value of F_1 averaged over the solar surface including region 2 as well as region 1. Kopp shows in his analysis that $a_1 P_1 \ll P_0$, and it follows from Equation (VII-53), with the P_2 term set to zero, that $F_0 \gg a_1 F_1$, i.e., $\langle F_0 \rangle_1 \ll F_0$. This result means that the conductive energy flux over the network area is markedly reduced below the flux derived for the spherically symmetric case. The analysis to this point, which is as far as Kopp carried it, says nothing about the magnitude of F_2. It is possible, however, to estimate the magnitude of F_2.

Along with the constraint that is imposed upon $P_0^2 F_0^{-1}$ by dividing the atmosphere into two components, there is the added constraint imposed from continuum data that

$$\langle n_e \rangle = a_1 n_1 + a_2 n_2.$$

If we then make the reasonable supposition that the temperature is approximately the same in regions 1 and 2, this latter constraint on $\langle n_e \rangle$ becomes

$$\langle P_0 \rangle = a_1 P_1 + a_2 P_2. \tag{VII-54}$$

It follows from Equations (VII-54) and (VII-53) that

$$\langle P^2 F^{-1} \rangle = a_1 P_1^2 F_1^{-1} \left(1 + \frac{a_1 F_1}{a_2 F_2} \right) + \langle P_0 \rangle^2 F^{-1} \left(\frac{1}{a_2} - \frac{2a_1}{a_2} \frac{P_1}{\langle P_0 \rangle} \right) \tag{VII-55}$$

and if we use Kopp's conclusions that $a_1 \ll a_2$, $a_1 P_1 \ll \langle P_0 \rangle$ and $a_1 F_1 \ll F_0$ we are left with

$$\langle P_0^2 F_0^{-1} \rangle \approx \langle P_0 \rangle^2 F_2^{-1},$$

which implies that F_2 is of the order of F_0. Since the magnetic field is expected to have a strong horizontal component throughout most of the supergranule cell inside of the network, the result $F_2 \approx F_0$ implies that the conductive flux over the cell itself is reduced substantially below the flux derived for the spherically symmetric model. Thus, the total conductive flux is reduced substantially in this model.

Kopp's analysis is based on the observation by Reeves and Parkinson (1972) that the network features persist through the transition region. Their observations are made with only 35″ spatial resolution and show an actual contrast between network and supergranule cell of approximately 2 to 1. By assuming that the network width is 5 ″, they conclude that the true intensity contrast is more nearly 10 to 1. This conclusion is supported by initial results from ATM data with 5″ resolution (Reeves *et al.*, 1974). The ATM data show strong contrast between network and cell for lines formed in the transition region. Maximum contrast occurs at a temperature near 2×10^5 ° (O IV) and is observed to reach typical values of 10 to 15 at this temperature. Somewhat lower contrast is observed near 1×10^5 ° (C III) and 3×10^5 ° (O VI). As noted earlier, the network is still evident at 5×10^5 ° (Ne VII), but is largely indistinct at 1.5×10^6 ° (Mg X).

It is of interest at this point to evaluate $P_0^2 F_0^{-1}$ from radio data using the same assumptions as were used in the analysis of XUV data, i.e., $P_0 = $ const and $F_0 = $ const. The brightness temperature of the Sun at disk center at cm wavelengths is given by

$$T_b = \int_0^\infty T e^{-\tau_\lambda} \, d\tau_\lambda.$$ (VII-56)

This follows from the fact that the source function for the free-free radio continuum (Chapter VI) is given by the Rayleigh-Jeans law. The integral in Equation (VII-56) may be divided into two components, a coronal component and a transition region component. For the coronal component, we take $T_c = $ const $= 1.6 \times 10^6$ K and $n_e = 4 \times 10^8 \, e^{-h/H_c}$. To obtain H_c, we set

$$H_c = \frac{kT_c}{\mu m_H g} = 8 \times 10^9 \text{ cm}.$$ (VII-57)

Equations (VI-32), (VI-33) and (VI-34) give

$$K_{\lambda c} = 2.4 \times 10^{-22} n_e^2 T^{-3/2} \lambda^2 = 1.9 \times 10^{-14} e^{-2h_0/H_c} \lambda^2.$$ (VII-58)

At the base of the corona, $h_0 = 0$, and

$$\tau_{\lambda c} = 1.9 \times 10^{-14} \frac{H_c}{2} \lambda^2 = 7.6 \times 10^{-5} \lambda^2.$$ (VII-59)

Thus, for $\lambda \ll 100$ cm, $\tau_{\lambda c} \ll 1$, and we may set

$$T_{bc} = \int_0^{\tau_{\lambda c}} T_c \, d\tau_{\lambda c} = T_c \tau_{\lambda c} = 122 \, \lambda^2.$$ (VII-60)

The transition region contribution to T_b is given by

$$T_{bt} = \int_{\tau_{\lambda c}}^\infty T_t \, e^{-\tau_{\lambda t}} d\tau_{\lambda t},$$ (VII-61)

which may be approximated by

$$T_{bt} = \int_{\tau_{\lambda c}}^{1} T_t \, d\tau_{\lambda t}, \tag{VII-62}$$

i.e., by setting $e^{-\tau_{\lambda t}} = 1$ for $\tau_{\lambda t} \leqslant 1$ and $e^{-\tau_{\lambda t}} = $ for $0 \; \tau_{\lambda t} > 1$.

Within the transition region the gaunt factor, g, given by Equation (VI-33) varies from about 6.1 to 7.8. We adopt a mean value of 7.0, which gives

$$K_{\lambda t} = 1.9 \times 10^{-22} n_e^2 T^{-3/2} \lambda^2. \tag{VII-63}$$

Equation (VII-62) then becomes

$$T_{bt} = -\int_{\tau_{\lambda c}}^{1} 1.9 \times 10^{-22} n_e^2 T^{-\frac{1}{2}} \lambda^2 \, dh$$

$$= -\int_{T_c}^{T(T_\lambda = 1)} 1.9 \times 10^{-22} \lambda^2 P_0^2 F_0^{-1} \, dT$$

$$= 1.9 \times 10^{-22} \lambda^2 P_0^2 F_0^{-1} [T_c - T(\tau_\lambda = 1)]. \tag{VII-64}$$

Sufficiently deep into the transition region that $T(\tau_\lambda = 1) \ll T_c$, which occurs for $\lambda \lesssim 30$ cm, Equation (VII-64) becomes

$$T_{bt} = 3.0 \times 10^{-16} \lambda^2 P_0^2 F_0^{-1}. \tag{VII-65}$$

Equations (VII-59) and (VII-65) combine to give

$$T_b / \lambda^2 = 122 + 3.0 \times 10^{-16} P_0^2 F_0^{-1}. \tag{VII-66}$$

From the data in Figure V-27 we note that T_b/λ^2 is approximately 270 at $T_b = 10^5 \, °$. This corresponds to $P_0^2 F_0^{-1} = 4.9 \times 10^{17}$ in Equation (VII-66), which is very close to the value 4×10^{17} inferred from Dupree's results shown in Figure VII-13. For $T_b < 10^5 \, °$, the assumptions that P_0 and F_0 are constant are not valid and, for T_b much above $10^5 \, °$, T_{bt} is not negligible in comparison with T_c. Other authors, notably Pottasch (1967) and Dupree and Goldberg (1967) have shown that the radio brightness temperatures shown in Figure V-27, in fact, are compatible with a value of $P_0^2 F_0^{-1} \approx 4 \times 10^{17}$ at all values of $T_b \gtrsim 10^5 \, °$.

The smaller value of $A_{Si}^{-1} F_0 = 6.67 \times 10^{15}$ found by Elwert and Raju (1972) corresponds to $P_0^2 F_0^{-1} = 1.5 \times 10^{18}$ for $A_{Si} = 3.5 \times 10^5$ and $P_0 = 6 \times 10^{14}$ and would give $T_b/\lambda^2 = 570$, which is about a factor of two above the observed value near $T_b = 10^5 \, °$. Thus, the Elwert and Raju value for F_0 appears to be too small to be compatible with the radio data if we adopt $A_{Si} = 3.5 \times 10^{-5}$. On the other hand, for $A_{Si} = 10^{-4}$, as favored by Pottasch (1967) for the corona, his results give $P_0^2 F_0^{-1} = 5.4 \times 10^{17}$, in good agreement with radio data. Thus, we cannot really choose between the two sets of results for $P_0^2 F_0^{-1}$ on the basis of radio data until the silicon abundance in the corona is more accurately known. Relative abundances of elements within the corona derived by the methods described in the preceding discussion are not precisely determined. It appears, in fact, that the uncertainties in abundances are of the order of a factor two or three. So long as this is the case, there is no compelling reason to conclude that relative abundances in the corona are different from those in the photosphere. On the other hand, there is no compelling reason either to assume that the abundances are the same in the corona and in the photosphere or, for that matter, that the photosphere abundance determinations are necessarily correct to within a factor of two or three.

The values of $A_{Si}^{-1} F_0$ obtained from the work of Elwert and Raju (1972) and Dupree (1972) are presumably independent of the silicon abundance. They are not independent, however, of abundances of other elements relative to silicon, and they therefore are abundance dependent to some extent. Also, they are dependent upon the adopted value for P_0. Most workers simply adopt a value of P_0 near 6×10^{14}, which is inferred from the model of the low corona. Munro et al. (1971) derive P_0 by using line ratios for the beryllium isoelectronic sequence ions to obtain electron densities simultaneously with the temperature. They confirm a value of P_0 near 6×10^{14}.

As a working value for F_0, we adopt a value of $A_{S,}^{-1} F_0 = 1.6 \times 10^{16}$, intermediate to the values found by Elwert and Raju (1972) and Dupree (1972), and we adopt $A_{Si} = 3.5 \times 10^{-5}$. This gives $F_0 = 5.6 \times 10^{11}$. The temperature structure of the transition region is then given directly by integration of F_0. We have

$$T^{5/2} \frac{dT}{dh} = 5.6 \times 10^{11}, \deg^{7/2} cm^{-1},$$

or

$$T^{*5/2} \frac{dT^*}{dh^*} = 0.176, \ (10^5 \deg)^{7/2} km^{-1}, \tag{VII-67}$$

where $T^* = 10^{-5} T$ and h^* is in km. Integration of equation (VII-67) gives

$$T^{*7/2} - 1 = 0.62 \ h^*. \tag{VII-68}$$

Values of h^* are given in Table VII-5 for selected values of T^*. Results are given for $T < 1 \times 10^5 \ °$ even though Equation (VII-68) is not valid in this range. These will be of use in later discussions.

TABLE VII-5

Temperature structure of the transition region from Equation (VII-68)

$T (\times 10^5)$	0.5	1	2	4	8	10
$h (\times 10^5)$	−1.5	0	17	200	2300	5100

If instead of Equation (VII-67) we use the relation

$$T^{*-4} \, dT^*/dh^* = \text{const}$$

for $T^* < 1$, and evaluate the constant to obtain the same value of dT^*/dh^* at $T^* = 1$ as is required for Equation (VII-67), we obtain

$$T^{*-4} \frac{dT^*}{dh^*} = 0.176. \qquad\qquad (\text{VII-69})$$

Integration now gives

$$1 - T^{*-3} = 0.528 \, h^*, \qquad\qquad (\text{VII-70})$$

and for $T^{**} = 0.5$ and 0.25 we obtain $h^* = -13$ and $h^* = -120$, respectively. It may be seen from these results together with those in Table VII-5 that the details of the model for $T^* < 1$ are very sensitive to the form adopted for dT^*/dh^* but that in neither case is there evidence for an extended temperature plateau between $T = 2.5 \times 10^4$ ° and 1.0×10^6 °. We conclude, therefore, that within this entire temperature regime, the atmosphere is truly in a state of transition.

The conclusion that F_0 is constant, or nearly so, through the upper transition region has several interesting implications. If the Sun were free of magnetic fields, or if the fields were vertical, the inferred value of F_0 could be converted directly to an energy flux carried by thermal conduction. The conductive flux is given for the solar mixture of elements, with hydrogen and helium both ionized, by (Ulmschneider, 1969)

$$F_c = 1.1 \times 10^{-6} T^{5/2} \, dT/dh$$

$$= 1.1 \times 10^{-6} F_0. \qquad\qquad (\text{VII-71})$$

For the adopted value of F_0, $F_c = 6 \times 10^5$ erg cm^{-2} s^{-1}. This is a large energy flux, comparable to the radiant energy flux from the upper chromosphere, transition region and corona combined. It therefore must be considered carefully, and we will have more to say about this aspect of the problem in Chapter IX.

At this point, we note only that F_c is entirely dependent upon the magnetic field configuration. The picture of the field geometry developed in Chapter IV is one in which bundles of lines of force at certain locations in the network in the chromosphere and lower transition region diverge rapidly in the upper transition region to form a more or less uniform field in the corona. In such a model, the field lines over much of the solar surface necessarily have a strong horizontal component in the upper transition region.

Thermal conduction across magnetic field lines is miniscule in comparison to the conduction along the field lines. Thus, one may view the conduction as being directed along the field lines. This will lead to a focussing and enhancement of the conducted flux in the network regions. The average conductive flux, however, will be reduced over the whole sun by a value $\cos \theta$ where θ is the average angle of inclination of the field lines to the vertical. As noted, this average angle could be quite small.

In addition to the influence of the magnetic field orientation on the conduction itself, the magnetic field undoubtedly influences the temperature gradient, as was suggested by Kopp and Kuperus (1968). It seems premature, however, to attempt to develop specific models including the magnetic geometry until more is known about the magnetic fields in the transition region.

The transition region is penetrated by spicules. Spicules are estimated to cover about 1% of the solar surface (Chapter II) and their mean height is estimated at some five times their diameter. Their total surface area exceeds their base area by an amount given approximately by

$$\frac{\pi dh}{\pi d^2/4} = \frac{4h}{d} \approx 20.$$

Thus, the surface area of spicules is less than but possibly comparable to the surface area of the Sun. This leads to the possibility that much of the XUV line emission arises in a 'transition region' between the corona and the spicules, i.e., in a sheath surrounding spicules. If this were the case, the transition region over the rest of the Sun would necessarily be even thinner and F_0 would be larger. This conclusion is readily verified from Equations (VII-53), (VII-54) and (VII-55). Thus, if we let a_1 and a_2 each equal unity and let $P_1 = P_2$ and $F_1 = F_2$, P_0 is increased by a factor 2 (Equation (VII-54)). It follows from Equation (VII-53) that F_0 must be increased by a factor of 4, and from Equation (VII-55) that F_1 and F_2 are each a factor 2 over the original value of F_0. Again, however, we know essentially nothing about the temperature structure between spicules and the corona so there is little merit to speculating about the influence of the spicules at this point.

Attempts have been made to derive models of the transition region allowing for a hydrodynamic flow of matter through the transition region. No completely consistent treatment has been given to this problem, however. The original attempt at such a model by Lantos (1972) failed to take into account work done against the gravitational field by the expanding gas. At velocities less than the escape velocity (615 km s^{-1}), the rate of doing work against gravity exceeds the kinetic energy flux, and, by omitting the gravitational term one obtains flow velocities that are much too large.

A more recent attempt to derive a dynamic model by Chiuderi and Riani (1974) includes the gravitational term in the momentum equation. However, in this latter work the temperature structure of the transition region is derived from an assumed functional form for the observed radio brightness temperature as a function of wavelength, which is then inverted to obtain $T(h)$. The results of such a procedure depend critically upon the exact form of the temperature curve, and, as a result, the inversion contains large uncertainties. A number of authors have attempted such inversions, but none has produced a model that even approximately satisfies the XUV line fluxes. The model derived by Chiuderi and Riani suffers from this difficulty.

All of the models based exclusively on radio data have values of dT/dh that are much too large to be compatible with the line data. On the other hand, the line flux models with much steeper values of dT/dh have been shown to be consistent with the radio data, and it is clear that these models are strongly preferred over the radio models. The difficulty with the radio models is that the contribution functions of any one wavelength are very broad, and, as a result, the true temperature structure is obscured.

Thus, the two major attempts at dynamic models of the transition region contain major deficiencies — the Lantos model because of the neglect of the gravitational term and the Chiuderi-Riani model because of the neglect of the XUV data. The Lantos model gives large flow velocities and the Chiuderi-Riani model gives relatively low flow velocities. Neither model is reliable, however, and nothing definite can be said, at this time, concerning the importance of the dynamic effects.

8. The Corona

Eclipse observers during the late 19th and early 20th century established the basic nature of coronal structure and the nature of the coronal spectrum. Both aspects of the corona presented serious challenges that remained unresolved for many years. The large radial extent of the corona was a puzzle for it was widely believed that the coronal temperature could not possibly exceed 6000°. Also, the several coronal emission lines that were known in the visual spectrum defied identification. For a while these were attributed to a new element 'coronium', but as laboratory workers filled out the periodic table of elements no room was left for such an element.

Major break-throughs that eventually led to a successful explanation of both the coronal spectrum and the radial extent of the corona began near 1930. By this time it was recognized that the K and F continua were physically distinct. Both were believed to arise from scattered photospheric light, and they had been attributed correctly to scattering by electrons in the solar atmosphere and to scattering by dust particles in interplanetary space.

At an eclipse in 1929, Grotrian claimed to have observed the H and K lines of Ca II superposed on the K-corona as broad shallow absorption lines. In 1931, he proposed as an explanation that the lines, and the K-continuum, were scattered by electrons whose mean velocity was 7.5×10^8 cm s^{-1} (which is a factor of 13.6 above the mean electron velocity at a temperature of 6000 K), and he raised the question as to whether the velocities might be thermal.

As is so often the case in astrophysics, developments in one area of research are boosted by those in another area, often seemingly remote. Studies of novae in the late 1920's led to the discovery of forbidden lines of multiply ionized atoms in the spectra of these objects. Five of the lines present in the coronal spectrum were observed in the spectrum of R. S. Ophiuchi, a recurrent nova, during its 1933 brightening. Following this discovery, Grotrian suggested that the coronal lines might also be forbidden lines of multiply ionized elements, and he noted that one of the coronal lines (λ 6374) was very close in wavelength to a transition in Fe X identified in laboratory spectra by Edlén. Edlén, in turn, followed this lead and successively identified more and more lines. By

1942, some 19 coronal lines had thus been identified with ions of iron, nickel, calcium and argon.

The Lyot-coronagraph was first used successfully in 1932 to observe the corona outside of eclipse. Lyot soon showed that the coronal lines were broad even though they were relatively faint. The green line at λ 5303 [Fe XIV] was shown consistently to have a width of 0.9 Å. Lyot suggested that the width was due to thermal motions, but was unable to derive a temperature because the lines were still unidentified.

Once the coronal lines were successfully identified the evidence for a high temperature corona was overwhelming and several pieces of the puzzle soon fell into place. Fe XIV, which produced the strongest coronal line, could not exist without a very high temperature. The high temperature, in turn, immediately accounted for, in a crude way, the widths of the coronal lines, the high velocity of electrons (as evidenced by the near absence of the H and K absorption lines in the K-continuum) and the large lateral extent of the corona. This successful, if somewhat crude, explanation for several outstanding coronal problems has withstood the test of time. However, efforts to bring all of the observations into quantitative agreement with a simple coronal model continue to challenge solar physicists. This attempt at a synthesized model is converging towards a common solution but several outstanding difficulties remain.

A. DENSITY MODELS

Analyses of the coronal K-continuum lead directly to the determination of the coronal electron density distribution. Most such analyses, however, have been based on 'average' intensity data and require that the geometrical form of the distribution of matter be specified *a priori*. It is normally assumed that the average corona is spherically symmetric or, stated in another way, that the average brightness can be interpreted in terms of an average electron density in an assumed spherically symmetric distribution. Hydrogen and helium in the corona are known to be nearly fully ionized so that the electron density is directly proportional to the total density. Then, since the scattered light intensity is linearly related to the electron density (hence total density), the brightness and density should average in the same way. Thus, even though the corona is known to be highly structured (see Chapter II) it is still meaningful to define an 'average' density model.

Even though one assumes spherical symmetry in analyzing the continuum data, the assumption is needed only for performing line-of-sight integrations. Mean isophotes of coronal brightness are dependent on time and solar latitude. The sunspot minimum corona differs from the sunspot maximum corona and the equatorial corona differs from the polar corona, particularly near sunspot minimum. Thus, some 'mean' coronal models distinguish between sunspot maximum and minimum and between pole and equator.

The K-corona is polarized in a known way since Thompson scattering provides the great majority of the continuum emission and since the distribution of intensity over the solar disk is well known. The F-corona is believed to be unpolarized, and it follows that observations of coronal brightness coupled with observations of the degree of polarization permit a separation of the K and F components. Also, the F coronal spectrum contains the

Fraunhofer lines with very nearly the same width and relative depth as in photospheric light, whereas in the K-corona the Fraunhofer lines are broadened beyond recognition. This provides an additional check on the separation of the radiation from the K and F coronas. Very near the Sun (in angular position) the K corona is dominant and bright enough that observations at eclipse are unhampered by sky light. Beyond about $6\,R_\odot$, however, the sky brightness at total eclipse becomes a serious problem and is comparable to or greater than the coronal brightness. Sky brightness can be reduced by observing from high altitude balloons, rockets or satellites.

Coronal electron densities are derived using Equations (VI-2) to (VI-5), or a variation of them. The procedures are well standardized and straightforward and will not be repeated here. (The interested reader should consult Billings, 1966.) Mean models derived by van de Hulst (1953) and Newkirk (1967) are given in Table VII-6. Newkirk's model for the minimum-equatorial corona is very similar to van de Hulst's sunspot maximum model. In van de Hulst's models, the minimum equatorial model has densities that are lower than the sunspot maximum model by a factor of 1.8. The minimum-polar model has a steeper density gradient as well as lower densities.

TABLE VII-6
Electron densities[a] in the 'average' corona

$r/R_\odot$	Maximum Equator and Pole	Minimum Pole	Minimum Equator	(Newkirk)
1.03	8.50	8.10	8.25	8.51
1.06	8.37	7.94	8.12	–
1.1	8.20	7.73	7.95	8.15
1.2	7.85	7.21	7.60	7.84
1.4	7.36	6.44	7.11	7.36
1.6	7.00	5.93	6.75	7.00
1.8	6.70	5.57	6.45	–
2.0	6.45	5.30	6.20	6.45
2.5	5.92	4.70	5.67	5.95
3.0	5.50	4.23	5.25	5.60
3.5	5.21	3.87	4.96	–
4.0	4.95	3.60	4.70	5.08
5.0	4.65			–
6.0	4.46			4.49
8.0				4.11
10				3.99
15				3.40
20				3.11
30				2.66
50				2.00
100				1.30
215 (1 AU)				0.40

[a]The first three columns give $\log N_e$ from van de Hulst (1953). The fourth column gives the minimum equatorial corona from Newkirk (1967).

A word of caution needs to be stated in connection with the polar model. Because the equatorial corona is quite bright and extends far into space, observations made near the pole will inevitably show some brightness from long streamers lying at low latitude that happen to project into the plane of the sky at the solar pole. Ney *et al.* (1961) found, for example, that the polar corona observed in 1959 (near sunspot maximum) could be satisfactorily explained with the projection of features lying below 70° latitude and, hence, that no electrons were required above 70° latitude. Similarly, Gillet *et al.* (1964) found that the corona observed in 1952 and 1963 (near sunspot minima) required no electrons above 60° latitude. Although these results do not prove that the polar corona is non-existent, they do cast serious doubt on the meaning of the mean polar density models.

The polar corona does at times show moderately strong emission in both the Fe X red line at $\lambda\,6374$ and the Fe XIV green line at $\lambda\,5303$. The polar red line emission is particularly evident at sunspot minimum (cf. Billings, 1966). Line emission from the corona decreases much more rapidly with height than does the continuum emission, and it can be shown readily that the observed polar emission in the lines cannot result from low latitude projection effects. Thus, there is no doubt that a polar corona does exist, on the average, but the density model must be regarded as uncertain.

Electron densities derived from the brightness of the K-corona are reliable out to about 8 to $10\,R_\odot$ only. From about 10 to $100\,R_\odot$ the densities are best established by observations of the coronal occultation of radio sources. Near the Earth, satellites have measured the density *in situ*. Newkirk's model beyond $10\,R_\odot$ in Table VII-6 has been deduced from these sources. This part of the model is included in Table VII-6 for completeness but will not be discussed further.

An 'average' model for a medium as highly structured as the corona is almost by definition a model of something that never really exists. The corona is clearly dominated by discrete features on both the large and small scale. Near sunspot maximum, coronal condensations over active regions and streams associated with activity tend to dominate the coronal picture on the large scale. Near sunspot minimum, the polar corona becomes much less prominent while the equatorial corona remains extended. The total mass of the corona at sunspot maximum may well exceed that at sunspot minimum by a factor of four or more.

Morphological changes in the corona during the course of the sunspot cycle and with specific centers of activity are crucial to any thorough study of the corona. However, in this text we are dealing specifically with the so-called quiet Sun and to digress into a study of the active corona, however enticing that may be, would require a similar digression into the study of all forms of solar activity and would greatly enlarge upon the intended scope of this book.

On the small scale, the corona appears to consist of a complex of arches, rays and polar plumes that are clearly associated with magnetic lines of force. The gas pressure at the base of the corona is approximately 0.16 dyn cm^2, and it decreases to approximately 0.001 dyn cm^{-2} at $2\,R_\odot$. Comparable pressure results from magnetic fields of 2 G and 0.16 G, respectively. Such fields are to be expected frequently in local areas if not as an average condition. In regions where the magnetic pressure dominates, the structure of the corona yields to the field configuration. The field configuration, in turn, is controlled by forces and currents lying much deeper in the atmosphere, possibly in the photosphere or subphotospheric layers.

Within the immediate vicinity of arch and curved ray structures, the electron density gradient may reverse to give an outwards increase of density in a local region. An obvious example where this is true is near the bottom borders of prominences. It is equally true, although less conspicuously so, underneath some arches and curved ray structures.

The construction of 'average' coronal density models is an act requiring some faith that such a model will be meaningful. When one uses the average density model to infer other properties of the corona, faith perhaps gives way to foolhardiness. In this vein, it is frequently assumed that the density gradient in the average corona can be related to a mean temperature in the corona. Thus, one assumes that the corona is near to hydrostatic equilibrium at a constant temperature. The hydrostatic temperature is then derived from the density scale height given by the average model.

Because the scale height in the corona is non-negligible in comparison to the solar radius, the gravitational acceleration cannot be taken as constant. For this case, we write

$$dp = -\mu m_H \, ng \, dr, \tag{VII-72}$$

$$dp = kT \, dn \tag{VII-73}$$

and

$$g = g_0 (R_\odot / r)^2, \tag{VII-74}$$

from which we obtain

$$\frac{dn}{n} = -\frac{\mu m_H g_0}{kT} \left(\frac{R_\odot}{r} \right)^2 dr$$

$$= -H_c^{-1} \left(\frac{R_\odot}{r} \right)^2 dr. \tag{VII-75}$$

Integration of Equation (VII-75) with H_c constant and with $n = n_0$ at $r = R_0$ yields

$$n = n_0 \exp -\frac{r}{H_c} \left[\left(\frac{R_\odot}{r} \right) - \left(\frac{R_\odot}{r} \right)^2 \right]. \tag{VII-76}$$

It follows from Equation (VII-75) that

$$d \ln n_e = H_c^{-1} R_0 \, d \left(\frac{R_\odot}{r} \right) \tag{VII-77}$$

and, hence, that

$$T = \frac{\mu m_H g_0 R_\odot}{k} \left(\frac{d \ln n_e}{d(R_\odot / r)} \right)^{-1}. \tag{VII-78}$$

Figure VII-14 shows plots of $\log n_e$ vs $R_\odot/r$ for the van de Hulst maximum corona (crosses) and the polar minimum corona (pluses). (The two end points at $R_\odot/r$ equal 0.125 and 0.1 are taken from Newkirk's equatorial model.) A constant temperature is represented on the plot in Figure VII-14 by a straight line. The sunspot minimum polar corona seems well represented by a temperature of 1.0×10^6 K. For the sunspot maximum corona, two straight lines are drawn in the figure, one for $r/R_\odot \leqslant 2.5$ with $T = 1.4 \times 10^6$ K and one for $r/R_\odot > 2.5$ with $T = 0.97 \times 10^6$ K.

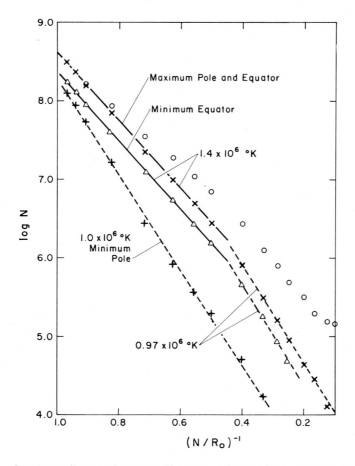

Fig. VII-14.   Density gradients in the corona. The slope of the straight lines is inversely proportional to temperature. See text for comment on the open circles.

The rather striking tendency for the plots in Figure VII-14 to be fitted by a single value of T over height intervals of a few solar radii is somewhat surprising in view of the rather obvious departure of the corona from the simple spherically symmetric, hydrostatic model assumed. Specifically, we note the following considerations: (1) the corona probably is not strictly isothermal over distances of the order of a few radii; (2) the pressure due to magnetic fields probably is not negligible, in the inner corona particularly; and (3) departures from spherical symmetry probably alter the mean density model so as to change the gradient in $\log n_e$. In connection with this latter point, it should be noted that if most of the matter at a given height is concentrated in radial rays occupying a fraction x^{-1} of the solar surface, then the mean density should be multiplied at each height by x. Brandt *et al.* (1965) adopt a compilation of estimated values of x from Allen (1963), which they represent by the expression

$$x = 1.6r/R_\odot - 0.7. \tag{VII-79}$$

A correction of this magnitude applied to the sunspot maximum data in Figure VII-14 yields the open circles plotted in the Figure. These points can again be fitted by two straight lines similar to those drawn through the uncorrected points but with slopes corresponding to $T \approx 1.8 \times 10^6$ K and $T \approx 1.3 \times 10^6$ K.

Although any such correction as the one indicated by Equation (VII-79) is of questionable validity, it is evident that the mean density structure of the corona out to about $2.5 R_\odot$ is not incompatible with an isothermal, hydrostatic corona (strictly speaking, of course, any departure from spherical symmetry is, in itself, incompatible with an isothermal, hydrostatic model). The additional complications arising from magnetic pressure and from temperature gradients, both horizontal and radial, are undoubtedly present, but there is no obvious way of assessing their importance from the density model.

The corona is known not to be in static equilibrium but to be undergoing a general expansion. Under such conditions, continuity conditions require that the relation

$$n_e v r^2 = \text{const} \tag{VII-80}$$

be satisfied at all values of r. Satellite observations at 1 AU place the value of $n_p v$ as approximately 2×10^8 cm^{-2} s^{-1}. Assuming that the coronal electrons and protons expand with the same velocity, we may use this value of $n_p v$ at $r = 215 R_\odot$ to obtain

$$n_e v \left(\frac{r}{R_\odot}\right)^2 = 9.25 \times 10^{12} \text{ cm}^{-2} \text{ s}^{-1} \tag{VII-81}$$

The combinations of n_e and $r/R_\odot$ for the sunspot maximum corona in Table VII-6 then give expansion velocities of 0.28, 2.1, 8.2, 33, and 93 km s^{-1} at $1.02 R_\odot$, $1.4 R_\odot$, $2 R_\odot$, $3 R_\odot$, and $10 R_\odot$ respectively.

In an expanding corona, the equation of hydrostatic equilibrium should be replaced by (Parker, 1960)

$$\frac{d}{dr} nkT = -\mu n m_H g_0 \left(\frac{R_\odot}{r}\right)^2 - \mu n m_H \frac{dv}{dr} \tag{VII-82}$$

If we designate the two terms on the left as the static and dynamic terms, the preceding values of v and r yield the ratio of the dynamic term to the static term as less than 0.01 out to about $2\,R_\odot$ and less than 0.1 out to about $5\,R_\odot$. Thus, the expansion of the corona appears to be of negligible consequence to the equilibrium conditions within a distance of about $5\,R_\odot$ of the center of the Sun.

In a fully ionized plasma, such as the corona, permeated by strong magnetic fields, the matter is effectively 'frozen' to the lines of force in directions normal to the lines of force but is free to move in directions that lie along the lines of force. This permits, in theory at least, the possibility for a static atmosphere that is not spherically symmetric. Along any given flux tube in the magnetic field the distribution of matter will satisfy the hydrostatic condition, i.e., density will decrease exponentially with vertical height within each flux tube. It is not necessary, however, that the gas pressure be constant along surfaces of constant height since only the total of the gas pressure plus magnetic pressure need be constant. The magnetic field geometry may well be determined by gas motions and currents in the photosphere or chromosphere leaving the coronal fields in rather complex array. Also, the pressure at the base of the corona as well as the coronal temperature may be expected to vary from one flux tube to another. This leads to the distinct possibility that nearby flux tubes may have quite different pressure and density distributions. Such a picture is consistent with the observed structural properties of the corona.

Although a static, aspherical corona is theoretically possible in the presence of strong magnetic fields, it seems unlikely that the highly structured features observed in the corona are static in reality. The magnetic field pressure will not everywhere dominate the gas pressure and the field configuration itself is unlikely to be everywhere static.

In spite of all the reasons one can state for the corona at any point in time to depart from the simple spherically symmetric, hydrostatic model, the mean corona at a given solar latitude does appear to be broadly consistent with such a picture. The systematic pole to equator trends and the obvious violations of the simple model in coronal condensations (also prominences), loops, rays and filaments negate the model locally in space and time, however.

The rather pronounced tendency for the sunspot maximum coronal densities shown in Figure VII-14 to show a change in the gradient d log n_e/d$(1/r)$ near $2.5\,R_\odot$ is of interest. It could be associated with a decrease in coronal temperature or with changes in the geometrical and magnetic structure of the corona. If coronal structure is categorized in two broad classes of 'open' and 'closed' structures, where 'closed' means loops and arches and 'open' means diverging rays, then nearly all of the closed structures lie below $2.5\,R_\odot$ (Chapter II) and the corona above $2.5\,R_\odot$ is nearly always open. Open structures do exist below $2.5\,R_\odot$, of course, so the structure below $2.5\,R_\odot$ is a mixture of open and closed. Nevertheless, one can, in a general way, categorize the region below $2.5\,R_\odot$ as closed and the region beyond as open.

Magnetic field models of the corona that are computed from photospheric fields agree best with the observed closed and open structures of the corona when a zero-potential magnetic surface is placed near $2.5\,R_\odot$ (Altschuler and Newkirk, 1969 see also Chapter

IV). The effect of such a surface on the magnetic model is to provide a predominantly radial magnetic field beyond $2.5\,R_\odot$ while the field below $2.5\,R_\odot$ is predominantly closed. Such a magnetic model is consistent with the requirement of solar wind models that the field become radial beyond some fixed distance from the solar surface. Pneuman (1968), for example, finds that closed arches can exist in helmet streamers out to one-half the distance from the solar surface to the critical point where the solar wind becomes supersonic. For a coronal temperature near 1.6 to 1.8×10^6 K, which is the most generally acceptable range of possible temperatures, the limit on the closed structures given by Pneuman is $2.6\,R_\odot$ (measured from solar center).

It appears, therefore, that the evident 'break' in the curves in Figure VII-14 is associated with the change in the geometery of coronal magnetic fields from closed to open as the solar wind pulls the magnetic lines of force into interplanetary space. Whether there is also an associated decrease in temperature near $2.5\,R_\odot$ is not known, but such a result appears likely to be the case.

B. IONIZATION TEMPERATURE

One of the long standing methods for determining coronal temperatures is through the ionization equilibrium. In order to use this method two steps must be successfully accomplished. Firstly, the relative concentrations of ions in the corona must be deduced from observational data. Secondly, a correct theory of the ionization equilibrium coupled with accurately known atomic parameters must be used to synthesize the deduced ionic distribution.

The first problem is hampered by the long path length of integration. As we saw in Chapter I, the effective path length in the corona is of the order of $1\,R_\odot$ or greater at the limb and of the order of $0.1\,R_\odot$ on the disk. Many observed coronal features are small by comparison to these path lengths. Thus, one can never be certain that spectral lines observed for different ions are emitted from the same geometrical regions. Indeed, there is every reason to suppose that different ions will be distributed differently in space due to the presence of temperature gradients.

If, for example, one uses observations of the red coronal line ($\lambda\,6374$) to deduce the average concentration of Fe X and the green coronal line ($\lambda\,5303$) to deduce the average concentration of Fe XIV, is there any real justification for concluding that the average ratio of concentrations is given by the ratio of their averages? If the two spectral lines are formed systematically in different regions, the answer is no. In the first place, the average ionic concentrations are incorrectly determined because of the unknown geometry, and, in the second place, the ratio of average concentrations contains an unknown ratio of the fractional volumes where the two ions are concentrated.

The type of difficulty mentioned in the preceding paragraph has long been evident in the comparison of different ionic concentrations as deduced from visual forbidden lines. Such studies have shown, for example, that Fe XIV is the most populous iron ion whereas the second most populous is Fe X. Both Fe XI and Fe XIII (Fe XII is not observed) are, on the average, less abundant (cf. van de Hulst, 1953). This is inconsistent with any known ionization equilibrium distribution for a single temperature. All of the ionization equilibrium computations (see Figure VII-11) give an Fe X concentration considerably below the Fe XI and Fe XIII concentrations at temperatures near that for which Fe XIV is most abundant. One is forced to conclude, therefore, that Fe X and Fe XIV emissions do, in fact, arise in volume elements that are systematically different.

The second step mentioned in connection with temperature determinations using relative concentrations of ions is that of using a correct ionization theory and correct atomic parameters. Incorporation of di-electronic recombinations (Burgess, 1964) and, to a lesser extent, the incorporation of auto-ionizations (Goldberg *et al.*, 1965) into the ionization equilibrium model has corrected most of the difficulties that plagued earlier attempts to compute an equilibrium distribution that gave ionization temperatures consistent with temperatures deduced by other means (see the comparison of House's results with Jordan's results in Figure VII-11). This is not to imply that the last word has been said about the ionization equilibrium. Refinements are still necessary in determining just which particular energy levels are important in the ionization equilibrium and in accurately assigning appropriate excitation, ionization and recombination cross sections.

Jordan's (1969) ionization equilibrium computations for iron shown in Figure VII-11 give the maximum concentration of Fe XIV ($m = 13$) and Fe X ($m = 9$) at temperatures of 1.8×10^6 K and 1.0×10^6 K, respectively. Under the assumption that Fe XIV is commonly the most abundant ion, which is consistent with the mean intensity of the λ 5303 line relative to other coronal lines, we would therefore place the most common ionization temperature of the corona at 1.8×10^6 K. At the same time, however, we must recognize that ionization temperatures near 1.0×10^6 K (Fe X) are relatively common also in the normal corona.

At an ionization temperature of 1.8×10^6 K the most abundant ions of elements other than iron include C XII, N VII–VIII, O VII–VIII Ne IX, Mg XI, Si XII–XIII, S X–XI–XII and Ni XII–XIII. Thus, for O, Ne, Mg and Si the helium-like ions are populous in the corona. Lighter elements are hydrogen-like, or fully stripped, and heavier elements are not ionized as far as the lithium sequence. Ions in the helium and hydrogen sequences are not very useful for determining ionization temperatures because these ions exist in high relative proportions through wide ranges of temperature. It follows that the most useful ions are those with atomic weight greater than that of silicon. Iron is the most abundant element in this class and is observed in several stages of ionization. However, it has the disadvantage that near an ionization temperature of 2×10^6 K Fe XIII, XIV, XV, XVI and XVII are all quite populous and there is little chance of determining an accurate ionization temperature. Nickel suffers similar disadvantages and other elements either have been observed in an insufficient number of ionization stages to accurately locate the ionization maximum or their ionization equilibrium has not been accurately determined.

In active centers, particularly in association with flares, ionization temperatures much above 2×10^6 K are present. Iron has been observed in X-ray spectra, for example, in ionization stages up to Fe XXVI, which is hydrogenic. Ca XV, with an ionization potential of 898 eV, has a forbidden line in the visual spectrum that is not uncommonly observed in association with flares and loop prominences over sunspots. According to Billings (1966) observation of the CA XV ion requires ionization temperatures of 5.6×10^6 K and according to Jordan (1970) Fe XXVI occurs in abundance only for temperatures greater than about 5×10^7 K.

The occurrence of such highly excited conditions (including Ca XV), however, is always associated with transitory events, such as flares, that are known to be associated with energetic particles. It is debatable, therefore, whether the ionization equilibrium is to be associated with the characteristic kinetic temperature of the gas or with the energies of non-maxwellian particles. Evidence from XUV and radio data suggests that kinetic temperatures up to a few million degrees exist in some active centers. For apparent ionization temperatures in excess of 10^7 K, however, there is little evidence that the excitation is thermal.

C. LINE WIDTH TEMPERATURES

Coronal forbidden lines in the visual spectrum have been extensively studied with moderate spatial and spectral resolution. The half-widths of the forbidden lines provide an alternative means for estimating the coronal kinetic temperature. Observations made with coronagraphs outside of eclipse suffer from superposed skylight and instrumentally scattered light, both of which contain the Fraunhofer spectrum. Unfortunately the green line (λ 5303) is blended with a Fraunhofer line and the profile must be corrected for this blend. Observations made at eclipse, however, do not suffer from this effect to the same degree and these observations serve to verify the coronagraph data.

The corona is optically very thin in the forbidden lines even at the limb, and the lines are nearly Gaussian in shape. For a Gaussian line that is thermally broadened, the intensity distribution in the line is given by

$$I_{\Delta\lambda} = I_0 \, e^{-(\Delta\lambda/\Delta\lambda_D)^2}$$

and the half intensity point, $I_{\Delta\lambda} = I_0/2$, occurs at $\Delta\lambda_{1/2} = 0.833 \, \Delta\lambda_D$. (The full width at half maxmum occurs $\Delta\lambda_{1/2}^* = 1.67\Delta\lambda_D$.) The Doppler width, in turn, is related to T through the usual conditions

$$v = \left(\frac{2kT}{m_i} \right)^{1/2}$$

and

$$\Delta\lambda_D = \lambda \frac{v}{c}.$$

It follows that

$$T = \frac{m_i}{2k} \left(\frac{c\Delta\lambda_{1/2}^*}{1.67\lambda} \right)^2 \tag{VII-83}$$

Billings (1966) summarizes the kinetic temperatures determined in this way as having the mean values of 2.5×10^6 K for Fe XIV (λ 5303), 1.8×10^6 K for Fe X (λ 6375) and 4.0×10^6 K for Ca XV(λ 5694). The dispersion in temperatures about the mean values is approximately $\pm 2 \times 10^5$ K for Fe XIV and $\pm 5 \times 10^5$ K for Ca XV. Billings further reports that the line widths for λ 5303 and λ 6374 show little variation within or near active regions. Quiet areas of the corona, the borders of active regions and bright condensations within active centers all show similar line widths. This supports the notion that a given ion radiates mostly in regions of a rather well defined temperature. Again, however, we caution that the long path length of integration may effectively obscure most of the intrinsic changes in line width.

For both Fe XIV and Fe X the line width temperatures exceed the ionization temperatures. This suggests the possibility that part of the line broadening is of non-thermal origin. A non-thermal, random velocity, of magnitude, ξ, increases the mean velocity to

$$\bar{v} = \left(\frac{2kT}{m_i} + \xi^2 \right)^{1/2}.$$

This leads to an apparent broadening temperature, T_a, of

$$\bar{v} = \left(\frac{2kT_a}{m_i} \right)^{1/2}.$$

Thus, the true broadening temperature is related to T_a by the equation

$$T = T_a \left(1 - \frac{m_i \xi^2}{2kT_a} \right). \tag{VII-84}$$

For the Fe XIV line we set $T = 1.8 \times 10^6$ and $T_a = 2.5 \times 10^6$ so that

$$\frac{m_i \xi^2}{2kT_a(\text{Fe XIV})} = 1 - \frac{T}{T_a} = 0.28,$$

similarly, for Fe X we obtain with $T = 1.0 \times 10^6$ and $T_a = 1.8 \times 10^6$

$$\frac{m_i \xi^2}{2kT_a(\text{Fe X})} = 0.44.$$

The former result gives $\xi = 14$ km s^{-1} and the latter gives $\xi = 15$ km s^{-1}. By comparison, the mean thermal velocity for $T = 1.8 \times 10^6$ K is 23 km s^{-1} for iron atoms and 172 km s^{-1} for hydrogen atoms (protons).

Non-thermal, line broadening velocities in the chromosphere increase from a few km s^{-1} in the low chromosphere to 10 to 15 km s^{-1} in the upper chromosphere. The indicated values of 14 to 15 km s^{-1} in the corona seem somewhat mild by comparison. In fact, the velocities in the upper chromosphere appear to be near the sonic velocity whereas the coronal velocities appear to be only about one-tenth of the sonic velocity. If one is to be surprised, by the corona velocities, therefore, the surprise is that they are so low. This result will be of interest in connection with later discussions of the possible heating mechanisms in the corona.

As yet, there are no observations of coronal line widths or line shifts on the disk of the Sun. Needless to say such data are of great interest and therefore highly desirable. Satellite observations within the next few years should provide much valuable data of this type.

D. LINE INTENSITY RATIOS

With the advent of space data from rockets and satellites, it is now possible to make use of the full coronal spectrum in order to determine the physical properties of the corona. The most notable contributions added by these new techniques involve the excitation properties inferred for individual ionic species from the relative intensities of different lines of that particular specie. Relative intensities of certain combinations of lines are sensitive to temperature and, in some cases, electron density. Since a given ionic specie gives rise to each of the lines, there is little doubt that the lines originate within nearly the same volume elements. These methods are useful in that they can often be used in cases where other data are not sufficient and, in any case, they provide a consistency check on other methods.

Interpretation of line ratios does, at times, require knowledge of atomic parameters that are not accurately known. Also, in many cases the line intensity ratios are less sensitive to density and temperature than are the absolute intensities of the individual lines. However, because these methods provide direct information on the local values of temperature and density in the regions from whence the emission arises, the interpretation can be made independently of the geometry of the emitting region. This is a strong advantage of this type of analysis, particularly in the case of active events in which the geometry is often unknown and for which only limited data may be available. On the other hand, the line intensity ratio methods have not yet added substantially to our understanding of the quiet corona even though the potential for doing so is present.

In lithium-like ions the 2p and 3p levels differ in excitation energy by a large factor. The excitation of such levels is primarily by electron collisions with ions in the ground state. For ease of notation let 1, 2 and 3 denote the 2s, 2p and 3p levels, respectively. Equation (VI-68) then gives the relative line intensities as

$$\frac{I_{12}}{I_{13}} = \frac{\nu_{12} C_{12}}{\nu_{13} C_{13}} = \frac{x_{12} C_{12}}{x_{13} C_{13}} . \qquad \text{(VII-85)}$$

In the Bethe approximation, c_{ij} is of the form

$$c_{ij} \propto f_{ij}(x_{ij}/kT)\bar{g} \exp\left(-\frac{x_{ij}}{kT}\right). \qquad \text{(VII-86)}$$

where $\bar{g}$ is an effective gaunt factor. Thus, we have

$$\frac{I_{12}}{I_{13}} = \frac{f_{12}x_{12}(x_{12}/kT)\bar{g}_{12}}{f_{13}x_{13}(x_{13}/kT)\bar{g}_{13}} \exp\left(\frac{x_{13}-x_{12}}{kT}\right). \tag{VII-87}$$

Obviously, the usefulness of this method depends upon the magnitude of x_{13} and x_{12} at the values of T where the ions are abundant. Usually, x_{12}/kT is small so that the temperature sensitivity of the line ratio enters through $\exp(x_{13}/kT)$. Values of x_{13}/kT, for T equal to the temperature of maximum ion density, for typical lithium-like ions are as follows C IV – 4.6, N V – 3.9, O VI – 3.4, Mg X – 2.2 and Si XII – 1.8. It is clear from these expected magnitudes of x_{13}/kT that the method is not very accurate for the corona where changes in temperature of 20% change the relative intensities of the lines for Mg X and Si XII by a factor of about 1.5 only. The temperature sensitivity could be increased by using lines of higher excitation energy in place of the 2S – 3P line. However, the lines then lie further in the X-ray spectrum and are weakened by the reduced collision rate.

The above method was proposed by Henoux (1964) and has been applied to O VI, Ne VIII and Mg X by Flower (1971).

A method essentially similar to that discussed above has been developed for hydrogen-like ions by McWhirter and Hearn (1963), Beigman and Vainshtein (1967) and Jacobs (1968) and has been applied to the Lyman-α and Lyman-β lines of O VIII in the corona by a number of authors, including Jacobs (1968) and Evans and Pounds (1968). The energy separation between the $n = 2$ and n = 3 levels in O VIII is 121 eV, and the corresponding value of x_{23}/kT at $T = 10^6$ K is 1.4. Most of the temperature sensitivity of the Lyman-α to Lyman-β ratio again enters through the factor $\exp(x_{23}/kT)$, which is again only a slow function of temperature for temperatures in excess of 1×10^6 K. Thus, coronal temperatures are not accurately determined by this method and different authors have found quite different results. By allowing for two-photon decay of the 2s level, the need for which was pointed out by Beigman and Vainshtein (1967), Jacobs (1968) found a quiet coronal temperature of 1.3×10^6 K to 2.0×10^6 K in agreement with the generally accepted range of values.

Still a third method of estimating temperatures from line ratios suggested by Gabriel (1971) utilizes the different excitation requirements of the di-electronic satellite recombination lines and the resonance lines. This method, as the previous two discussed, is also relatively inaccurate for conditions found in the quiet corona. Each of the three methods is useful for hotter regions of the corona, however, where other techniques may be inapplicable.

For helium-like ions in the corona the intercombination line $2^3P - 1^1S$ and the forbidden line $2^3S - 1^1S$ are both present with appreciable intensity. The system of levels 1^1S, 2^3S and 2^3P may be thought of as a closed loop of three levels each connected to the other two by radiative and collisional transitions. Under certain conditions of temperature and density (Gabriel and Jordan, 1969) the transition rate from level 2^3S to 2^3P by electron collisions will exceed the radiative rates from 2^3S to both 2^3P and 2^1S.

Under these conditions the population of the 2^3S level, and hence the intensity of the forbidden line, is proportional to electron density. This is most likely to be true, of course, at higher densities so this method again becomes of most interest in connection with active region phenomena. At normal coronal densities the line ratios are only weakly dependent upon density and the method is not particularly useful. Di-electronic recombination satellite lines of the lithium-like ions lie in close proximity to the intercombination and forbidden lines in the helium-like ions and may contribute significantly to the apparent fluxes in the helium-like lines (Walker and Rugge, 1971). This may account for the unusually high values of n_e sometimes obtained from the use of this method.

E. RADIO TEMPERATURES

Brightness temperatures measured at the center of the solar disk at radio wavelengths (see Figure V-27) provide the most direct of all methods for obtaining the coronal temperature. However, the method is complicated by a number of factors. At short wavelengths the corona is transparent and is observed against the chromosphere and transition region radiation. At long wavelengths the complex index of refraction bends the rays away from the higher density regions of the low corona, as shown in Figure VII-15, and the corona again becomes transparent (cf. Smerd, 1950). In both of these regions the observed brightness temperature of the corona, which is given by

$$T_{b\lambda} = T(1 - e^{-\tau\lambda}),   \text{(VII-88)}$$

is less than the true coronal temperature since τ_λ is less than unity.

For wavelengths of the order of 1 to 5 m, the ray path penetrates deeply enough that τ_λ may exceed unity and $T_{b\lambda}$ should approach the true coronal temperature. From observations at 178 cm Leblanc and LeSqueren (1969) obtain coronal temperatures of

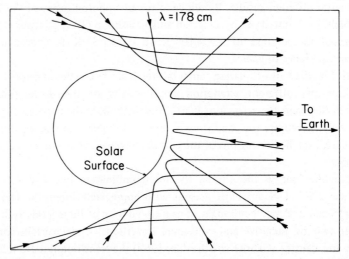

Fig. VII-15.   Ray trajectories in the corona as computed at λ 178 cm by Newkirk (1967).

1.1×10^6 K at sunspot minimum and 1.8×10^6 K for the 'quiet' corona at sunspot maximum.

Although the determination of coronal temperature from disk center brightness temperatures is straightforward in principle, in practice it is difficult. Since the radio wavelengths lie on the Rayleigh-Jeans end of the Planck curve, the brightness temperature is linearly related to the observed flux. This means that an error in the *absolute* calibration of the antenna radiation temperature reflects as an equal error in the inferred temperature. It is difficult to reduce such errors below about 10%. Nevertheless, the radio brightness temperature measurements remain as one of the more accurate means for determining the coronal temperature.

A further difficulty associated with radio measurements of central disk temperatures is that of obtaining the necessary angular resolution. Attempts to infer the central disk brightness from the integrated brightness face the difficulty that the Sun is an irregular object at radio wavelengths so that the distribution of brightness over the disk at any given time may differ appreciably from the average distribution (see Figure V-28). Antennas with fan-beam patterns offer considerable improvement, but still face the difficulty of an unknown brightness distribution.

Observations of integrated disk intensities as the Moon systematically occults the Sun during solar eclipse afford the possibility of recovering the brightness distribution within the occulted areas. This technique has been used successfully a number of times. However, the opportunities for having sizeable antenna arrays at suitable eclipse sites are relatively rare.

By far the best method for obataining central brightness temperature is through direct observations with pencil beam antennas. However, the pencil beam must have a resolution of a few arc min and, at meter wavelengths, this requires large, complex antenna arrays. The results reported by Leblanc and LeSqueren (1969) are based on observations made with the Nancay interferometer. This antenna has two crossed interferometer arrays giving an east-west resolution of 2.8 arc min and a north-south resolution of 8 to 12 arc min depending upon the solar declination angle. The two arrays combine to produce a pattern of pencil beams each of which is elliptical in shape with axes of 3.8 arc min and 8–12 arc min in the east-west and north-south directions, respectively. This resolution is adequate for determining central brightness temperatures. Even so, it is necessary to combine data from many days of observation to determine the true 'quiet' Sun conditions.

F. TEMPERATURE DISTRIBUTION

A variety of evidence indicates that the corona is approximately isothermal over a considerable distance from the solar surface. On the other hand, we know that the coronal temperature must reach a maximum at some height since the temperature must ultimately drop going in either direction from the corona. For purposes of determining the local energy balance in the corona, which is one of the major problems of solar physics, however, we must know the detailed temperature structure. An isothermal corona at 1.8×10^6 K has a far different energy balance than a corona with a mean temperature of 1.8×10^6 K but which has a definite maximum temperature.

Thermal conductivity in the corona is proportional to $F_0 = T^{5/2} dT/dh$. For the transition region we found a value for F_0 of 5.6×10^{11} cm^{-1}. If we adopt $T = 1.8 \times 10^6$ for the corona, a value of $F_0 = 5.6 \times 10^{11}$ cm^{-1} corresponds to $dT/dh = 13$ deg km^{-1}. Such a gradient extending for only one density scale height ($\approx 0.1\ R_\odot$) would produce a temperature change of 9×10^5 K. In the mean corona, at least, such gradients (with reverse sign) cannot be present beyond the temperature maximum. Otherwise, the corona would not extend so far from the Sun and the approximation to isothermal, hydrostatic equilibrium would not extend nearly so far as is indicated by Figure VII-14. In fact, there is no reason to expect thermal conduction outwards from the corona to be as large as the inward conduction. The large inward conduction results from a temperature gradient forced upon the atmosphere by the combination of mechanical heating and radiant cooling. The rapid drop in temperature inwards is compensated by a rapid rise in density so that pressure equilibrium can be achieved with an arbitrarily large temperature gradient. Also, since density increases inward the conducting layer can be terminated by absorbing the conducted energy in the lower atmosphere. In the outer corona, however, there are no such restraints. A rapid outwards decrease of temperature would be associated with a rapid drop in pressure. If the temperature gradients in the outer corona were as large as in the inner corona, the resulting pressure gradient in the outer corona would be intolerably large, and the corona would simply expand to relieve the pressure gradient. Also, if the corona is considered to be static and non-radiating then the outwards thermal conduction must be constant to infinity, i.e., the conducted energy flux cannot be terminated at any finite distance from the Sun. This latter condition carries with it the restriction that the temperature cannot decrease to zero at any finite distance from the Sun and this severely limits the temperature structure. It is tantamount, in fact, to allowing only one possible value of the conducted flux for a given coronal maximum temperature.

This point was first recognized by Chapman (1957) who solved the coupled heat conduction and hydrostatic equilibrium equations to obtain the temperature and density profiles in the outer corona. We now know that the corona is not static, but rather that it expands. Furthermore, the corona loses energy by radiation and receives energy from mechanical dissipation. Nevertheless, it is of interest to examine the hydrostatic, conduction model.

Since the corona is extended in space, the plane parallel approximation must be dropped. The conducted energy flux then becomes

$$F_c = k \left(\frac{r}{R_\odot} \right)^2 T^{5/2} \frac{dT}{dr} = \text{const.} \tag{VII-89}$$

we integrate this equation from r_0 to r to obtain

$$\frac{2}{7} K(T^{7/2} - T_0^{7/2}) = F_c R_\odot^2 \left(\frac{1}{r_0} - \frac{1}{r} \right). \tag{VII-90}$$

As a boundary condition, we set $T = T_\infty$ at $r = \infty$. Whence we find

$$T^{7/2} = T_\infty^{7/2} + \left(\frac{r_0}{r} \right) (T_0^{7/2} - T_\infty^{7/2}). \tag{VII-91}$$

Near the Sun, we may set $T_\infty \ll T$ and $T < T_0$ so that

$$T = \left(\frac{r_0}{r} \right)^{2/7} T_0. \tag{VII-92}$$

The corresponding conducted energy flux is

$$F_c = \frac{2}{7} K \frac{r_0}{R_\odot^2} T_0^{7/2}, \tag{VII-93}$$

and the temperature gradient is

$$\frac{dT}{dh} = \frac{dT}{dr} = -\frac{2}{7} \frac{r_0^{2/7} T_0}{r^{9/7}}. \tag{VII-94}$$

For $r = xr_0$ and $r_0 = 1.05\,R_\odot$, $dT/dr = -0.70x^{-9/7}$ deg km^{-1}. Thus, at $x = 1$, dT/dr is -0.7 deg km^{-1} and, at $x = 10$, $dT/dr = -0.036$ deg km^{-1}. The outward conductive flux for $K = 1.1 \times 10^{-6}$, $r_0 = 1.05\,R_\odot$ and $T_0 = 1.8 \times 10^6$ K is 3.7×10^4 erg m^{-2} s^{-1}. By comparison the inward conductive flux for $F_0 = 5.6 \times 10^{11}$ is 6.2×10^5 erg cm^{-2} s^{-1}, which is some 17 times greater.

We note from Equation (VII-92) that for $r = 2r_0$ and $4r_0$, T has the values $0.82\,T_0$ and $0.67\,T_0$, respectively. Thus, if the corona near the Sun is truly dominated by conduction the expected decrease of temperature outwards is so slow as to make it very difficult to observe by any existing techniques. Such a model is, in fact, consistent with present data for the corona and with current theoretical models of the corona, even those allowing for coronal expansion via the solar wind (Chapter IX).

In the case where the corona is expanding, most of the energy of expansion close to the Sun is used up in overcoming the gravitational potential. Thus, a more accurate representation of the corona close to the Sun is obtained by setting (Parker, 1965)

$$F = - Kr^2 \frac{dT}{dr} - nvr^2 m_H \frac{GM_0}{r} = \text{const.} \tag{VII-95}$$

This equation can be combined with the Equations (VII-80) and (VII-82) and integrated to obtain $T(r)$. The boundary conditions are $T = T_0$ at the position r_0 of maximum temperature and $T \approx 0$ at $r = \infty$. Parker (1965) characterizes the solution as being approximately of the form $T(r) \propto r^{4/7}$ for $r_0 < r \leqslant r_2$ and $T(r) \propto r^{2/7}$ for $r_2 < r \leqslant r_c$, where r_c is the location of the critical point where the wind becomes supersonic. The location of r_2 is given by

$$r_2 = \frac{r_0}{\alpha - 1},$$ (VII-96)

where

$$\alpha = \frac{4}{7} \frac{kT_0^{7/2}}{n_0 v_0 m_H GM_0}.$$ (VII-97)

The condition $\alpha = 1$ corresponds to the case of all the conductive energy flux being used to lift the particles out of the gravitational field. In this limit, which is reached for large n_0 or small T_0, the expansion remains subsonic at all distances. On the other hand, for $\alpha \geqslant 2$, r_2 is less than r_0, which means physically that the expansion does not change the form of $T(r)$ from the pure conductive case. For $n_0 v_0 = 9.2 \times 10^{12}$ cm^{-2} s^{-1}, as inferred from solar wind studies at 1 AU, $K = 1.1 \times 10^{-6}$ and $T_0 = 1.8 \times 10^6$ K, α has the value 2.4. However, this estimate of α based on average values of nv at the orbit of Earth is clearly uncertain by as much as 50%. This means that we cannot rule out the possibility that the temperature gradient between $r = r_0$ and $r = 2 R_\odot$ is closer to $(r_0/r)^{4/7}$ than to $(r_0/r)^{2/7}$. Even for $T(r) \propto (r_0/r)^{4/7}$, however, T decreases to only $0.67T_0$ at $2 R_\odot$ and the gradient is still relatively small.

In the above discussions, it has been assumed that conduction is not seriously inhibited by coronal magnetic fields, that radiation losses are negligible and that there is negligible dissipation of mechanical energy. None of these assumed conditions is very probable. It is implausible that the coronal temperature continues to rise after the mechanical energy dissipation stops and equally implausible that the energy dissipation stops abruptly where the temperature maximum occurs. The most likely situation is that the mechanical dissipation has some more or less smooth decrease with height and that the maximum temperature occurs where there is still appreciable mechanical energy dissipation. It follows that near the temperature maximum, and possibly to some distance much beyond the maximum, mechanical dissipation still plays an important role.

The temperature maximum probably occurs within $r = 1.1 R_\odot$ ($h = 70\,000$ km) where the density is still relatively large. Coronal emission lines in the XUV radiate a combined energy flux of the order of 1×10^5 erg cm^{-2} s^{-1}, which is larger than the outward conductive flux estimated earlier by about a factor of three. Only a fraction of the radiation loss arises near the temperature maximum, but, clearly, radiation losses cannot be ignored completely in this region.

That region of the corona where closed magnetic field structures evidently play a large role extends well beyond the region of maximum temperature. Since the field lines are often closed in this region much of the energy that is carried by conduction will be trapped within the corona rather than flow radially outwards.

The effect of adding mechanical energy above the temperature maximum, of trapping conducted energy in the corona itself and of including radiation losses from the lower layers of the corona all seems to act in the direction of reducing the average temperature gradient, i.e., of making the corona more nearly isothermal. Thus, we have little reason to expect temperature gradients of any easily detectable magnitude in the corona below about $2 R_\odot$.

We turn next to the empirical evidence for or against temperature gradients in the corona. Two types of data for this purpose are available: line half-widths and equivalent widths. The first type of data gives information on the velocity distribution with height and the second gives information on the ionization distribution as a function of height. Unfortunately, data of either type are sparse.

The basic difficulties inherent in attempts to infer temperature gradients from gradients in line widths, of course, are that part of the line width is evidently due to macroscopic motions and thus only part of the width is due to thermal motions. In discussing temperatures determined from line widths in Section 8C we concluded that the mean macroscopic velocity may be as high as 14 km s^{-1} as compared to a mean thermal velocity for iron ions of 23 km s^{-1}. If we allowed the macroscopic motion to decrease to zero and kept the thermal velocity constant, the line width would decrease by about 17%. Hence, any change of line width of this magnitude, or less, has an ambiguous interpretation. It could be due either to changes in thermal motion or to changes in macroscopic motion or some combination of the two.

Billings (1966, p. 247) concludes from an earlier study by Billings and Lilliequist (1963) using the Fe XIV (λ 5303) line that the temperature gradient in the inner corona observed in 1959 is negative and has a magnitude of a few degrees per km. Their observations were made with a coronagraph outside of eclipse and are restricted to the height range $1.03 R_\odot$ to $1.13 R_\odot$, i.e., $\Delta h \approx 70\,000$ km. The 'few degrees' per km that they infer therefore corresponds to a total change of perhaps 2×10^5 to 3×10^5 K, which is about 10% of the mean temperature inferred. This total change of temperature is considerably less than the scatter in temperatures observed at any one height and is therefore of doubtful statistical significance.

Jarrett and von Klüber (1961) report on the widths of the Fe XIV (λ 5303) and Fe X (λ 6374) lines observed at the 1958 eclipse (sunspot maximum) using a Fabry-Perot interferometer. They give results for several heights along 9 solar radii and for three specific heights (1.15, 1.30 and $1.45 R_\odot$) along some 30 solar radii. For the 9 radii studied in detail, 5 show a net increase in line width of about 20% in the interval $1.1 R_\odot$ to $1.5 R_\odot$, and 4 show a net decrease in width of about 10% in the same interval followed by a region from $1.5 R_\odot$ to $1.8 R_\odot$ where no further change in width occurs. For the 3 heights studied in detail, there is a slight decrease in average line width ($\sim$10%) between 1.15 and $1.30 R_\odot$ and no detectable change in average width between $1.30 R_\odot$ and $1.45 R_\odot$ Jarrett and von Klüber conclude that no definite average trend in line widths between $1.1 R_\odot$ and $1.8 R_\odot$ can be established from their data. This appears to be the most plausible conclusion at this time, viz., that no definite trend has been established.

Equivalent widths of Fe XIV and Fe X lines as a function of height have been reported by a few authors. Billings and Cooper (1957) give average height gradients for the equivalent widths of the lines λ 6374 (Fe X) and λ 5303 (Fe XIV) between approximately 1.03 $R_\odot$ and 1.1 $R_\odot$. Within this interval the average gradients obtained from many observations indicate that the ratio of intensities in the two lines remains constant, on the average, to within better than 20%. Allen (1946) gives data for the eclipse of 1940 (2 years past sunspot maximum) showing the relative equivalent width of the λ 5303 and λ 6374 lines as a function of height between approximately 1.4 and 2.0 $R_\odot$. Within this interval the ratio of mean equivalent widths, $F(\lambda\ 6374)/F(\lambda\ 5303)$, is nearly constant with a weak tendency for a maximum near 1.6 to 1.8 $R_\odot$. However, the scatter in the data is such that a constant ratio of equivalent widths would fit the data just as well. One concludes from these data that the relative concentrations of Fe X and Fe XIV are nearly constant with height.

In Figure V-23 data are presented that show the relative fluxes in λ 6374 and the continuum between about 1.8 and 6.5 $R_\odot$. Over this height interval, the ratio of the line flux to the continuum flux is constant to within a factor of two. At distances from the solar surface of this order the electron density has decreased enough that the λ 6374 line is excited mainly by scattering of photospheric radiation rather than collisions. Thus, the line flux varies linearly with the local density of Fe X and the local intensity of photospheric radiation. The continuum flux varies linearly with local electron density and the local intensity of photospheric radiation. Thus, the relatively constant ratio of λ 6374 flux to continuum flux implies that the ratio of Fe X density to electron (hence, hydrogen) density is relatively constant as well.

If the corona were in static equilibrium, the ionization equilibrium computed by Jordan (1969) (Figure VII-11) would be valid over a wide range of heights. However, the corona is undergoing hydrodynamic expansion with the expansion velocity increasing with radial distance from the Sun. At some point in the corona the 'expansion time' will become short enough that the ionization equilibrium will become independent of local density and temperature, i.e., the equilibrium will be frozen in by the expansion (Hundhausen *et al.*, 1968).

To estimate the freezing in depth for Fe X in an approximate way, we set the characteristic collisional ionization time, τ_c, for Fe X equal to the characteristic expansion time, τ_{ex}, through one density scale height. The characteristic collision time is

$$\tau_c = (n_x C_{x1})^{-1} \tag{VII-98}$$

and the characteristic expansion time is

$$\tau_{ex} = H_c/v, \tag{VII-99}$$

where H_c is the coronal density scale height and v is the expansion velocity. As before we obtain the expansion velocity from the continuity equation and the mean particle flux observed at Earth. This gives

$$v = \frac{9.2 \times 10^{12}}{n_e(r/R_\odot)^2} \text{ cm s}^{-1}.$$

The coronal scale height can be approximated by

$$H_c \approx \frac{kT}{\mu m_H g_0} \left(\frac{r}{R_\odot}\right)^2 = 7 \times 10^9 \left(\frac{r}{R_\odot}\right)^2 \text{ cm}$$

so that τ_{ex} is given approximately by

$$\tau_{ex} \approx 7.6 \times 10^{-4} n_e (r/R_\odot)^4 \text{ s.} \tag{VII-100}$$

The collision rate C_{XI} given by equation (VII-48) for $T = 1.8 \times 10^6$ K is

$$C_{XI} \approx 3.6 \times 10^{-10} n_e.$$

Thus, if we set the Fe X concentration equal to $a_x n_e$, the collision time is

$$\tau_c \approx \frac{2.8 \times 10^9}{a_x n_e^2} . \tag{VII-101}$$

Equations (VII-100) and (VII-101) give

$$n_e (r/R_\odot)^{4/3} = \frac{1.5 \times 10^4}{a_x^{1/3}} . \tag{VII-102}$$

For a coronal temperature of 1.8×10^6 K, a_x is approximately $10^{-2} A_{Fe}$, or approximately 2×10^{-7}. Thus, at the height where the Fe X concentration is frozen in by the coronal expansion, we expect $n_e (r/R_\odot)^{4/3}$ to be less than or approximately 2.6×10^6 cm^{-3}. For Newkirk's equatorial model, or for van de Hulst's sunspot maximum model in Table VII-6, this relation is satisfied for $r/R_\odot$ greater than or approximately 2.5 and for van de Hulst's sunspot minimum equatorial model it is satisfied for $r/R_\odot$ greater than or approximately 2.25. At $r/R_\odot = 1.5$, however, τ_{ex} exceeds τ_c for Fe X by nearly a factor of ten. Similar computations for Fe XIV yield a collision time

$$\tau_c \approx \frac{7 \times 10^9}{a_{XIV} n_e^2} . \tag{VII-103}$$

The value of a_{XIV} is $0.15 A_{Fe}$, or 3×10^{-6} at $T = 1.8 \times 10^6$ K. Equations (VII-104) and (VII-107) give $n_e (r/R_\odot)^{4/3} \lesssim 1.5 \times 10^6$ cm^{-3} at the value of $r/R_\odot$ where $\tau_c \gtrsim \tau_f$. This point is somewhat further out in the corona than the point where the Fe X concentration becomes frozen in.

It appears from the above arguments that the constant relative concentration of Fe X can be accounted for above $2.5 R_\odot$ solely in terms of the coronal expansion, but that the constancy in the relative concentrations of iron ions below $2.5 R_\odot$ must be due to local conditions of temperature in the corona.

The ionization equilibrium for iron computed by Jordan (1969) (see Figure VII-11) shows that between temperatures of 2×10^6 K and 1.6×10^6 K the ratio of Fe XIV to Fe X concentrations decreases by a factor of 27 whereas the ratio of Fe X to hydrogen concentration increases by a factor of 24. The evidence from the ratios of equivalent widths of λ 6374 and λ 5303 and λ 6374 flux and continuum flux is inconsistent with relative changes in Fe XIV to Fe X concentrations of this magnitude. Since the relative iron ionization appears to be nearly constant over heights of several solar radii and since the Fe X concentration is frozen in near $2.5\,R_\odot$ a temperature change greater than about 1×10^5 K between 1 and $2.5\,R_\odot$ seems incompatible with the data.

The conduction model of the corona giving $T(r) \propto r^{-2/7}$ yields a drop in T to $0.77\,T_0$ at $2.5\,R_\odot$. For $T_0 = 1.8 \times 10^6$ K, therefore $T \approx 1.4 \times 10^6$ K at $2.5\,R_\odot$. Thus, even this modest rate of decrease of temperature seems to be too large to be compatible with the equivalent width data. This suggests that mechanical heating or conduction along closed field lines serves to increase the coronal temperature above the open field conduction value out to at least $2.5R_\odot$. Alternatively, radiation losses from the low corona could cool the lower regions thereby reducing the temperature gradient. It seems unlikely, however, that radiation cooling can influence any but the very lowest layers near 1.0 to $1.5\,R_\odot$. Conduction along closed field lines could play a substantial role out to about $2.5\,R_\odot$, and this may be sufficient to explain the result that T is nearly constant out to $2.5\,R_\odot$. In this connection, it should be noted that the data on which we have based the conclusion that T is constant were obtained at low solar latitudes where the magnetic fields below $2.5\,R_\odot$ are predominantly closed. The same conclusion may not be valid in open field regions. Also, of course, the conclusions apply only to the average corona. Specific active regions may, and obviously do, violate these average conditions.

Mechanical heating in the corona undoubtedly plays a role in determining the temperature structure of the corona near the Sun and may well influence the temperature structure out to several solar radii. We can do little more than speculate about this point, however.

The coronal holes are observed in the continuum, and must, as a result, have lower density than the normal corona. Munro and Withbroe (1972) find from a study of one coronal hole observed on OSO-4 that the coronal density is decreased below the average coronal density in the region of Mg X (1.2×10^6 K) emission by a factor of three. They further concluded from the relative intensities of spectral lines from different ions that the temperature in the hole is reduced by about 600 000° from the average coronal temperature. This places the temperature of the corona in the hole at a value near 1×10^6 K.

Pneuman (1973) suggests that the coronal holes result from an open structure of the magnetic field in the corona. This allows energy to be carried away from the corona by thermal conduction and by the solar wind. In this interpretation, the coronal hole is similar to the polar corona and would be expected to have a steeper density gradient as well as reduced density (see Figure VII-14). The factor of three reduction in density found by Munro and Withbroe (1972) is consistent with this interpretation.

References

Allen, C. W.: 1946, *Monthly Notices Roy. Astron. Soc.* **106**,137.

Allen, C. W.: 1963, in J. W. Evans (ed.), *The Solar Corona*, Academic, New York, p. 1.
Allen, J. W. and Dupree, A. K.: *Astrophys. J.* **155**, 27.
Altschuler, M. D. and Newkirk, G., Jr.: 1969, *Solar Phys.* **9**, 131.
Athay, R. G.: 1958, *Ann. Astrophys.* **67**, 475
Athay, R. G.: 1965, *Astrophys. J.* **142**, 755.
Athay, R. G.: 1966, *Astrophys. J.* **145**, 784.
Athay, R. G.: 1969, *Solar Phys.* **9**, 51.
Athay, R. G. and Bessey, R. J.: 1964, *Astrophys. J.* **140**, 1174.
Athay, R. G. and Canfield, R. C.: 1969, *Astrophys. J.* **156**, 695.
Athay, R. G. and Canfield, R. C.: 1970, in H. G. Groth and P. Wellman (eds.), *Extended Atmosphere Stars*, Dept. of Commerce, Washington, NBS Spec. Publ. 332.
Athay, R. G. and Johnson, H. R.: 1960, *Astrophys. J.* **131**, 413.
Athay, R. G. and Lites, B. W.: 1972, *Astrophys. J.* **176**, 809.
Athay, R. G. and Menzel, D. H.: 1956, *Astrophys. J.* **123**, 285.
Athay, R. G. and Skumanich, A.: 1967, *Solar Phys.* **4**, 176.
Athay, R. G. and Skumanich, A.: 1968, *Astrophys. J.* **152**, 141.
Avery, L. W.: 1970, *Solar Phys.* **13**, 301.
Avery, L. W. and House, L. L.: 1969, *Solar Phys.* **10**, 88.
Avery, L. W., House, L. L. and Skumanich, A.: 1969, *J. Quant. Spectrosc. Radiat. Transfer* **9**, 519.
Beckers, J. M.: 1972, *Ann. Rev. Astron. Astrophys.* **10**, 73.
Beigman, I. L. and Vainshtein, L. A.: 1967, *Astron. Zh.* **44**, 668. (*Soviet Astron. AJ.* **11**, 531.)
Billings, D. E.: 1966, *A Guide to the Solar Corona*, Academic, New York.
Billings, D. E. and Cooper, R. H.: 1957, *Z. Astrophys.* **43**, 218.
Billings, D. E. and Lilliequist, C. G.: 1963, *Astrophys. J.* **137**, 16.
Brandt, J. C., Michie, R. W., and Cassinelli, J. P.: 1965, *Icarus* **4**, 19.
Burgess, A.: 1964, *Astrophys. J.* **139**, 776.
Burgess, A.: 1965a, *Astrophys. J.* **141**, 1588.
Burgess, A.: 1965b, *Proc. Second Harvard-Smithsonian Conf. Stellar Atmos.* p. 47.
Burgess, A. and Seaton, M. J.: 1964, *Monthly Notices Roy. Astron. Soc.* **127**, 355.
Chapman, S.: 1957, *Smithsonian Contr. Astrophys.* **2**, No. 1.
Chipman, E.: 1971, Thesis Harvard Univ.
Chiuderi, C. and Riani, I.: 1974, *Solar Phys.* **34**, 113.
Coates, R. J.: 1968, *Astrophys. J.* **128**, 83.
Cuny, Y.: 1971, *Solar Phys.* **16**, 293.
Dunn, R. B.: 1960, Thesis, Harvard Univ.
Dupree, A. K.: 1972, Solar Satellite Prog. Rep. TR33, Harvard College Obs. Cambridge.
Dupree, A. K. and Goldberg, L.: 1967, *Solar Phys.* **1**, 229.
Elwert, G.: 1952, *Z. Nat.* **7a**, 703.
Elwert, G. and Raju, P. K.: 1972, *Solar Phys.* **25**, 319.
Evans, K. and Pounds, K. A.: 1968, *Astrophys. J.* **152**, 319.
Flower, D. R.: 1971, *Highlights of Astronomy* **2**, 544.
Gabriel, A. H.: 1971, *Highlights of Astronomy* **2**, 486.
Gebbie, K. B. and Thomas, R. N.: 1970, *Astrophys. J.* **161**, 229.
Gabriel, A. H. and Jordan, C.: 1969, *Monthly Notices Roy. Astron. Soc.* **145**, 241.
Gillet, F. C., Stein, W. A., and Ney, E. P.: 1964, *Astrophys. J.* **140**, 292.
Gingerich, O. and de Jager, C.: 1968, *Solar Phys.* **3**, 5.
Gingerich, O., Noyes, R. W., Kalkofen, W., and Cuny, Y.: 1971, *Solar Phys.* **18**, 347.
Giovanelli, R. G.: 1967a, *Australian J. Physics* **20**, 81.
Giovanelli, R. G.: 1967b, in J. N. Xznthakis (ed.), *Solar Phys.*, p. 353.
Giovanelli, R. G., Hall, D. N. B., and Harvey, J. W.: 1972, *Solar Phys.* **22**, 53.
Goldberg, L. Dupree, A. K., and Allen, C. W.: 1965, *Ann. Astrophys.* **28**, 589.
Henoux, L.: 1964, *Proc. Phys. Soc. London* **83**, *121*.
Henze, W., Jr.: 1969, *Solar Phys.* **9**, 56.
Hiei, E.: 1963. *Publ. Astron. Soc. Japan* **15**, 277.
Hinteregger, H. E., Hall, L. A., and Schmidtke, G.: 1965, *Space Res.*, Vol. 5, North Holland, Amsterdam.
Hirayama, T.: 1971, *Solar Phys.* **19**, 384.
House, L. L.: 1964, *Astrophys. J. Suppl.* **8**, 307.
Hulst, H. C. van de: 1953, in G. P. Kuiper (ed.), *The Sun*, Univ. of Chicago, Chicago.
Hundhausen, A. J., Gilbert, H. E., and Bame, S. S.: 1968, *Astrophys. J.* **152**, L3.

Jacob, A.: 1968, *Solar Phys.* **5**, 359.
Jarrett, A. H. and von Klüber, H.: 1961, *Monthly Notices Roy. Astron. Soc.* **122**, 223.
Jordan, C.: 1969, *Monthly Notices Roy. Astron Soc.* **142**, 501.
Jordan, C.: 1970, *Monthly Notices Roy. Astron. Soc.*. **148**, 17.
Kawabata, K. and Sofue, T.: 1972, *Publ. Astron. Soc. Japan* **24**, 469.
Kopp, R. A.: 1972, *Solar Phys.* **27** 373.
Kopp, R. A. and Kuperus, M.: 1968, *Solar Phys.* **4**, 212.
Krat, V. A. and Krat, T. V.: 1971, *Solar Phys.* **17**, 355.
Lantos, P.: 1972, *Solar Phys.* **22**, 387.
Leblanc, Y. and LeSqueren, A. M.: 1969, *Astron. Astrophys.* **1**, 239.
Linsky, J. L. and Avrett, E. H.. 1970, *Publ. Astron. Soc. Pacific* **82**, 169.
Linsky, J. L. and Shine, R. A.. 1972, *Solar Phys.* **25**, 357.
Lites, B. W.: 1972, Thesis, Univ. of Colorado, Boulder. (National Center for Atmospheric Research
    Coop. Thesis No. 28.)
Makita, M.: 1972, *Solar Phys.* **24**, 59.
Mankin, W. G., Eddy, J. A., Lee, R. H., and MacQueen, R. M.: 1974, *Proc. of the Soc. of
    Photo-optical Inst. Engrs.* **44**, 133.
McCrea, W. H.: 1929, *Monthly Notices Roy. Astron. Soc.* **89**, 718.
McWhirter, R. W. P. and Hearn, A. G.: 1963, *Proc. Phvs. Soc. London* **82**, 641.
Milkey, R. W. and Mihalas, D.: 1974, *Astrophys. J.* **192**, 769.
Morton, D. C. and Widing, K. G.: 1961, *Astrophys. J.* **133**, 596.
Munro, R., Dupree, A. Y., and Withbroe, G. L.: 1971, *Solar Phys.* **19**, 347.
Munro, R. and Withbroe, G. L.: 1972. *Astrophys. J.* **176**, 511.
Newkirk, G., Jr.: 1967, *Ann. Rev. Astron. Astrophys.* **5**. 213.
Ney, E. P., Huch, W. F., Kellogg, P. J., Stein, W., and Gillet, F.: 1961, *Astrophys. J.* **133**, 616.
Noyes, R. W. and Kalkofen, W.: 1970, *Solar Phys.* **15**, 120.
Parker, E. N.: 1960, *Astrophys. J.* **132**, 821.
Parker, E. N.: 1965, *Space Sci. Rev.* **4**, 666.
Parkinson, W. H. and Reeves, E. M.: 1969, *Solar Phys.* **10**, 342.
Pasachoff, J. M., Noyes, R. W., and Beckers, J. M.: 1968, *Solar Phys.* **5**, 131.
Pneuman, G. W.: 1968, *Solar Phys.* **3**, 578.
Pneuman, G. W.: 1973, *Solar Phys.* **28**, 247.
Poland, A., Skumanich, A., Athay, R. G., and Tandberg-Hanssen, E.: 1971, *Solar Phys.* **18**, 391.
Pottasch, S. R.: 1964, *Space Sci. Rev.* **3**, 816.
Pottasch, S. R.: 1967, *Bull. Astron. Inst. Neth.* **19**, 113.
Praderie, F. and Thomas, R. N.: 1972, *Astrophys. J.* **172**, 485.
Reeves, E. M. and Parkinson, W. H.: 1972, *Solar Phys.* **24**, 113.
Reeves, E. M., Foukal, P. V., Huber, M. C. E., Noyes, R. W., Schmahl, E. J., Timothy, J. G., Vernazza,
    J. E. and Withbroe, G. L.: 1974, *Astrophys. J.* **187**, L27.
Sandlin, G. D. and Widing, K. G.: 1967, *Astrophys. J.* **149**, L129. (See also Widing, K. G., Purcell, J.
    D., and Sandlin, G. D.: 1970, *Solar Phys.* **12**, 52.)
Seaton, M. J.: 1964, *Planetary Space Sci.* **12**, 55.
Simon, M. and Zirin, H.: 1969, *Solar Phys.* **9**, 317.
Smerd, S. F.. 1950, *Australian J. Sci. Res.* **3**, 34.
Tanaka, K. and Hiei, E.: 1972, *Publ. Astron. Soc. Japan* **24**, 323.
Thomas, R. N. and Athay, R. G.: 1961, *Physics of the Solar Chromosphere*, Interscience, New York.
Tousey, R.: 1967, *Astrophys. J.* **149**, 239.
Ulmschneider, P. H.: 1969, *The Chromosphere-Corona Transition Region*, National Center for
    Atmospheric Research, Boulder.
Vernazza, J. E.: 1972, Thesis, Harvard University, Cambridge.
Vernazza, J. E., Avrett, E. H., and Loeser, R.: *Astrophys. J.* **184**, 605.
Vernazza, J. E. and Noyes, R. W.: 1972, *Solar Phys.* **22**, 358.
Walker, A. B. C. and Rugge, H. R.: 1971, *Astrophys. J.* **164**, 181.
White, O. R. and Suemoto, Z.: 1968, *Solar Phys.* **3**, 523.
White, O. R., Altrock, R. C., Brault, J. W., and Slaughter, C. D.: 1972, *Solar Phys.* **23**, 18.
Wilson, R.: 1967, *Plasmas in Space and in the Laboratory*, P373 (ESRO-SP-20).
Withbroe, G. L.. 1970, *Solar Phys.* **11**, 208.
Zirin, H. and Howard, R.: 1966. *Astrophys. J.* **146**, 367.
Zirker. J. R.: 1962, *Astrophys. J.* **136**, 250.

CHAPTER VIII

CHROMOSPHERIC STRUCTURE INFERRED FROM SPECTRAL LINES

1. Chromospheres and Coronas as Phenomena of Stellar Atmospheres

The chromosphere-corona complex plays such an important role in our understanding of the Sun and its space environment that it behooves us to ask how general these phenomena are in other stars. We shall turn to this question at several places in the remaining chapters. At this point, we consider just what evidence we would have of the solar chromosphere and corona if the Sun were at a stellar distance. The answer is: very little!

Throughout the visual and photographic spectrum of the Sun there is no detectable evidence of the solar corona and very little evidence for the existence of the chromosphere. Only the faint emission reversals in the Ca II H and K lines and the shallow depression in λ 10 830 of He I exist as directly identifiable indicators of the chromospheric temperature rise, and even these, on face value, completely belie the importance of the phenomena. On the other hand, numerous other subtle features of the visual spectrum betray the presence of the chromosphere. Most of the stronger lines have too much emission at line center to be properly explained without a chromosphere. Other strong lines, such as the hydrogen lines, would have smaller equivalent width without the chromosphere and the faint λ 10 830 line of He I would be absent. Given the present quality of stellar spectra, it is likely that none of these indicators of the chromosphere would have been noticed. The Sun, seen as a star, would be a placid object, presumably in radiative equilibrium and presumably showing a monotonic outwards decrease in temperature.

The Sun, at one parsec distance, would not have been seen as a radio source. Neither would it have been seen in the XUV. The Lyman-α flux, for example, would be the equivalent of about 5 photons $cm^{-2} s^{-1}$, which is too faint to have been observed with present day space telescopes. This is not to say, of course, that future instruments would not detect the XUV and radio emission. Nevertheless, the conclusion that the Sun observed at a distance of one parsec would not be recognized at the present time as having either a corona or a chromosphere seems inescapable.

It is true, also, perhaps that the features now recognized in the solar disk spectrum as being of chromospheric origin would not be so recognized were it not for the fact that the chromosphere and corona are so prominent at eclipse and in the XUV and radio spectrum of the Sun. Most of the effects of which we speak, after all, are rather subtle. They most likely exist to a greater or lesser extent in the spectra of multitudes of stars and have gone undetected because, until very recently, we have not had the requisite skill in spectrum diagnostics to recognize those features of the visual and photographic spectrum that are indicative of chromospheres.

In the remainder of this chapter, we consider the analysis of several strong Fraunhofer lines whose central cores are formed either wholly or partially in the chromosphere. The problem is of dual interest in that it provides an additional powerful means for studying chromospheric phenomena on the Sun and, in addition, serves as a useful guide to the nature of the effects to be expected in the spectra of other stars.

The reader is perhaps aware, at this point, that little use was made in the preceding discussion of the low and middle chromosphere of spectral line data. Continuum data have provided the primary basis for the model. This is a matter of choice only, and was done essentially for the reason that continuum data are more readily analyzed and discussed than are line data. Alternatively, we could have used the spectral lines observed on the disk to construct the basic model and reserved the continuum for corroboration. Such an approach would have been somewhat more difficult, however, and the model would be somewhat less well defined. Since stellar astronomers typically do not have access to the type of continuum data used in constructing the model of the low and middle chromosphere, they must rely heavily upon the line data to infer chromospheric properties. It is of interest, therefore, to examine, in some detail, the chromospheric model from the standpoint of the spectral lines observed on the disk.

2. Influence of the Chromospheric Temperature Rise

It was shown in Chapter VI (Equation (VI-55)) that if the chromospheric temperature rise is mimicked by an equation of the form

$$B = B_0(1 + A e^{-\gamma N}),$$ (VIII-1)

then the source function for a two level atom in an atmosphere with constant Doppler width and constant destruction probability, p_d, is increased for $\gamma N < 1$ by a factor

$$x = 1 + \frac{A}{\left(1 + \dfrac{\gamma}{p_d}\right)^{\frac{1}{2}}}$$ (VIII-2)

over the value the source function would have if A were zero. The parameter γ is here defined in terms of the mean scattering depth, N, near the base of the chromosphere. Since the chromospheric temperature rise sets in near $\gamma N = 1$, γ may be replaced by the reciprocal of the characteristic scattering depth N_c at the base of the chromosphere. In all cases of interest in the visual spectrum, the optical thickness of the chromosphere is small enough that the absorption coefficient may be approximated by the Doppler form. For this case, N_c is given approximately by (Equation (VI-136))

$$N_c \approx 10 \, \tau_{0c}.$$

Thus, the increase in the source function due to the chromospheric increase in B is given approximately by

$$x = 1 + \frac{A}{\left(1 + \dfrac{1}{10\tau_{oc}p_d}\right)^{1/2}} = 1 + \frac{(10\tau_{oc}p_d)^{1/2}}{(1 + 10\tau_{oc}p_d)^{1/2}} \qquad \text{(VIII-3)}$$

The asymptotic limits on x are $x = 1$ for $\tau_{oc} \ll 1$ and $x = 1 + A$ for $\tau_{oc} \gg (10p_d)^{-1}$.

For some XUV lines, the chromosphere is thick enough that the damping portion of the absorption coefficient dominates the transfer problem near the base of the chromosphere. In that case, $N_c \approx 2(\tau_{oc}/a)^{1/2}$, where a is the damping parameter, and x becomes of the form

$$x = 1 + \frac{(2\tau_{oc}p_d)^{1/2}A}{(a^{1/2}\tau_{oc}^{1/2} + 2\tau_{oc}p_d)^{1/2}} \qquad \text{(VIII-4)}$$

The asymptotic upper limit of $1 + A$ on x now occurs for $\tau_{oc} \gg 0.25\, ap_d^{-2}$.

The continuum optical depth at the base of the chromosphere is of the order of 10^{-4}. Thus, any line for which the chromosphere plays a major role must have a line center opacity that exceeds the continuum opacity at that wavelength by over a factor of 10^4. This means that the continuum destruction probability is small (see Equation (VI-159)). A typical collisional destruction probability for a line with transition probability of 10^8 s^{-1} is of the order of

$$P_d = \epsilon = \frac{C_{21}}{A_{21}} = 10^{-8}\pi a_0^2\, n_e v_e.$$

For $n_e = 2 \times 10^{11}$ and $v_e = 10^8$ cm s^{-1}, we find $P_d \approx 2 \times 10^{-5}$.

The temperature rise in the low chromosphere from $4200°$ to, say, $5200°$ corresponds to values of A of 3.8 at λ 5000 and 4.2 at λ 4000. As a typical set of values, then, we pick $A = 4$, $P_d = 2 \times 10^{-5}$ and $a = 10^{-3}$. Values of x obtained from Equations (VIII-3) and (VIII-4) are given as a function of τ_{oc} in Table VIII-1.

It is of interest now to estimate the value of τ_{oc} for the Ca II K line. For this purpose, we adopt a hydrogen density near the base of the chromosphere of 1×10^{14}, a relative calcium abundance (assumed to be all Ca II) of 1.4×10^{-6}, a broadening velocity of 3 km s^{-1} and a density scale height of 10^7 cm. The absorptivity at line center is then given by

$$\alpha_0 = \frac{\pi e^2}{mc} \frac{f}{\pi^{1/2}v} A_{Ca}N_H = 1.8 \times 10^{-4}\ \text{cm}^{-1},$$

and the optical depth, τ_0, is 1.8×10^3. To the extent that the results in Table VIII-1 are applicable, therefore, the K line source function should be increased by somewhat over a factor of 2 above its non-chromospheric value.

This, of course, is only a crude estimate of the chromospheric effect in the K line, but
it does serve to illustrate that the Ca II H and K lines are expected to show an emission
peak that will perhaps be marginally present and, in any case, will not be overwhelmingly
strong. The IR lines of Ca II, by comparison, have optical depths that are less than those
in the K line by a factor of 10^2 or more, and the results in Table VIII-1 correctly predict
that only a modest chromospheric effect is present. On the other hand, magnesium is
more abundant than calcium in the Sun by about a factor of 15. Thus, the chromospheric
optical thickness in the stronger member of the Mg II resonance doublet should be more
nearly 3×10^4. Also, the characteristic value of A is larger for the Mg II lines. The results
in Table VIII-1 are seen to be consistent, therefore, with large emission peaks in the Mg II
lines. These lines, and others, will be discussed in more detail in the remainder of this
chapter.

Table VIII-1
The chromospheric increase in S_L due to a
temperature rise corresponding to $A = 4$ and for
$P_d = 2 \times 10^{-5}$ and $a = 10^{-3}$

τ_{oc}	X	
	Equation (VIII-3)	Equation (VIII-4)
10	1.1	–
10^2	1.4	–
10^3	2.2	1.6
10^4	3.8	2.2
10^5	4.8	2.9
10^6	5	3.8

Figure VIII-1 illustrates schematically the influence of the chromosphere on the source
function for three lines: one of very large optical depth, one of moderately large optical
depth and one of only modest optical depth. The former two will exhibit self-reversed
emission features and the latter will show only a partially filled in line core.

3. Influence of the Chromospheric Increase in Doppler Width

One of the most severe restrictions placed on Equation (VIII-1) is that of constant
Doppler width. In the solar chromosphere the Doppler widths of the lines of metal ions
increase quite markedly with height. Calcium atoms, for example, have a mean thermal
velocity of only 1.6 km s^{-1} at a temperature of 6000 K. The results in Figure III-7
indicate that throughout the low and middle chromosphere the non-thermal component
of line broadening increases at the approximate rate of 5 km s^{-1} per 1000 km of height
starting from a value near zero at height zero. Thus, near the base of the chromosphere
where the temperature is, say, 4500°, the broadening of Ca II lines is mainly thermal with
a velocity amplitude equal to 1.4 km s^{-1}. Some 500 km above the base of the

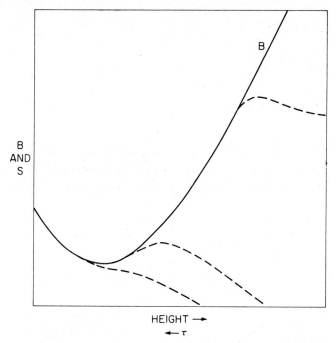

Fig. VIII-1.   A schematic illustration of the coupling between the line source function, dashed lines, and the chromospheric rise in B for three lines of different optical thickness. The H and K lines of Ca II are representative of the intermediate case.

chromosphere the rms broadening velocity is of the order of $(1.6^2 + 2.5^2) = 3.0$ km s^{-1}, i.e., about twice the broadening velocity at the base of the chromosphere.

This relatively rapid outwards increase in the Doppler width of the absorption coefficient has a very marked influence on the optical depth scale of the atmosphere at wavelengths in the edge of the Doppler core (see Chapter I). To illustrate this effect in a simplistic way, we write

$$\tau_{\Delta\lambda} = -\int_k^\infty \alpha_0 n_1 e^{-(\Delta\lambda/\Delta\lambda_D)^2} dt, \qquad\qquad \text{(VIII-5)}$$

$$\alpha_0 = \text{const},$$

$$n_1 = n_0 e^{-\beta h}, \qquad\qquad\qquad\qquad \text{(VIII-6)}$$

$$\Delta\lambda_D = \Delta\lambda_{D0}(1 + \delta h), \qquad\qquad\qquad \text{(VIII-7)}$$

$$y_0 = \frac{\Delta\lambda}{\Delta\lambda_{D0}},$$

and

$$f = \left[\beta h + \left(\frac{y_0}{1 + \delta h} \right)^2 \right]$$ (VIII-8)

Whence, we obtain

$$\tau_{y0} = -\alpha_0 n_0 \int_k^\infty e^{-f} dh.$$ (VIII-9)

To be consistent with the rate of increase of $\Delta \lambda_D$ with height as discussed previously, we adopt $\delta = 0.6$ per 100 km. Also, the value of β for Ca II obtained from Figure II-2 is 0.7 per 100 km. Plots of f vs height are shown in Figure VIII-2 for $\beta = 0.7$, $\delta = 0$ and 0.6 and $y_0 = 2$ and 3.

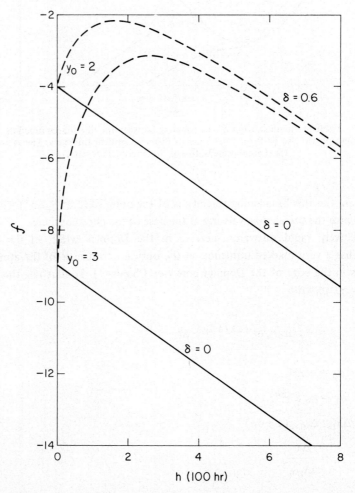

Fig. VIII-2.   Plots of f (Equation (VIII-8)) vs height for selected values of S and y.

Although τ_{y0} is proportional to the integral of $\exp(-f)$ over height, the behavior of τ_{y0} can be inferred from the plots of f vs height. Thus, for $\delta = 0$, for example, it is readily seen that τ_{y0} increases by a factor e each 140 km decrease in height. For $\delta = 0.6$, however, τ_{y0} increases rapidly down to a height h_0 after which it increases only very slowly. At $y_0 = 3$, h_0 is approximately 300 km and, at $y_0 = 2$, h_0 is approximately 200 km. Note, also, that τ_{y0} is much larger for $\delta = 0.6$ than for $\delta = 0$ and that there is relatively less difference between τ_2 and τ_3 for $\delta = 0.6$ than for $\delta = 0$. This is because $y_0 = 3$ corresponds approximately to $y = 1.5$ at 500 km whereas $y_0 = 2$ corresponds approximately to $y = 1.0$ at 500 km.

The effects on line profiles arising from distortions of the $\tau_{\Delta\lambda}$ scale of the type illustrated by Equation (VIII-9) are quite severe. Figure VIII-3 illustrates, again schematically, the variation of the line source function with optical depth at three

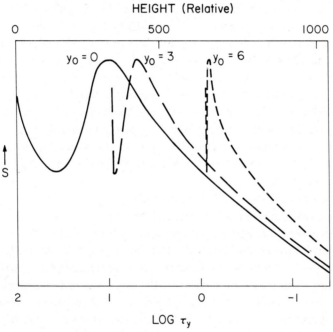

Fig. VIII-3. Plots of S vs log τ_y at three wavelengths in the line profile for $\beta = 0.7$ and $\delta = 0.6$ per 100 km (see Figure VIII-2 and Equations (VIII-8) and (VIII-9)).

wavelengths: $y_0 = 0$, $y_0 = 3$ and $y_0 = 6$. In this plot, the source function has the same value at all wavelengths; only the τ_y scale has been changed. The values of τ_y were obtained by numerical integration of Equations (VIII-8) and (VIII-9) with $\beta = 0.7$ and $\delta = 0.6$ per 100 km interval. For the illustration chosen, the line source function maximizes at $\tau_0 = 10$, $\tau_3 = 7$ and $\tau_6 = 0.8$. (Note that $y_0 = 6$ refers to units of the Doppler width at the base of the chromosphere. At 500 km, the corresponding value of y

is approximately 3 Doppler units.) To a crude approximation the emission maximum in the line profile will occur near the value of y_0 where $\tau_{y0} = 1$ at the depth where S is a maximum. We note, however, that, in the illustration chosen, the full variation in S between $\tau_0 = 100$ and $\tau_0 = 1$ is compressed to the range $\tau_6 \approx 0.7$ to $\tau_6 \approx 0.3$, i.e., the atmospheric region in which the source function maximizes is optically thin at the wavelengths where the line profile maximizes. This has the consequence that the emission maximum will be less intense than the maximum in S; the emission peak, in other words, is weakened by the gradient in Doppler width. Similarly, the minimum in S will map into a profile minimum (k_1 in Ca II) that is less deep than the minimum in S. This distortion between S_λ and its mapping into I_λ can be of major importance, and apparently is in the solar case.

Streaming motion in the vertical direction in the chromosphere combines with the gradient in Doppler width in an interesting way to further distort the emission peaks in the line profile. Suppose, for the sake of illustration, that the chromosphere illustrated in Figure VIII-3 is streaming outwards for $\tau_0 < 10$ and is stationary for $\tau_0 \geqslant 10$. The outwards streaming will reduce the optical thickness on the red side of the line and increase the optical thickness on the violet side. Thus, the red emission peak will be seen at a smaller value of y_0 than will the violet emission peak. Since the compression effect illustrated in Figure VIII-3 increases as y_0 increases, the effect of the streaming will be to enhance the reduction of the violet emission maximum and to diminish the reduction of the red emission maximum resulting from the gradient in Doppler width. The emission peaks will then be asymmetric; shifted slightly to the violet but with the red peak enhanced. Conversely, downward motion above $\tau_0 = 10$ will produce a slight red shift and an enhanced violet peak.

If, now, we allow the region of the chromosphere at $\tau_0 > 10$ in Figure VIII-3 to be in vertical motion with the rest of the atmosphere fixed, the asymmetry effects on the emission peaks reverse, i.e., outwards motion for $\tau_0 > 10$ enhances the violet peaks and inwards motion enhances the red peak. This latter effect is not easily illustrated in graphical terms but is clearly evident in numerical computations (Athay, 1970).

In illustrating these effects of gradients in Doppler width and of differential streaming motion, we have considerably oversimplified the real situation. Nevertheless, the examples cited properly represent the particular effects that we have drawn attention to.

The gradient in Doppler width produces effects other than distorting the mapping of S into I_ν. An increase in Doppler width in the chromosphere, as we have seen, greatly increases τ_y in the low chromosphere in the edges of the Doppler core. This, in turn, decreases the escape probability for line photons from the low chromosphere and increases the coupling between the Planck function and the line source function. Thus, the positive gradient in Doppler width in the chromosphere enhances S over the value predicted by Equation (VIII-1). Of the two effects, enhancement of S and compression of the τ_y grid, the latter is the more pronounced in terms of the resulting line profile. The net effect of a positive gradient in Doppler width reduces the observed emission peaks in the line profile.

A second severe restriction on Equation (VIII-1) is that it is derived assuming that the chromosphere is plane parallel. The main significance of this restriction is that photons are allowed to escape only in the vertical direction. When lateral structure is present and has a characteristic dimension of the same order of or less than the chromospheric thickness, photon diffusion laterally will play a large role in the determination of the line source functions. Chromospheric structures observed in K_2 and K_3 are, in fact, comparable in diameter to the chromospheric thickness, and the plane-parallel approximation is probably not adequate for the interpretation of the line profile. Solution of the radiative transfer problem for an inhomogeneous atmosphere is vastly more difficult than for a plane parallel atmosphere. Substantial progress has been made on this difficult problem (cf. Rybicki, 1968; Jones and Skumanich, 1968; Jones, 1970; Avrett and Loeser, 1971), but the techniques of analysis are not sufficiently developed to permit ready application to chromospheric problems.

Still a third restriction on Equation (VIII-1) is that S is assumed to be independent of frequency. As was noted in Chapter VI, this leads to considerable error in certain cases where coherent scattering produces a marked change between the source function near line center and the source function in the line wings (cf. Figure VI-12).

It is clear from the preceding discussion in this section and in Section 2 that attempts to infer chromospheric properties from observed line profiles cannot be based upon a simplistic approach. Emission reversals in lines such as the H and K lines of Ca II are reliable indicators of chromospheric structure. However, the source function giving rise to the emission peaks is not the Planck function and the emergent intensity in the emission peaks is not simply related to the source function. The source function depends, sensitively, upon both the magnitude and rate of the chromospheric temperature rise, the optical thickness of the chromosphere, the destruction probability, the lateral geometrical structure and the gradient in Doppler width. The resultant line profile in a given feature, in turn, depends upon the source function and the optical thickness but, additionally, it depends upon the gradient in Doppler width and upon streaming motions. Hence, the line profile cannot be properly inverted to obtain the line source function without knowing the gradients of velocity in addition to the velocity amplitude. Furthermore, the line source function cannot be related back to the Planck function without simultaneously evaluating the destruction and escape probabilities. Interlocking effects in multilevel atomic models further complicate the determination of the escape and destruction probabilities. Needless to say, the problem is complicated, but it is nevertheless tractable.

Throughout the preceding illustrations we have used the Ca II lines as reference. The problem is not less complicated for other lines. Chromospheric emission is more prominent in the XUV, but the basic problems are the same.

4. Ca II Lines

Emission reversals in the H and K lines of Ca II are a common phenomena in stars of spectral type G0 and later. These emission features can be described, somewhat crudely, by their total emission flux, F_2, and their total width, W_2. In practice, W_2 is defined by the outer limits of the observed emission and is therefore somewhat larger than the total separation of the K_2 peaks and somewhat smaller than the total separation of the K_1 minima. For the Sun W_2 is approximately 0.4 Å.

Wilson (1954) and Wilson and Bappu (1957) have shown that W_2 correlates with the visual absolute magnitude, M_v, of the star. The correlation between ln W_2 and M_v is linear for $10 \geqslant M_v \geqslant -6$, and, for W_2 expressed in km s^{-1}, the relation

$$\ln W = 4.28 - \frac{1}{6} M_v \qquad \qquad \text{(VIII-10)}$$

is approximately obeyed. Stars subdivided according to spectral type or according to the intensity F_2 obey the same relation. It has been shown further by Wilson and Skumanich (1964) that F_2 is greatest in stars that are near the zero-age main sequence and is diminished in stars that are evolved.

These rather remarkable correlations have prompted a great deal of interest in both the K_2 phenomena and in the associated study of stellar chromospheres. It is generally assumed that W_2 is proportional to the line broadening from macroscopic motions in the stellar chromosphere, and, hence, that the macroscopic motion is correlated with the visual magnitude. Also, it is generally assumed that F_2 is proportional to the mechanical energy flux fed into the chromosphere from the stellar convection zone. Thus, the correlation of F_2 with stellar age is interpreted as a decrease of mechanical energy flux with stellar age. The lack of correlation between F_2 and W_2 implies a lack of correlation between the mean velocity of macroscopic motions and the mechanical energy flux.

The Sun, as a star, obeys the Wilson-Bappu relation. When examined in detail, however, both F_2 and W_2 vary markedly over the solar surface and with time at a given location on the solar surface. Detailed studies of these spatial and temporal changes in the profiles have been carried out by several authors. The most systematic and least ambiguous study appears to be that of Liu (1972). Following is a summary of his results for the quiet Sun.

From a study of approximately 4000 profiles showing K_2 emission observed near disk center, Liu finds that 63% show emission peaks on both sides of the line, 25% show only a violet peak and 12% show only a red peak. For the majority of doubly peaked profiles, the violet emission is stronger than the red emission. From a more systematic study of 374 line profiles obtained from a limited area of the Sun, Liu finds that K_2 emission covers about 80% of the surface area of the Sun whereas 20% of the Sun shows no discernible emission, i.e., the intensity increases montonoically from line center to continuum.

Network features always show both the violet and red emission peaks. Bright points within the network cells show both doubly and singly peaked profiles. Violet single peaks occur twice as often as red single peaks, but the relative frequencies of doubly peaked and singly peaked profiles has not been determined for the cell points. Thus, doubly peaked profiles occur both in the network and within the cells, whereas singly peaked profiles occur exclusively within the cells.

These characteristics of the K_2 emission are illustrated photographically 'in the composite spectroheliogram shown in Figure II-10. In this figure the wavelength of observation changes steadily from left to right across the solar image in such a way that the left hand half of the solar image is made in the violet side of the line and the right hand side of the solar image is made in the red side of the line. K_3 is at disk center (left to right) and the K_2 emission is seen as bright bands either side of line center. Note that the violet K_2 (K_{2V}) emission shows many more small bright points than does the red K_2 (K_{2R}) emission. Note, also, that the network is apparently less evident in K_{2V} than in K_{2R}. Closer inspection shows, however, that the apparent decrease in network structure in K_{2V} is partly an illusion caused by the added bright points inside the network cells in K_{2V}.

Obviously, there is a great deal to be learned about the nature of structures within the chromosphere and their associated velocity fields through a careful, systematic study of the K line profile and its interpretation. Most studies to date have been restricted to plane-parallel, static atmospheres, i.e., lateral structure and streaming motions are ignored. This results, inevitably, in profiles that are symmetric about line center. Such models can predict the gross center-limb effects but say relatively little about statistical variations in the profile.

A number of authors have synthesized the mean profiles of the H and K lines using chromospheric models similar to the HSRA model. An extensive review of the work prior to 1970 is given by Linsky and Avrett (1970) together with their synthesis of the profiles of both the H and K doublet and the infrared triplet. The temperature model used by Linsky and Avrett is cooler than the HSRA in the low chromosphere by amounts ranging from 250 to 450 K. As a result of the lower temperatures in the chromosphere, the electron densities are significantly lower than in the HSRA model. Microturbulent velocities used by Linsky and Avrett are shown in Figure III-7. Although precise computations have not been carried out for the HSRA, the higher temperature and electron density in the HSRA model would appear to require a steeper gradient in the microturbulent velocity than that adopted by Linsky and Avrett.

Synthesized profiles of the H and K doublet and the IR triplet of Ca II obtained by Linsky and Avrett (1970) are shown in Figures VIII-4a and 4b. The synthesis, though not perfect, is quite satisfactory. In this modelling of the line emission the K_2 peaks are formed at line center optical depths ranging from about 10^2 to 10^3, i.e., the peaks occur in the edge of the Doppler core. The geometrical heights where the K_2 peaks occur range from about 900 km to 650 km. K_3 is formed at a height of about 1600 km (cf. Figure II-2) and K_1 forms in the temperature minimum at the base of the chromosphere. Alternative representations of the profiles by other authors result in model chromospheres that are somewhat different from the Linsky-Avrett model. However, each of the models has the general character of the HSRA and differ in detail only.

There has been no detailed treatment of the effects of coherent scattering on the Ca II profile, although the problem is under investigation by Milkey and Mihalas (private communication). The previous work of these authors on the Mg II lines and on Lyman-α

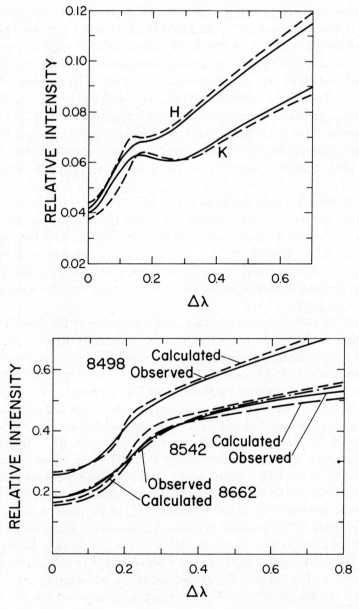

Fig. VIII-4.   Profiles of Ca II synthesized by Linksy and Avrett 1970 (courtesy *Astrophys. J.*).

(Milkey and Mihalas, 1973, 1974) clearly suggests that coherent scattering may significantly influence the profiles of the H and K lines. This would render the previous syntheses of the profiles as being somewhat inaccurate, but would not change the general conclusions.

Athay and Skumanich (1968) have investigated the influence of various atomic and chromospheric parameters on W_2 in an effort to isolate the particular influence that leads to the Wilson-Bappu effect. They find that W_2 depends sensitively upon each of the parameters λ, a and $\Delta\lambda_D$. For the first two of these, the increase in W_2 results primarily from the effect of shifting the K_2 peak further into the line wings, whereas for the latter it is the overall widening of the profile that increases W_2.

If the Wilson-Bappu effect were due to changes in either γ or a, there should be both a strong correlation between F_2 and W_2 and a marked change in the character of the profile as W_2 increased. This latter conclusion follows from the fact that as the K_2 emission moves into the wings where Φ_V is a slow function of frequency the K_2 emission feature becomes broad and indistinct. Changes in W_2 resulting from changes in $\Delta\lambda_D$ are, to a first approximation, uncorrelated with changes in F_2 provided that the gradient in $\Delta\lambda_D$ does not change markedly. Also, the general character of the profile is unaltered by changes in $\Delta\lambda_D$. It appears therefore that the most consistent explanation of the Wilson-Bappu effect is that invoking changes in $\Delta\lambda_D$.

For the stars in which K_2 emission is observed, W_2 changes by approximately a factor of ten. In the case of the Sun, which is near the lower end of the W_2 scale, the broadening is mainly non-thermal and corresponds to a mean microturbulent velocity at the depth where K_3 and K_2 are formed of about 5 km s^{-1}. The established Wilson-Bappu relation, if similarly interpreted, would require a range of mean microturbulent broadening velocities of approximately 2.5 to 25 km s^{-1}.

An appreciable amount of work has been done on the influence of the network structure on the overall properties of the H and K profiles, whereas very little has been done on the influence of the cell points. It appears to be premature to attempt any generalization of these results at the present time.

5. Mg II Lines.

The most complete set of observations of the Mg II doublet at λ 2795 and λ 2802 is those published by Lemaire and Skumanich (1973). Spectra were obtained near disk center with a balloon borne spectrograph yielding a spectral resolution of 25 mÅ and a spatial resolution of 7″. This spatial and spectral resolution is sufficient to illustrate the characteristic profile changes from the cell to network (as shown in Figure VIII-5) and to give statistical variations of the profiles for each of these features. It is not sufficient for the study of small scale features such as the cell points.

Figure VIII-6 is a reproduction of one of the spectrograms obtained by Lemaire. It is readily apparent from this illustration that the Mg II emission features cover almost all of the solar surface and that both emission peaks are virtually always present. The Mg II emission is much stronger than the Ca II emission. Thus, the areas of the Sun that have either no emission peaks or only one peak in the Ca II K line seem to show both emission peaks in the Mg II lines. As in the case for the Ca II lines, however, the violet emission peaks in the Mg II lines are brighter, on the average, than are the red peaks; the intensity

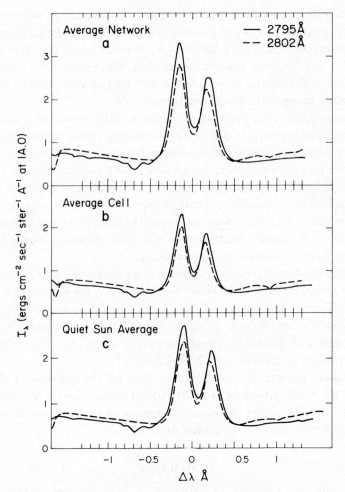

Fig. VIII-5.   Profiles of the Mg II lines for average network features and average supergranule cells
(Lemaire and Skumanich, 1973).

of the emission varies markedly from point to point on the solar surface; and the
wavelength separation of the two peaks and the full width of the emission at
half-maximum vary by about ±10% over normal quiet areas of the Sun. Larger variations
occur in active phenomena.

The average separation of the emission peaks in the Mg II lines is 0.30 Å corresponding
to a value of $\Delta\lambda/\lambda$ of about 1.1×10^{-4}. For the Ca II lines, the corresponding values are
$\Delta\lambda = 0.33$ and $\Delta\lambda/\lambda \approx 0.84 \times 10^{-4}$. These comparative results are consistent with the
interpretation that the emission peaks, in both Ca II and Mg II, are formed near the outer
edge of the Doppler core. For the Doppler core, we may write

$$\tau_\lambda = \tau_0 \exp - (\Delta\lambda/\Delta\lambda_D)^2 = \tau_0 \exp - \left(\frac{\Delta\lambda}{\lambda}\frac{c}{v}\right)^2. \qquad \text{(VIII-11)}$$

Fig. VIII-6.   A portion of the solar spectrum near λ 2800 showing the two resonance lines of Mg II at λ 2795 and λ 2803 across the solar disk. Note the fluting of the line cores near the two limbs. These spectra were photographed from a high altitude balloon by P. Lemaire, Laboratorire de Physique Stellaire et Planetaire, Verriéres-Le-Buisson.

Also, the emission peaks are expected to form near the depth where $\tau_\lambda \approx 1$ and where τ_0 equals the thermalization length τ_{th}. If we now assume that the thermalization lengths are the same for Ca II and Mg II, which is a reasonably good assumption, we find from Equation (VIII-11) that $(\Delta\lambda/\lambda)(c/v)$ should be approximately the same for the Mg II lines as for the Ca II lines. For a mean value of $v = 5$ km s^{-1} and half the observed value of $\Delta\lambda = \lambda_R - \lambda_V$, $(\Delta\lambda/\lambda)(c/v)$ for Ca II has a value of 2.5. The same value is obtained for the observed half-value of $\Delta\lambda/\lambda$ for Mg II by increasing v to 6.6 km s^{-1}. In view of the general outwards increase of v in the chromosphere (Figure III-7) and the high abundance of Mg relative to Ca in the Sun, it is to be expected that the mean value of v for the Mg II lines is somewhat greater than for the Ca II lines.

The opacity in the Mg II lines exceeds that in the Ca II lines by about a factor 15. Since the density scale height in the middle chromosphere is of the order of 150 km, the density drops by a factor of 15 in a distance given by $15 = e^{\Delta h/150}$, which gives $\Delta h = 400$ km. An increase in v from 5 km s^{-1} to 6.6 km s^{-1} in a distance of 400 km corresponds to a mean gradient of $dv/dh = 4$ s^{-1}. This is in good agreement with the average trend indicated by Figure III-7.

If, on the other hand, the emission peaks occurred either in the line wings or near one Doppler width from line center the mean velocities required to explain the peaks would be inconsistent with the results in Figure III-7, i.e., the required velocities would be, respectively, much lower or much higher than those given in the figure.

The prominence of the Mg II emission relative to the Ca II emission results from a stronger coupling of the line source function to the chromospheric Planck function in the case of Mg II and to the relatively stronger increase in the chromospheric Planck function at shorter wavelengths. The former influence arises from a relative increase in τ_0, and therefore decrease in γ, (Equations (VIII-1) and (VIII-2)) and the latter from a relative increase in A for the Mg II lines. Thus, the stronger emission peaks for Mg II are a natural consequence of the outwards increase in temperature in the chromosphere, coupled with the higher abundance of Mg II.

Figure VIII-7 shows the comparison between the quiet sun average profile for $\lambda\,2795$ and profiles computed by Lemaire and Skumanich (1973) for the HSRA and Linsky-Avrett model. The Linsky-Avrett model gives relatively good agreement with the observed profile whereas the HSRA predicts too much emission. However, the

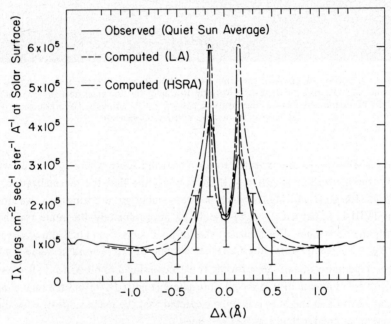

Fig. VIII-7.   Profiles of the $\lambda\,2795$ line of Mg II computed for the HSRA and the Linsky-Avrett model (Lemaire and Skumanich, 1973).

microturbulent velocity model assumed for the HSRA computation is the same as that given in the Linsky-Avrett model. A steeper velocity gradient, with the same mean velocity, would bring the HSRA results into better agreement with the observations. Thus, we emphasize that the thermodynamic model needed to explain the lines is closely coupled to and dependent upon the velocity model.

The Mg II profile synthesis by Lemaire and Skumanich is based on non-coherent scattering. A more accurate synthesis of the Mg II profiles has been achieved by Milkey and Mihalas (1974) allowing for coherent scattering (partial redistribution). Their results are shown in Figure VIII-8. Note especially the improved fit to the observed profiles in the K_1 and H_1 minima and at K_2 and H_2. Note, also, that the wing of the H line lies systematically above the wing of the K line, which is a feature of the observed profiles that is not properly predicted by the non-coherent scattering (complete redistribution) models.

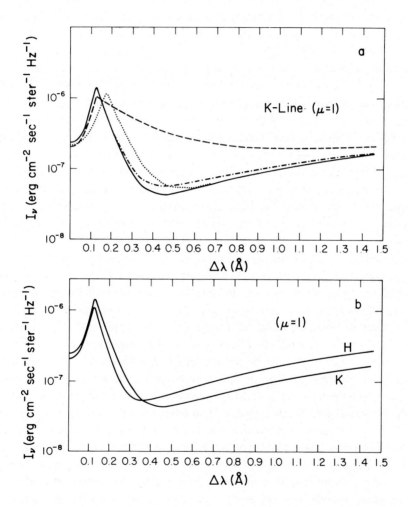

Fig. VIII-8. Profiles of the Mg II lines computed by Milkey and Milhalas (1974) allowing for partial redistribution in the photon scattering (solid lines). The dashed line in part a shows the computed profile assuming complete redistribution. The two remaining profiles in part a allow for partial redistribution bot have increased values of microturbulence velocity (dotted lines) and van der Waal's interaction constant (dot-dash line). Part b of the figure illustrates the relative profiles of the two lines in the doublet. The observed profiles exhibit the same relative behavior.

6. Neutral Metal Lines

Many of the stronger lines in the visual Fraunhofer spectrum are formed by neutral atoms of the more abundant species of metals. Notable lines in this category are the sodium D doublet, the magnesium b triplet (multiplet 2) and magnesium multiplet 3, the calcium resonance line (multiplet 2), the aluminum resonance doublet (multiplet 1), and the stronger lines in multiplets 4, 5, 20, 21, 23, 41, 43 and 45 of iron. In each of these cases the lines have strongly developed wings, and the line cores have chromospheric contributions. Attempts to synthesize these lines using realistic model chromospheres have been limited so far to the Na D lines, the Mg b lines and some of the stronger iron lines.

The problems related to the synthesis of the profiles of neutral metal lines are basically similar to those encountered in the synthesis of the Ca II and Mg II lines. The principal difference encountered is that the metals are predominantly singly ionized in the solar atmosphere so that the neutral atoms represent a minor constituent of their particular specie of element. Relative concentrations of the neutral components are very sensitive to the model atmosphere and vary markedly with depth in the atmosphere (cf. Table I-2). A practical result of this is that, in the case of the neutral atoms, the ionization equilibrium of each element must be solved simultaneously with the solutions for the line source functions. On the other hand, for the singly ionized metals it is frequently sufficient to assume that the concentration of each specie relative to hydrogen remains fixed with depth and is equal to the relative abundance of the element.

The sensitivity of the neutral metal concentration to the model atmosphere makes the strong, low excitation lines of neutral metals very valuable as diagnostic tools for the low chromosphere. None of the lines of the neutral metals are sufficiently strong to have much contribution from the middle or high chromosphere and none of the lines have emission reversals. The lack of emission reversals results simply from the relatively large values of γ (Equation (VIII-2)) for these lines. On the other hand, some of the stronger lines of neutral metals are near the strength at which emission reversals begin to occur, and relatively modest changes in the chromospheric model, such as moving the temperature minimum deeper in the atmosphere, do in fact lead to predicted profiles that have emission reversals. In this regard, it should be noted that the emissions reversals in the Ca II and Mg II lines are deceptively weak as a result of the strong increase in Doppler width in the chromosphere coupled with the location of the emission peak near the outer edge of the Doppler core. An emission peak near line center where ϕ_v is a relatively slow function of v is influenced to a much lesser degree by the gradient in Doppler width. Thus, emission reversals may appear more readily in those lines with intermediate values of γ than in those with small values of γ, contrary to the expectations from Equation (VIII-2). A possible example of such behavior can be found in results reported by Shine and Linsky (1972) who find that the weakest member of the infrared triplet of Ca II (λ 8498) tends to exhibit emission reversals in plages more readily than does the strongest member of the triplet (λ 8542).

The heights of formation of some of the strong lines of neutral metals are summarized in Table V-1. For the group of these formed in the height interval 600 to 800 km, the optical depth at line center at the base of the chromosphere ($h \approx 250$ km) lies between about 20 and 100. From equation (VIII-3) and for $\tau_{oc} = 100$, we find $x \approx 1 + 0.22\, A$ for $P_d = 10^{-4}$ and $x = 1 + 0.07\, A$ for $P_d = 10^{-5}$. Lines in the violet part of the spectrum where A is of the order of 4 or greater may reasonably be expected to show emission reversals for $x \gtrsim 1 + 0.2\, A$. Thus, we see that a number of stronger Fraunhofer lines are formed at depths where the idealized two-level results represented by Equation (VIII-3) predict the marginal apppearance of emission reversals. The actual appearance of such reversals depends, of course, on a number of complicating factors, such as multilevel effects and the smearing effects of both the macro- and micro-scale mass motions.

Efforts to synthesize the profiles of strong lines of neutral metals using currently acceptable model chromospheres have been limited to the Na D lines the Mg b lines and several of the stronger Fe lines. Several authors have treated the Na D lines with quite similar results. The work summarized here is that of Athay and Canfield (1969) and is selected because it combines the discussion of the Na D lines and the Mg b lines using a single model chromosphere. Although the model chromosphere and upper photosphere used by Athay and Canfield is significantly different from the HSRA, it bears enough similarity to the HSRA to predict the type of changes in the computed profiles that would be expected for the HSRA. The Athay-Canfield model has a higher minimum temperature and a somewhat hotter low chromosphere than the HSRA. The higher minimum temperature and the hotter chromosphere are, in fact, of little consequence to the computations. Both tend to elevate the source functions and, hence, to increase the core intensities. However, the source functions are already beginning to uncouple from the Planck function at the temperature minimum. Reducing the minimum temperature does not reduce, correspondingly, the line source function. The changes induced in the source functions by changing the temperature model could be retrieved by increasing the gradient in microturbulence or by increasing the collisional excitation cross sections by a small factor. Current uncertainties in the collision cross sections and in the gradient of microturbulence permit either course as a valid alternative.

Illustrative source functions for the b_1 and D_2 lines are shown in Figure VIII-9, the resultant synthesis of the observed lines profiles is shown in Figure VIII-10. For both the Na D and Mg b lines the outer parts of the Doppler cores are formed in the upper photosphere and are sensitive to the photospheric gradient in microturbulence. Athay and Canfield found it necessary to introduce a depth-dependent, anisotropic microturbulence in the photosphere, as have a number of other authors. The photospheric micro-turbulence has a horizontal component of about 1.5 km s^{-1} and a vertical component of about 1 km s^{-1}. Both components decrease slowly outwards reaching values less than 0.5 km s^{-1} near $\tau_5 = 10^{-3}$. The microturbulence increases again, rapidly, in the low chromosphere and is sufficiently well represented by an isotropic distribution.

In the case of both sodium and magnesium the line source functions are reasonably well represented by an equivalent two-level atom. The source functions for different

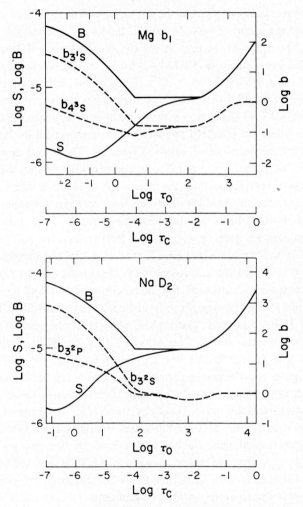

Fig. VIII-9.   Source functions computed for the D_2 line of Na I and the b_1 line of Mg I by Athay and Canfield (1969).

members of the multiplets involved are closely equal and their amplitude is determined mainly by the collisional excitation rates, i.e., interlocking effects are of little consequence. However, interlocking effects are important in the case of sodium in establishing the close equality of the D_1 and D_2 source functions. The equality, in this case, is due largely to the interchange of electrons between the two $3p^2P^0$ states via transitions to and from higher levels of excitation in sodium. Also, the ionization equilibrium in sodium and magnesium depends quite strongly upon the ionization from and recombinations to energy levels other than those directly involved in the transitions. Thus, the two-level approximation is not adequate for computing the opacities in the lines.

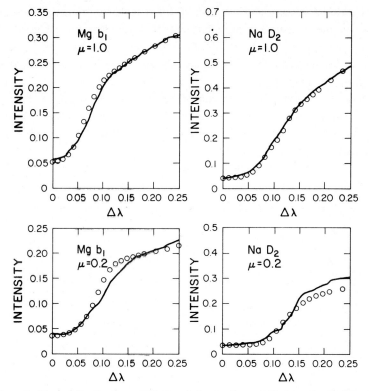

Fig. VIII-10.   Profiles of the D_2 line of Na I and the b_1 line of Mg I synthesized by Athay and Canfield (1969). The circles are computed points and the solid lines are the observed profiles.

Departures from LTE have two notable effects upon the formation of the Na D and Mg b lines: (a) they decrease the line opacity in the temperature minimum region and increase the line opacity in the chromosphere, and (b) they partially decouple the line source function from the chromospheric and temperature minimum Planck functions. Neglect of either effect would seriously alter the computed profiles.

The energy level structure in the neutral sodium and magnesium atoms is sufficiently simple that the D and b lines can be treated with a modest number of energy levels (< 6) in the model atom. Attempts to synthesize these lines with improved values for atomic collision cross sections would provide a valuable guide to the temperature, density and microturbulence model of the low chromosphere.

Computations of the Fe I equilibrium in the photosphere and chromosphere have been carried out by Athay and Lites (1972) and a synthesis of the Fe I lines was obtained with a model atmosphere identical to the HSRA up to the temperature minimum but with a model of the low chromosphere somewhat hotter than the HSRA. The hotter chromosphere was invoked in order to raise the core intensities in the lines. Again, however, similar results could be obtained with larger collisional excitation cross sections or a sharper gradient in microturbulence.

As is the case for sodium and magnesium, the departures from LTE decrease the iron line opacity in the temperature minimum region and increases it in the chromosphere. They further serve to decouple the source function from the Planck function. In the chromosphere the line profiles are again quite insensitive to the actual temperature near the temperature minimum region.

In contrast to sodium and magnesium the iron atom is complex and simple atomic models are insufficient for determining the ionization equilibrium. The computations by Athay and Lites (1972) and Lites (1972) include up to 18 atomic energy levels and about an equal number of spectral lines.

Lites' (1972) synthesis of some of the iron lines is illustrated in Figure VIII-11. For

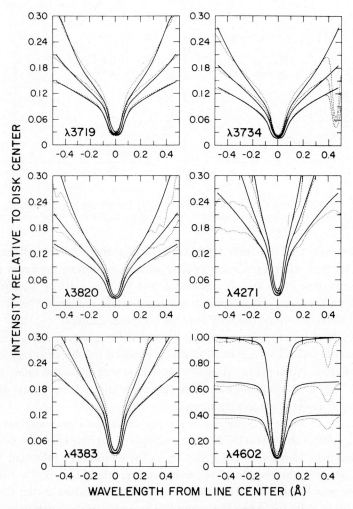

Fig. VIII-11.   Profiles of the Fe I lines synthesized by Lites (1972). Profiles are shown at $\mu = 1, 0.5$ and 0.2. Observed profiles are shown as dots.

comparison, a profile computed for the λ 3719 line using the same model atmosphere and assuming LTE is shown in Figure VIII-12.

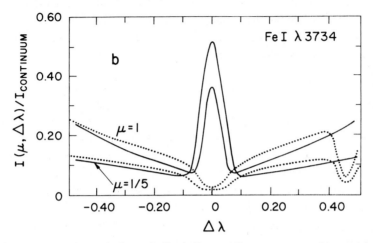

Fig. VIII-12.   A comparison of observed (dots) Fe I profiles at two μ positions with computed profiles assuming LTE and the HSRA (Lites, 1972).

More work is needed in the synthesis of the neutral metal lines both for determining reliable heights of formation and for determining the thermodynamic and hydrodynamic structure of the low chromosphere. Of particular interest in this regard is the resonance line of Ca I. Calcium is the only element for which the resonance lines of both the neutral and first ion stage occur in the visual spectrum and, in addition, extend into the chromosphere. Also, the atomic structure is not exceedingly complex and can be treated with relatively simple model atoms.

7. Hydrogen Lines

The solar hydrogen lines present a special challenge to solar physicists. A wealth of chromospheric structural detail is visible in the Hα line (see Chapter II), and this line has become more or less the 'standard line' for monitoring chromospheric activity. It may seem somewhat surprising, therefore, that a satisfactory diagnostic for this line has not yet been developed. However, unlike other strong chromospheric lines the Hα line is formed by energy levels that have large excitation potentials and that are connected to the ground state of the atom by strongly permitted transitions (Lyman α and β). In the case of all other strong lines (excluding other Balmer and Paschen lines and a few lines in the extreme uv) the lower level of the line transition is either the same as the ground state of the atom or is a low excitation metastable level. The subordinate character of the Balmer lines makes them difficult to treat from the radiation transfer standpoint.

This difficulty with the treatment of the Balmer lines is further enhanced by the fact that the Lyman-α line itself is very difficult to treat in the middle and low chromosphere where Hα is formed. At these depths the line center optical depths are very large in Lyman-α so that the photon transfer occurs in the line wings well outside the Doppler core. Within the line wings the frequency redistribution of photons is neither fully non-coherent nor fully coherent, and, in the case of Lyman-α, these two extremes yield very different solutions to the hydrogen equilibrium (Vernazza et al., 1973; Milkey and Mihalas, 1973).

Vernazza et al. (1973) approximate the redistribution function by a weighted sum of a fully coherent component and a fully non-coherent component. They then adjust the weighting of the two components until the computed Lyman-α wing agrees with the observed wing for a model chromosphere very similar to the HSRA. The best fit is obtained with a redistribution function that is 93% coherent and 7% non-coherent. Subsequent computations were made by Milkey and Mihalas (1973) using a redistribution function that accounts for both broadening and splitting of the 2s and 2p levels in hydrogen plus Doppler redistribution (see Warwick's (1955) R_{II} function, Jefferies' (1968) g_{II} function or Athay's (1972) R_{II} function). Their results show that, indeed, the approximation of 7% non-coherent scattering plus 93% coherent scattering is quite good for Lyman-α.

The wings of the Lyman-α line are superimposed on free-bound continua of C I. Figure VIII-13 shows the fit to the Lyman-α wings and the free-bound continua from λ 1900 to λ 700 obtained by Vernazza et al. (1973). The C I, Si I and H free-bound continua are treated allowing for departures from LTE whereas the Fe I and Mg I continua are approximated by LTE. The general agreement between the observed and predicted continuum distribution is about as good as the observational data justify. Note that the temperature minimum region corresponds to a wavelength of approximately λ 1550. At longer wavelengths the free-bound continua have absorption edges, and at shorter wavelengths they have emission edges.

The model atmosphere used by Vernazza et al. (1973) is shown in Figure VIII-14. It differs significantly from the HSRA in the temperature range common to both models ($T < 9000°$) only in that the pressure scale height is extended in the model in Figure VIII-14 by the addition of pressure due to microturbulence as discussed in Chapter VII, Section 1. The influence of the added pressure term extends the scale height. Its effect is most evident in the model by Vernazza et al. above a height of 1000 km. The temperature jump beginning near 8500 K is some 350 km higher in the model by Vernazza et al. than in the HSRA. This similarity of the two models (other than in the scale height) is to be expected since the models are based on essentially the same data and computed in essentially the same way.

Since the chromospheric opacity in the Hα line depends upon the population of the $n = 2$ level of hydrogen, it follows that the Hα opacity is directly dependent upon the Lyman-α source function (hydrogen is mainly neutral below the temperature jump near 8500 K). As we have just seen, the Lyman-α source function depends upon the redistribution function, and as may be seen from Figure VIII-14 the Lyman-α wings form

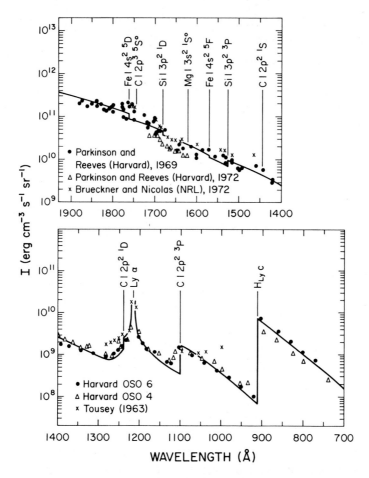

Fig. VIII-13.   Synthesis of the continuum spectrum from λ 1900 to λ 700 by Vernazza *et al.* (1973).

at depths that overlap with the region where the core of Balmer-α forms. This result casts serious doubt on the validity of all previous computations of the Balmer-α profile since in each of these earlier computations Lyman-α was considered to be either in detailed balance or in complete redistribution, both of which are incorrect.

Vernazza (1972) obtained computed profiles for the Hα, Hβ, Pα, Lyman-α and Lyman-β lines that agree well with the observed profiles when allowances are made for partial redistribution in the Lyman lines and for broadening of the lines with macroturbulence. Figures VIII-15 and VIII-16 illustrate the agreement obtained for the Hα, Hβ, Lyman-α and Lyman-β lines. The assumed values of macroturbulence velocity are indicated in the figure captions. The regions of the Lyman-α and Lyman-β profiles shown are formed entirely in the upper chromosphere near the 20 000 K temperature plateau. This plateau is responsible for most of the self-reversal in the cores of the Lyman-α and

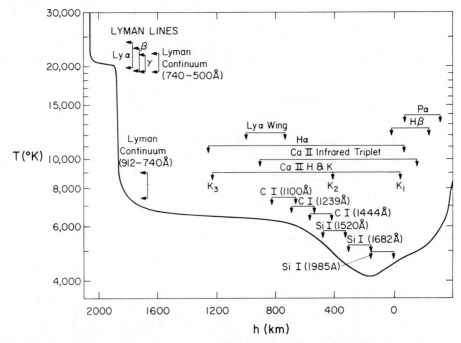

Fig. VIII-14.   The model atmosphere adopted by Vernazza *et al.* (1973), together with the locations
where different spectral features form.

Lyman-β lines. Without the plateau the Lyman-β line, in particular, is not sufficiently
self-reversed.

The existence of a temperature plateau near 20 000 K is the first feature of the
temperature model of the chromosphere that is due mainly to line data. Inference of the
temperature in the plateau and of the thickness of the plateau is based on the assumption
that there are no similar plateaus at temperatures that could influence the Lyman-α and
Lyman-β lines. However, preliminary analyses of the He II resonance line at λ 304
(Milkey, private communication) suggest the presence of an additional plateau near
40 000 K. Should such a plateau prove to exist it would contribute to the Lyman-α and
Lyman-β profiles and this, in turn, would modify the requirements on the 20 000 K
plateau. Thus, the 20 000 K plateau should not be regarded as being as firmly established
as is the lower plateau ending near 8500 K.

Network and supergranule cell structure shows evidence of being much more
pronounced in the Lyman lines than in any of the lines discussed thus far. Indeed, the
majority of the Lyman-α and Lyman-β flux may come from the network regions (Reeves
et al., 1974), and the 'mean' profiles of the Lyman lines may not be very representative
of the average region of the solar surface. In addition, the profile data for these lines are
still somewhat crude. The observed profile data are clearly influenced by the finite
spectral resolution of the observing instruments and they have not been carefully

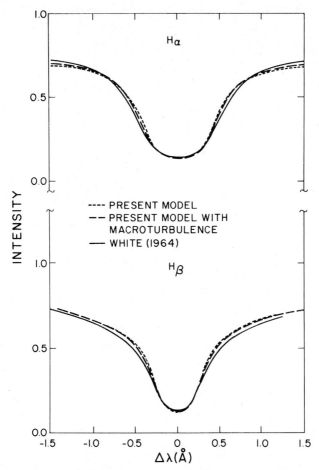

Fig. VIII-15.   Synthesis of Hα and Hβ by Vernazza (1972) using a microturbulent velocity of 10 km s^{-1}. The observed profiles are shown as solid lines.

corrected for this effect. Because of the lack of sufficient spatial and spectral resolution in the profiles, and because of the evidence for large spatial variations in the profiles, a detailed synthesis of the profiles is not warranted. In this sense, the computed profiles in Figure VIII-16 agree sufficiently well with the observed profiles.

8. C II and O I Lines

Although profile data for many XUV lines have been obtained in recent years, little of the data is available in quantitative form and relatively little use has been made of the data. This is particularly regrettable since some of the XUV lines provide our most effective diagnostic tool for the upper chromosphere in just those regions where the chromospheric model is least understood.

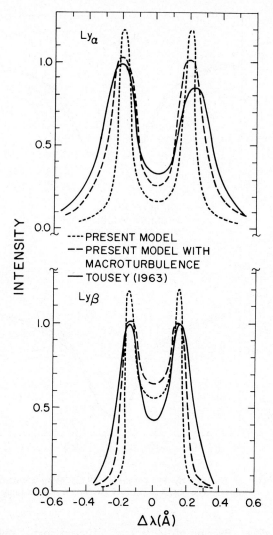

Fig. VIII-16.   Synthesis of Lyman-α and Lyman-β profiles by Vernazza (1972) using a microturbulence of 10 km s^{-1}. The observed profiles are shown as solid lines.

Profiles of the O I lines at λ 1302, 1304 and 1306 together with the profile of the C II
line at λ 1336 are shown in Figure V-37. The O I profiles are heavily saturated and show
evidence of reversal at line center similar to that observed in the Lyman lines. The C II
profile, however, is more nearly Gaussian in shape, but shows evidence of saturation near
line center.

An analysis of the C II lines by Chipman (1971) suggests that these lines are formed
primarily in a temperature plateau in the upper chromosphere rather similar to the upper
plateau in Figure VIII-14. The best fit to the C II lines is obtained with a plateau

temperature near 16 000 K and a plateau width of about 100 km. A plateau temperature near 20 000 K is not excluded, however. Because the lines are closely Gaussian in shape and too broad to be explained by thermal motions, their width is determined primarily by the assumed microturbulence velocity. A value of 18 km s^{-1} is required to fit the profiles. As noted by Chipman, however, much of the broadening could be due to larger scale motions. As in the case with Lyman-α and Lyman-β, the C II emission is concentrated quite strongly in the network regions. Since spicules are prominent features of the network in the upper chromosphere, broadening due to spicule motion is to be expected.

The most extensive analysis of the O I lines is that due to Chipman (1971). Most of the emission in these lines occurs in the damping wings formed in the low chromosphere. Thus, both the line shapes and intensities are strongly influenced by the damping constants as well as by the chromosphere model. The Doppler cores of the O I lines are formed primarily in the middle chromosphere in approximately the same layers as cores of the Mg II lines. However, the temperature plateau near 20 000 K contributes a little to the central intensities of the lines and reduces the amount of reversal in the core regions. This helps to remove an earlier difficulty found by Athay and Canfield (1970) in which the computed profiles were too strongly reversed at line center.

The very pronounced effects of coherency between absorbed and emitted photons in the wings of Lyman-α suggest that similar effects may be present in the wings of the O I lines. Neither the analysis by Chipman (1971) nor the analysis by Athay and Canfield (1970) produced a satisfactory fit to the line wings with an acceptable model chromosphere.

The large values of microturbulence and/or macroturbulence in the upper plateau region indicated by the C II lines is of further help in reducing the magnitude of the reversal in the computed profiles, a difficulty encountered by Chipman as well as by Athay and Canfield. Both the C II and O I lines merit additional careful study. Improvements are needed in both spatial and spectral resolution in the observed profiles.

9. Summary

Profiles of the stronger Fraunhofer lines and of the XUV emission lines are potentially our most powerful diagnostic tools for probing and studying the solar chromosphere. Within the intensity map of the line profile is contained information on the temperature, density, velocity and magnetic field structure of the atmosphere. To be sure, the information is present in ways that are sometimes subtle and obscure. It is present, nevertheless, and can be extracted given sufficiently good data and sufficiently detailed analyses.

Because of the strong localization of emission in the network structure in the upper chromosphere, it is questionable whether many profiles averaged over large spatial regions can be uniquely interpreted with a single component chromospheric model. What seems to be needed is a two-component model together with the average network profiles and average cell profiles. Even then there may be too much diversity within each class to provide a full explanation for all the lines with a two-component (network and cell) model atmosphere. However, the evident success of the one-component model in Figure VIII-14 in providing an approximate fit to a wide class of profiles ranging from C II to Lyman-α and the Na D lines clearly suggests that a two-component model may work very well indeed for a wide class of spectral lines. Given our present capabilities in space observations, such data should be available in the near future.

Analytical techniques need sharpening, particularly with respect to the proper handling of the velocity field. As we have demonstrated, the velocity field influences strongly both the amplitude of the source function for the line and the mapping of the source function into the line profile. Both aspects of this problem provide serious challenges in the development of proper diagnostic techniques. The rewards for a successful solution to these problems, however, are large. With a proper description of the velocity field, the mode and magnitude of mechanical energy transport can be identified and evaluated. Similarly, failure to properly diagnose the velocity field cannot help but lead to an erroneous picture of the mechanical energy processes.

As little as a decade ago, one would scarcely have dared to suggest a large scale analysis of chromospheric line profiles including a full treatment of departures from thermodynamic equilibrium and a detailed description of the velocity field. Today, however, such a program is not only possible but is necessary if we are to progress markedly in our understanding of the energy and momentum balance in the chromosphere. Line profile data adequate for such a program are still lacking, but this condition should be shortly relieved by data obtained by the Apollo Telescope Mount and by the experiments planned for Orbiting Solar Observatory-I. The future looks to be promising, indeed, for line profile studies.

References

Athay, R. G.: 1970, *Solar Phys.* **11**, 347.

Athay, R. G.: 1972, *Radiation Transport in Spectral Lines*, D. Reidel Publishing Co., Dordrecht, Holland.

Athay, R. G. and Canfield, R. C.: 1969, *Astrophys. J.* **156**, 695.

Athay, R. G. and Canfield, R. C.: 1970, in H. G. Groth and P. Wellman (eds.), *Extended Atmosphere Stars*, Dept. of Commerce, Washington, NBS Spec. Publ. No. 332.

Athay, R. G. and Lites, B. W.: 1972, *Astrophys. J.* **176**, 809.

Athay, R. G. and Skumanich, A.: 1968, *Astrophys. J.* **152**, 141.

Avrett, E. H. and Loeser, R.: 1971, *J. Quant. Spectrosc. Radiat. Transfer* **11**, 559.

Brueckner, G. E. and Nicolas, K.: 1972, *Bull. Am. Astron. Soc.* **4**, 378.

Chipman, E.: 1971, Thesis, Harvard Univ., Cambridge.

Jefferies, J. T.: 1968, *Spectral Line Formation*, Blaisdell, Waltham.

Jones, H. P.: 1970, Thesis, Univ. of Colorado, Boulder. (National Center for Atmospheric Research Cooperative Thesis No. 21.)

Jones, H. P. and Skumanich, A.: 1968, *Resonance Lines in Astrophysics*, National Center for Atmospheric Research, Boulder.

Lemaire, P. and Skumanich, A.: 1973, *Astron. Astrophys.* **22**, 61.

Linsky, J. L. and Avrett, E. H.: 1970, *Pub. Astron. Soc. Pacific* **82**, 169.

Lites, B. W.: 1972: Thesis Univ. of Colorado, Boulder. (National Center for Atmospheric Research Cooperative Thesis No. 28.)

Liu, S. Y.: 1972, Thesis, Univ. of Maryland, College Park.

Milkey, R. W. and Mihalas, D.: 1973, *Astrophys. J.* **185**, 709.

Milkey, R. W. and Mihalas, D.: 1974, *Astrophys. J.* **192**, 769.

Parkinson, W. H. and Reeves, E. M.: 1969, *Solar Phys.* **10**, 342.

Parkinson, W. H. and Reeves, E. M.: 1972, Quoted by Vernazza, J. E., Avrett, E. H., and Loeser, R.: 1973, *Astrophys. J.* **184**, 605.

Reeves, E. M., Foukal, P. V., Huber, M. C. E. Noyes, R. W., Schmahl, E. J., Timothy, J. G., Vernazza, J. E., and Withbroe, G. L.: 1974, *Astrophys. J.* **188**, L27.

Rybicki, G. B.: 1968, Results quoted by Wilson, P. R.: 1968, *Astrophys. J.* **151**, 1029.

Shine, R. A. and Linsky, J. L.: 1972, *Solar Phys.* **25**, 357.

Tousey, R.: 1963, *Space Sci. Rev.* **2**, 3.

Vernazza, J. E.: 1972, Thesis, Harvard University, Cambridge.

Vernazza, J. E., Avrett, E. H., and Loeser, R.: 1973, *Astrophys J.* **184**, 605.

Warwick, J. W.: 1955, *Astrophys. J.* **121**, 190.

Wilson, O. C.: 1954, *Proc. National Sci. Foundation Conf. on Stellar Atmospheres*, Indiana Univ., Bloomington.

Wilson, O. C. and Bappu, M. K. U.: 1957, *Astrophys. J.* **125**, 661.

Wilson, O. C. and Skumanich, A.: 1964, *Astrophys. J.* **140**, 1401.

CHAPTER IX

ENERGY AND MOMENTUM BALANCE

1. Energy Balance

The outwards increase of temperature in the chromosphere and corona is associated with dissipation of mechanical energy presumably carried upwards from the photosphere as wave motion. This mechanical energy, dissipated as heat, ultimately finds its way out of the Sun primarily in the form of radiation. A small percentage of the energy is carried outwards by conduction and convection in the solar wind. This is not to say, however, that convection and conduction are not locally important within the atmosphere. The sharp increases in temperature between temperature plateaus suggest that thermal conduction is important in these regions. In the case of the transition region between the chromosphere and corona where we have been able to form an estimate of the conductive energy flux we find that the flux indeed is large. The average value of $T^{5/2} dT/dh$ for the transition region corresponds to a conductive flux of $F_c \approx 6 \times 10^5$ erg cm^{-2} s^{-1} in regions of vertical magnetic field and in a spherically symmetric atmosphere. Radiation flux from the corona amounts to $1-3 \times 10^5$ erg cm^{-2} s^{-1} depending on the level of solar activity. Similarly, the radiation flux from the upper chromosphere, which is almost entirely in the Lyman-α line, amounts to $2-4 \times 10^5$ erg cm^{-2} s^{-1}, again depending on the level of solar activity. The radiation losses from the upper chromosphere and corona occur in emission lines that are produced by collisional excitations. As such, they represent a direct loss of energy from the thermal reservoir and must, therefore, be balanced by an equivalent energy input. As we noted earlier, the basic heat input comes from the mechanical energy carried upwards. It seems clear, however, that thermal conduction through the transition region may represent a strong energy drain on the inner corona and may provide a strong source of energy in the upper chromosphere. The actual value of the conductive flux is uncertain because of the complex magnetic field structure in the transition region.

Convective energy flux of appreciable magnitude is clearly evident in the case of spicules. The matter density in spicules is approximately 1.7×10^{-13} g cm^{-3}, corresponding to a hydrogen density of 1×10^{11} cm^{-3}. Spicule matter moves upwards at an average velocity of 25 km s^{-1}. The resultant flux of energy is given by

$$F_k = \tfrac{1}{2} \rho v^3 + \frac{5\rho kT}{m_H} v + \rho g_0 R_0 v.$$

The first term is the kinetic energy transport, the second is the enthalpy flux (Parker,

1963, Chapter VII) and the third is the potential energy flux, i.e., the rate of doing work against the gravitational field. At a spicule velocity of 2.5×10^6 cm s^{-1}, a density of 10^{-13} g cm^{-3} and temperature of 15 000 K, the terms on the right hand side of the expression for F_k have the values 8×10^5, 3×10^6 and 5×10^8, respectively, in units of erg cm^{-2} s^{-1}. The last term is by far the largest, and, in addition, is about three orders of magnitude larger than the conductive and radiative losses from the corona. At the site of a spicule, therefore, the great bulk of the energy flux is due to the convective motion of the spicule.

If we dilute the spicule energy flux by a factor 10^{-2} to obtain the surface average of the spicule energy flux, we see that the average convective flux is still an order of magnitude larger than the radiative and conductive fluxes.

Although we now recognize that both conduction and convection have strong local consequences in the upper chromosphere and low corona, it is of interest to consider the characteristics of the model atmosphere that results from the overall energy balance between mechanical heating and radiation loss. In order to do this, it is necessary to apply the energy balance locally, of course, and we are in danger of misrepresenting the true model because we are overlooking important terms. Thus, we proceed with these reservations in mind.

It will be assumed in the following that a local dissipation of mechanical energy, $M = \nabla \cdot F_{\text{mech}}$, is balanced by a radiation loss, $L = \nabla \cdot F_{\text{rad}}$. Little is known about the nature of M other than what can be deduced from empirical evidence and from crude theoretical arguments. The available evidence (see Section IX-6) suggests that M is proportional to the total density n_H, at least in the upper chromosphere and corona. M probably depends upon temperature, also, but the nature of this dependence is unknown other than that M is evidently not an extreme function of temperature. With some generality, M could be written in the form

$$M = m(T)n_H^{\alpha}. \tag{IX-1}$$

For the sake of illustration, however, we will take $\alpha = 1$ and $m(T) = $ const. This choice is somewhat arbitrary. However, as we shall soon show, L is a strong function of temperature and is proportional to n_H^2. Thus, the choices of $m(T)$ and α are not critical as long as α is near unity and $M(T)$ is not a strong function of T.

2. Radiation Loss Rates

Radiant energy loss from the chromosphere and corona occurs mainly in a few spectral lines of the more abundant elements. Free-bound continua contribute only a small fraction of the energy loss, with the possible exception of Balmer continuum losses from the middle chromosphere and H$^-$ losses from the low chromosphere. Estimates of the rate of energy loss by radiation from an optically thin gas of solar composition have been made by a number of authors.

For the case of a two-level atom excited by electron collisions, the energy loss rate is (cf. Chapter VI)

$$L = h\nu C_{12} n_1,$$ (IX-2)

where C_{12} is the collisional excitation rate and n_1 is the population density of the ground state. As in Chapter VI, we substitute

$$C_{12} = Q_0 f(T) n_e$$

$$= Q_0 f(T) g(T) n_H.$$ (IX-3)

Here, n_H is the total hydrogen (neutral plus ions) density and $g(T)$ gives the ratio n_e/n_H. We represent $g(T)$ as a function of temperature only, which is not strictly true but which is adequate for the present purposes. Also, we set

$$n_1 = x(T) A n_H,$$ (IX-4)

where A is the abundance of the element relative to hydrogen and $x(T)$ gives the fractional population of the element in energy level 1 of the ion producing the line. Equations (IX-2), (IX-3) and (IX-4) yield

$$L = h\nu Q_0 A n_H^2 x(T) g(T) f(T),$$ (IX-5)

or

$$L = \text{const } n_H^2 Q(T).$$ (IX-6)

The total loss rate of course is made up of a sum of terms each of which has the form of Equation (IX-6). In order to understand the approximate form of $Q(T)$, consider the three functions $x(T)$, $g(T)$ and $f(T)$. The function $x(T)$ for ions is proportional to factors of the form

$$x(T) \propto e^{-a_1/T}(1 - e^{-a_2/T}), \qquad a_2 > a_1.$$ (IX-7)

The first factor represents the increase in concentration of the ion under consideration, and the second represents the decrease in the ion concentration as the next stage becomes populous. For neutral atoms, the first factor is unity. The function $g(T)$ has a factor of the form

$$g(T) \propto e^{-a_3/T}$$ (IX-8)

representing mainly the increase in electron concentration due to ionization of hydrogen. Also, $f(T)$ has a factor of the form

$$f(T) \propto e^{-a_4/T},$$ (IX-9)

which arises from the exponential term in the Maxwellian velocity distribution for electrons. Each of the functions $x(T)$, $g(T)$ and $f(T)$ contains other factors involving T^α where α is a number of order unity and may be positive or negative. The precise value of α depends upon the details of the ionization equilibrium, and the energy dependence of the electron collision cross sections for ionization and excitation. From the forms of $x(T)$, $g(T)$ and $f(T)$, it may be seen that $Q(T)$ has terms of the approximate form

$$Q(T) \propto T^\beta (e^{-b_1/T} - e^{-b_2/T}), \qquad b_2 > b_1 . \tag{IX-10}$$

This function has a well defined maximum when b_2/b_1 and β are not too large. Typically both b_2/b_1 and β are of order unity rather than of order ten.

The compostie $Q(T)$ function for an optically thin solar mixture of elements has been computed by a number of authors. The most extensive and most reliable computations are those of Cox and Tucker (1969). Figure (IX-1) illustrates a plot of $L/n_e n_H$ from their results. Through the range of temperature considered n_e is approximately equal to n_H and is at most a very slow function of temperature.

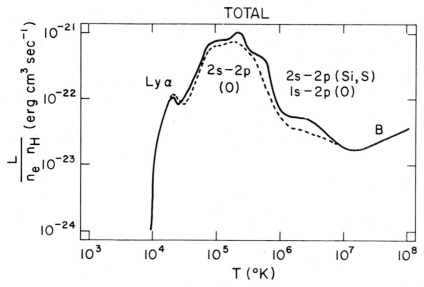

Fig. IX-1. Radiative power for an optically thin, cosmic mixture of elements as a function of temperature. Near 10^5 K most of the radiation arises from 2S–2P transitions in oxygen ions. Other important contributors are indicated at other temperatures. The solid curve includes dielectronic recombinations, but the dashed curve does not (Cox and Tucker 1969; courtesy *Astrophys. J.*).

Cox and Tucker's computations are based on a specific set of assumptions, viz., ionization and excitation by electron collisions, recombination and de-excitation by radiative processes, and optical thinness. These assumptions may be criticized from several standpoints. For example, proton collisions may be effective in exciting some spectral lines of importance. More importantly, however, the solar atmosphere is not optically thin in the principal radiations at temperatures below about 5×10^4 to 10^5 K. This casts doubt on the validity of the curve shown in Figure IX-1 for $T_e < 10^5$ K. The condition of effective thinness, which is less restrictive than the condition of optical thinness, and which is valid in much of the temperature range between about 10^4 and

10^5 K, still allows all of the photons generated by collisions to escape as energy loss. However, in the effectively thin case photons escape only after a number of scatterings and the mean radiation density inside the atmosphere builds up. This buildup in the radiation density produces added excitation of atoms and ions and leads to additional possibilities for energy loss. In the case of hydrogen, for example, a buildup of the Lyman-α radiation increases markedly the population of the second quantum level in hydrogen. When the population of the second quantum level is sufficiently high, the Balmer series of hydrogen becomes an efficient source of energy loss through collisional excitations from the second level. This possibility, as well as related ones, is omitted from the Cox and Tucker curve, and their results are particularly inadequate for temperatures below about 2×10^4 K where the Lyman-α loss maximizes. At these lower temperatures radiation losses due to the Balmer series of hydrogen, the Mg II and Ca II resonance lines, strong lines of Fe II and H$^-$ greatly exceed the results in Figure IX-1.

These added sources cannot be included in a generalized L function in a meaningful way since they depend critically upon the details of the optical depth, temperature and density relationships in the model atmosphere. However, for the solar case there is a strong likelihood that the correct $Q(T)$ function has one or more additional maxima for $T < 2 \times 10^4$ K.

The major peak in the energy loss rate in Figure IX-1 near 2×10^5 K results primarily from a composite emission in lines of oxygen ion stages II, III and IV. Each of these ions has several strong transitions in the $2s - 2p$ class that provide the dominant energy loss.

For purposes of illustration in the following section, we choose an analytic form for $Q(T)$ that has two maxima. The form chosen is

$$Q(T) = \frac{1}{T}(24e^{-1.6/T} - 30e^{-3/T}$$

$$+ 120e^{-8/T}),$$

(IX-11)

where T is measured in units of 10^4. In the following section, we investigate the consequences of this particular form for $Q(T)$ in combination with Equation (IX-6) and a simple form for M.

3. Thermal Stability

Equation (IX-6) in combination with Equation (IX-11) gives

$$L = \left(\frac{\rho}{\rho_0}\right)^2 \frac{1}{T}(24e^{-1.6/T} - 30e^{-3/T} + 120e^{-8/T}),$$

(IX-12)

where n_H is replaced by ρ and the constant in Equation (IX-6) is replaced by ρ_0^{-2}. As indicated in the discussion following Equation (IX-1), the form adopted for M is

$$M = \frac{\rho}{\rho_0}, \tag{IX-13}$$

and ρ/ρ_0 is given, in turn, by

$$\frac{\rho}{\rho_0} = e^{-h/H}, \tag{IX-14}$$

where H is a constant. Equations (IX-12), (IX-13) and (IX-14) can be solved formally by setting $L = M$ to yield

$$e^{h/H} = Q(T). \tag{IX-15}$$

However, $e^{h/H}$ is a monotonic function whereas $Q(T)$ is multivalued. At heights less than, say, h_1, $Q(T)$ has two roots that satisfy Equation (IX-15). Between h_1 and h_2 there are four roots to Equation (IX-15) and between h_2 and h_3 there are again two roots. Above h_3 there are no roots that satisfy the equation. Thus, we have deliberately, but not unrealistically, chosen L and M in such a way as to make the condition $L = M$ insufficient for specifying $T(h)$.

Before proceeding further with the model resulting from the specific choices of L and M, we caution again that we are ignoring energy transport by both conduction and convection. Also, the difficulties with multiple roots or non-existent roots could be avoided by appropriate changes in L and M. On the other hand, the forms chosen for L and M are believed to be representative of the true situation, in broad outline, if not in detail. Also, thermal conduction and convection will not necessarily remove all of the ambiguities in Equation (IX-15) even though they may remove some and modify others.

Any atmosphere does in fact, have restraints in addition to energy balance. The second law of thermodynamics requires that the energy balance be achieved in the configuration that minimizes the inernal energy, i.e., that minimizes the temperature. Thus, when multiple roots of Equation (IX-15) exist only the root giving the lowest stable value of T is physically acceptable. This condition removes all of the ambiguity in the multiple roots but it does not remove the difficulty of having no root above h_3.

Still a third restraint on the energy balance in an atmosphere is that the equilibrium must be stable against small perturbations. The concept of thermal stability in a gaseous medium has been discussed by a number of authors following Parker's (1953) original suggestion that bound-bound and free-bound emissions are unstable in some temperature ranges. Extended discussions of the conditions leading to thermal instability are given by Field (1965) and Defouw (1970a, b, c). The rigorous discussions are rather involved, and we will first consider a somewhat superficial but much simpler approach introduced by Athay and Thomas (1956) and later amplified by Thomas and Athay (1961). Even though this initial approach is overly simplified it serves to illustrate the basic character of the problem.

In the particular problem we are considering, a perturbation in T by ΔT, at constant pressure, will perturb both L and M. If $dL/dT > dM/dT$, the resultant changes in L and M

from a positive increment ΔT will lead to $\Delta L > \Delta M$. Thus, a net cooling will occur and the increase in T will be resisted by the excess loss of energy. Conversely, if $dL/dT < dM/dT$ a positive ΔT will lead to $\Delta L < \Delta M$ and the disequilibrium between L and M will further increase T. In this latter case the perturbation in ΔT will grow rather than be damped. For the atmosphere to be thermally stable, therefore, it is necessary that

$$\frac{dL}{dT} > \frac{dM}{dT} .$$

(IX-16)

In fairness to the reader, it should be pointed out again that this stability criterion is incomplete and oversimplified. Even in the restricted problem we are considering there are internal energy sinks and sources that are not represented explicitly in L and M. Thus, an increase in T raises both the internal kinetic energy and the potential energy (increased ionization) of the gas. We are treating these effects as being implicit in the loss rate L. In a rigorous treatment, they should be handled explicitly. Also, as Defouw (1970a, b, c) has pointed out, thermal instability implies convective instability and the stability criterion should include the effects of convection. The onset of instability further implies the presence of a large temperature gradient and energy flow by conduction. This can be seen by noting (Defouw, 1970c) that the condition of marginal stability is $d(L - M)/dh = 0$ since $L - M$ is zero at all heights. It follows from this condition that

$$\frac{\partial(L - M)}{\partial T} \frac{dT}{dh} + \frac{\partial(L - M)}{\partial \rho} \frac{d\rho}{dh} = 0,$$

or that

$$\frac{dT}{dh} = - \frac{\dfrac{\partial(L - M)}{\partial \rho} \dfrac{d\rho}{dh}}{\partial(L - M)/\partial T} .$$

(IX-17)

The derivative $\partial(L - M)/\partial \rho$ is always positive whenever $\alpha < 2$ in Equation (IX-1). However, at marginal stability $\partial(L - M)/\partial T$ goes to zero. Thus dT/dh becomes infinite at the onset of instability. This implies that the temperature jumps discontinuously to a higher value where the atmosphere is again stable. Actually, of course, dT/dh cannot be infinite. Thermal conduction becomes increasingly important as dT/dh increases and provides an upper limit on dT/dh that is determined by the amount of energy available for conduction. Obviously, the conducted energy flux cannot exceed the total mechanical energy flux. The conducted flux will serve also to modify the stability criterion itself.

In spite of the conclusion that the onset of instability leads to both convection and conduction and therefore invalidates the simple assumption that the mechanical energy dissipation is balanced by radiative energy losses, it is still instructive to proceed with the simple version of the problem.

The condition (IX-16) when combined with Equations (IX-12), (IX-13) and (IX-14) leads to the result

$$\frac{2}{\rho}\left(\frac{\rho}{\rho_0}\right)^2 \frac{d\rho}{dT} Q(T) + \left(\frac{\rho}{\rho_0}\right)^2 \frac{\partial Q(T)}{\partial T} > \frac{1}{\rho_0} \frac{d\rho}{dT} \tag{IX-18}$$

Temperature perturbations could occur in the solar atmosphere under a variety of conditions, such as constant density or constant pressure. The constant density case leads immediately to the result $\partial Q(T)/\partial T > 0$. On the other hand for the constant pressure case $d\rho/dT = -\rho/T$ and the inequality (IX-18) reduces to

$$\frac{T}{Q(T)} \frac{\partial Q(T)}{\partial T} > 2 - \frac{\rho_0}{\rho} \frac{1}{Q(T)}. \tag{IX-19}$$

According to Equations (IX-14) and (IX-15), which are still valid, the last term on the right hand side of (IX-19) is just unity. Thus, the inequality reduces to

$$\frac{\partial \ln Q(T)}{\partial \ln T} > 1. \tag{IX-20}$$

Since this condition is violated at a lower value of T than is the condition $\partial Q(T)/\partial T > 0$, the inequality (IX-20) is the appropriate one for finding the allowed limits on the stable temperature regimes. Note that the result expressed by Equation (IX-20) depends only on the assumption that $L/M \propto \rho$ plus the condition of constant pressure for perturbations in temperature. It does not depend upon either the form of $Q(T)$ or the form of $\rho(h)$. Using a somewhat more complex equation for M that is representative of energy dissipation by periodic shock waves, Weymann (1960) finds $\partial \ln Q(T)/\partial \ln T > 5/4$ as the stability criterion. For the set of conditions chosen here, Equation (IX-15) gives $\ln Q(T) = h/H$. Thus, the inequality (IX-20) is equivalent to

$$\frac{1}{H} \frac{\partial h}{\partial \ln T} > 1. \tag{IX-21}$$

Figure IX-2 shows a plot of $\log Q(T)(= h/2.3H)$ vs $\log T$. The regions of stability are indicated by the solid segments of the curve. The lines made with long dashes give $h/2.3H$ vs $\log T$ as determined by the stability criterion in combination with Equation (IX-15). Note that the temperature versus height curve consists of two relative plateaus in temperature bordered by sharp temperature increases. The sharp temperature increases occur for $T \approx 6300°$ and $T \approx 35\,000°$.

The criterion (IX-20) applied to the curve in Figure IX-1 again indicates two regions of stability: one for $T < 2 \times 10^4$ K, and one for 3×10^4 K $< T < 8 \times 10^4$ K. Thus, the run of T with height predicted by the curve in Figure IX-1 consists of a relative plateau up to 2×10^4 K, then a sharp rise in T to 3×10^4 K followed by a second plateau reaching to 8×10^4 K and, finally, a second rapid rise in T beginning at 8×10^4 K. (All of these

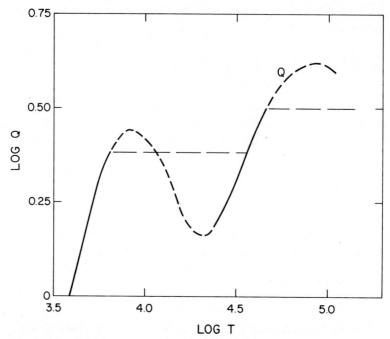

Fig. IX-2.   A plot of log Q vs log T (see text for explanation). The solid segments are regions of
stability and the dashed horizontal lines indicate regions of sharp temperature rise.

values of T are approximate.) These results differ from those shown in Figure IX-2 only
in the temperature limits on the plateaus.

Although the $Q(T)$ curves in both Figures IX-1 and IX-2 predict a plateau type of
structure for the temperature model, neither curve leads to temperature plateaus that
correspond closely with the chromosphere-corona model with but two exceptions. The
Cox and Tucker curve in Figure IX-1 does give the correct approximate temperature
($\approx 1 \times 10^5$ K) at the base of the last major temperature jump, i.e., at the base of the
chromosphere-corona transition region, and it gives a reasonable temperature for the top
of the Lyman-α plateau ($\approx 2 \times 10^4$ K). As we have already noted, however, the Cox and
Tucker curve is not really to be trusted below 1×10^5 K. In particular, we note that the
addition of Balmer series losses at temperature below about 10^4 K will alterate the $Q(T)$
curve markedly at low temperatures and will therefore reduce the derivative $\partial \ln Q(T)/$
$\partial \ln T$ with the likely result that an additional temperature plateau will occur below
10^4 K as is found empirically. Also, we note that the detailed shape of the $Q(T)$ curve
between 10^4 K and 10^5 K depends upon a number of factors, such as relative abundances
and collision cross sections, which are not known to high accuracy. It is not unlikely,
therefore, that the true solar $Q(T)$ curve contains regions between 10^4 K and 10^5 K
where $\partial \ln Q(T)/\partial \ln T < 1$ in addition to the indicated interval between 2×10^4 K and
3×10^4 K. (See Canfield and Athay, 1974.)

Defouw (1970c) has carried out a more complete investigation of the thermal stability conditions for the chromosphere than the simplified picture presented here. He includes the effects of ionization as an energy sink and the effects of convection but, as in the case treated above, he does not include the effects of conduction. Defouw defines a net cooling function $\mathscr{L}$ that is the same as $L - M$ in our notation. He further defines a net ionization rate, $\mathscr{D}$, as the rate of ionization minus the rate of recombination, and he considers both $\mathscr{L}$ and $\mathscr{D}$ to be functions of T, n_e and x, where x is he fractional ionization. The resultant criterion for stability against perturbations at constant pressure is

$$\left(\frac{\partial \mathscr{L}}{\partial x} - \frac{\rho}{1+x} \frac{\partial \mathscr{L}}{\partial \rho} \right)\left(\frac{\partial \mathscr{D}}{\partial T} - \frac{\rho}{T} \frac{\partial \mathscr{D}}{\partial \rho} \right) -$$

$$- \left(\frac{\partial \mathscr{L}}{\partial T} - \frac{\rho}{T} \frac{\partial \mathscr{L}}{\partial \rho} \right)\left(\frac{\partial \mathscr{D}}{\partial x} - \frac{\partial}{1+x} \frac{\partial \mathscr{D}}{\partial \rho} \right) > 0 \qquad \text{(IX-22)}$$

In the same notation, the criterion (IX-20) is

$$\frac{\partial \mathscr{L}}{\partial T} - \frac{\rho}{T} \frac{\partial \mathscr{L}}{\partial \rho} > 0.$$

However, in the derivation of the criterion (IX-20) it is assumed that the dependence of $\mathscr{L}$ on x is incorporated in $Q(T)$ and in the dependence of L on ρ. Thus, the two criterion are not as different as they appear to be. The reader should note, also, that for most applications of solar interest $\partial \mathscr{D}/\partial \rho = 0$. Furthermore, the term $\partial \mathscr{D}/\partial x$ is negative, and the criterion (IX-22) is equivalent to

$$\frac{\partial \mathscr{L}}{\partial T} - \frac{\rho}{T} \frac{\partial \mathscr{L}}{\partial \rho} - \left(\frac{\partial \mathscr{L}}{\partial x} - \frac{\rho}{1+x} \frac{\partial \mathscr{L}}{\partial \rho} \right)\frac{\partial \mathscr{D}/\partial T}{\partial \mathscr{D}/\partial x} > 0. \qquad \text{(IX-23)}$$

Defouw (1970c) has applied his stability criterion only for the case of Lyman continuum radiation, under the assumption that the chromosphere is effectively thin and that ionization occurs by direct collisions from the ground state. Thermal instability occurs, in this case, for $T > 17\,500$ K, which is not very different from the instability point for Lyman-α obtained using the criterion (IX-20). The conditions for dynamic stability obtained by Defouw depend upon the density and temperature gradient in the thermally unstable regions and on the presence or absence of magnetic fields. In a strong magnetic field, the condition (IX-22) is necessary and sufficient for monotonic instability. In a weak magnetic field, monotonic instability occurs only if dT/dh is sufficiently small and ρ is sufficiently large. Large values of dT/dh lead to overstability and low values of ρ lead to stability. Defouw does not give the numerical limits of H, ρ and dT/dh under which these conditions apply, however.

Again, we caution that even in the more complete analysis of Defouw thermal conduction is ignored. The effects of thermal conduction could easily equal or exceed those of convection and ionization.

A point noted by both Weymann (1960) and Defouw (1970c), and which is demonstrated by Equation (IX-17), is that the rapid rise in temperature at the top of a plateau begins while the solution is still stable because of the decrease in $\partial(L - M)/\alpha T$ as the point of marginal stability is approached. Thus, in identifying the steep rises in temperature in the empirical models with the intervals of instability it is important to keep in mind that the beginning of instability occurs after the steep rise in temperature begins rather than immediately at the base of the transition.

In the Athay and Thomas (1956) and Thomas and Athay (1961) discussions of thermal instabilities in the chromsophere and corona, they associated the corona with the last major region of stability. It now appears that such a conclusion may be a misconception that arose from an overestimate of $Q(T)$ near 10^6 K and an underestimate of $Q(T)$ near 10^5 K. The Cox and Tucker curve in Figure IX-1 suggests that the last major region of stability for a balance between L and M ends in the upper chromosphere. It could be, of course, that the Cox and Tucker curve is in error near 1.8×10^6 K and that dQ/dT is positive at this temperature. Also, it may be that the assumption that M retains the same functional form throughout the chromosphere, transition region and corona is invalid. Either of these two possibilities could lead to a possibility for a stable balance between L and M. In this connection, we note that the Cox and Tucker (1969) curve does not include energy losses due to radiation by ions of iron. The recent upward revision of the iron abundance makes iron comparable in abundance to silicon and magnesium and a possible major source of radiation at higher temperatures.

The radiation loss rate predicted by the results in Figure IX-1 for $T = 1\ 8 \times 10^6$ K is approximately

$$L = 6 \times 10^{23} n_e n_H \text{ erg cm}^{-3} \text{ s}^{-1} \tag{IX-24}$$

Observed energy losses for the corona give the average loss rate as

$$LH = 1-3 \times 10^5 \text{ erg cm}^{-2} \text{ s}^{-1}.$$

For the inner corona, $n_e \approx 3 \times 10^8$, $n_H \approx 3 \times 10^8$ and $H \approx 10^{10}$ cm. Thus, the predicted value of LH from Equation (IX-24) is 5×10^4 erg cm^{-2} s^{-1}. This is lower than the observed flux by a factor of 2 to 6. In view of the uncerainties in the various quantities involved, this is not a conclusive argument, but it does conform with the suggestion that $Q(T)$ may be underestimated near 1.8×10^6 K. In the following, however, we consider the possibility that the coronal equilibrium is not established by the condition $L = M$ and ask what other equilibrium conditions might exist.

4. Coronal Energy Loss by Thermal Conduction and Evaporation

As we have shown earlier in Chapter VII, the conductive flux from the corona inward to the chromosphere, if the average magnetic field is nearly vertical within the transition region, is of the same order, or perhaps greater than, the radiation flux from the corona. Thus, thermal conduction is an obvious contender for the dominant energy loss from the corona. The amount of energy carried by thermal conduction does limit, in a particular way, the temperature of the corona near the Sun. It was shown in Chapter VII that the value of $T^{5/2} \, dT/dh$ inferred for the transition region from the observed line fluxes is such that a temperature of 2×10^6 K is reached about 60 000 km above the base of transition region. This estimate was obtained in the plane-parallel approximation, which is inappropriate in the case of the corona. The requirement for constant conductive flux is

$$T^{5/2} \frac{dT}{dr} = \frac{C}{R_\odot} \left(\frac{r}{R_\odot} \right)^2 ,$$
(IX-25)

which results in

$$\tfrac{2}{7} (T^{7/2} - T_0^{7/2}) = \frac{C}{3} \frac{r^3 - R_\odot^3}{R_\odot^3} .$$
(IX-26)

If we use as reference $T = T_0 = 10^5$ K at $r/R_\odot = 1$, we find

$$\frac{r}{R_\odot} = \left\{ \frac{6}{7C} T_0^{7/2} \left[\left(\frac{T}{T_0} \right)^{7/2} - 1 \right] + 1 \right\}^{1/3}$$
(IX-27)

In these units $6 \, T_0^{7/2}/7C$ is approximately 1.2×10^{-5}.

Values of $r/R_\odot$ obtained from Equation (IX-27) are 1.1 for $T = 2 \times 10^6$ K, 1.8 for $T = 4 \times 10^6$ K and 3.8 for $T = 8 \times 10^6$ K. Thevalues of $r/R_\odot$ would be even larger if the quantity $T^{5/2} \, dT/dh$ were permitted to decrease outwards. It is clear therefore that the inferred value of $T^{5/2} \, dT/dh$ in the transition region limits the average temperature within $r/R_\odot < 2$ to about 4×10^6 K. Note that the value of $r/R_\odot$ associated with a given value of T varies only as the one-third power of $T^{5/2} \, dT/dh$, so the above result is not very sensitive to errors in this quantity.

The magnitude of the inward conductive flux tells us, of course, only the values of temperature in the corona that are consistent with such a flux. Higher coronaι temperatures would imply a higher conductive flux and, necessarily, a higher input of mechanical energy. For a given amount of mechanical energy, therefore, the coronal temperature for $r < 2 R_\odot$ is clearly limited by the conductive flux inwards to the chromosphere.

Energy flux is carried outwards also by thermal conduction. The magnitude of this outwards flux is again determined by the maximum temperature of the corona. For this case, we write

$$T^{5/2} \frac{dT}{dr} = -\frac{C}{R_\odot} \left(\frac{R_\odot}{r} \right)^2$$
(IX-28)

and integrate with the boundary condition $T = 0$ at $r = \infty$. This yields

$$C = \frac{2}{7} T_{max}^{7/2} \left(\frac{r}{R_\odot} \right)_{T_{max}} \tag{IX-29}$$

The numerical value of the conductive flux is given by $1.1 \times 10^{-6} T^{5/2} (dT/dr)$ or from Equations (IX-28) and (IX-29) by

$$F_c = 3 \times 10^{-7} T_{max}^{7/2} \left(\frac{R_\odot}{r^2} \right)_{T_{max}} \tag{IX-30}$$

The mechanical energy flux at $r = R_\odot$ that is required to sustain this conducted flux is given by $4.3 \times 10^{-18} T_{max}^{7/2}$. Thus, the values of mechanical flux required (in units of erg cm^{-2} s^{-1}) are 4.8×10^4 for $T_{max} = 2 \times 10^6$ K, 5.4×10^5 erg cm^{-2} s^{-1} for $T_{max} = 4 \times 10^6$ K and 6.1×10^6 erg cm^{-2} s^{-1} $T_{max} = 2 \times 10^6$ K. In the case of the solar corona, where the empirical value of $T_{max} \lesssim 2 \times 10^6$ K, the loss of energy due to conduction into space is evidently not the limiting factor. Both the radiation loss and the inward conductive flux exceed the outward conductive flux by a significant factor.

A hot corona loses additional energy through expansion of the coronal gases into space. In a later section of this chapter we discuss the coronal expansion in some detail. Here we note only that in the extreme case of a corona whose temperature is so high that the thermal velocity of protons exceeds the gravitational escape velocity the corona will evaporate rapidly into space carrying its internal energy with it. This condition has been used by Durney (private communication) to limit the coronal temperature to a value given by

$$v_{ther}^2 < v_{escape}^2$$

or

$$\frac{2kT_{max}}{m\rho} < \frac{GM_\odot}{r_{max}}$$

and, finally,

$$T_{max} < \frac{m_p GM_\odot}{2kr_{max}}. \tag{IX-31}$$

For the Sun, this gives $T_{max} < 1 \times 10^7$ K for $r_{max} > 1.1 R_\odot$. A more detailed discussion and derivation of the limiting value of T due to evaporation is given by Roberts and Soward (1972).

The evaporative restriction on the coronal temperature appears to be less imposing than the conductive restriction. For temperatures less than the evaporative limit set by inequality (IX-31), the corona undergoes hydrodynamic expansion in the form of the solar wind (see Section IX-9). Near the Sun, however, outwards thermal conduction still carries more energy away from the corona than does the convective current of the solar wind.

Inasmuch as the mean temperature of the corona between about $1.1\,R_\odot$ and $R_\odot$ is near 1.8×10^6 K, it seems clear that the two dominant sources of energy loss are inwards thermal conduction through the transition region and radiation. The thermal conduction flux depends, as we have previously noted, on the configuration of the magnetic field. The field configuration in turn, depends upon the hydrodynamic motions in the chromosphere and corona. In the low corona, however, field strengths of the order of 2 G and greater are sufficient to have the magnetic pressure dominate over the gas pressure. The average photospheric field strength is near 2 G so the expected average coronal field stength is less than 2 G. In active regions and over the borders of supergranules, coronal fields in excess of 2 G are to be expected. Here the field geometry will govern the hydrodynamic flow rather than the other way around. Over the supergranule cells away from active regions, the hydrodynamic flow of the corona may carry the magnetic field along. Without knowing the hydrodynamic flow field in the corona above the supergranule cells it is impossible to predict the magnetic field geometry in detail. Nevertheless, it seems clear that the field will not everywhere be vertical and that over much of the solar surface the field lines in the inner corona and in the transition region may have a strong horizontal component. This means that the inwards conductive flux will be reduced below the vertical field value of 6×10^5 erg cm^{-2} s^{-1} and may even be less than the radiative flux of $1-2 \times 10^5$ erg cm^{-2} s^{-1}. This point has been emphasized by (Kopp, 1972) and will be discussed in more detail in Section IX-5.

Three alternatives seem to remain for the explanation of a mean coronal temperature near 1.8×10^6 K: (1) Either thermal conduction inwards is removing the bulk of the mechanical energy dissipation thereby limiting the coronal temperature, or (2) thermal conduction (inwards or outwards) modifies the stability conditions in such a way as to result in a stable regime near 1.8×10^6 K even though $d\phi/dT$ is negative, or (3) the correct radiative cooling function for the corona has a positive derivative with respect to T near 1.8×10^6 K. Given the present uncertainties in the coronal magnetic field geometry and in the radiative cooling function for the corona, none of the three alternatives can be eliminated and none can be pointed to as more probable than others. The problem is not likely to be settled until we have reliable models for the magnetic field geometry in and around the supergranule cells and until we· have a more accurate radiative loss function.

5. Energy Balance Within the Transition Region

Certain questions relative to the detailed energy equilibrium in the solar atmosphere are brought into particularly sharp focus in the case of the chromosphere-corona transition region. It was noted in Chapter VII that the spatial average of the quantity $F_0 = T^{5/2}\,dT/dh$ is very large in the transition region, and, further, that the average value of the heat energy carried by thermal conduction implied by F_0 is very large if the magnetic field is predominantly vertical through the transition region. For such a case, the conductive flux is $F_c = 1.1 \times 10^{-6} F_0 \approx 6 \times 10^5$ erg cm^{-2} s^{-1}. It was noted also, however, that the regions of dominantly vertical fields may be limited to the network regions and their immediate surroundings and that the average value of F_c inferred from the average of F_0 may be very misleading. Nevertheless, the conductive energy flux is obviously of importance in the transition region.

Earlier in this chapter (Section IX-1), the energy flux carried by an average spicule was estimated at 5×10^8 erg cm^{-2} s^{-1}. The spicules lie over and emanate from the network. They arise somewhat below the transition region and penetrate, on the average, to somewhat above the level where $T = 10^6$ K, i.e., they penetrate through the transition region. Since spicules at a height of 3000 km cover about 0.1% of the solar surface, the spatial average of their kinetic energy flux is still large. Within the network regions, especially, the energy transport due to spicules is obviously of paramount importance.

The importance of radiation losses from the transition region varies strongly with temperature. For example, the average value of the emission measure, η, given by

$$\eta = \int_{T_1}^{2T_1} n_e^2 \, dh \tag{IX-32}$$

is approximately equal to

$$\eta \approx \frac{p^2}{T_1^2} (h_{2T_1} - h_{T_1}).$$

For $T^{5/2} \, dT/dh = $ const and $P = $ const, the height difference $h_{2T_1} - h_{T_1}$ is proportional to $T_1^{7/2}$ so that $\eta \propto T_1^{1.5}$. Thus, as T_1 varies from 10^5 to 5×10^5 η increases by about a factor of 11. Over this same range of temperature, $Q(T)$ is essentially constant (see Figure IX-1), and η is a good measure of the relative radiating efficiency of the transition region for fixed increments in $\Delta \log T$. Kopp (1972) has deduced from the observed intensities of lines in the XUV spectrum that the spatially averaged radiation loss from that portion of the transition region lying between about 3×10^4 K and 1×10^6 K is 1.4×10^5 erg cm^{-2} s. The solar spectrum is not sufficiently well observed in the XUV nor are line identifications sufficiently complete to allow a ready division of the observed energy fluxes into different characteristic temperature regimes. However, in accordance with the preceding arguments on the relative radiating efficiency at different temperatures, one concludes that less than 10% of the total transition region radiation comes from the low temperature regime, say, $1 \times 10^5 < T < 2 \times 10^5$ K, and that the large bulk of the radiation comes from the higher temperature regime.

Since the network structure persists into the transition region, and since the radiation intensity within the network at temperatures near 1×10^5 K (C IV, N V, O IV) is some 10 times higher than the ambient background (Reeves et al., 1974), the radiation loss from the network in the temperature range 1×10^5 to 2×10^5 K may be quite comparable to the total spatially averaged loss rate of 1.4×10^5 erg cm^{-2} s^{-1} inferred by Kopp (1972).

Our preceding estimates of conduction and convection of energy represent the total energy flux F. In the case of the radiation losses, however, the quantity we have estimated is $\int \nabla \cdot F \, dh$ over a limited height range. Thus, the radiation losses should be compared with the net gain or loss of conductive and convective energy within the same height range. The nature of the conductive and convective fluxes is not known in sufficient detail to permit such a comparison. Nevertheless, it can be seen that the radiation losses are sufficiently high that they are of major importance in the overall energy balance.

Little is known about mechanical heating in the transition region or its variation from network to cell features. Strong refraction and reflection of hydrodynamic shock waves is expected to occur both in the chromosphere and in the transition region because of the sharp density decrease. Similarly fast-mode Alfvén waves are strongly refracted and reflected both in the chromosphere and in the transition region (Osterbrock, 1961; Kopp, 1968). Slow-mode Alfvén waves, however, are not so strongly refracted and reflected in the chromosphere and in the transition region and may provide the needed mechanical heat source as suggested by Osterbrock (1961) and more recently by Kopp (1972). The slow-mode waves have the added attraction that they are expected to be of most importance in regions of strong vertical magnetic fields i.e., in the network regions. On the other hand, the theory of energy transport and dissipation by slow-mode waves is not in a satisfactory state, and, in the long run, it may turn out that the most important heating mechanisms are yet to be identified. At this point little more can be said than that whatever heats the corona most likely traverses the transition region and most likely loses some of its energy in the transition region itself.

Although a number of authors have constructed theoretical models of the transition region, none has taken into account all of the relevant physical processes. Early models either balanced radiation against conduction or radiation against mechanical heating using somewhat *ad hoc* expressions to represent the mechanical heating terms. More recent models have drawn attention to the importance of the network magnetic field structure and to the necessity of including the convective processes.

Sufficient data with spatial resolution adequate to resolve the network features and to deduce changes in transition region structure from network to cell are not available at present. It is of interest, however, to consider empirical evidence for changes in transition region structure for large scale surface features associated with solar activity and to consider some of the recent attempts to discuss theoretically the nature of the spatial changes in transition region structure.

Withbroe (1972) has shown that the spatial intensity of the Mg X line at $\lambda 625$ correlates strongly with spatial maps of coronal electron density and that, to a good approximation, $I_{MgX} \propto P^2$, where P is the coronal and/or transition region pressure. This is not a surprising result since Mg X is a coronal ion formed near temperatures of $1-2 \times 10^6$ K and, as such, is unaffected by the thickness of the transition region. Spectral lines of lower excitation ions, such as N V or O VI, for example, will be influenced both by changes in pressure in the transition region and/or corona and by the thickness of the transition region itself. Indeed, Withbroe and Gurman (1973) find from a correlation of the intensity in the N V line at λ 1239 with the intensity of the Mg X line at λ 625 (as shown in Figure IX-3) that two distinctive correlation regimes exist: one for $I_{MgX} \leqslant 250$ erg cm^{-2} s^{-1} sr^{-1} and a second for $I_{MgX} > 250$ erg cm^{-2} s^{-1} sr^{-1}. To interpret this result they set

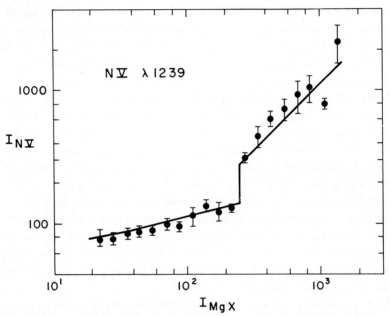

Fig. IX-3.   The correlation between the intensities of the λ 1239 line of N V and the λ 625 line of Mg X (after Withbroe and Gurman, 1973).

$$I_{Mg\ X} \propto P^2$$

and

$$I_{NV} \propto \frac{P^2}{F_c}.$$

They then conclude from the correlations that for $I_{MG\ X} < 250$ erg cm^{-2} s^{-1} sr^{-1}

$$\Delta \log F_c \approx 1.7 \, \Delta \log P \qquad\qquad\qquad (IX\text{-}33)$$

and for $I_{Mg\ X} > 250$ erg cm^{-2} s^{-1} sr^{-1}

$$\Delta \log F_c \approx \text{const.} \qquad\qquad\qquad (IX\text{-}34)$$

These results are consistent with correlations found for lines of O VI, Ne VIII, Na IX, Al XI, and Si XII as well. (It should be noted here that F_c is simply a measure of dT/dh, i.e., the inverse thickness of the transition region, and does not take into account magnetic field effects.) The average quiet Sun value of I_{MgX} is near 100 erg cm^{-2} s^{-1} sr^{-1}.

The results expressed by Equation (IX-33) indicate that near quiet Sun conditions and for large scale features an increase in P is associated with an increase in F_c such that $I_{NV} \propto P^2/F_c \propto P^{0.3}$. Thus, an increase in pressure is nearly cancelled, so far as the radiation loss from the transition region is concerned, by a decreased thickness of the transition region (or steepening of the temperature gradient). On the other hand, bright active regions show no evidence for either steepening or flattening of the temperature gradient in the transition region.

If the relation $I \propto P^{0.3}$ were valid in the network-cell contrast, the observed radiative flux increase in the network by a factor of 10 relative to the cell would require a relative increase in P in the network of a factor 10^3. It follows from Equation (IX-33) that F_c increases by a factor 10^5.

There is, however, no evidence for such large changes in pressure, and, in fact, there is no evidence at present that the pressure in the network exceeds the pressure within the cells by more than a small factor. It appears, therefore, that the network-cell contrast is due either to a relative *decrease* in the temperature gradient in the network regions or to a small increase in pressure in the network or to both effects combined. In either case, it is clear that the network-cell contrast does not obey the correlation given by Equation (IX-33). This is rather surprising in view of the fact that both the network and the large scale bright features are thought to result from the enhanced magnetic field strengths in these areas.

For the spatial average transition region, there is some justification for the conclusion that the major factors in the energy equilibrium are radiation loss and heat conduction. Because the average heat conduction is so large in the absence of a magnetic field, only a relatively small divergence in the heat flux is required in order to balance the radiation losses. Under such conditions, the radiation loss per unit volume may be expressed in the form

$$\nabla \cdot F_{rad} \propto n_e^2 Q(T),\qquad\qquad(\text{IX-35})$$

and the energy gain due to heat conduction is given by

$$\nabla \cdot F_{cond} \propto \frac{d}{dh}\left(T^{5/2}\frac{dT}{dh}\right)$$

$$\propto \frac{5}{2}T^{3/2}\left(\frac{dT}{dh}\right)^2 + T^{5/2}\frac{d^2T}{dh^2}\qquad\qquad(\text{IX-36})$$

If we ignore the term in d^2T/dh^2, which is likely to be small, this latter proportionality may be rewritten as

$$\nabla \cdot F_{cond} \propto \frac{F_c^2}{T^{7/2}}.\qquad\qquad(\text{IX-37})$$

Now, since the radiation from a particular ion always occurs within the same temperature limits, the proportionalities (IX-35) and (IX-37) yield

$$n_e^2 \propto F_c^2,$$

or, alternatively

$$\Delta \log P = \Delta \log F_c. \tag{IX-38}$$

Also, we note that the intensity observed for a given line is given by Equation (VI-76) as

$$I \propto \frac{P^2}{F_c}. \tag{IX-39}$$

This, together with Equation (IX-38) gives

$$\Delta \log F = \Delta \log F_c = \Delta \log P \tag{IX-40}$$

Models of this type and correlation have been proposed independently by Reimers (1971a,b) and by Moore and Fung (1972). The correlation between $\Delta \log F_c$ and $\Delta \log P$ given by Equation (IX-40) is very different from that found empirically by Withbroe and Gurman and is not valid, therefore, for the large scale features. In the case of the network, Equation (IX-40) requires that the increased brightness of the network be accompanied by a steepening of the temperature gradient by a factor of 10 and an increase in pressure by a factor of 10.

   Although one cannot completely rule out this possibility, it is not a very plausible one. The absence of network structure in the corona must be taken to mean that, at coronal heights, there is no large pressure disequilibrium from network to cell. This conclusion follows from the fact that the coronal line fluxes are proportional to P^2. If we accept this conclusion, then it follows that there can be no large pressure disequilibrium between the network and cells at transition region heights since there is practically no change in pressure from the transition region to the low corona. Logic would seem to require, therefore, that the network-cell contrast in the transition region is due to changes in the temperature gradient that are largely uncorrelated with changes in pressure. Thus, the cells are relatively dark because the temperature gradient is relatively steep, and the network is bright because the temperature gradient is reduced. Such a model is consistent with the magnetic field model in which the regions of vertical field, and hence high vertical conductivity, lie in the network. Intuitively, one expects low temperature gradients where the conductivity is high and high temperature gradients where the conductivity is low. Such a model implies, however, that the energy equilibrium is not simply a balance between radiation and conduction.

   A model with these latter characteristics has been proposed by Kopp (1972), and was discussed to some extent in Chapter VII, Section 7. As noted there, Kopp considers a network structure of fractional surface area $a(h)$, and, although he allows $a(h)$ to change with height in deriving appropriate analytical expressions, he later takes a as constant. The magnetic field strength is assumed to obey the relation $B(h) = a^{-1}(h)$, and divergence of the magnetic field lines is taken into account in determining the divergence of the conductive flux. Kopp assumes that all of the observed radiation flux arises from the network and uses the observed line fluxes to derive a 'reduced pressure', ap, and subsequently a conductive flux. He shows that the conductive flux is proportional to $(aP)^2$, and obtains a value that is about a factor of 15 lower than the usual value of 6×10^5 erg cm^{-2} s^{-1} obtained for the plane-parallel, vertical field models. Since this is an important conclusion, if correct, it should be examined in some detail.

Kopp's analysis of the XUV data follows the standard procedure with the following exceptions. He sets

$$T^{7/2} = \eta \qquad \text{(IX-41)}$$

so that

$$F_0 = \frac{d\eta}{dh}, \qquad \text{(IX-42)}$$

and he assumes $F_0 = $ const to integrate Equation (VI-76). He then follows the customary analytical procedure of plotting $\log aP^2/F_0$ vs $\log \eta$ to obtain an empirical relation giving aF_0^{-1} as a function of η. Following Kopp, we set

$$\nabla \cdot F_c = - \frac{2}{7} kB \frac{\partial}{\partial l}\left(\frac{1}{B}\frac{\partial \eta}{\partial l}\right) \qquad \text{(IX-43)}$$

and

$$\nabla \cdot F_{rad} = Q(T)n_e^2 , \qquad \text{(IX-44)}$$

where l is measured along the magnetic lines of force. Also, we set $B(h) = a^{-1}(h)$ and equate $\nabla \cdot F_c = - \nabla \cdot F_{rad}$ to obtain

$$\frac{d^2\eta}{dh^2} + \frac{1}{a}\frac{da}{dh}\frac{d\eta}{dh} = \frac{7}{2k}\frac{P^2 O(\eta)}{4/7} \qquad \text{(IX-45)}$$

Next, we set

$$f = a P^2 (d\eta/dh)^{-1} \qquad \text{(IX-46)}$$

and eliminate the height derivatives, assuming $p = $ const, to obtain

$$\frac{d}{d\eta}\left(\frac{a^2}{f}\right) = \frac{7}{2\kappa P^2}\frac{f Q(\eta)}{\eta^{4/7}} . \qquad \text{(IX-47)}$$

The quantity $Q(\eta)$ is then given by

$$Q(\eta) = - \frac{2}{7}\frac{k(aP)^2}{f^2\eta^{3/7}}\frac{d\ln f}{d\ln \eta} - 2\frac{d\ln a}{d\ln \eta} . \qquad \text{(IX-48)}$$

Values of f and $d \ln f/d \ln \eta$ can be obtained, in principle, at least, from the observed fluxes in the XUV lines. Kopp's determination of $f(\eta)$ is shown in Figure IX-4. If we

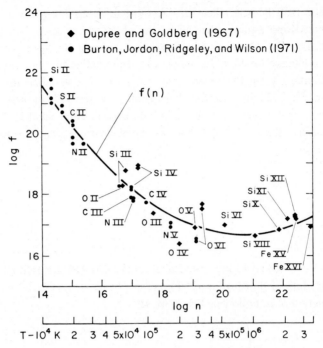

Fig. IX-4.   The relationship between f and η in the transition region (see Equations (IX-41) and (IX−46)) (Kopp, 1972; courtesy *Astrophys. J.*).

consider that P is known, the remaining unknowns on the right hand side of Equation (IX-48) are a and d ln a/d ln η. Kopp evaluates $Q(\eta)$ by setting a = const, which corresponds to a uniform vertical field in the network. He then finds the value of ap for which $Q(\eta)$ is everywhere less than or equal to the corresponding function given by Cox and Tucker (1969). A comparison of the two curves for three values of aP is shown in Figure IX-5.

Kopp's argument for requiring that the empirical $Q(T)$ be everywhere less than or equal to the Cox and Tucker values for $Q(T)$ is based on the belief that d ln a/d ln η is positive and therefore, if taken into account, would reduce the value of $Q(\eta)$ deduced from Equation (IX-48). The relative importance of the term d ln a/d ln η in determining the empirical $Q(\eta)$ is hard to estimate without some direct observational measure of $a(\eta)$, which is lacking. However, we note from Figure IX-4 that even a modest dependence of a upon η will make the term 2d ln a/d ln η competitive with the term d ln f/d ln η for temperatures above about 2×10^5 K where d ln f/d ln η is small. For example, Kopp's results give d ln f/d ln η approximately -0.5 at 2×10^5 K, -0.15 at 5×10^5 K and $+0.08$ at 1×10^6 K. By comparison, a dependence of a upon η of the form $a \propto \eta^{1/10}\,(a \propto T^{7/20})$, which is perhaps a conservative estimate, gives 2d ln a/d ln $\eta = +0.2$. This value is sufficiently large to strongly increase the empirical $Q(\eta)$ for all temperatures above about 3×10^5 K and for fixed values of ap.

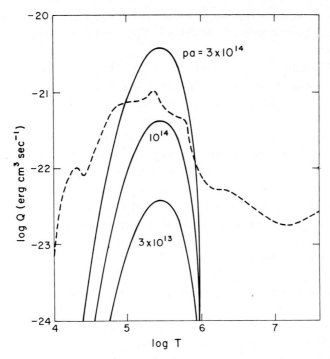

Fig. IX-5. The quantity $Q(T)$ (see Equation (IX-44)) derived for the transition region. Cox and Tucker's radiation loss curve is shown as a dashed line.

Kopp recognizes that the unknown value of $d \ln a/d \ln \eta$ poses a difficulty but argues that this derivative is positive and concludes, as a result, that $ap \, (= an_e T)$ must be less than 1.2×10^{14} K cm^{-3}. An added difficulty with this analysis, and which causes some concern about the validity of this upper limit on ap is that the derivative $d \ln f/d \ln \eta$ is poorly determined in the temperature range of interest. The curve drawn in Figure IX-4 is a least squares quadratic fit to the data. However, there is no particular reason to select a quadratic form. One could, as well, use either a higher order polynomial or two linear segments without increasing the residuals. A fit of either type would substantially alter the derivative $d \ln f/d \ln \eta$ for temperatures above 1×10^5 K. Given the scatter in the data shown in Figure IX-4, it is difficult to argue, in fact, for any particular value of $d \ln f/d \ln \eta$ other than the broad conclusion than $d \ln f/d \ln \eta$ is small, possibly zero, for values of $T \geqslant 1 \times 10^5$ K. Thus, the inferred value of $Q(\eta)$ from Equation (IX-48) might be determined entirely by $d \ln a/d \ln \eta$ for $T \geqslant 1 \times 10^5$ K and might have value considerably smaller than those indicated in Figure IX-5. If this were the case, then Kopp's evaluation of ap is not an upper limit after all. Thus, one should perhaps not pay too much attention to the exact numbers resulting from Kopp's analysis but should look, instead, at the trends in his results.

Finally it should be noted that the assumption of constant F_0 on which the plot in

Figure IX-4 is based is not valid for temperatures below or near 1×10^5 K and this again casts doubt on the evaluation of d ln f/d ln η. This point is discussed in more detail in the next section.

The general conclusion from the results in Figure IX-5 is that the reduced pressure is less than the average pressure of 6×10^{14} ($= n_e T$) and, hence, that $a < 1$. Since, in the proposed model, the conductive flux is zero outside the network and is proportional to a p^2/f within the network, the conductive flux within the network is less by a factor a than the average Sun value for the plane-parallel, vertical field model. Such a conclusion merely states that the enhanced line flux in the network regions is due mainly to an increase in the thickness of the transition region, which, in turn, requires a decrease in F_c within the network. Thus, the conclusions are not particularly surprising. By the same logic, however, one wonders about the thermal conduction in the cell regions. Presumably the radiation flux in these regions is suppressed by steepening the temperature gradient by a significant factor. Since the magnetic field is not likely to be completely horizontal in all parts of the cells, the enhanced temperature gradient over the cells may well lead to conduction in excess of that inferred for the network (see Chapter VII, Section 7).

The preceding analysis is based upon the assumption that radiation losses in the transition region are balanced by the divergence of the heat conduction. However, the heat conduction is reduced in Kopp's model to a value considerably lower than the observed radiation losses from the transition region, and Kopp concludes from this that mechanical heating must play an important role. Again, this is a logical conclusion that follows from the apparent lower temperature gradient in the network transition region. Any attempts at numerical estimates of the heating required, however, are again dependent upon the uncertain derivatives d ln f/d ln η and d ln a/d ln η. They are dependent, additionally, upon our ability to assign the proper distribution of the radiation losses as a function of temperature in the transition region, which we are still unable to do empirically. Nevertheless, the conclusion that mechanical heating plays an important role in the energy balance, based both on Kopp's analysis and on the predicted correlations $\Delta \log F_c$ and $\Delta \log P$ for the radiation-conduction models, seems almost inescapable.

6. The Base of the Transition Region

As may be seen from the plots in Figures IX-4 and VII-13, the quantity $P^2 F_0^{-1}$ increases rapidly with decreasing temperature below 1×10^5 K and at 1×10^4 K has a value some four orders of magnitude higher than at 1×10^5 K. Such an increase cannot be due to an increase in P^2. All recent chromospheric models (see Figures VII-1 and VII-2) give the gas pressure at the 1×10^4 K level within a factor 2 of the gas pressure in the corona. It is clear, therefore, that F_0 has decreased markedly at lower temperatures. Again, this is not surprising since the linear distance between temperatures of 1×10^4 K and 1×10^5 K for F_0 constant at the mean value found in the transition region is only about 4 km. The conclusion that F_0 is changing rapidly below about 1×10^5 K vitiates the assumption that F_0 is constant, which is basic to the plots in Figures VII-13 and IX-4. It is not clear what assumptions, if any, are appropriate in the regime below 1×10^5 K. Perhaps, here, one should simply follow the lead of Pottasch (1964) and introduce the approximation

$$\int n_e^2 Q(T) \, dh = \langle Q(T) \rangle \int n_e^2 \, dh \tag{IX-49}$$

where $\int n_e^2 \, dh$ is the so-called emission measure. This has the advantage of not introducing any spurious effects through faulty assumptions, but it has the disadvantage that the emission measure is of less immediate interest than is the temperature gradient.

Pottasch's (1964) results indicate that the quantity $d \ln \int n_e^2 \, dh / d \ln T$ has a mean value of about -1.6 for $T < 1 \times 10^5$ K. If we interpret the emission measure as

$$\int_{h(T_1)}^{h(2T_1)} n_e^2 \, dh \approx P_0^2 \int_{h(T_1)}^{h(2T_1)} \frac{dh}{T^2} \approx \frac{P_0^2}{T_1^2} \, \delta h, \tag{IX-50}$$

where P_0 is constant, T_1 is the temperature at which the emission from a given line maximizes and δh is the height difference $h(2T_1) - h(T_1)$, then we infer that

$$\frac{d \ln \int n_e^2 \, dh}{d \ln T} = \frac{d \ln \delta h}{d \ln T} - 2. \tag{IX-51}$$

Thus, Pottasch's results give $d \ln \delta h / d \ln T \approx 0.4$. Since $d \ln T = \frac{2}{7} d \ln \eta$ the corresponding equality is $d \ln \delta h / d \ln \eta \approx 0.11$. The same interpretation of the results in Figure IX-4 corresponds to the approximation

$$\frac{d\eta}{dh} = \frac{7}{2} T^{5/2} \frac{dT}{dh} \approx \frac{7}{2} T^{7/2} \delta h^{-1}$$

$$\approx \frac{7}{2} \eta \, \delta h^{-1}, \tag{IX-52}$$

or

$$\frac{d \ln (d\eta/dh)^{-1}}{d \ln \eta} = \frac{d \ln \delta h}{d \ln \eta} - 1. \tag{IX-53}$$

The mean value of $d \ln (d\eta/dh)^{-1}/d \ln \eta$ from Figure IX-4 for $T < 1 \times 10^5$ K is about -1.17. Thus, the mean value of $d \ln \delta h / d \ln \eta$ is about -0.17. This has the opposite sign from the results obtained using Pottasch's analysis. However, both results indicate that δh varies only slowly with temperature. An estimate of the value of δh can be obtained using Pottasch's results together with Equation (IX-50). For $T = 3 \times 10^4$ K, Pottasch's results give $\int n_e^2 \, dh \simeq 1 \times 10^{27} \text{ cm}^{-5}$, and for $P_0 = 6 \times 10^{14} \text{ cm}^{-3}$ Equation (IX-50) gives $\delta h \approx 25$ km. A similar evaluation of δh from Equation (IX-52) and the results in Figure IX-4 gives $\delta h \approx 15$ km. Both of these results are consistent with the value $\delta h \approx 20$ km for the height interval between 2×10^5 K to 1×10^5 K as obtained in Chapter VII. The small values of δh together with the evidence that δh varies only slowly with temperature strongly indicates, as noted by Kopp (1972), that there are no extensive temperature plateaus between temperatures of 1×10^4 and 1×10^5 K.

The evidence for a rather abrupt decrease in F_0 beginning near $T = 1 \times 10^5$ K led Kuperus and Athay (1967) to suggest that thermal conduction deposits a large amount of heat in a thin layer at the top of the chromosphere. They further argued that this energy could not be radiated away, since, if that were possible, the corona would not exist. This conclusion is based on a prior conclusion that the coronal temperature jump occurs only because the upper chromosphere is radiating at maximum efficiency. Thus, the existence of the corona was taken to mean that the chromosphere could not radiate more energy and, hence, could not radiate away the heat flux from the transition region. As an alternative, they proposed that the heat flux was carried back to the corona by spicules and was responsible for the acceleration of the spicules. This conclusion is inconsistent, however, with Kopp's conclusion that the heat conduction is less in the network than was inferred from the spherically symmetric model and the large energy flux in spicules discussed in Section 1.

The conclusion of Kuperus and Athay that thermal conduction drives the spicules has been challenged by a number of authors. Defouw (1970a) has pointed out that the Kuperus-Athay model is based on the assumption that $\partial Q(T)/\partial T = 0$ at the top of the chromosphere whereas the correct criterion is given by Equation (IX-20), viz., $\partial \ln Q(T)/\partial \ln T > 1$. This latter criterion is violated while $\partial Q(T)/\partial T$ is still positive. Thus, the upper chromosphere is not radiating at maximum efficiency and may, indeed, radiate away the energy deposited by heat conduction. In a more detailed treatment, Moore and Fung (1972) conclude that while much of the conducted heat energy is radiated away some portion must still be carried by convective motions, possibly spicules.

In each of the cases discussed in the preceding paragraph, the upper chromosphere and the transition region are assumed to be plane parallel Kopp's (1972) model of the network transition region leads to both increased radiating ability in the network (due to decreased temperature gradient) and decreased thermal conductivity. In his model, in fact, the heat conduction is insufficient to produce the observed radiation and an additional mechanical energy source is required. As noted in the preceding section, Kopp's specific numerical conclusions are open to question and the relative roles of heat conduction and convection are still uncertain. Mechanical heating probably does play a significant role, especially within the network. Convection seems to be of importance, also, but our knowledge of the upper chromosphere, lower transition-region and spicule structure is still too rudimentary to allow a proper evaluation of these effects.

Obviously, the question of energy balance at the base of the transition region is still to be resolved. Within the resolution of this problem lies valuable clues both as to the origin of spicules and as to the origin of the corona itself.

7. The Effect of Motions

The stability criterion (IX-20), is based on the assumption that the atmosphere is at rest. Defouw (1970c) has shown, however, that thermal stability and dynamic stability are interrelated in the sense that the onset of thermal instability is accompanied by dynamic instability. If, therefore, we regard the transition region as being associated with thermal instability, we should expect that it is dynamically unstable as well. The actual dynamic instability depends upon the magnetic field strength and geometry and no detailed theoretical models have been derived.

Lantos (1972) and Delache (1973) have noted that models allowing for a large radial flow velocity in the transition region will strongly modify the structure and energy balance within the transition region. Transport of enthalpy by the outwards flow becomes a major factor in the energy balance for flow velocities of the order of 20 km s^{-1} or higher. Lantos (1972) has derived a model of the lower transition region based on cm wavelength radio data together with the assumption that downward conductive flux is removed by radiation and expansion. His interspicular model shows no expansion for $T \leqslant 1 \times 10^5$ K but rapid acceleration for $T \gtrsim 1.1 \times 10^5$ K. At $T = 2 \times 10^5$ K the expansion velocity is already about 22 km s even though the height separation between temperatures of 1×10^5 and 2×10^5 K is only about 30 km. At a temperature of 3×10^5 K, which is reached in an additional 50 km, the expansion velocity is 43 km s^{-1}. Delache (1973) has further amplified the model and has shown that it gives general agreement with XUV data. (Few details are given in Delache's paper, however.) However, as we noted in Chapter VIII, Lantos (1972) omitted the gravitational term in the energy equation. Since this is the largest term by a few orders of magnitude, the velocities inferred by Lantos are much too high. The dynamic model derived by Chiuderi and Riani (1974) has lower expansion velocities, but also has a temperature gradient that is much too small to satisfy the XUV line fluxes. Thus, it is difficult to assess the validity of such models other than to note that they call attention to an important facet of the problem and demonstrate that the effects of expansion may be significant. The physical reality of such models can be judged only when more data are available, in particular when accurate profiles and wavelengths are established for lines formed in the transition region.

Preliminary studies of the profiles of lines of C IV ($T_0 = 1 \times 10^5$), Si III ($T_0 = 4 \times 10^4$), C II ($T_0 = 2 \times 10^4$) and Si II ($T_0 = 1 \times 10^4$) by Boland et al. (1973) reveal that the observed line widths are considerably in excess of the values predicted for thermal motions in an optically thin atmosphere. Under the assumption that the excess broadening is produced by sound waves propagating radially outwards, they obtain an estimate of the mechanical energy flux in the waves of about 5×10^5 erg cm^{-2} s^{-1}. This is in excellent agreement with the estimated total energy loss from the corona and transition region and suggests that the interpretation in terms of propagating waves may be correct. Much more data of precision character are needed, however, before a complete description of the broadening motions and their propagation characteristics can be given. The problem is both interesting and challenging and deserves considerable effort in the future.

8. Energy Losses from the Chromosphere

A. UPPER CHROMOSPHERE

In the chromosphere proper, i.e., in the temperature regime below about 5×10^4 K, energy loss by thermal conduction is believed to be relatively unimportant. It is true that some regions of sharp temperature rise exist. However, at the densities found in the chromosphere radiation losses tend to be large, which makes the conductive losses relatively less important. It is to be noted, for example, that the energy flux in the chromospheric Lyman-α line exceeds, by itself, all of the energy flux in the coronal spectrum.

For purposes of discussing the energy losses, it is convenient to divide the chromosphere into three regimes characterized by their dominant sources of radiation. Thus, we here define the upper chromosphere as the region where Lyman-α provides the dominant radiation loss, the middle chromosphere as the region where line emissions of neutral and singly ionized elements, other than Lyman-α of hydrogen, provides the dominant radiation loss and the low chromosphere as the region where the free-bound and free-free emission from H^- provides the dominant radiation loss. This division of the chromosphere into layers distinguished by their dominant emission mechanisms is a somewhat different division than our earlier one based on temperature structure, but the two alternate divisions represent approximately the same geometrical layers.

In this section, we are concerned with the upper, or Lyman-α, layer of the chromosphere. The Lyman-α line carries an average energy flux of about 6 erg cm^{-2} s^{-1} at Earth or about 2.7×10^5 erg cm^{-2} s^{-1} at the surface of the Sun. As we remarked earlier, this energy flux is comparable to the combined radiation loss from the corona and the transition region. It is comparable, also, to the estimated flux of thermal conduction in the spherically symmetric transition region. Spectroheliograms of the Sun in the Lyman-α line show a pronounced network structure. Reeves *et al.* (1974) find that the network in Lyman-α is 3 to 10 times brighter than the cells. Since the relative area of the network in the upper chromosphere is of the order of 10%, the total radiation flux from the network may exceed that from the cells even though the network area is relatively small.

In Chapter VI, Equation (VI-68), it was shown that for the case of a two-level atom the net flux contributed to the line by the effectively thin layer of the atmosphere is just equal to the total number of collisional excitations within that layer. Since the collisional excitations followed by photon emission represent an energy flow from the thermal reservoir of the gas to the radiation field, it follows that the radiant energy flux contributed by the effectively thin layer represents a direct energy loss from that layer.

For an effectively thick layer, most of the collisional excitations are nullified by collisional de-excitations and do not produce photons that escape the atmosphere. The radiation arising from such a layer still represents an energy loss, but the energy loss is no longer equal to the total collisional excitations.

In the case of Lyman-α the chromosphere is effectively thin in the layers where the

Doppler core of the line forms and becomes effectively thick in the layers where the line wings form. Of the total flux of 2.7×10^5 erg cm^{-2} s^{-1}, nearly all is in the line core formed in the effectively thin layers above about 10^4 K. Thus, the observed Lyman-α flux represents a direct energy loss from the upper chromosphere.

Spicule matter flowing upwards from the network regions provides an additional source of energy loss from the upper chromosphere. In Paragraph IX-1, the energy flux within a typical spicule was estimated to be 5×10^8 erg cm^{-2} s^{-1} and the enthalpy flux was estimated to be 3×10^6 erg cm^{-2} s^{-1}. The fraction of the network area involved in spicule eruption at a given time is not accurately known. However, the estimated fraction of the total solar surface covered by spicules is near 10^{-3} (see Chapter II). In the upper chromosphere, the network covers about 0.1 of the solar surface. Thus, the fraction of the network area involved in spicule eruptions at a given time appears to be about 0.01. This is a crude estimate only and could easily be in error by a factor of three in either sense. (Note that the estimate given represents only the cross sectional area of the spicules normal to the spicule axis. Observations of spicules on the disk but off disk center show the spicules projected as elongated fibers and give the impression that the relative spicule area is much larger than 1% of the network area. This is misleading, however.)

It appears from the above estimate that the average energy flux in spicule flow in the network is near 5×10^6 erg cm^{-2} s^{-1}, which is larger than the Lyman-α flux in the network areas. The energy fluxes averaged over the entire solar surface show an approximate equality between the Lyman-α flux and the spicule energy flux.

The upper chromosphere gains energy from the transition region via thermal conduction. Again, the largest energy flux due to conduction appears to be in the network regions. As Kopp (1972) has shown, however, the increased radiation flux in the network implies that the transition region over the network is thicker (lower temperature gradient) than for the average Sun and, hence, that the thermal conduction in the transition region network is less than is inferred for the average Sun. Again, accurate quantitative estimates are not available. It appears, however, that the energy gain in the upper chromosphere from thermal conduction is insufficient to produce the Lyman-α flux either in the network or in the cells.

Evidently the upper chromosphere, even though relatively thin in vertical extent, has a rather complex energy balance. The mechanical energy dissipation required to produce the observed Lyman-α flux has an average Sun value of about 2.7×10^5 erg cm^{-2} s^{-1} and within the network is about a factor of ten greater. The average Sun mechanical energy dissipation needed to sustain the upper chromosphere radiant energy loss is comparable to the average mechanical energy dissipation needed to sustain the corona and transition region combined. Possible mechanisms for providing the required mechanical energy supply will be discussed in a following section.

B. MIDDLE CHROMOSPHERE

In the middle and low chromosphere the temperature gradients are too moderate for thermal conduction to play an important role in the energy balance. Also, discrete geometrical structures in these layers of the atmosphere appear to be associated with low translational velocities. No middle or low chromospheric features with motions similar to spicules have been identified. Thus, convective transport of energy similar to that associated with spicules does not appear to be of major importance.

This is not to argue, of course, that convective transport of energy is not important generally. Because of the relatively high densities found in the middle chromosphere, large energy fluxes may occur in low amplitude wave motions. A wave train with velocity amplitude v_a and propagation velocity v_p carries an energy flux

$$F_{conv} = \frac{1}{2}\rho v_a^2 v_p. \tag{IX-54}$$

In the middle chromosphere the density, ρ, is of the order of 10^{-11} g cm^{-3}. Thus, a wave train of velocity amplitude 5 km s^{-1} and propagation velocity 10 km s^{-1} carries a convective flux of 1.2×10^6 erg cm^{-2} s^{-1}. Motion of this type would be difficult to detect in the middle chromosphere, partly because of its complex effect upon the profiles of spectral lines and partly because of the complex pattern of geometrical structures observed in the middle chromosphere. The lack of direct observational evidence for convective flow of energy of an amount comparable to that needed to heat the upper chromosphere and corona is therefore not too surprising and can not be used as an argument against the existence of such motions.

Radiation losses from the middle chromosphere are complicated by two factors: (1) the chromosphere is effectively thick in some lines, and (2) the major radiation losses occur in lines that are seen on the disk as absorption lines. A complete analysis of the radiation losses from the middle chromosphere has not yet been attempted. What has been done, as a substitute, is to estimate the losses contributed by the major suspect lines.

It can reasonably be argued that one effect of the chromospheric temperature rise is to distend the atmosphere of the Sun with the result that the geometrical layers occupied by the chromosphere are more opague in some spectral lines than they would otherwise be. The added opacity of the atmosphere scatters photons incident from the lower layers. These scattered photons are observed as chromospheric emission, but they do not constitute an energy loss. In addition to the scattered photons, there are other photons present in the line that have arisen from thermal excitation in the chromospheric layers. It is only this latter group of photons that should be included in the chromospheric energy budget as radiation loss. Before discussing the means of separating the scattered photons from the thermally excited photons, however, it is useful to consider the total radiation budget of the chromosphere.

There are two alternate approaches to assessing the total energy radiated by the middle chromosphere. One approach is simply to add up all of the emission observed in the chromosphere at the limb during total solar eclipse and to convert this to a mean disk value. A second approach is to add up the residual intensities observed on the disk in those portions of the stronger spectral lines that are formed in the chromosphere. The former method has the advantage that there is no ambiguity in the level of origin of the radiation and the disadvantage that eclipse data are not accurately calibrated in absolute intensity. The second approach has the advantage that the intensities are well calibrated but the disadvantage that it is not easy to identify the regions of the spectral lines that are of chromospheric origin.

As an attempt at the second approach, consider all of the lines listed in Table V-1 as being formed, at their centers, at heights greater than 600 km. For the lines other than H and K and Hα, a crude approximation to the chromospheric contribution is obtained by taking a bandwidth equal to $\pm \Delta\lambda_D$ centered on the line. The value of $\Delta\lambda_D$ for metal atoms in the middle chromosphere is given approximately by $\Delta\lambda_D = 1.6 \times 10^{-5}\lambda$, corresponding to a mean broadening velocity of 5 km s^{-1}. For Hα, the broadening velocity is more nearly the mean thermal velocity of hydrogen atoms, which is about 10 km s^{-1}, and yields a Doppler width of 0.22 Å. Since the K_2 and H_2 emission peaks are known to form near the top of the initial steep temperature rise at the base of the chromosphere, the middle chromospheric contribution, of the H and K line is given by the residual intensities inside of the K_2 and H_2 emission peaks. Similar arguments hold for the emission cores in the resonance lines of Mg II, which we add to the list in Table V-1. The net outgoing fluxes represented by the residual intensities in several lines are given in Table IX-1.

TABLE IX-1

Estimates of πH from the middle chromosphere in units of 10^5 erg cm^{-2} s^{-1}

Spectral features	10 lines of Fe I (Table V-1)	Na I D$_2$	IR triplet Ca II	H and K Ca II	Hα	Mg II	Balmer cont. (Henze, 1969)
Residual disk intensities	4.8	0.8	7.0	1.8	6.4	2.3	–
Eclipse (Dunn et al., 1968)	1.5	0.14	3.7	3.8	6.0	–	10

Emission fluxes observed at eclipse (Dunn et al., 1968) and converted to equivalent disk fluxes over the outgoing hemisphere are given in Table IX-1 in the last row. The two sets of data for the estimated chromospheric emission are in reasonable agreement considering the uncertainties involved in each set. It is readily seen from the results in Table IX-1 that the total middle chromospheric emission is about an order of magnitude greater than the emission from the upper chromosphere or from the corona and transition region. As was noted earlier, however, not all of the middle chromospheric emission is the result of thermal excitation and therefore does not represent an energy drain from the middle chromosphere.

The net loss of energy from a layer of thickness $d\tau_0$ is given by Equation (VI-65). This equation reduces in the effectively thin case and for a two level atom to Equation (VI-68). Athay (1966) used an equation of this latter form to determine the energy loss from the middle chromosphere due to the Balmer series of hydrogen. He obtained a value for $4\pi H$ of 1.8×10^6 erg cm^{-2} s^{-1}.

For many of the lines in Table V-1 it is possible to obtain a direct evaluation of

$$2\pi H = 2\pi^{3/2} \int_0^{\tau_0} \Delta\nu_D \rho S \, d\tau_0 \qquad (IX-55)$$

using self consistent computations of ρ, S and τ_0. The ρ used here is the excape coefficient defined in Chapter VI. These quantities have been computed for the Fe I lines by Lites (1972), for the Ca II lines by Avrett (1973) and for the Hα and Hβ line by Vernazza et al. (1973). The three sets of computations unfortunately are not based on the same model chromosphere. Each of the three models differs in detail from the HSRA but are very similar in their general properties. The Hα and Hβ line, especially, are sensitive to the model chromosphere. Energy losses in these lines are proportional to n_2 (the population of the second quantum level in hydrogen), which is quite sensitive to the temperature structure of the middle chromosphere.

Energy losses obtained from Equation (IX-55) using the computations cited above are given in Table IX-2. The computed values of $2\pi H$ for Fe I and Ca II which should represent the true energy loss, are near the eclipse values in Table IX-1. It should be emphasized again, however, that the computed values are quite sensitive to the model chromosphere, particularly for Hα, and the computations have not all been done using the same model chromosphere. Thus, the results in Table IX-2 should not be regarded as being accurate to more than about a factor of three.

TABLE IX-2
Computed values of $2\pi H^a$ from Equation (IX-55)
in units of 10^5 erg cm^{-2} s^{-1}

Lines	Fe I (Table V-1)	IR Triplet Ca II	H and K Ca II	Hα	Hβ
$2\pi H$	1	6	2	8	-1

[a]Except for Hα and Hβ in which case $4\pi H$ is tabulated.

In the case of both Fe I and Ca II, the low chromosphere and upper photosphere have high opacity and the radiation loss occurs almost exclusively in the outward direction. Thus, $2\pi H$ is a valid measure of the total energy loss. For Hα and Hβ, however, the low chromosphere and upper photosphere have low opacity and radiation flows out of the chromosphere in both inward and outward directions. For this case, the combined energy losses in the two directions is approximately $4\pi H$, which is the value given in Table IX-2. The energy loss for Hβ is negative, indicating that this line heats the chromosphere rather than to cool it. This happens for a few subordinate lines because of photon exchanges (interlocking) with other lines but is never of major importance.

The Balmer continuum energy flux from the chromosphere is seen from Table IX-1 to be moderately large. No detailed computations of the net energy loss or gain in the free-bound continua have been carried out. However, in the middle chromsophere all of the bound energy levels in hydrogen are over populated with respect to their LTE values (i.e. $b_j > 1$). This is a direct result of the radiative imbalance between the bound levels and the continuum, and the overpopulation of the bound levels results from an excess of photorecombinations over photoionizations. Thus, the middle chromosphere does experience a net loss of energy due to the free-bound continua of hydrogen.

The Lyman continuum flux from the Sun is of the order of 2×10^4 erg cm^{-2} s^{-1} (Vernazza and Noyes, 1972). This is small in comparison to the Balmer continuum emission seen at eclipse. Also, the Paschen continuum at eclipse is weak in comparison to the Balmer continuum. We may safely conclude, therefore, that most of the free-bound emission from hydrogen in the middle chromosphere is in the Balmer continuum. However, to a rather good approximation the equilibrium of the continuum states is fixed relative to the bound states by the balance between photoionizations and photorecombinations. Collisional ionizations and three-body collisional recombinations are relatively unimportant. The majority of the Balmer continuum flux from the chromosphere, is scattered photospheric radiation, and only a relatively small fraction of the flux is to be counted as a net energy loss. It seems likely, then, that the Hα losses exceed the Balmer continuum losses.

Energy loss due to the resonance lines of Mg II represents another potentially important cooling mechanism. Computations for Mg II have not been published in sufficient detail to allow an integration of Equation (IX-55). However, in the case of Mg II most of the energy flux listed in Table IX-1 is in the emission peaks, which are of chromospheric origin and are due directly to the chromospheric temperature rise. For this reason, it is likely that most of the flux listed in Table IX-1 is to be considered as direct energy loss from the middle chromosphere.

The sum of the energy losses in Table IX-2 plus the Mg II flux in Table IX-1 yields a total energy loss rate for the middle chromosphere of 2×10^6 erg cm^{-2} s^{-1}. Additional energy loss occurs in the Balmer continuum and in numerous other spectral lines not considered in the preceding discussion. The total contribution from these added sources is unknown but probably does not exceed the value 1×10^6 erg cm^{-2} s^{-1}. We emphasize, again, that the computed losses in Table IX-2 are uncertain to perhaps a factor of two. It seems clear, however, that the radiation losses from the middle chromosphere exceed by a substantial factor those from the upper chromosphere and from the transition region and corona as well.

C. LOW CHROMOSPHERE AND T_{min} REGION

In the lowest layers of the chromosphere, energy loss from emission by H$^-$ ions is the dominant loss term. With increasing height, the line emission becomes increasingly important relative to the H$^-$ emission, and at a height of 500 km in the chromosphere the line emission has taken over.

From Equations (VII-23) and (VII-27), we note that the energy loss due to H$^-$ emission is proportional to the difference, δT, between the actual chromsopheric temperature and the radiative equilibrium temperature. A completely satisfactory radiative equilibrium model does not yet exist. On the other hand, the existing models are moderately good and merit discussion.

In radiative equilibrium the radiation flux integrated over frequency is constant. The transfer equation is

$$\mu \frac{dI_\nu}{d\tau_\nu} = I_\nu - S_\nu. \tag{IX-56}$$

Integration of Equation (IX-56) over angle and frequency with $d\tau_\nu = \phi_\nu \, d\tau$ yields

$$\frac{1}{4\pi} \iint \mu \frac{dI_\nu}{d\tau} \, d\omega \, d\nu = \frac{dH}{d\tau} = \int \phi_\nu (J_\nu - S_\nu) \, d\nu. \tag{IX-57}$$

Since $dH/d\tau = 0$ and since J_ν approaches an asymptotic value at small values of τ, it follows that

$$\int \phi_\nu S_\nu \, d_\nu = \text{const}.$$

For the H^- ion, ϕ_ν is only a slow function of frequency in the spectral regions where most of the solar flux is found. Thus, to a good approximation

$$\int S_\nu \, d_\nu = \text{const} = \frac{\sigma T^4}{\pi b} \tag{IX-58}$$

The b coefficient for H^- is unity near the temperature minimum and increases to a value 2 in the low chromosphere (Gebbie and Thomas, 1970). Since b is increasing with height, T must increase with height also. For a factor 2 increase in b, T increases a factor $2^{1/4} = 1.19$ over its minimum value. The empirical rise in T in the first 500 km of the chromosphere is by a factor near 1.24. On this basis, it would appear that most of the temperature rise in the low chromosphere is consistent with radiative equilibrium, as was first proposed by Cayrel (1966) and subsequently verified by a number of authors. This conclusion is incorrect, however, and arises from the neglect of the spectral lines.

In order to demonstrate the effect of spectral lines on the radiative equilibrium model, we rewrite Equation (IX-57) as

$$\frac{dH}{d\tau_c} = \int \left(1 + \frac{\phi_\nu}{r_0}\right) (J_\nu - S_\nu) \, d\nu, \tag{IX-59}$$

where ϕ_ν is now the shape of the line absorption coefficient and $r_0 = d\tau_c/d\tau_0$. The continuum absorption coefficient has been assumed to be independent of frequency. In the case of a single spectral line whose source function is S, we may write

$$S_\nu = \frac{r_0}{\phi_\nu + r_0} B_\nu + \frac{\phi_\nu}{\phi_\nu + r_0} S, \tag{IX-60}$$

where B_ν is the continuum source function. Equations (IX-59) and (IX-60) yield

$$\frac{dH}{d\tau_c} = \int (J_\nu - B_\nu)\, d\nu + \frac{1}{r_0} \int \phi_\nu (J_\nu - S)\, d\nu$$

$$= \int (J_\nu - B_\nu)\, d\nu + \frac{\Delta\nu_0 \pi^{\frac{1}{2}}}{r_0} \left[\int J_\nu \Phi_\nu\, d\nu - S \right]. \tag{IX-61}$$

From Equation (VI-58), we find

$$\int J_\nu \Phi_\nu\, d\nu - S = \epsilon(S - B_0).$$

Thus,

$$\frac{dH}{d\tau_c} = \int (J_\nu - B_\nu)\, d\nu + \frac{\Delta\nu_D \pi^{\frac{1}{2}}}{r_0}\ \epsilon(S - B_0), \tag{IX-62}$$

where B_0 denotes the value of B_ν at the spectral line.

We next add and subtract the term $\int J_\nu^c\, d\nu$ on the right hand side of Equation (IX-62). This yields

$$\frac{dH}{d\tau_c} = \int (J_\nu^c - B_\nu)\, d\nu - \int (J_\nu^c - J_\nu)\, d\nu - \frac{\Delta\nu_D \pi^{\frac{1}{2}}}{r_0}\ \epsilon(S - B_0) \tag{IX-63}$$

If we interpret J_ν^c as the value of the continuum intensity, then $J_\nu^c - J_\nu = 0$ everywhere except at the spectral line. The value of J_ν^c at the line is obtained by equating J_ν^c to J_ν near to, but outside of, the line. Thus, the integral $\int (j_\nu^c - J_\nu)\, d_\nu$ is just $J_0^c W_j$, where J_0^c is the value of the continuum near the spectral line and W_j is the equivalent width of the line measured in the mean intensity. The first term on the right hand side of Equation (IX-63) is recognized as being just the gradient of the continuum flux, $dH_c/d\tau_c$. Since $dH/d\tau_c = 0$ in radiative equilibrium, we find

$$\frac{dH_c}{d\tau_c} = J_0^c W_j + \frac{\Delta\nu_D \pi^{\frac{1}{2}}}{r_0}\ \epsilon(S - B_0) \tag{IX-64}$$

Note that in the absence of a spectral line each of the terms in Equation (IX-64) vanish. The effect of the line, therefore, is to induce a gradient in the continuum flux. Since there is now a gradient in H_c, the solution to Equation (IX-64) is necessarily different from the solution to the equation

$$\int (J_\nu^c - B_\nu)\, d\nu = 0,$$

which is obtained when H_c is constant.

The two terms on the right hand side of Equation (IX-64) are readily interpreted. The first term represents the imbalance between continuum absorptions and emissions at the wavelength of the line, and the second term represents the imbalance in collisional excitations and de-excitations in the line transition. As a result of the line, the continuum absorption rate is changed from being proportional to J_ν^c to being proportional to J_ν. The continuum emission rate, is unchanged, however. Similarly, the rate of collisional de-excitations is proportional to ϵS whereas the rate of collisional excitations is proportional to ϵB.

Note that W_j is positive when the line is in absorption. This tends to produce an increase in H_c with τ_c, which requires a greater value for the temperature gradient than in the case where $W_j = 0$. One result of this increased temperature gradient is the familiar back-warming effect near $\tau_c = 1$.

When $S < B_0$, which is the most common result, the second term on the right hand side of Equation (IX-64) is negative. The magnitude of this term increases in proportion to r_0^{-1} i.e., as the line becomes stronger. The first term increases as r_0 decreases through the increase in W_j. However, W_j does not increase nearly as rapidly as r_0^{-1} because of saturation effects in the lines. It follows that for sufficiently small values of r_0 the second term on the right hand side of Equation (IX-64) has a larger absolute value than the first term. Since the difference between S and B is greatest at small values of τ, the effect of this second term is to produce surface cooling. Thus, the back-warming effect arises mainly from lines on the linear portion of the curve-of-growth and the surface cooling effect comes mainly from lines on the shoulder and wing portions of the curve-of-growth.

The blanketing effect produced by several spectral lines is just the additive effects of the individual lines. Thus, it is convenient to define a quantity β as

$$\beta = \frac{1}{W_0 H_c^0} \frac{dH_c}{d\tau_c}, \tag{IX-65}$$

where W_0 is now the combined equivalent width of all of the spectral lines measured in the net flux, H_c^0, at optical depth zero. For individual lines

$$\beta_i = \frac{1}{W_{i0} H_{ci}^0} \frac{dH_{ci}}{d\tau_{ci}} \tag{IX-66}$$

and, hence

$$\beta = \frac{1}{W_0 H_c^0} \sum_{i=1}^{N} \beta_i W_i H_{ci}^0 . \tag{IX-67}$$

Values of β_i have been computed (Athay, 1970) for twelve of the strongest Fraunhofer lines as well as for many lines represented by rather simplified atomic models representing Fe I and Ti II. The model atoms used for the twelve selected lines of other ions were sufficiently complex to give a good synthesis to the observed solar profiles for these lines. For Fe I and Ti II, the model atoms consisted of only three bound levels plus a continuum. The position of the bound levels relative to the continuum distinguished the Fe I and Ti II atoms.

No attempt was made to synthesize observed lines of Fe I and Ti II. Instead, a large number of cases were computed with different spacing of energy levels, different relative abundances, different combinations or r_0, and different atomic cross sections. This resulted in blanketing functions for lines with a wide range in wavelength and equivalent width with varying degrees of departure from LTE. Average values of β_i were than determined separately for lines on the linear, shoulder and wing portions of the curve-of-growth for Fe I. The same was done for Ti II. This entire process was repeated for four model atmospheres. Three of the model atmospheres had temperatures decreasing monotonically to a boundary value, T_0. Values of T_0 and the continuum optical depth where T_0 is reached were selected as follows: $4600°$, $\tau_5 \leqslant 2 \times 10^{-2}$; $4300°$, $\tau_5 \leqslant 2.5 \times 10^{-3}$; and $4000°$, $\tau_5 \leqslant 5 \times 10^{-4}$. The fourth model included a chromosphere similar to, but somewhat different from, the HSRA. Composite values of β were determined for each of the four model atmospheres from Equation (IX-67).

Figure (IX-6) shows the values of β obtained for the model including a chromosphere for the six classes of Fe I and Ti II lines. Note that the backwarming ($\beta < 0$) near $\tau_5 = 1$ comes mainly from the simulated Fe I lines on the linear portion of the curve-of-growth. Near $\tau_5 = 10^{-4}$, however, the strongest surface cooling comes from the simulated Ti II lines on the shoulder of the curve-of-growth. All of the lines produce heating at large τ_5 and all produce cooling for $\tau_5 \lesssim 3 \times 10^{-3}$.

The surface cooling effects due to Hα, the Mg II resonance lines and the strong Ca II lines for the chromospheric model are shown in Figure IX-7. At $\tau_5 < 10^{-5}$, the cooling

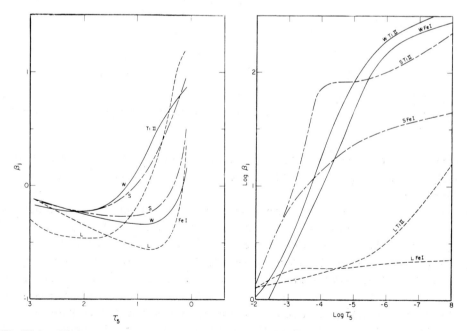

Fig. IX-6.   Blanketing functions for Fe I and Ti II lines on the linear, shoulder and wing portions of the curve-of-growth (courtesy *Astrophys. J.*).

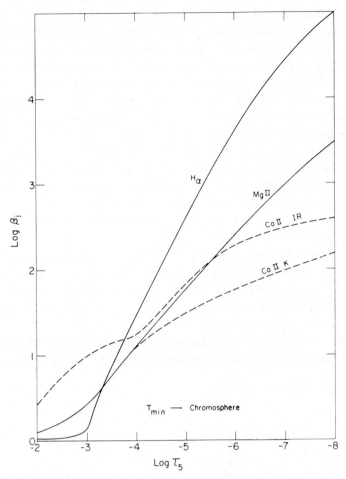

Fig. IX-7.   Surface cooling (blanketing) functions for selected strong lines (courtesy *Astrophys. J.*).

effect due to Hα alone exceeds the combined cooling effect of all of the metal lines. This result is model dependent, however, and needs to be computed for a range of chromospheric models.

Composite values of β for the four models are shown in Figure IX-8. One feature of the results of immediate interest is the rather marked increase in β as the assumed value of T_0 increases. Thus, a low assumed value for T_0 results in a low value of β and a high assumed value for T_0 results in a high value of β. Since it is the value of β that determines T_0, and since a large value of β produces a relatively low value of T_0, the results in Figure IX-8 show that there is a unique pair of values of T_0 and β that satisfy Equation (IX-65).

In order to obtain a self-consistent value of T_0, it is necessary to compute β for the values of $T(\tau)$ that satisfy Equation (IX-65). However, an estimate of the surface value of T_0 for a given value of β can be obtained from Equation (IX-57) by noting that a change

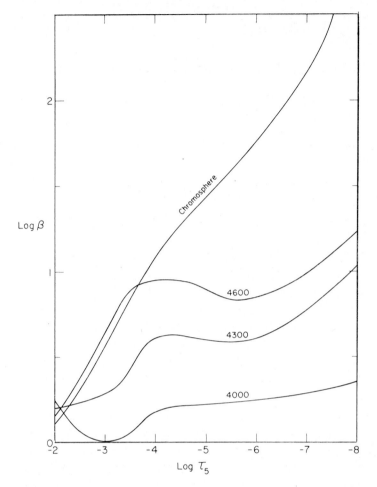

Fig. IX-8.   Composite blanketing (cooling) functions for four model atmospheres (see text) (courtesy *Astrophys. J.*).

in the Planck function near $\tau_c < 10^{-4}$ has little or no effect on the mean intensity. Thus, to a reasonably good approximation, we may write:

$$0 = (\textstyle\int J_\nu \, d\nu)_0 - B_{ou}$$

for the unblanketed case, and

$$\left(\frac{dH_c}{d\tau_c}\right)_0 \approx (\textstyle\int J_\nu \, d\nu)_0 - B_{ob}$$

for the blanketed case. Thus,

$$\left(\frac{dH_c}{d\tau_c}\right)_0 \approx B_{0u} - B_{ob},$$

and, since $H_c^0 = B(T_{eff})$ and $B(T) \propto T^4$, we find

$$\frac{1}{H_c^0}\left(\frac{dH_c}{d\tau_c}\right)_0 \approx \frac{T_{0u}^4 - T_{ob}^4}{T_{eff}^4}.$$ (IX-68)

From Equations (IX-65) and (IX-68), we finally obtain

$$T_{ob} \approx T_{0u}\left(1 - \beta W_0\left(\frac{T_{eff}}{T_{0u}}\right)^4\right)^{\frac{1}{4}}.$$ (IX-69)

The results in Figure (IX-8) can be approximated, for $\tau_5 \lesssim 10^{-4}$ by $\beta = 1.5, T_0 = 4000°$; $\beta = 3.3$, $T_0 = 4300°$ and $\beta = 8$, $T_0 = 4600°$. For the Sun $W_0 \approx 0.12$, $(T_{eff}/T_{0u})^4 = 2$, and $T_{0u} = 4860$. These values give $T_{0u} = 4350°$ for $T_0(\text{assumed}) = 4000°$ and $T_{ob} 3300°$ for $T_0(\text{assumed}) = 4300°$. In order to interpolate between these temperatures we note that the above values of β are approximately proportional to $T_0^{1/2}$. Thus, for $T_0(\text{assumed}) = 4100°$, we obtain $\beta = 2.1$ and $T_{ob} = 4100°$. Thus, on the basis of Equation (IX-69), the self-consistent solution occurs at $T_0 = 4100°$. A more accurate solution to Equation (IX-65) gives the self-consistent solution near $T_0 = 4300°$. Uncertainties in the values of β and W_0 suggest an uncertainty in T_0 of about $\pm 100°$ (Athay, 1970). This result suggests that the minimum temperature in the HSRA is somewhat too low.

Above $\tau_5 = 10^{-4}$, the chromospheric rise in temperature results in a rapid increase in β. This produces a strong local cooling in these layers that resists the outwards increase in temperature predicted by the radiative equilibrium model without lines. The effect of the lines is so strong as to negate any appreciable temperature rise by the Cayrel mechanism. Thus, we will assume that in the radiative equilibrium model, the boundary temperature stays constant at $4300°$.

It was shown in Chapter VII, Equations (VII-23) and (VII-27), that the energy loss due to H^- emission is given by

$$4\pi H = 16\sigma T^3 \delta T\tau_{min}$$ (VII-23)

when $b = 1$ for H^- and by

$$4\pi H = \frac{4\sigma T^3}{b} \delta T\tau_{min}$$ (VII-27)

for $b \propto T^3$.

Values of b near the temperature minimum are not sufficiently well known to justify the use of Equation (VII-27) at the present time.

Values of $4\pi\delta H$ obtained from the HSRA and Equation (VII-23) sum to approximately 2×10^6 erg cm^{-2} s^{-1}. Most of this loss arises just above the temperature minimum where τ_1 is relatively large. As may be seen from inspection of Equation (VII-23) the losses are very sensitive to the model atmosphere adopted. Early models of the chromosphere, such as the Bilderberg continuum atmosphere, in which the minimum temperature occurred near $\tau = 10^{-2}$ would require much greater energy to sustain them. At $\tau_1 = 10^{-2}$, for example, a value of δT of only 10^0 would require a mechanical energy supply of 6.6×10^6 erg cm^{-2} s. For this reason it is important that the actual solar temperature distribution and the theoretical radiative equilibrium temperature distribution in the upper photosphere, T_{min} region and low chromosphere be determined as accurately as possible. At the present time, we can do little more than estimate the loss from the region just above T_{min} as about 2×10^6 erg cm^{-2} s^{-1}. The losses in the atmosphere at greater depth could be either larger or smaller than this value.

The preceding discussion is based upon the assumption that b is unity and, hence, that $\delta b / \delta T = 0$. A more accurate analysis should be done with more realistic values of b. However, the value of b needs to be computed self-consistently with the model atmosphere. Until computations of b are carried out for the HSRA a more realistic evaluation of the effects of b on the value of $4\pi\delta H$ computed near the temperature minimum cannot be made. Since the preceding estimates of $4\pi\delta H$ are already strongly model dependent and are obtained from a model in which δT is not well known, we adopt the values obtained with $b = 1$.

Table IX-3 contains a summary of the radiation losses from the different layers of the chromosphere and corona together with other salient physical characteristics. It is to be emphasized that each of the entries in the table represents only an approximate value. The estimates of total electron and mass densities are uncertain both because of the uncertainties in the representative physical thickness of each layer and because of uncertainties in the densities themselves.

TABLE IX-3
Summary of radiation losses and physical characteristics

	Low chromosphere and T_{min} region	Middle chromosphere	Upper chromosphere	Transition region	Corona
Radiation loss (erg cm^{-2} s^{-1})	2×10^6	2×10^6	3×10^5	2×10^5	1×10^5
Physical thickness (km)	200	1000	200	10 000	100 000
$\int n_e \, dh$(cm^{-2})	3×10^{18}	8×10^{18}	1×10^{18}	1×10^{18}	3×10^{18}
$\int \rho \, dh$(cm^{-2})	2×10^{-2}	2×10^{-3}	2×10^{-6}	2×10^{-6}	5×10^{-6}

Even with these reservations in mind, one sees clearly that the radiation loss is much more nearly proportional to $\int n_e \, dh$ than to either $\int \rho \, dh$ or Δh. In the regions of the atmosphere under consideration $n_p \approx n_e$ and $\int n_e \, dh \approx \int n_p \, dh$. The approximate proportionality between radiation loss and $\int n_e \, dh$ and $\int n_p \, dh$ suggests that the mechanical energy dissipation may be associated with the charged component of the atmosphere rather than the neutral component. This suggestion is purely speculative, however, and the seeming proportionality between radiation loss and the charged component may be fortuitous.

9. The Quiet Sun Solar Wind

A. AN INCONSISTENCY OF HYDROSTATIC EQUILIBRIUM

The gaseous atmosphere surrounding a star is pulled inward by the gravitational attraction of the star and pushed outward by the thermal gas pressure in the atmosphere itself. If the temperature of the atmosphere is low, the gravitational field confines the atmosphere; and the atmosphere contracts until pressure equilibrium is reached. On the other hand, if the temperature of the atmosphere is sufficiently high the gravitational pull is unable to contain the atmosphere and expansion occurs.

A simple illustration of the tendency for a hot atmosphere to expand is obtained from an inconsistency resulting from the assumption of hydrostatic equilibrium. Consider, for example, an isothermal atmosphere of temperature T, particle density n and constant mean molecular weight μ. Hydrostatic equilibrium requires that

$$dp = -\mu m_H n(r) g(r) \, dr, \tag{IX-70}$$

where

$$g(r) = g_0 \left(\frac{R_\odot}{r} \right)^2. \tag{IX-71}$$

Let

$$P = n(r) kT \tag{IX-72}$$

so that

$$dp = kT \, dn(r) \tag{IX-73}$$

Equations (IX-70) and (IX-73) yield

$$\frac{dn(r)}{n(r)} = -\frac{\mu m_H g_0}{kT} R_\odot^2 \frac{dr}{r^2}, \tag{IX-74}$$

whose solution is

$$n = n_0 \, e^{R_\odot^2 / H(1/r - 1/R_\odot)}, \tag{IX-75}$$

where the boundary condition $n = n_0$ at $r = R_\odot$ is imposed and where the scale height, H, is given by

$$H = \frac{kT}{\mu m_H g_0}. \tag{IX-76}$$

At $r = \infty$, Equation (IX-75) gives

$$n_\infty = n_0 e^{-R_0/H}. \tag{IX-77}$$

The corresponding gas pressure at $r = \infty$ is

$$P_\infty = P_G \, e^{-R_0/H}. \tag{IX-78}$$

In order for this hypothetical, isothermal atmosphere to be confined to the star it is necessary that P_∞ be less than the pressure in interstellar space. Interstellar pressure results mainly from magnetic fields of the order of 10^{-4} G and has an approximate value of $B^2/8\pi = 4 \times 10^{-10}$ dyn cm^{-2}. In the solar corona the approximate conditions are: $P_0 = 0.2$ dyn cm^{-2}, $r_0 = 7 \times 10^{10}$ cm, $g_0 = 2.7 \times 10^4$ cm s^{-2}, $\mu = 0.6$ and $T = 1.6 \times 10^6$ K. These values give $H = 8.2 \times 10^9$ cm, $R_\odot/H = 8.5$ and $P_\infty = 4 \times 10^{-5}$ dyn cm^{-2}. Thus, the combination of the high value of T with the low value of μ in the solar corona leads to a relatively large gas pressure at large distance from the Sun, assuming that the corona is static. Since there is no resisting pressure, expansion is implied.

B. ISOTHERMAL AND POLYTROPIC EXPANSION

The preceding argument is inadequate, of course, to demonstrate the actual expansion of the corona. Both the assumption of an isothermal corona and a hydrostatic corona are invalid. As a next step, consider an isothermal corona expanding with velocity $v(r)$. The assumed isothermal structure implies a particular mechanical energy input and obviates the need to consider thermal conduction. Radiation losses may be considered as being combined with the mechanical energy input to yield the isothermal condition. The model is further restricted to be spherically symmetric and to have equal numbers of protons and electrons with the same temperature (single-fluid model).

Mass conservation requires that

$$\frac{1}{r^2} \frac{d}{dr} (r^2 n_p v) = 0 \tag{IX-79}$$

or

$$n_p v r^2 = \text{const.} \tag{IX-80}$$

Also, momentum conservation requires

$$n_p m_H v \frac{dv}{dr} = -\frac{dp}{dr} - n_p m_H g_0 \frac{R_\odot^2}{r^2} \tag{IX-81}$$

or

$$n_p m_H v \frac{dv}{dr} = -2kT \frac{dn_p}{dr} - n_p m_H \frac{g_0 R_\odot^2}{r^2}, \tag{IX-82}$$

where n_p is the proton density. Elimination of n_p from Equations (IX-80) and (IX-82) leads to the result

$$\frac{1}{v}\frac{dv}{dr}\left(v^2 - \frac{2kT}{m_H}\right) = \frac{4kT}{m_H r} - \frac{g_0 R_\odot^2}{r^2}.$$
(IX-83)

For values of T such that the right hand side of Equation (IX-83) is negative at $r = r_0$ (in the case of the Sun $T < 6 \times 10^6$ K), both sides of the equation must vanish at some critical value of $r = r_c$. In order for the left hand side to vanish, either $dv/dr = 0$ or $v_c = (2kT/m_H)^{1/2}$. At the coronal value of $T = 1.8 \times 10^6$ K, $v_c = 170$ km s^{-1} and $r_c/R_0 = \frac{1}{2}(g_0 R_0/v_c^2) = 3.2$.

Equation (IX-83) has two unique solutions for which $v_c = (2kT/m_H)^{1/2}$ and two families of solutions for which $dv/dr = 0$, as illustrated in Figure IX-9. One family of solutions has minima in v near $r = r_c$, and v is everywhere greater than v_c. Since v_c is just

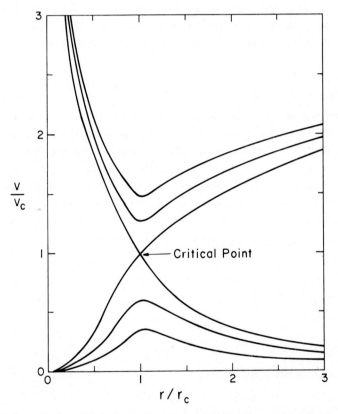

Fig. IX-9.   A schematic plot of the solar wind expansion velocity. See text for definition of v_c and r_c.

the mean thermal velocity, this class of solutions requires that the corona near the Sun is expanding superthermally at speeds much in excess of 170 km s^{-1}, which is inconsistent with observations. One of the unique solutions that satisfies the condition $v = v_c$ has $dv/dr < 0$ everywhere and suffers from the same objections. The second family of solutions has maxima in v near $r = r_c$ and v is everywhere less than v_c. This family of solutions permits low expansion velocities near the Sun but has the difficulty that the low expansion velocity still leads to finite gas pressure at infinity. It is physically unacceptable for that reason. We are left with one unique solution that satisfies the conditions: $v = v_c$ at $r = r_c$, v is small near $r = R_\odot$ and the gas pressure approaches zero at large distances from the Sun. This solution is physically acceptable and has $dv/dr > 0$, i.e., the corona accelerates outwards. For the model under consideration, $v \to \infty$ as $r \to \infty$, but this result arises from the assumption that the corona expands isothermally.

Parker (1958) was the first to treat the problem of the expanding corona. He assumed a polytropic model for the corona in which

$$P = P_0 \left(\frac{n_p}{n_p(0)} \right)^\alpha \tag{IX-84}$$

This is equivalent to

$$T = T_0 \left(\frac{n_p}{n_p(0)} \right)^{\alpha - 1} \tag{IX-85}$$

and allows T to decrease outwards. Parker's results for two model coronas, one with $\alpha = 3/2$ and one with an isothermal corona out to a distance b beyond r_c followed by a region at $r > b$ with $\alpha = 5/3$ (adiabatic expansion), gave the same types of solutions as the isothermal case. Unlike the solutions for the isothermal corona, however, the class of solutions for the polytropic corona for which v has maxima near $r = r_c$ do lead to vanishingly small pressure at infinity. Hence, this class of solutions cannot be immediately rejected as being inconsistent with the boundary conditions. Nevertheless, Parker rejected all but the one unique solution with $dv/dr > 0$ and $v = v_c$ at $r = r_c$. On this basis, he predicted in 1958 the existence of a general solar wind blowing throughout the inner solar system. As evidence for the existence for such a wind, he cited Biermann's (1951) contention that the acceleration of matter in comet tails near the Sun, as well as the ionization of cometary material, implied such a wind. Parker further cited the more or less continual agitation of the terrestrial magnetic field in polar regions as evidence that the field lines were being buffeted by a high velocity solar wind plasma.

C. EXPANSION COMBINED WITH THERMAL CONDUCTION

Equations (IX-79) and (IX-81) are inadequate for describing a realistic expansion of the corona in that they implicitly assume that there are no energy sinks or sources in the corona. The energy conservation equation is

$$\nabla \cdot (F_{KE} + F_H) + W_g + \nabla \cdot F_S = 0, \tag{IX-86}$$

where F_{KE}, F_H and F_S are, respectively, the fluxes of kinetic energy, enthalpy and source energy and W_g is the work done against gravity. Equation (IX-86) may be written more explicitly as

$$\frac{1}{r^2} \frac{d}{dr} \left[n_p m_H v r^2 \left(\frac{1}{2} v^2 + 5 \frac{kT}{m_H} + \frac{F_S}{n_p m_H v} \right) \right] = -n_p m_H v g_0 r_0^2 / r^2. \tag{IX-87}$$

A full treatment of the solar wind problem should include in F_S the radiation, conduction and mechanical energy terms. however, little or nothing is known about the mechanical energy term and one usually attempts to simulate this term by imposing an isothermal corona within a given distance from the Sun. This, of course, is only a crude approximation and there is no firm basis, either observationally or theoretically, for extablishing the boundary of the isothermal corona.

Radiation losses are important in the inner corona. Since they are proportional to the square of the density, however, they become increasingly less important with increasing distance from the Sun. By assuming an isothermal corona near the Sun, one approximates, again in a crude way, the effect of the radiation terms in the domain in which they are important.

Thermal conduction in a collision-dominated plasma, is a well understood physical process. It is known to be important in the high temperature corona and can properly be included in the energy equation. In fact, Chapman's (1957) conduction model for a static corona preceded the solar wind models. In the absence of (or in the direction of) a magnetic field the thermal conductivity is given by

$$F_c = - K_0 T^{5/2} \, dT/dr. \tag{IX-88}$$

Across magnetic field lines the conduction is greatly inhibited and is normally taken as zero. Thus, the thermal conduction terms in the energy equation are critically involved with the magnetic field geometry. The field geometry is not well known. When the field becomes sufficiently weak, however, the solar wind carries the field into a spiral pattern opening into interplanetary space, and there is substantial evidence (see Chapters IV and VII) that the field is predominantly open above about $2.5 \, R_\odot$

The complexities of radiation losses, unknown mechanical energy terms and unknown magnetic field geometry in the corona near the Sun render the solar wind problem as presently intractable in these regions. What is required to solve the solar wind equations is an appropriate set of boundary conditions. Although one can hopefully by-pass the uncertainties of the corona near the Sun by specifying the boundary conditions at a sufficiently large distance from the Sun to insure that radiation losses, mechanical heating and closed magnetic field lines are no longer important, such a procedure leaves much to be desired. At what distance from the Sun, for example, does the mechanical heating cease to be important? Similarly, where do closed field lines cease to restrict the flow of the wind and the flow of thermal conduction? Even if we could answer these questions with certainty, we would still face the difficulty that the coronal temperature structure is only crudely known. We know with some degree of confidence only the average temperature for the corona inside the region where closed magnetic field lines are perhaps dominant and where mechanical heating and radiation loss are still important. We do not know sufficiently well the coronal temperature near, say, $2.5 \, R_\odot$ to use it with certainty in predicting solar wind conditions. More often than not, in fact, one used observed properties of the solar wind near 1 AU in combination with solar wind models to infer the conditions in the outer corona.

Conduction models of the solar wind have both marked similarities to and marked differences from the polytropic models considered by Parker. In the absence of any expansion, Equations (IX-87) and (IX-88) reduce to

$$\frac{d}{dr}\left(r^2 T^{5/2} \frac{dT}{dr}\right) = 0,$$

(IX-89)

which is equivalent to

$$T^{5/2} \, dT = C \frac{dr}{r^2},$$

(IX-90)

where C is a constant. This latter equation integrates to give

$$T \propto r^{-2/7}.$$

(IX-91)

Such a dependence of T on r will tend to occur whenever thermal conduction is the dominant energy term in Equation (IX-86).

Near the sun where the expansion velocity is low the combined thermal conduction and gravitational terms are of more importance than the enthalpy and kinetic energy terms (Parker, 1965). Under these conditions Equation (IX-87) becomes

$$\frac{1}{r^2}\frac{d}{dr}(r^2 F_c) = -n_p m_H v \frac{g_0 R_\odot^2}{r^2}$$

$$= -\frac{c_1}{r^4},$$

(IX-92)

where c_1 is constant. This follows from the constancy of $n_p v r^2$. Equation (IX-92) reduces to

$$\frac{d}{dr}\left(r^2 F_c - \frac{c_1}{r}\right) = 0,$$

(IX-93)

or to

$$F_c = \frac{F}{r^2} + \frac{c_1}{r^3},$$

(IX-94)

where F is the total energy flux.

Substituting Equation (IX-88) for F_c and integrating, we find

$$\frac{2}{7} T^{7/2} = \frac{F}{r} + \frac{1}{2}\frac{c_1}{r^2} + c_2.$$

(IX-95)

For the boundary condition $T = 0$ at $r = \infty$, we then find $c_2 = 0$ and

$$T = T_0 \left(\frac{R_\odot}{r}\right)^{4/7} \left(\frac{r + r_2}{R_\odot + r_2}\right)^{2/7},$$

(IX-96)

where

$$r_2 = \frac{c_1}{2F} = \frac{n_p m_H v r^2 g_0 R_\odot^2}{2F}$$

$$= \frac{f_p m_H g_0 R_\odot^2}{2F} \tag{IX-97}$$

and where f_p is the constant quantity $n_p v r^2$. According to Equation (IX-96) T is proportional to $r^{-4/7}$ for $r \ll r_2$ and to $r^{-2/7}$ for $r \gg r_2$. This result is associated with the fact that, near the Sun ($r \ll r_2$), work done against the gravitational field by the moving particles exceeds the gain in energy by thermal conduction.

The quantity r_2 can be estimated from the observed solar wind flux at the orbit of Earth. At this distance most of the energy flux is in the kinetic energy of the protons and is given by $\tfrac{1}{2} n_p m_H v^3$. Using this value for F/r^2, we find from Equation (IX-97)

$$r_2 = \frac{g_0 R_\odot^2}{(v^2)_{1\,\text{AU}}}. \tag{IX-98}$$

The quiet Sun wind velocity at 1 AU is approximately 400 km s^{-1}, which gives $r_2 = 1.2\ R_\odot$. Since this value is so near $R_\odot$, it is questionable whether the neglect of mechanical heating and whether the assumption of open magnetic field lines is valid. Thus, we cannot be certain that a region of $r \ll r_2$ where $T \propto r^{-4/7}$ actually exists. If such a region did exist over an appreciable distance in the corona, a temperature drop of several hundred thousand degrees could occur. Such a large temperature drop is inconsistent with the results discussed in Chapter VII. Also, Parker (1965) noted in his original discussion of the results expressed by Equation (IX-96) that the predicted temperature decrease near the Sun leads to a wind velocity at 1 AU of 330 km s^{-1} when combined with the momentum equation. At the time of Parker's work, this value appeared to be too low and Parker concluded, on this basis, that the temperature of the corona was sustained by mechanical heating to an appreciable distance into space. Although Parker's specific argument is not justified using current solar wind data, his conclusion that the temperature falls off less rapidly than predicted by Equation (IX-96) is probably correct. Whether this is due to mechanical heating or thermal conduction is not presently known.

Thus far, we have discussed only the solution of the energy equation near the Sun. General solutions to the couple energy, momentum and mass conservation equations include a number of different classes of solutions. As in the isothermal and polytropic cases, we will consider as acceptable only the solutions passing through the critical point $v = v_c$ at $r = r_c$ and with $dv/dr > 0$. In the isothermal models this solution is unique, but in the thermal conduction model three classes of solutions meet these conditions. The three classes are distinguished by their asymptotic properties as $r \to \infty$ and by the density and temperature in the corona. Hundhausen (1972) discusses these three classes of solutions in terms of the ratio of enthalpy flux to heat conduction flux as $r \to \infty$. This ratio is given by

$$\epsilon = \frac{5 n_p v r^2 kT}{K_0 r^2 T^{5/2} \, dT/dr},$$ (IX-99)

$$= \text{const} \; \frac{T}{r^2 T^{5/2} \, dT/dr}$$

and the three classes of solutions correspond to: Class I, $\epsilon \to 0$; Class II, $\epsilon \to$ constant (finite); and Class III, $\epsilon \to \infty$ as $r \to \infty$. The first class of solution results when the coronal density is low, the second class results at a single intermediate coronal density and the third class results when the coronal density is high.

Parker's (1958) polytropic solution is of the Class I type. These solutions have the property that $T \to 0$ as $r \to \infty$ (enthalpy flux $\to 0$) while $r^2 T^{5/2} \, dT/dr \to$ constant finite value $(T \propto r^{-2/7})$ as $r \to \infty$. The Class II solution has been discussed by Whang and Chang (1965). It has the property that both T and $r^2 T^{5/2} \, dT/dr \to 0$ as $r^{-2/5}$ for $r \to \infty$. Thus, both the enthalpy flux and the heat conduction flux tend to zero at the same rate. Class III solutions have been discussed by Durney (1971) and by Roberts and Sowards (1972). These solutions have $T \propto r^{-4/3}$ and $r^2 T^{5/2} \, dT/dr \propto r^{-11/3}$ at large r. Thus, the enthalpy flux tends to zero less rapidly than the thermal conduction flux. Physically, these latter solutions correspond to a polytropic model with $\alpha = 5/3$, which is the adiabatic case.

The behavior of v at $r \to \infty$ in the three classes of solutions can be deduced from the relative behavior of the enthalpy and conductive fluxes. Most of the energy flux at large r is in the kinetic energy term. Thus, we have

$$F_{KE}^{\infty} = (n_p v r^2 \tfrac{1}{2} m_H v^2)_{\infty}$$

$$= F_p \tfrac{1}{2} v^2,$$ (IX-100)

where F_p is the constant mass flux. It follows that

$$v_{\infty} = \left(\frac{2 F_{KE}}{F_p} \right)^{1/2}.$$ (IX-101)

Since the total energy flux is conserved, $F = F_{KE}^{\infty} + F_H^{\infty} + F_C^{\infty}$. Then, since both F_H^{∞} and $F_C^{\infty} \to 0$ for solutions in Classes II and III, $F_{KE}^{\infty} \to F$. For Class I solutions, however, $F_C^{\infty} \to$ finite constant and $F_H^{\infty} \to 0$. Thus, $F_{KE}^{\infty} \to F - F_C^{\infty}$.

D. TWO-FLUID MODELS

Sturrock and Hartle (1966) have pointed out that the low particle densities near 1 AU in the solar wind $(n_p \approx 10 \text{ cm}^{-3})$ imply such long collisional interaction times between particles that the single fluid description of the solar wind plasma is inadequate. Instead, one expects that the electron fluid and proton fluid will be largely uncoupled. Since thermal conduction is important in the energy equilibrium and since the primary conduction is in the electron fluid, the protons are likely to have a much lower temperature near 1 AU than the electrons.

In the two-fluid treatment of the solar wind, Equation (IX-87) is replaced by two similar equations: one for protons and one for electrons. The two equations are coupled by an energy flow between the two fluids induced by collisional interactions. In the Sturrock-Hartle model the collisions are assumed to be of the Coulomb type.

The solution obtained by Sturrock and Hartle has $T_e \propto r^{-2/7}$ and $T_p \propto r^{-6/7}$ as $r \to \infty$. Thus, so far as conduction is concerned, their solution falls in Class I. Hundhausen (1972) has raised the interesting question of whether corresponding solutions of Class II and Class III exist in this case. Unfortunately, the only single-fluid model that has been carried out rigorously is the Whang and Chang Class II model, and it is not clear that the results from their model are directly comparable to the results of the Sturrock-Hartle model. In fact, however, the two models predict quite similar flow speeds at 1 AU. The predicted densities at 1AU are about a factor of two greater in the two-fluid model. The greatest difference in the two models at 1 AU is in the predicted electron and proton temperatures, as might be expected. The single-fluid model predicts $T_e \approx T_p \approx 1.6 \times 10^5$ at 1 AU whereas the two-fluid model predicts $T_e \approx 3.4 \times 10^5$ and $T_p \approx 4.4 \times 10^3$. In effect, the enthalpy flux is nearly the same in the two models but is carried entirely by the electrons in the two-fluid model.

Observations place the electron temperature at 1 AU near 1.5×10^5 and the proton temperature near 4×10^4. Since the two-fluid model based on Coulomb collisions gives a proton temperature that is too low and an electron temperature that is too high at 1 AU there is obviously some trouble with the model. A further difficulty with the model is that the predicted expansion velocity at 1 Au is too low (250 km s^{-1}) to agree with the observed quiet Sun velocity (302 to 325 km s^{-1}). Sturrock and Hartle interpreted these difficulties as being due to too rapid a temperature decrease near the sun, and they concluded, as Parker had earlier, that mechanical heating maintained nearly constant coronal temperatures out to approximately $r = 2 R_\odot$. This conclusion may be criticized from two standpoints. Firstly, the coronal temperature may be maintained by thermal conduction along closed magnetic field lines, and secondly, coupling between protons and electrons other than via Coulomb collisions may be important. In the latter case, the model adopted by Sturrock and Hartle is inadequate.

Among the refinements added to the relatively simple models discussed so far are the following: (1) proton-electron interactions through waves generated by plasma instabilities in the solar wind, (2) viscous effects due to the radial velocity gradient, (3) reduction in heat conduction flux by non-radial magnetic field lines, and (4) reduction in heat conduction flux by a failure of Equation (IX-88). This latter condition arises in the low density solar wind at large values of r where the use of Equation (IX-88) gives a larger conductive flux than could be supplied by transporting the entire internal energy of the gas with the mean thermal speed of electrons (Hundhausen, 1972) (note that this is considerably higher than the electron drift velocity in the solar wind). Since the electrons cannot transport energy more rapidly than allowed by this latter condition, the conductive flux has the limit

$$F_c < \tfrac{3}{2}\, n_e k T_e \left| \left(\frac{2kT_e}{m_e} \right)^{1/2} \right.$$

$$= \tfrac{3}{4} n_e m_e \left(\frac{2kT_e}{m_e} \right)^{3/2}. \tag{IX-102}$$

The ratio of this limiting conductive flux to the value given by Equation (IX-88) is

$$\beta = \frac{3 n_e m_e}{4 k_0} \left(\frac{2k}{m_e} \right)^{3/2} \frac{1}{T\, dT/dr}. \tag{IX-103}$$

At large values of r, v approaches a constant and, as a result, $n_e r^2$ approaches a constant also. It follows from Equation (IX-103) that β is constant for $T \propto r^{-1/2}$, and $d\beta/dr > 0$ when T decreases less rapidly than $r^{-1/2}$. The two-fluid model of Sturrock and Hartle gives $\beta = 1$ somewhat inside of 1 AU. A somewhat tighter restriction on the classical conductive flux is imposed by the condition that the characteristic scale length of the temperature gradient be long compared to the mean free path between collisions. This is necessary in order to insure that the velocity distribution function be in equilibrium with the local temperature. For further discussion of these points and those related to the effects of viscosity, non-radial magnetic fields and plasma instabilities the reader is referred to the excellent treatise on the solar wind by Hundhausen (1972).

E. EVAPORATIVE MODELS

In the preceding discussion, the solar wind plasma is treated either as a single fluid or as two interacting fluids. The fluid approach has been the subject of some criticism. Chamberlain (1961) proposed as an alternative to Parker's fluid model of the coronal expansion an evaporative model. In the evaporative model particles act as individual particles moving under the influence of gravity, Coulomb collisions and a polarization electric field induced by the preferential evaporation of electrons. The expansion properties of such a model depend rather critically upon the collisional interaction between particles and the polarization electric field. In Chamberlain's (1961) initial analysis an expansion speed of approximately 10 km s^{-1} was found at 1 AU. This prompted Chamberlain to suggest that since the evaporative model was more consistent with the subsonic family of solutions to the hydrodynamic equations, the subsonic solutions were indeed the proper ones. Subsequent modifications to the evaporative theory, however, give solutions that are nearer the supersonic hydrodynamic solutions passing through the critical point.

In his initial treatment of the evaporative problem, Chamberlain introduced a critical level above which no collisional interactions occurred. To estimate this level he assumed a single collision time for all particles. Brandt and Cassinelli (1966) modified this by introducing a collision time that depended on the Maxwellian velocity distribution function. Fast moving ions have longer collision times and escape preferentially from lower levels in the corona than do slow moving ions. This effect alone increased the expansion velocity at 1 AU to 140 km s^{-1}. A second modification proposed by Brandt and Cassinelli was to limit the escaping particles to regions of the corona where the magnetic field lines opened into interplanetary space. No particles were allowed to escape from closed field regions. Assuming that half the corona was 'closed' near $r = R_\odot$, they obtained a further increase in expansion velocity to 220 km s^{-1}.

Perhaps a more important modification of the evaporative model is that proposed by Jockers (1970). Earlier authors used the Pannekock-Rosseland polarization field which is based on a static solution and which does not properly ensure charge neutrality when expansion is present. Jockers modified the polarization field to include momentum and charge conservation. For the case of a single collision time with the collisionless limit at $2.5\,R_\odot$ and a coronal temperature of 1.32×10^6 K at $2.5\,R_\odot$, he obtained a coronal expansion velocity at 1 AU of 170 km s^{-1}. The additional effect of including different collision times for protons and electrons raised the expansion velocity to the value 253 km s^{-1} for a model in which $T \propto r^{-1/2}$ for $r > 9\,R_\odot$ and $T = 1.32 \times 10^6 r \leqslant 9\,R_\odot$. Extending the isothermal region to $25\,R_\odot$ further increased the expansion velocity at 1 AU to 322 km s^{-1}, and reducing the temperature gradient to $T \propto r^{-1/3}$ beyond $9\,R_\odot$ increased the expansion velocity at 1 AU to 288 km s^{-1}.

Further extensions of the evaporative model by Hollweg (1970) move the collisionless surface to a distance beyond the Parker critical point and otherwise include the essential revisions of Jockers and Brandt and Cassinelli. This gave an expansion velocity at 1 AU in excess of 300 km s^{-1}.

It seems clear that the hydrodynamic and particle approaches to the solar wind theory are converging toward a common solution that is consistent with the supersonic expansion of the corona predicted originally by Parker. If the physics is done correctly, the two alternative approaches should be completely equivalent and should give the same results. The two-fluid hydrodynamic model already incorporates some aspects of the particle approach.

F. THE OBSERVATIONAL SOLAR WIND

In situ observations of the solar wind were made possible by artificial satellites probing interplanetary space beyond the bounds of the Earth's magnetosphere ($\sim$ 10 Earth radii in the solar direction). Russian deepspace probes on missions to the Moon and Venus launched between 1959 and 1961 first detected positive ion fluxes in excess of 10^8 cm^{-2} s^{-1}. The U.S. Explorer 10 spacecraft succeeded in measuring the speed of the ion flow at approximately 280 km s^{-1} and a proton temperature of $3-8 \times 10^5$ K. Since these early detections of the solar wind, a number of increasingly sophisticated experiments have been carried out on U.S. satellites in the Pioneer, Mariner and Vela series and in the Russian Kosmos and Luna series.

The accumulated data for the solar wind reveal wide variations in the physical characteristics of the wind on a variety of time scales. Flow speeds vary from near 250 km s^{-1} to in excess of 600 km s^{-1} and proton densities vary from less than 0.5 cm^{-3} to in excess of 50 cm^{-3}. Among the recognized features of the wind are 27-day recurrent high speed streams associated with long lived solar features; shock waves associated with flares and hydromagnetic discontinuities associated with Alfvén waves propagating outwards in the solar-interplanetary magnetic field. These latter phenomena occur very frequently in contrast to the flare associated shocks, which are relatively rare.

The composite picture of the solar wind is not so much one of a steady background upon which one sees occasional superposed disturbances as it is one of more or less constantly varying properties. This makes the identification of the 'quiet-Sun' solar wind questionable in both concept and definition. Figure IX-10 displays a histogram of flow speeds obtained from approximately 28 months of observations obtained during the rising period of the solar cycle beginning in 1965. Although the histogram shows a clear preference for speeds near 350 km s^{-1}, the wind speed is between 300 and 400 km s^{-1} only about 50% of the time. On the other hand, the wind speed falls below 300 km s^{-1} less than 6% of the time. This suggests that there is, in fact, a quite well defined value below which the wind speed seldom falls. On this basis, Hundhausen (1972) defines the quiet solar wind as having speeds between 300 and 325 km s^{-1}.

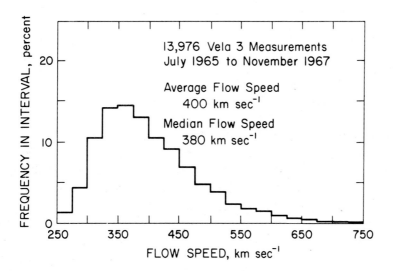

Fig. IX-10.   A histogram of observed flow speeds in the solar wind at 1 AU (Hundhausen, 1972; courtesy Springer-Verlag).

Unfortunately there is not a strong correlation between wind speed and proton flux. Hundhausen notes, for example, that the standard deviation of proton densities for the cases on Vela 3 for which the wind speed lies between 300 and 325 km s^{-1} show as much range as they do for the entire set of observations. Thus, even for the quiet solar wind defined in a restricted range of speeds only average properties can be given. Hundhausen gives for these average properties: speed 300 to 325 km s^{-1}, $n_p = 8.2$ cm^{-3}, $T_e = 1.5 \times 10^5$ and $T_p = 4 \times 10^4$. The associated energy fluxes in erg cm^{-2} s^{-1} at 1 AU are: kinetic energy = 0.22, enthalpy = 0.008, gravitational = 0.004, magnetic = 0.003 and thermal conduction = 0.007. The total at 1 AU is 0.242 erg cm^{-2} s^{-1}. Note that the average conditions indicated by Figure IX-10 give a speed of 400 km s^{-1}, which, for the same proton density, increases the kinetic energy flux by about a factor of two.

Presumably the only energy sink (or source) between Sun and Earth is the solar gravitational field. Thus, the energy drain from the corona due to the solar wind is given by

$$F = F_g + F_1 \left(\frac{AU}{R_\odot} \right)^2 \tag{IX-104}$$

where F_1 is the solar wind energy flux at 1 AU and F_g is the gravitational energy flux at at $R_\odot$. The value of $F_1 (AU/R_\odot)^2$ is 1.2×10^4 erg cm^{-1} s^{-1}.

To evaluate F_g, we note that

$$F_g = \frac{n_p m_H v g_0 R_\odot^2}{r}, \tag{IX-105}$$

and that

$$F_p = n_p m_H v r^2 = \text{const.} \tag{IX-106}$$

Hence,

$$F_g = F_p g_0 \frac{R_\odot^2}{r^3} = \frac{F_p g_0}{R_\odot}, \qquad (r = R_\odot). \tag{IX-107}$$

F_p can be evaluated at 1 AU from the observed proton flux $n_p v = 2.4 \times 10^8$ cm^{-2} s^{-1}. This corresponds to $F_p = 9 \times 10^{10}$ gs^{-1} sr^{-1}. Whence, we find that $F_g = 3.5 \times 10^4$ erg cm^{-2} s^{-1}, and a total flux from Equation (IX-99) of 4.7×10^4 erg cm^{-2} s^{-1}. Between Sun and Earth, therefore, about three-fourths of the wind energy is expended against the gravitational potential.

The total energy loss rate in the solar wind is less than the coronal radiation loss by a factor of 3 to 6, and is less than the energy loss by thermal conduction to the chromosphere by a factor of about 6. However, the thermal conduction loss to the chromosphere is based on the assumption of spherical symmetry and radial magnetic fields and is almost certainly smaller than the adopted value. Also, since the radiation loss is proportional to n_e^2 most of this loss occurs in the low corona. By contrast much of the solar wind energy flux is generated in the outer corona. At $r = 2 R_\odot$, for example, the electron density has decreased by about two orders of magnitude and the radiation loss rate by about four orders of magnitude. The solar wind flux, however, is $F_g/8 + F_1 (AU/2 R_\odot)^2 = 7.4 \times 10^3$ erg cm^{-2} s^{-1}, which is only a factor of 6.4 below the value at $r = R_\odot$. For the outer corona, therefore, solar wind losses are dominant.

The total mass flux in the solar wind is $4\pi F_p = 1.1 \times 10^{11}$ g s^{-1} or about 1.7×10^{-15} solar masses per year. The total mass of the corona is estimated at approximately 3×10^{17} g. Thus, the time for emptying the corona by the solar wind is approximately 33 days. Mass cycling in the corona occurs via spicules and prominences as well as via the solar wind. In Section 11 of this chapter, the characteristic cycling time for one coronal mass is estimated at approximately 1 h for spicules and 10 h for prominences. It is evident, therefore, that the solar wind flow is only part of a more complicated pattern of mass flow in the corona.

10. Spicule Mechanisms

A. SUMMARY OF SPICULE CHARACTERISTICS

Spicules are observed as jet-like features penetrating the chromosphere-corona transition region and low corona. Thus, they are moving upward through a region of strong positive temperature gradient that would normally be considered as being convectively stable. The mean spicule velocity of 25 km s^{-1} corresponds to the sound velocity at a temperature of $22\,600°$, which is close to the currently accepted temperature for the upper chromosphere Lyman-α plateau. The spicule densities, also, are characteristic of this same region.

Most spicules elongate to a height of 10 000 km or higher with little reported evidence for a marked change in velocity. In assessing the meaning of this apparent constancy of velocity, however, it is necessary to consider how the velocities are determined. The value of 25 km s^{-1} for the mean velocity is derived from the rate of elongation of the spicules as seen at the limb. Typically, one plots the observed height of the spicule above some reference point as a function of time, then draws one or more straight lines, through the data points. This is illustrated by the dashed line in Figure IX-11 for a spicule that reaches a maximum height of 12 500 km and that spends approximately one third of its lifetime

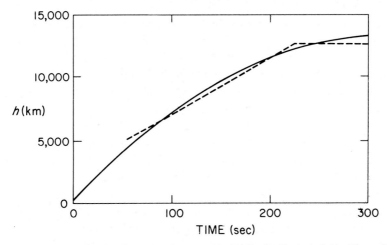

Fig. IX-11.   A schematic height-time plot for a typical spicule (dashed line). The solid curve represents a ballistic trajectory approximating the spicule plot (after Thomas and Athay, 1961).

near maximum height. The spicule is first identified as a separate feature near a height of 5000 km. A typical scatter of the data points about such a plot is ± 300 to 500 km. The parabolic curve drawn in Figure IX-11 is the locus of height vs time for a projectile given an initial upward velocity of 75 km s^{-1} at the top of the chromosphere ($h \approx 2000$ km) and thereafter decelerating in the gravitational field. Within the accuracy of the observational data, either interpretation is equally acceptable.

A second method for obtaining spicule velocities is through the Doppler shift of spectral lines emitted by spicules at the limb. Such Doppler shifts arise from two sources; bodily motion of the spicule in the line of sight and streaming motion along the spicule axis. Lateral bodily motions of spicules are observed, but are believed by most observers to be of the order of 5 km s^{-1} or smaller (Beckers, 1972). If we assume that spicules at the limb with streaming velocity v are, on the average, lying in a cone whose axis is perpendicular to the limb and whose apex angle is 2θ and that the spicules are distributed randomly in the cone, then the average absolute Doppler velocity in the line of sight is given by

$$\langle v_D \rangle = \frac{2}{\pi} V \sin \theta \int_0^{\pi/2} \sin \phi \, d\phi$$

$$= \frac{2}{\pi} v \sin \theta, \tag{IX-108}$$

where ϕ is the apparent angle of the spicule measured in the plane of the sky from the axis of the cone. For $V = 25$ km s^{-1}, we see that $\langle v_D \rangle = 16 \sin \theta$. Not very much is known about average orientations of spicules, but the average value of approximately 20° from the vertical found by Beckers (1964) seems reasonable. This would give an average Doppler velocity of about 5 km s^{-1}. The maximum Doppler velocity would be $25 \sin \theta \approx 9$ km s^{-1}.

Figure IX-12 shows a plot of the number of spicules with a given observed Doppler shift. While the average is close to 5 km s^{-1}, an appreciable number of spicules have Doppler velocities in excess of 10 km s^{-1} and a few have velocities in excess of 20 km s^{-1}. The spectroscopic data from which the Doppler shifts of lines are measured are inherently restricted in spatial resolution. The finite slit width in the spectrograph and atmospheric seeing both contribute to averaging over a considerable range of heights. Individual spicules are not clearly resolved in such observations at heights below about 4000 km. Thus, the observed Doppler shifts refer systematically to the spicules as seen in projection above 4000 km and contain a significant averaging over height. It could not be argued, therefore, that the observed Doppler shifts are inconsistent with the parabolic plot in Figure IX-11.

Our inability, at this time, to differentiate between a constant velocity model and a gravitational deceleration model on the basis of limb observations alone is unfortunate. The two alternatives lead to very different conclusions about the nature of the spicule acceleration mechanism. In the constant velocity model, the accelerating force acts

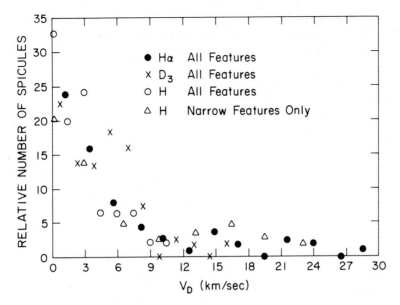

Fig. IX-12.   The observed frequency distribution of spicule Doppler shifts (Athay and Bessey, 1964, courtesy *Astrophys. J.*).

continually and steadily through the life of the spicule, whereas in the gravitational deceleration model the accelerating force ceases once the spicule motion is established. Evidence presented in Section 10B will tend to discount the ballistic model, but, unfortunately, the results are not absolutely certain.

Spicule densities are some six orders of magnitude lower than photospheric densities and are relatively constant above a height of 5000 km. Some authors have been tempted by the similarity of diameter and lifetime between spicules and granules to speculate that the spicules and granules are one and the same phenomenon. The vast differences in densities, however, rule this out as a plausible association. We hasten to add, however, that a direct association between spicules and granules via some hydrodynamic or hydromagnetic coupling mechanism is not ruled out.

As noted in Chapters II, III and IV spicules occur preferentially in the network regions. The accepted magnetic model of the network-cell complex is one in which the radial outflow in the supergranule cell sweeps the cell virtually free of magnetic field and compresses the field into a collection of vertical magnetic filaments in the network. A positive correlation has been established between the brightness in the core of Hα and magnetic field strength. Similarly, a correlation has been claimed between the bright coarse mottles of the network and the spicules bushes. Thus, there is an implied correlation between spicules and network regions of strong vertical magnetic field.

In summary, then, we note the following basic associations and properties of spicules:

(1) Spicules are associated with regions of the network in which the vertical magnetic field is strong and predominantly of one sign.

(2) Spicules grow primarily by elongation along a nearly vertical axis with little evidence of lateral expansion. The average elongation rate is near the speed of sound in the upper chromosphere temperature plateau. Lateral body motions of a few km s^{-1} may be present.

(3) The average matter density in spicules is near the chromospheric matter density in the upper temperature plateau. Within the spicule, density decreases outwards with an approximate scale height of 4000 km, which is much greater than the isothermal hydrostatic scale height for the spicule temperature.

(4) Spicule temperatures are not well known, but the temperature appears to be near 16 000 K. Substantial temperature gradients may exist both along the spicule axis and perpendicular to it. The most commonly accepted directions of temperature increase are outwards from the spicule axis and upwards along the axis.

B. SUMMARY OF THEORETICAL MODELS

The proposed mechanisms for giving rise to spicules fall into three broad categories: (1) Shocked matter flowing behind vertically propagating shock fronts. (2) Jets driven upwards along magnetic field lines by hydrostatic and hydrodynamic pressure forces. (3) Matter accelerated by magnetic forces.

Within each category of mechanisms several variations exist. Thus, the shock wave models include both hydrodynamic and Alfvén shocks; the pressure forces driving the jet models include gas pressure, radiation pressure and a combination of gas pressure and sound wave pressure; the magnetic models include the lifting of matter frozen to moving magnetic field lines and the expulsion of diamagnetic clouds by induction forces.

Pikel'ner (1971a) has argued against the class of jet models driven by gas pressure on the basis that the empirical models of spicules yield a low pressure gradient. In fact, the empirical models indicate that the downward gravitational force exceeds the pressure gradient force, and hence that the net force from gravity and pressure gradient is downward. This result clearly implies that within the spicule gas pressure gradient forces are not dominant and other mechanical forces are present. It appears that the only escape from this conclusion lies in the ballistic interpretation of spicule motion, for in this picture, pressure gradient forces need be present only at the base of the spicule. The observed spicule itself then consists of matter carried aloft by its initial momentum, as in a water fountain. Such a jet will rise even though the pressure gradient vanishes.

Although the ballistic picture of spicule motion cannot be ruled out purely on the basis of limb observations, disk observations provide rather strong evidence against such an interpretation. A spicule reaching an altitude of 12 000 km in a ballistic trajectory requires a velocity of 73.5 km s^{-1} at a height of 2000 km. Within the height interval between 2000 km and 3000 km the material within the spicule will decelerate from 73.5 km s^{-1} to 50.3 km s^{-1}. Thus, for an interval of some 1000 km the spicule material will have an average velocity of about 62 km s^{-1}. The corresponding Doppler shift at Hα is 1.36 Å. If most spicules indeed did have initial velocities of the order of 75 km s^{-1}, then the combined effects of Doppler shift together with the intrinsic width of the Hα

line should show clear evidence of spicule structure in the blue wing of the Hα line to values of Δλ in excess of 1.8 Å. There is little or no evidence that this is the case. Most of the evidence for spicule structure is lost beyond Δλ ≈ 1.2 Å, which is inconsistent with spicule velocities in excess of about 25 km s^{-1}. In view of these difficulties, it is very difficult to accept a ballistic model for spicule motions, and the objections raised by Pikel'ner against the jet models driven by gas pressure must be taken seriously.

A more subtle argument against ballistic spicule motions can be advanced by considering the motions of surge prominences. It is not uncommon to observe a surge prominence with a velocity near 200 km s^{-1} over a height range of some 100 000 km. A height vs time plot for such a prominence leaves little possibility for a ballistic interpretation. The height-time plot, in this case, is too well determined and the slope is too constant with height to permit the possibility of fitting the data with a ballistic parabola. The surges establish, therefore, the existence in the Sun of material ejections of a non-ballistic nature. There is, in fact, a definite avoidance of ballistic type motions among the majority of prominence motions, and one gains the impression that the non-ballistic motions are the rule rather than the exception. Surges and spicules are not greatly different in appearance except for their sizes. Both occur in association with magnetic fields and both exhibit predominantly radial motion. The material densities are of the same order, also. It is not altogether unrealistic, therefore, to draw a parallel between spicules and surges, and, hence, to conclude that spicule motions are most likely of a non-ballistic nature, as in the case of surges.

The preceding objections to the jets driven by gas pressure do not apply specifically to jets driven by radiation pressure. However, a similar type of objection arises because the radiation pressure seems again to be insufficient to provide the spicule motion. This point will be discussed in more detail in Section 10E.

A very recent spicule model proposed by Unno et al. (1974) postulates specific mechanical forces to help support the spicule matter. This model will be discussed in Section 10E.

The shock wave theory for spicules appears to encounter difficulties in the energy equilibrium. Those authors, including the present one, who have advocated either the pressure driven jet or shock wave theory for spicules generally have failed to realize that the vast majority of spicule energy is expended as work against the gravitational field and that the energy requirements for a spicule are very large. A periodic shock train of period, P, mach number M, and mean matter density, ρ, carries an energy flux (Kuperus, 1965) of

$$F_{\text{Sh}} = \frac{4\,\tau}{3\,p}\,\frac{\rho v_{\text{S}}^3(M^2 - 1)^2}{(\gamma + 1)^2 M},$$

(IX-109)

where τ is the width of the shocked region in each period, v_s is the velocity of sound and γ is the ratio of specific heats. For $\gamma = 5/3$, $\tau/P = \frac{1}{2}$ and $M \gtrsim 2$, Equation (IX-109) reduces to

$$F_{Sh} \approx 9.4 \times 10^{-2} \rho v_{Sh}^3 \qquad\qquad (IX-110)$$

where v_{Sh} is the shock velocity. In the shock wave models, the shock first passes through the chromosphere near the sound speed then accelerates to a much higher speed in the corona. The spicule motion is accelerated as the shocked matter in the chromosphere and follows behind the very fast moving coronal shock as a contact discontinuity. To estimate the energy flux of the shock wave that is responsible for carrying the spicule matter, we may set $\rho_{Sh} = \rho_{Sp}$ and $v_{Sh} = v_{Sp}$. We then note that the energy flux given by Equation (IX-110) is a factor of five below the kinetic energy flux in the spicule. The rate of doing work against gravity by the spicule is given by

$$F_g = \rho g v_{Sp} R_0, \qquad\qquad (IX-111)$$

and the ratio of

$$\frac{F_{Sh}}{F_g} = \frac{9.4 \times 10^{-2} v_{Sh}^3}{g v_{Sp} R_0} = 3 \times 10^{-4} \qquad\qquad (IX-112)$$

for $v_{Sh} = v_{Sp} = 25$ km s^{-1}. Note that this ratio depends only on the spicule velocity and the assumed ratio τ/P. The very small value found for F_{Sh}/F_g poses a fundamental difficulty for the hydrodynamic shock wave theory for spicules. It appears to be impossible for such shocks to provide the energy required to lift the spicule matter.

The above objection to the hydrodynamic shock wave theory of spicules applies equally as well to the magnetic shock theories. For a magnetic field of about 5 G or greater, which seems to be conservative for the regions in which spicules occur, the Alfvén velocity at the matter densities found in spicules exceeds spicule velocities. Thus, one identifies the spicule flow with the chromospheric slow mode MHD waves (Osterbrock, 1961). In the limit $v_{Sm} < v_A$, where v_{Sm} and v_A are the slow mode and Alfvén mode velocities, respectively, the slow mode wave has just the characteristics of the hydrodynamic wave. Thus, the slow-mode shocks, in this limit, are the same as the hydrodynamic shocks and suffer the same energy limitations.

It appears from the preceding arguments that neither the pressure driven jets nor the shock wave theories are quantitatively able to account for spicule motions. This leaves the theories for magnetically driven spicules as the only remaining candidates. Unfortunately, the only magnetic theories proposed thus far require bipolar magnetic fields. It is possible that bipolar field conditions occur at the sites of some network regions where spicules occur, but it seems totally at variance with the existing magnetic field observations to assume that bipolar fields exist everywhere where spicules are found. We are left, then, in the unfortunate situation where no spicule theory is free from strong objections of either theoretical or observational basis.

In view of the fact that there are strong objections to all current classes of spicule theories it does not seem appropriate to review any of the theories in detail. The following review, therefore, is intentionally somewhat superficial.

C. SHOCK WAVE MODELS

Thomas (1948) was the first to suggest that the spicule outflow is the wake of shocked material rising behind a vertically propagating shock front. At the time of this suggestion the network-supergranulation structure of the Sun had not been recognized, nor was anything known about the distribution of magnetic fields. Spicules were believed to occur randomly over the solar surface. Their temperature and density structure were largely a matter of conjecture, and there was little basis, therefore, from which quantitative models could be constructed. Although the idea of a spicule as a flow of shocked matter was relatively popular, little was done to test the hypothesis quantitatively for a number of years.

In 1961, Uchida readvocated the idea of the spicule as a flow of matter behind a hydrodynamic shock. At the same time, Osterbrock (1961) specifically identified the spicule flow with matter flowing behind a slow-mode MHD shock, which, for the case considered, was essentially the hydrodynamic shock arising from a vertically propagating wave.

Parker (1964) made the mechanism for generating the sound and shock waves associated with spicules more specific by suggesting that the photospheric motions of granules and supergranules would jostle the magnetic field lines and, as a result, produce magneto-acoustic waves propagating along the field lines with velocity amplitudes in the photosphere of the order of 1 km s^{-1}. In the regions of vertical field, these magneto-acoustic waves would propagate vertically.

It has long been recognized that for a sound wave propagating without dissipation the quantity ρv_w^2 is conserved, where $v\omega$ is the wave amplitude. For waves propagating vertically, ρ is decreasing exponentially and $v\omega$ will therefore increase. A wave of velocity amplitude 1 km s^{-1} at a photospheric depth where $\rho = 10^{-7}$ will grow to a velocity amplitude of 10 km s^{-1} at a height where $\rho = 10^{-9}$. This latter density is characteristic of the low chromosphere where the sound velocity is near 7 km s^{-1}. Thus, the wave is now of shock amplitude. This same basic idea is fundamental to each of the shock wave models for spicules. The uniqueness of Parker's (1964) suggestion lies in the direct generation of the magento-acoustic waves as monopole radiation generated by the motions of granules and supergranules. Previous theories had either left the mode of wave generation unspecified or had relied upon the quadrapole generation of sound waves in the solar convection zone as suggested by Lighthill (1967).

Quantitative discussions of the Parker mechanism have been given by Lust and Scholer (1966) and by Wentzel and Solinger (1967). The latter authors note that there is some ambiguity as to whether the shock waves that supposedly produce the spicules are to be treated as isothermal shocks or adiabatic shocks. They further note that the guiding influence of the field becomes unstable when $v_{Sh} > v_A > v_S$. This places limits on the magnetic field strengths in spicules of > 200 G for isothermal shocks and > 60 G for adiabatic shocks. The larger field strength required for the isothermal shock arises from the greater compression behind the isothermal shock as compared to the adiabatic shock.

The shock wave theory of spicule generation has been rather popular. None of the quantitative discussion of the theory, however, has specifically included the energy expended by the shock front in lifting the shocked matter in the gravitational potential. The large energy flux associated with the lifting of spicule matter, as noted in Section 10B, appears to provide an insurmountable obstacle to the shock wave theories.

D. JETS DRIVEN BY GAS PRESSURE

The concept of a spicule as a jet driven purely by gas pressure has taken a variety of forms, none of which has been treated quantitatively. Kuperus and Athay (1967) suggested that thermal conduction from the corona deposits heat energy in the upper chromosphere, which, in turn, leads to increased gas pressure and the eventual release of the excess pressure as spicule eruptions. They argued that the energy release could not be via radiation since the corona would not be expected to form unless the upper chromosphere were radiating near maximum efficiency. Kopp and Kuperus (1968) expanded the conduction model to include the expected magnetic field geometry associated with the network structure (see Figure IX-13).

As noted in Section 3 of this chapter, Defouw's analysis of the thermal stability associated with the upper chromosphere indicates that the onset of instability near the base of the chromosphere-corona transition region includes both a thermal instability and a hydrodynamic instability. This leads Defouw (1970c) to identify the spicule flow with the hydrodynamic instability.

Hollweg (1972) has suggested that the converging horizontal flows in the network between adjacent supergranules will force both a downflow and an upflow in the network. He identifies the upflow with spicule motion.

In none of the three cases suggested in the preceding has there been sufficient quantitative treatment of the energy and momentum balance to attach any meaningful plausibility to the suggested mechanisms. All three are of interest, but they each suffer from the general criticism raised by Pikel'ner and discussed in Section 10B.

Although none of the three mechanisms suggested here may be the total cause of spicules, all three mechanisms are undoubtedly present to some degree and may well play a role in initiating or aiding the spicule eruption.

E. JETS DRIVEN BY A COMBINATION OF GAS PRESSURE AND HYDRODYNAMIC FORCES

As we have noted, the difficulty with the jet driven purely by gas pressure is that the pressure gradient deduced from the empirical spicule models is very low. The deduced gradient, in fact, corresponds to a 'supporting temperature' of near 6.5×10^4 K if the spicule is assumed to be isothermal and in hydrostatic equilibrium. The actual spicule temperature appears to be near a quarter of this value.

In a very recent paper, Unno et al. (1974) propose a spicule model in which both upward deceleration of spicule matter and pressure excited by upward moving sound waves add to the support of spicule matter. Their suggestion follows an earlier suggestion by Meyer and Schmidt (1968) in which the latter authors pointed out that a magnetic arch connected at either end to the chromosphere would, in all probability, experience different gas pressures at a common height at the two ends of the arch. This difference in pressure was shown by Meyer and Schmidt (1968) to lead to a flow of matter along the arch, rising in one leg of the arch and falling in the other. Pikel'ner (1971b) considered the problem in somewhat more detail and applied the concept specifically to the syphoning of matter from the chromosphere into prominences craddled in magnetic field valleys located near the tops of magnetic arches. He concluded that the flow would be sufficient for supplying quiescent prominence material but would be insufficient for producing spicule flow.

Unno *et al.* (1974) modify these concepts by introducing the added effects of deceleration within the spicule and upwardly propagating sound waves. The full mathematical treatment by Unno *et al.* is somewhat tedious and complex, but the following abstract contains most of the essential ideas.

The model treated consists of a magnetic sheet of thickness σ. Matter is confined to flow along the magnetic sheet with the result that the mass flux

$$\Phi = \rho v \sigma \tag{IX-113}$$

is conserved. If the magnetic field is strong enough that the magnetic pressure exceeds the gas pressure, σ will increase rapidly with height as the field diverges. In this configuration the product $\rho \sigma$ decreases only slowly with height and may even increase. Thus, conservation of mass flux does not require a large increase of v with height. If, on the other hand, the magnetic pressure is weak relative to the gas pressure σ is more nearly constant with height and $\rho \sigma$ will decrease rapidly outwards. Conservation of mass flux then requires that v increase rapidly outwards. Thus, a small initial velocity at the base of the spicule disturbance, which Unno *et al.* assume to be the base of the chromosphere, grows to a large velocity at spicule heights. This is the primary mechanism in the model for producing the observed spicule velocity. The acceleration is assumed to occur primarily at heights within the chromosphere before the spicule is observed as a jet above the normal chromosphere. An initial velocity at the base of the chromosphere is assumed.

Even in the weak field case, the magnetic pressure will eventually dominate over the gas pressure and the field lines will diverge. When this occurs $\rho \sigma$ increases with height and the spicule decelerates. Along the spicule axis, which we assume to be vertical, the equation of motion is

$$\frac{1}{\rho} \frac{\partial p}{\partial h} + g + \frac{\partial}{\partial h} \left(\frac{v^2}{2} \right) = 0, \tag{IX-114}$$

which integrates in the isothermal case to give

$$P = P_0 e^{-[(g\rho h/p) + (\rho v^2/2)]}$$

Since $P/\rho = v_S^2$, we write the above expression in the form

$$P = P_0 e^{-[(gh/v_S^2) + (M^2/2)]}, \tag{IX-115}$$

where M is the spicule Mach number relative to the sound speed within the spicule. The scale height for the gas pressure is $(d \ln P/dh)^{-1}$ and is given by

$$H^{-1} = \frac{g}{v_S^2} + M\frac{dM}{dh},$$

or

$$H = \frac{v_S^2}{g}\left(\frac{1}{1 + \dfrac{v}{g}\dfrac{dv}{dh}}\right) \qquad\qquad (IX\text{-}116)$$

Since v_S^2/g is just the isothermal hydrostatic scale height, the dynamic scale height is increased when dv/dh is negative and approaches a value, say, $\tfrac{1}{4}g/v$. For a spicule velocity of 25 km s^{-1}, the required value of dv/dh is 2.7 km s^{-1} per 1000 km. This value if within the limits allowed by spicule observations. Note, however, that if we assume that

$$\frac{v}{g}\frac{dv}{dh} = \frac{1}{4}$$

at all heights then v decreases from its initial value to zero in a height interval given by

$$\frac{v^2}{2g} = \frac{1}{4}(h' - h),$$

and, for $v = 25$ km s^{-1}, $h' - h = 4600$ km. This is a marginal height for spicules. However, we should note that in this condensed summary of the model proposed by Unno *et al.* we have neglected force terms arising from the magnetic pressure gradient and the centrifugal force of the matter flowing along curved field lines. Thus, the preceding estimate of the spicule height should be taken only as being indicative of the types of effects that are present in the model.

As a further means of adding to the spicule support, Unno *et al.* (1974) compute v_S from an 'effective temperature'. Since H is proportional to v_S^2, which is proportional to T, a doubling of the temperature doubles H. For the spicules extension above 2000 km, the 'effective temperature' adopted by Unno *et al.* is 25 000°. To account for the high 'effective temperature' they postulate the existence of sound waves propagating along the field lines. The sound waves are supposed to be produced via the Parker mechanism.

In order for the sound waves to be effective in supporting the spicule they must be approaching shock amplitude but not have reached full shock strength, otherwise they would have dissipated much of their energy. Thus, Unno *et al.* adopt $0.3 < M_0 < 1$ for the waves at the base of the region where support is required. The energy flux carried in the waves at the base of the support region is

$$F_{w0} = \rho_0 v_{\omega 0}^2 (v_{S0} + v_{Sp0}),$$

which reduces to

$$F_{w0} \approx \rho_0 v_{S0}^2 v_{Sp0} = P_0 v_{Sp0}$$

since $v_{\omega 0} \approx v_{Sp}$ and $\rho_0 v_{S0}^2 = P_0$. Unno et $al.$ set $P_0 \approx 5 \times 10^3$ dyn cm^{-2} and $v_{Sp0} \approx v_{S0} = 6 \times 10^5$ cm s^{-1}. The required energy flux in the waves is then $F_{w0} \approx 3 \times 10^9$ erg cm^{-2} s^{-1}. The value of P_0 adopted corresponds to the temperature minimum region, which is near the depth where Parker's mechanism is effective.

The flux of energy estimated for the acoustic waves by Unno et $al.$ is high. The reason it is high stems from the adoption of the base of the supporting region at the temperature minimum. At heights when spicules are observed (above 2000 km) and are known to need supporting forces other than gas pressure, the value of P is near 0.2 dyn cm^{-2} and v_{Sp} is near 25 km s^{-1}. Thus, the required energy in the wave flux for effective support of the spicule is reduced to approximately 5×10^5 erg cm^{-2} s^{-1}, which, of course, is of the same order as the kinetic energy flux in the spicule.

For the weak field condition, Unno et $al.$ give $B_0 < 250$ G where B_0 is the field at upper photospheric levels. Since the spicules are associated with strong field regions where B_0 seems to exceed this value, Unno et $al.$ suggest that the tendency for most spicules to deviate from the vertical arises from the fact that the spicules occur preferentially near the borders of the strong magnetic knots where the fields are weaker and diverging from the radial direction.

The model proposed by Unno et $al.$ is too new, at the time of this writing, to adequately evaluate all of its ramifications and merits. Of the models proposed thus far, however, it is clearly one of the more promising.

F. JETS DRIVEN BY RADIATION PRESSURE

Radiation flowing outwards through an atmosphere exerts a pressure force whose direction parallels the direction of the net flux' of energy in the radiation. More specifically the pressure gradient due to radiation in a particular spectral line is given by

$$\frac{dp}{dh} = \frac{h\nu}{c} B_{ij} n_i \int H_\nu \Phi_\nu \, d\nu, \tag{IX-117}$$

where B_{ij} is the Einstein absorption coefficient and where stimulated emissions have been ignored.

In an atmosphere of a star, H_ν is directed outwards in the outermost layers and the radiation pressure force is oppositely directed from the gravitational force. Since the integral in Equation (IX-117) contains Φ_ν, which is sharply peaked near line center, the most effective flux is that near line center. The flux at line center tends to reach an upper limit near $\tau_0 = 1$. However, it is n_1 and not τ_0 that determines the pressure gradient (i.e. the outward force) and the most favorable condition for a large pressure gradient is to have $\tau_0 \approx 1$ for a relative large value on n_1. This, in turn, requires that the effective geometrical scale length for the optical thickness be small and this happens when there is a rapid outwards increase in temperature near the depth $\tau_0 = 1$. This condition occurs near the top of both of the chromospheric temperature plateaus. The abrupt temperature rises at the top borders of the two temperature plateaus occur near optical depths unity in the Lyman continuum (lower plateau) and in Lyman-α (upper plateau).

Without doing the specific computation needed to evaluate Equation (IX-117), it is still possible to form an estimate of the possible importance of the radiation force. Energy conservation requires that the mechanical energy flux produced by the action of radiation pressure cannot exceed the energy flux in the radiation itself. At velocities less than the escape velocity most of the energy flux is in the potential energy term. Hence, we may write

$$\int H_v^{max}\, dv > \rho g v R_\odot \tag{IX-118}$$

In the case of the Lyman-α line the emergent line flux for the average Sun is 3×10^5 erg cm^{-2} s^{-1}. Within the average network region the flux is observed to be some ten times higher than the average Sun value, and, locally, within the areas producing spicules it could possibly be some ten times higher than the average network value. If we choose a maximum flux of 5×10^7 erg cm^{-2} s^{-1}, as a possible value, the value of v predicted by Equation (IX-118) for a spicule density of 10^{-13} is < 2.6 km s^{-1}. Since we have been quite generous in estimating the maximum flux, it seems unlikely that the radiation pressure due to Lyman-α is capable, by itself, of producing a spicule.

A detailed evaluation of the pressure gradient due to Lyman-α radiation in the HSRA shows a net upward force of magnitude 0.06 ρg for the average Sun (Athay, 1974). Assuming that the increased Lyman-α flux in the network results in a similar increase in radiation pressure, we find an upward force in the network of 0.6 ρg. The radiation pressure due to Lyman-α appears, therefore, to be a substantial contributor to the force balance within the spicule even though the radiation flux does not appear to be sufficient to provide the spicule velocity. The actual radiation pressure influence within the spicule should be computed, of course, using a realistic spicule temperature model and including the spicule velocity. Since the pressure gradient given by Equation (IX-117) is quite sensitive to the temperature and density model of the spicule, an accurate evaluation of the pressure gradient requires a much better temperature and density model than is now available.

Without quantitative evaluation of the radiation pressure for a reliable spicule model it is difficult to assess the real importance of this mechanism as a supporting force and as a factor in the spicule motion. On the other hand, it seems evident that this mechanism very probably exerts a substantial influence in determining the spicule structure.

A bright feature in the solar atmosphere, such as the network and bright mottles, experiences a lateral radiation flux as well as a radial one. It seems quite probable, therefore, that much of the fibril structure of the network and rosettes may be due to Lyman-α radiation pressure. Since the fibrils are largely horizontal in character, the energy flux required to produce a given velocity is much lower than it is in the case of spicules. The shock wave and radiation pressure models, in this case, may be quite capable of producing the observed motions. A horizontal velocity of 5 km s^{-1} and density of

10^{13} g cm^{-3}, for example, requires an energy flux of only about 10^4 erg cm^{-2} s^{-1}, which is small compared to the Lyman-α energy flux and comparable to the energy flux in a shock front of mach number near unity.

Estimates of the value of the radiation pressure for a number of other spectral features, including the Lyman continuum, Balmer-α and the Mg II and Ca II resonance lines, indicate that the radiation pressure due to these sources is much weaker than the Lyman-α radiation pressure. This is a result, of course, of the particular characteristics of the chromosphere temperature structure.

G. MAGNETICALLY DRIVEN SPICULES

The energy and momentum requirements for spicules coupled with the known association between spicules and network regions of high magnetic field strength make it tempting to seek a magnetic acceleration mechanism as a means of producing the spicule motion. The two mechanisms suggested thus far are both tied to the Petschek mechanism (Petschek, 1964). Outflowing material in neighboring supergranule cells carries magnetic lines of force from each cell into a common network region. This leads to a field configuration at the network as shown schematically in Figure IX-13a. When the field concentration at the network begins two anti-parallel sets of lines of force are brought together. At their point of contact field annihilation occurs with reconnection of the field lines as shown in Figure IX-13b. The reconnected field lines move upward and downward, as suggested

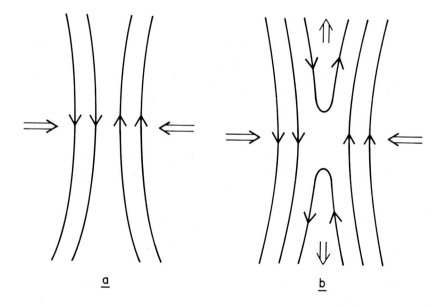

Fig. IX-13.   Illustration of spicule formation by field line reconnection (Petschek mechanism). Oppositely directed field lines brought together by fluid flow (a) reconnect and move away from the contact point (b).

by Petschek (1964), allowing the outlying field lines to continue to approach the neutral point.

In a model suggested by Uchida (1969) the magnetic annihilation occurs at a series of points along the axis of the converging fields. He then considers that bubbles of plasma with closed magnetic field lines form in the regions that separate adjacent neutral points where the field reconnection takes place. This is illustrated schematically in Figure IX-14. The enclosed bubbles of magnetized plasma are repelled by the surrounding fields as in the so-called 'melon seed' effect suggested by Schlutter (1957.).

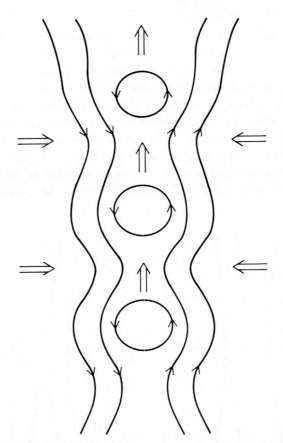

Fig. IX-14.   Diamagnetic bubbles moving outwards from regions of field line reconnection as suggested by Uchida (1969). .

Uchida identifies the spicule with a series of plasma bubbles pushed upward along the neutral line separating the two converging fields. He treats the problem of the motion of the plasma bubbles in an approximate way and shows that they may be expected to accelerate rather quickly to speeds of Mach 2 to 4 (relative to the sound speed within the spicule) then to decelerate to Mach 1. The acceleration to Mach 2 occurs within the first

2000 km above the level where the bubble forms and deceleration back to Mach 1 occurs some 7000 km higher up. Thus, the results are roughly consistent with spicule properties.

Pikel'ner (1969) proposed (concurrently with Uchida) a somewhat different version of the acceleration of matter by the Petschek mechanism. In Pikel'ner's model the spicule results from the reconnection process at a single neutral point, as shown in Figure IX-13. The field lines forming the 'V' in the trough above and below the neutral point move upward and downward in response to the continued advance of the neighboring field lines. The ionized matter in the trough has presumably been compressed somewhat by the approaching fields. This compressed matter is carried upwards by the rising magnetic field in the upper trough becoming the spicule when it crosses the upper chromospheric boundary.

Pikel'ner (1969, 1971b) postulates that the approaching fields will generate ionic sound waves in the reconnecting region and that the ionic sound speed, v_i, which is near 10 km s^{-1}, determines the rate at which the magnetic lines flow into the reconnecting region. He then argues that continuity requires

$$v_i B_\infty \approx B_0,$$ (IX-119)

where B_∞ is the field strength far from the reconnecting point, B_0 is the field strength in the reconnected region at the base of the trough and v is the velocity of the rising spicule matter. Pikel'ner estimates $B_\infty/B_0 \approx \frac{1}{2}$ and $v \approx 2v_i$, which gives a spicule velocity of about the correct value.

Neither the Uchida nor the Pikel'ner model has been worked out in sufficient detail to know whether such processes will indeed lead to spicule-like ejections of matter from the chromosphere. As in nearly all spicule models, much is left to conjecture. Both mechanisms have the advantage of identifying the spicule acceleration directly with the magnetic field. They both suffer the strong disadvantage, however, of requiring the presence of magnetic fields of opposite polarity.

The magnetic field data (Chapter IV) permit a relatively frequent juxtaposition of fields of opposite polarity but this appears to be the exceptional situation in the network rather than the common one. We seem forced to assume, at the present time, that in the typical situation neighboring supergranules are dominated by fields of the same general polarity, and, hence, that the local network fields created by compressing fields of the neighboring supergranules together are of one polarity. The spicule phenomenon appears to be much too universal to admit spicules only when fields of opposite polarity are forced together by the acion of supergranule flow.

References

Athay, R. G.: 1966, *Astrophys. J.* **146**, 223.
Athay, R. G.: 1970, *Astrophys. J.* **161**, 713.
Athay, R. G.: 1974, in R. G. Athay (ed.), *Chromospheric Fine Structure*, Reidel Publishing Co., Dordrecht, Holland.

Athay, R. G. and Bessey, R. J.: 1964, *Astrophys. J.* **140**, 1174.

Athay, R. G. and Thomas, R. N.: 1956, *Astrophys. J.* **123**, 299.

Avrett, E. H.: 1973, private communication.

Beckers, J. M.: 1964, Thesis, Univ. of Utrecht.

Beckers, J. M.: 1972, *Ann. Rev. Astron. Astrophys.* **10**, 73.

Biermann, L.: 1951, *Z. Astrophys.* **29**, 274.

Boland, B. C., Engstrom, S. F. T., Jones, B. B., and Wilson, R.: 1973, *Astron. Astrophys.* **22**, 161.

Brandt, J. C. and Cassinelli, J. P.: 1966, *Icarus* **5**, 47.

Burton, W. M., Jordon, C., Ridgeley, A., and Wilson, R.: 1971, *Phil. Trans. Roy. Soc. London* **A270**, 81.

Canfield, R. C. and Athay, R. G.: 1974, *Solar Phys.* **34**, 193.

Cayrel, R.: 1966, *J. Quant. Spectrosc. Radiat, Transfer* **6**, 621.

Chamberlain, J. W.: 1961, *Astrophys. J.* **133**, 675.

Chapman, S.: 1957, Smithsonian Contrib. to Astrophys. **2**, 1.

Chiuderi, C. and Riani, I.: 1974, *Solar Phys.* **34**, 113.

Cox, D. P. and Tucker, W. H.: 1969, *Astrophys. J.* **157**, 1157.

Defouw, R. J.: 1970a, *Astrophys. J.* **160**, 659.

Defouw, R. J.: 1970b, *Astrophys. J.* **161**, 55.

Defouw, R. J.: 1970c, *Solar Phys.* **14**, 42.

Delache, P.: 1973, in S. D. Jordan and E. H. Avrett (eds.), *Stellar Chromosphere*, National Aero. and Space Adm., Washington, NASA SP-317, p. 207.

Dunn, R. B., Evans, J. W., Jefferies, J. T., Orrall, F. Q., White, O. R., and Zirker, J. B.: 1968, *Astrophys. J. Suppl.* **15**, 275.

Dupree, A. K. and Goldberg, L.: 1967, *Solar Phys.* **1**, 229.

Durney, B.: 1971, *Astrophys. J.* **166**, 669.

Field, G. B.: 1965, *Astrophys. J.* **142**, 531.

Gebbie, K. B. and Thomas, R. N.: 1970, *Astrophys. J.* **161**, 229.

Hollweg, J. V.: 1970, *J. Geophys. Res.* **75**, 2403.

Hollweg, J. V.: 1972, *Cosmic Electrodyn.* **2**, 423.

Henze, W. Jr.: 1969, *Solar Phys.* **9**, 56.

Hundhausen, A. J.: 1972, *Solar Wind and Coronal Expansion*, Springer-Verlag, New York.

Jockers, K.: 1970, *Astron. Astrophys.* **6**, 219.

Kopp, R. A.: 1968, Thesis, Harvard Univ., Cambridge. (The Equilibrium Structure of a Shock Heated Corona, Air Force Cambridge Res. Lab. Scientific Report.)

Kopp, R. A.: 1972, *Solar Phys.* **27**, 373.

Kopp, R. A. and Kuperus, M.: 1968, *Solar Phys.* **4**, 212.

Kuperus, M.: 1965, *Rech. Astron. Obs. Utrecht* **17**, 1.

Kuperus, M. and Athay, R. G.: 1967, *Solar Phys.* **1**, 361.

Lantos, P.: 1972, *Solar Phys.* **22**, 387.

Lighthill, M. J.: 1967, in R. N. Thomas (ed.), *Aerodynamic Phenomena in Stellar Atmosphere*, Academic, London, p. 429.

Lites. B. W.: 1972, Thesis, Univ. of Colorado, Boulder. (National Center for Atmospheric Research Coop, Thesis No. 28.)

Lust, R. and Scholer, M.: 1966, *Z. Naturforsch.* **21a**, 1098.

Meyer, F. and Schmidt, H. U.: 1968, *Z. Angew. Math. Mech.* **48**, 218.

Moore R. L. and Fung, P. C. W.: 1972, *Solar Phys.* **23**, 78.

Osterbrock, D. E.: 1961, *Astrophys. J.* **134**, 347.

Parker, E. N.: 1953, *Astrophys. J.* **17**, 431.

Parker, E. N.: 1958, *Astrophys. J.* **128**, 664.

Parker, E. N.: 1963, *Interplanetary Dynamical Processes*, Interscience, New York.

Parker, E. N.: 1964, *Astrophys. J.* **140**, 1170.

Parker, E. N.: 1965, *Space Sci. Rev.* **4**, 666.

Petschek, H. E.: 1964, in W. N. Hess (ed.), *Symp. on the Physics of Solar Flares*, National Aero. and Space Adm., Washington, NASA SP-50, p. 425

Pikel'ner, S. B.: 1969, *Soviet Astron. AJ* **13**, 259.

Pikel'ner, S. B.: 1971a, *Comments Astrophys. Space Phys.* **3**, 33.

Pikel'ner, S. B.: 1971b, *Solar Phys.* **20**, 236.

Pottasch, S. R.: 1964, *Space Sci. Rev,* **3**, 816.

Reeves, E. M., Foukal, P. V., Huber, M. C. E., Noyes, R. W., Schmahl, E. J., Timothy, J. G., Vernazza, J. E., and Withbroe, G. L.: 1974, *Astrophys. J.* **188**, L27.

Reimers, D.: 1971a, *Astron. Astrophys.* **10**, 182.
Reimers, D.: 1971b, *Astron. Astrophys.* **14**, 198.
Roberts, P. H. and Soward, A. M.: 1972, *Proc. Roy. Soc. London* **328**, 185.
Schlutter, A.: 1957, in H. C. van de Hulst (ed.), *Radio Astronomy*, Cambridge Univ., Cambridge.
Sturrock, P. A. and Hartle, R. E.: 1966, *Phys. Rev. Letters* **16**, 628.
Thomas, R. N.: 1948, *Astrophys. J.* **108**, 130.
Thomas, R. N. and Athay, R. G.: 1961, *Physics of the Solar Chromosphere*, Interscience, New York.
Uchida, Y.: 1961, *Publ. Astron. Soc. Japan* **13**, 321.
Uchida, Y.: 1969, *Publ. Astron. Soc. Japan* **21**, 128.
Unno, W., Ribes E., and Appenzeller, I.: 1974, *Solar Phys.* **35**, 287.
Vernazza, J. E., Avrett, E. H., and Loeser, R.: 1973, *Astrophys. J.* **184**, 605.
Vernazza, J. E. and Noyes, R. W.: 1972, *Solar Phys.* **22**, 358.
Wentzel, D. G. and Solinger, A. B.: 1967, *Astrophys. J.* **148**, 877.
Weymann, R.: 1960, *Astrophys. J.* **132**, 380 (also 452).
Whang, Y. C. and Chang, C. C.: 1965, *J. Geophys. Res.* **70**, 4175.
Withbroe, G. L.: 1972, *Solar Phys.* **25**, 116.
Withbroe, G. L. and Gurman, J. B.: 1973, *Astrophys. J.* **183**, 277.

CHAPTER X

WAVE GENERATION AND HEATING

1. Wave Generation

A. EXPECTED WAVE MODES

The need in astronomy to investigate a large number of heterogeneous objects and phenomena has repeatedly led to oversimplification of the physical phenomena being studied. Nowhere in astronomy is this more evident than in the case of the astrophysical representation of macroscopic fluid motions within the atmospheres of stars and nebulae. The lines in stellar spectra are too broad to be explained entirely in terms of thermal motions and rotation. An obvious explanation for this is that fluid motions are present within the stellar atmosphere. In a rather general way, one can visualize three effects of the fluid motions on the spectral lines: (1) a Doppler shift of the lines, (2) a broadening of the lines, and (3) an asymmetric distortion of the lines. The first effect will occur if the motions are somewhat coherent over a range of depths comparable to the thickness, h_{cf}, of the line contribution function. Motions on a scale much smaller than h_{cf} will broaden the lines, and motions on a scale comparable to h_{cf} generally will Doppler shift the line, broaden it and make it asymmetric. The latter effect arises because the Doppler shift is presumably not the same at all depths over which the line is formed.

With moderate spectral and spatial resolution stellar and solar lines are relatively symmetric and any systematic Doppler shift due to internal motions is of little consequence in terms of the observed profiles. It was recognized by astronomers, however, that the motions that produced the line broadening in excess of the thermal width probably existed on scales both much smaller and much larger than h_{cf}. In the case of the Sun, h_{cf} was known to be of the order of, say, 200 km whereas the granulation structure on the Sun appeared to have a scale considerably larger than this. One visualized, therefore, that the spectral lines associated with a given structural element were both broadened and Doppler shifted. Within the spatial resolution element of a given spectral observation there might well be several such structural elements each with a different Doppler shift and different broadening velocity. It is understandable, therefore, that the astronomers first serious attempts to take the fluid motions into account was by the introduction of an average broadening velocity (microturbulence) and an average Doppler shift (macroturbulence). In looking back, however, it is doubtful whether this rudimentary, two parameter description of the velocity field is more of a help than a hindrance. It is helpful in the sense that it permits the extraction of two additional parameters from the line profiles. It is a hindrance to the extent that one tries to attach physical meaning to the micro- and macroturbulence parameters, and, more seriously, in that it lulled astronomers into a state of satisfaction where it was unnecessary to worry about the true nature of the motions.

The combination of high spatial and high spectral resolution available in many modern solar spectrographs led quickly to the discovery that small, granule-size features on the Sun do indeed have different Doppler shifts, but that, as suspected, the Doppler shifted elements still had line profile widths in excess of the thermal width (see Chapter IV). On the other hand, the discovery of the 5 min oscillations (Leighton, 1960) on a spatial scale much in excess of the granulation scale, and even much more in excess of h_{cf}, caught solar astronomers completely by surprise. The only excuse that can be offered for this is that, with one or two notable exceptions, astronomers were simply not thinking extensively about the nature of fluid motions to be expected in stellar atmosphere.

Several years prior to the discovery of the 5 min oscillations, it was suggested (Biermann, 1946; Schwarzschild, 1948) that the convective motions in the sub-photospheric convection zone would generate sound waves. It was further argued, correctly, that the sound waves would be of low amplitude and would tend to propagate upwards, initially without significant dissipation of energy. Thus, it was assumed that the upward energy flux $\rho v^2 v_s$ was constant. Since v_s is approximately constant, this requires that the wave amplitude, v, increase with height in proportion to $\rho^{-\frac{1}{2}}$. The density, ρ, decreases from about 10^{-6} in the upper convection zone to about 10^{-9} in the low chromosphere. Thus, a velocity amplitude of only 0.2 km s^{-1} for the sound wave in the convection zone grows to about the sound velocity (≈ 6 km s^{-1}) in the low chromosphere, assuming that the chromosphere is at least one wave-length from the convection zone. Presumably, then, most of the energy in such sound waves is dissipated in the low chromosphere when the wave reaches shock amplitude. Little or no thought was given at this time to the probable distribution of the sound energy with wavelength. Thus, it was simply assumed that the mean velocity amplitude in the sound waves would be some appreciable fraction of the granulation velocity of 0.5 km s^{-1}.

Biermann (1946) and Schwarzschild (1948) had proposed the sound wave theory in the hope that this would provide a mechanism for heating the corona. It was not generally recognized, at that time, that the chromosphere required mechanical heating also. Thus, the tendency of the waves to dissipate in the low chromosphere was regarded as an essentially fatal argument to the theory.

A specific mechanism for the production of sound waves in a convective medium was proposed by Lighthill in 1952. Quantitative estimates of the energy flux associated with sound waves produced by the Lighthill mechanism indicated that the sound wave flux was indeed sufficient to heat the corona if by some means the wave energy could be preserved to that height. It was not until after the discovery of the 5 min oscillations, however, that attention was turned seriously to the question of wave generation in the solar atmosphere.

A sound wave of 300 s period in the solar photosphere has a wavelength of approximately 1800 km, which is of the order of the thickness of the photosphere and chromosphere combined and encompasses a density range of some six to severn orders of magnitude. For a sound wave to have a wavelength comparable to the density scale height in the photospheres(150 km), the period must be of the order of 25 s. Since h_{cf} is of the order of the density scale height, the short period waves will manifest themselves mostly as 'microturbulence'. Only those waves whose periods exceed about 100 s will produce distinct Koppler displacements. The energy spectrum of sound waves generated in the convection zone (Stein, 1967, 1968) shows maximum energy near periods of 20 to 60 s.

The influence of the atmospheric stratification due to gravity is of importance to the nature of the sound waves generated in the convection zone. In addition to stratifying the atmosphere, gravity produces an additional restoring force (additional to the pressure force in sound waves) that modifies the properties of the wave motion itself. A third restoring force often of importance in the solar atmosphere is provided by solar magnetic fields. The forces due to magnetic fields become important whenever the Alfvén speed approaches the sound speed. The former is given by

$$v_a = \left(\frac{H^2}{4\pi\rho} \right)^{1/2}$$

and has a value near the sound speed for the combination of H and ρ shown in Table X-1. We note from this table that the magnetic forces are not often of importance in the photosphere and low chromosphere except near the magnetic knots in the network and in active regions. In the high chromosphere and transition region, however, the magnetic forces are probably of frequent importance.

TABLE X-1
Magnetic field strengths required to give $v_a = 6$ km s^{-1}

Approximate regior	T_{min}	Low Ch.	Mid. Ch.	Transition region
ρ	10^{-8}	10^{-10}	10^{-12}	10^{-14}
n_H	6×10^{15}	6×10^{13}	6×10^{11}	6×10^9
H (gauss)	200	20	2	0.8

When the magnetic restoring forces are unimportant, the wave character will be of the acoustic-gravity type. When the magnetic restoring force dominates over the gravitational forces, the waves will be predominantly of the magneto-acoustic type, but with an influence from gravity produced by the density stratification. At magnetic field strengths and wavelengths such that all three restoring forces are important, the wave character is trimodal (acoustic, Alfvén and gravitational) and of a complex nature. Waves of this character are still only partially understood and no attempt will be made here to fully summarize their properties. Instead we shall limit our discussion to a brief summary of the acoustic-gravity and magneto-acoustic modes followed by an equally brief summary of the magneto-acoustic-gravity modes. For a more complete review the reader is referred to the current literature dealing with wave motions. The discussions by McLellan and

Winterberg (1968), Bel and Mein (1971), Chiu (1971), Bray and Loughhead (1974) and Stein and Leibacher (1974) are particularly helpful.

B. WAVE EQUATIONS

The conservation equations for an inviscid, perfectly conducting, non-rotating fluid undergoing adiabatic perturbations may be written in the forms (Stein and Leibacher, 1974):

mass,

$$\frac{d\rho}{dt} + \rho \, \nabla \cdot \mathbf{v} = 0; \tag{X-1}$$

internal energy,

$$\frac{dp}{dt} + \gamma P \, \nabla \cdot \mathbf{v} = 0; \tag{X-2}$$

momentum,

$$\rho \frac{d\mathbf{v}}{dt} - \rho \mathbf{g} + \nabla P - \frac{1}{4\pi} \, (\nabla \times \mathbf{H}) \times \mathbf{H} = 0; \tag{X-3}$$

and

magnetic flux,

$$\frac{\partial \mathbf{H}}{\partial t} - \nabla \times (\mathbf{v} \times \mathbf{H}) = 0. \tag{X-4}$$

If we now set

$$\rho = \rho_0 + \rho_1,$$
$$P = P_0 + P_1, \tag{X-5}$$
$$H = H_0 + H_1,$$

and

$$v = v_1,$$

where subscript zero refers to the time independent state and subscript 1 refers to a small amplitude, time-dependent perturbation, the conservation equations, to first order in the perturbation terms, become:

mass,

$$\frac{\partial \rho_1}{\partial t} + \mathbf{v}_1 \cdot \nabla \rho_0 + \rho_0 \nabla \cdot \mathbf{v}_1 = 0; \tag{X-6}$$

internal energy,

$$\frac{\partial P_1}{\partial t} + \mathbf{v}_1 \cdot \nabla P_0 + \gamma P_0 \, \nabla \cdot \mathbf{v}_1 = 0; \tag{X-7}$$

momentum,

$$\rho_0 \frac{\partial \mathbf{v}_1}{\partial t} - \rho_1 \mathbf{g} + \nabla P_1 - \tfrac{1}{4}\pi (\nabla \times \mathbf{H}_1) \times \mathbf{H}_0 = 0; \tag{X-8}$$

and magnetic flux,

$$\frac{\partial \mathbf{H}_1}{\partial t} - \nabla \times (\mathbf{v}_1 \times \mathbf{H}_0) = 0. \tag{X-9}$$

The unperturbed magnetic flux $\mathbf{H}_0$ is taken to be constant in space. To obtain the wave equation, we differentiate Equation (X–8) with respect to t then eliminate the time derivatives $\partial\rho_1/\partial t$, $\partial P_1/\partial t$ and $\partial\mathbf{H}_1/\partial t$ by the use of Equations (X-6), (X-7) and (X-9). This yields

$$\frac{\partial^2 \mathbf{v}_1}{\partial t^2} = v_{\mathrm{s}}^2 \nabla(\nabla \cdot \mathbf{v}_1) + (\gamma - 1)g\,\nabla \cdot \mathbf{v}_1 + \nabla(\mathbf{v}_1 \cdot \mathbf{g}) -$$

$$- \frac{1}{4\pi\rho_0} \mathbf{H}_0 \times \{\nabla \times [\nabla \times (\mathbf{v}_1 \times \mathbf{H}_0)]\} \tag{X-10}$$

The energy density, E_{W}, in the wave and the energy flux, $\mathbf{F}_{\mathrm{W}}$, carried by the wave satisfy the relation

$$\frac{\partial E_{\mathrm{w}}}{\partial t} + \nabla \cdot \mathbf{F}_{\mathrm{w}} = 0. \tag{X-11}$$

By combining Equations (X-6) to (X-9) in this form it may be shown that

$$E_{\mathrm{w}} = \tfrac{1}{2}\rho_0 v_1^2 + \tfrac{1}{2}\rho_0 v_{\mathrm{s}}^2 \left(\frac{P_1}{\gamma P_0}\right)^2 + \tfrac{1}{2}\rho_0 N_{\mathrm{BV}}^2 \delta h^2 + \frac{H_1^2}{8\pi}, \tag{X-12}$$

and

$$\mathbf{F}_{\mathrm{w}} = P_1 \mathbf{v}_1 + \frac{1}{4\pi} \mathbf{H}_1 \times (\mathbf{v}_1 \times \mathbf{H}_0), \tag{X-13}$$

where

$$v_{\mathrm{s}}^2 = \gamma \frac{P_0}{\rho_0}, \tag{X-14}$$

and

$$N_{\mathrm{BV}}^2 = -\left(\frac{q^2}{v_{\mathrm{s}}^2} + \frac{q}{\rho_0}\frac{\mathrm{d}\rho_0}{\mathrm{d}h}\right). \tag{X-15}$$

The quantity δh in Equation (X-12) is the vertical displacement of a particle in the wave and N_{BV} is the Brunt-Vaisala frequency. Note that if we set

$$\frac{dP_0}{dh} = -\rho_0 g \tag{X-16}$$

and

$$P_0 = R\rho_0 T_0, \tag{X-17}$$

then

$$\frac{g}{\rho_0}\frac{d\rho_0}{dh} = \rho\frac{g^2}{P_0} - g\frac{d\ln T_0}{dh} \tag{X-18}$$

and

$$N_{BV}^2 = \frac{g}{T_0}\left[\frac{dT}{dh} - \left(\frac{dT}{dh}\right)_{ad}\right], \tag{X-19}$$

where the adiabatic temperature gradient is given by

$$\left(\frac{dT}{dh}\right)_{ad} = -\frac{(\gamma-1)gT}{v_s^2}. \tag{X-20}$$

Equation (X-19) does not require that N_{BV} be zero in the isothermal case since $(dT/dh)_{ad}$ is not zero.

The term in Equation (X-12) containing the Brunt-Vaisala frequency is the energy due to buoyancy of the displaced material in the wave. The second term in Equation (X-12) is the energy associated with the compression in the sound wave mode, and the remaining two terms are the kinetic and magnetic energy terms.

When the magnetic terms are included in the wave equation, the Alfvén speed $H_0/(4\pi\rho_0)^{1/2}$ appears as a coefficient. It is difficult to treat these terms unless the Alfvén speed is constant. In the following, therefore, we will assume that both H_0 and ρ_0 are constant. The latter assumption is the familiar Boussinesq approximation, and, strictly speaking, limits the applicability of the results to cases where the vertical wave number, k_{1z}, is much greater than unity. However, the extension to lower values of k_{1z} provides useful limits on the regions of propagation.

It is customary to discuss the wave equation in terms of its solutions in the k, ω plane, where k is the propagation vector (wave number) whose components are $2\pi/\lambda_i$ and ω is the circular frequency $2\pi\nu$. One assumes small amplitude, plane waves of the form

$$v_i = A_i e^{i(k_1 x_i - \omega t)}. \tag{X-21}$$

The dispersion relation between k and ω obtained from the full wave Equation (X-10) is of sixth order in both k and ω.. However, it can be factored into two separate equations of the form (McLellan and Winterberg, 1968)

$$\omega^2 = v_{\text{a}}^2 k^2 \cos^2 \theta_{\text{k}} \tag{X-22}$$

and

$$\omega^4 - \omega^2 (v_2^2 k^2 + ig\gamma k \cos \theta_{\text{k}}) + (\gamma - 1)g^2 k^2 \sin^2 \theta_{\text{k}} +$$

$$+ ig\gamma v_{\text{a}}^2 k^3 \cos \theta_{\text{H}} \cos \theta_{\text{kH}} + v_1^4 k^4 \cos^2 \theta_{\text{kH}} = 0 , \tag{X-23}$$

where

$$v_{\text{a}} = \frac{H}{(4\pi\rho)^{\frac{1}{2}}} , \tag{X-24}$$

$$v_1^2 = v_{\text{a}} v_{\text{s}} , \tag{X-25}$$

$$v_2^2 = v_{\text{a}}^2 + v_{\text{s}}^2 , \tag{X-26}$$

θ_{k} = angle between **k** and the vertical

θ_{H} = angle between **H** and the vertical

θ_{kH} = angle between **k** and **H**.

It is convenient to rewrite Equation (X-23) in terms of the dimensionless quantities

$$\omega_1 \equiv \frac{\omega}{\gamma g/2v_{\text{s}}} \equiv \frac{\omega}{v_{\text{s}}/2H} \equiv \frac{\omega}{N_{\text{ac}}} . \tag{X-27}$$

(in the subsequent discussion H is the density scale height of an equivalent isothermal atmosphere) and

$$k_1 \equiv 2Hk \equiv \frac{v_{\text{s}}^2}{\gamma g} k \equiv \frac{v_{\text{s}}}{N_{\text{ac}}} k. \tag{X-28}$$

We also make use of the relationship

$$\frac{N_{\text{ac}}^2}{N_{\text{BV}}^2} = \frac{\gamma^2 g^2/4v_{\text{s}}^2}{(\gamma - 1)g^2/v_{\text{s}}^2} = \frac{\gamma^2}{4(\gamma - 1)} , \tag{X-29}$$

which follows from Equations (X-14), (X-15) and (X-18) for the sothermal case, as well as the relationship

$$\gamma g \equiv 4H N_{\text{ac}}^2, \tag{X-30}$$

which follows from the definition (X-27). We then find that the fourth-order dispersion relation can be written in the form

$$\omega_1^4 - \omega_1^2 (A + iB) + C + iD = 0 \tag{X-31}$$

where

$$A = (1 + \beta^2)k_1^2 , \tag{X-31}$$

$$B = 2 \cos \theta_k k_1 , \tag{X-32}$$

$$C = k_1^2 \left(\frac{N_{BV}^2}{N_{ac}^2} \sin^{-2} \theta_k + \beta^2 k_1^2 \cos^2 \theta_{kH} \right) , \tag{X-34}$$

$$D = 2\beta^2 \cos \theta_H \cos \theta_{kH} k_1^3 \tag{X-35}$$

and

$$\beta = v_a/v_s . \tag{X-36}$$

Equation (X-31) has the solution

$$\omega_1^2 = \tfrac{1}{2} (A + iB) \left[1 \pm \left(1 - 4\frac{C + iD}{(A + iB)^2} \right)^{1/2} \right]. \tag{X-37}$$

Waves for which k_1 and ω_1 are imaginary are unstable and their amplitude either grows or decreases exponentially in time. Waves for which k_1 and ω_1 are real are stable and propagate. In the most general case, Equation (X-37) represents two separate classes of waves (one for each ω_1^2 in Equation (X-31)). Each of these has a fast mode, represented by the + sign, and each has a slow mode, represented by the minus sign. Thus, four wave modes are represented. Equation (X-22) represents a fifth mode readily identifiable as the usual Alfvén mode propagating along the magnetic field lines with phase velocity $\omega/k = v_a \cos \theta_k$. It can readily be shown that the group velocity is always v_a and is always along the field lines for these waves. These waves are separable from the others because they are pure shear waves in the magnetic field and are not affected by the compressibility of the medium or by buoyancy forces.

In the following discussion, we shall be mainly interested in the coupled waves whose amplitudes are either stable or which grow exponentially with height in proportion to $v_1 \propto \rho^{-1/2}$. This wave growth arises simply from the decrease of density with height and occurs in acoustic waves that are stable in a constant density atmosphere. The stable waves occur when ω_1 and k_1 are real. A general condition for the imaginary terms to vanish in Equation (X-31) derived by McLelland and Winterberg (1968) is given by

$$B = \pm\frac{1}{2^{1/2}} [A^4 + B^4 - 8A^2C + 8B^2C - 8ABD + 8D^2 + 16C^2 -$$

$$- A^2 + B^2 + 4C]^{1/2} \tag{X-38}$$

C. ACOUSTIC-GRAVITY WAVES

When the magnetic field strength is small enough that $v_a \ll v_s$ the magnetic terms in Equation (X-31) may be neglected. These are the terms containing β. Thus, we may set $D = 0$ and omit the second terms in C. Note that the qualitative effect of setting $\beta = 0$ is the same as setting $\cos \theta_{kH} = 0$, the only difference being in the magnitude of A. Thus, the character of the coupled waves when the magnetic forces are small is qualitatively the same as when the wave propagation is at right angles to the magnetic field lines.

For $\beta = 0$, the dispersion relation (X-31) is quadratic in κ_1 and of the form

$$\omega_1^4 - 2i\omega_1^2 \cos \theta_k k_1 + \left(\frac{N_{BV}^2}{N_{ac}^2} \sin^2 \theta_k - \omega_1^2 \right) k_1^2 = 0. \tag{X-39}$$

We write Equation (X-39) in the form

$$k_{1z}^2 = 2ik_{1z} + \left(\omega_1^2 - \frac{N_{BV}^2}{N_{ac}^2} \right) \frac{k_{ix}^2}{\omega_1^2} - \omega_1^2, \tag{X-40}$$

where k_{1z} and k_{1x} are the vertical and horizontal components of k_1. The solution for k_{1z} is

$$k_{iz} = -i \pm \frac{1}{2} \left[-4 + 4\omega_1^2 - 4 \left(\omega_1^2 - \frac{N_{BV}^2}{N_{ac}^2} \frac{k_{ix}^2}{\omega_1^2} \right) \right]^{1/2}$$

or,

$$k_{iz} = -i \pm \left[(\omega_1^2 - 1) - \left(\omega_1^2 - \frac{N_{BV}^2}{N_{ac}^2} \right) \frac{k_{ix}^2}{\omega_1^2} \right]^{1/2}. \tag{X-41}$$

The i term represents the exponential growth in amplitude of the acoustic waves with the scale height $2H$ corresponding to the requirement that $v_1 \propto \rho^{-\frac{1}{2}}$.

In order that the term under the radical sign be positive, it is necessary that ω_1 not lie within the interval $\omega_1 = 1$ and

$$\omega_1 = \frac{N_{BV}}{N_{ac}} = \frac{2(\gamma - 1)^{\frac{1}{2}}}{\gamma}.$$

We note, also, that for $\omega_1 \gg 1$ it is necessary that $\omega_1 > k_{ix}$. Conversely, when $\omega_1 \ll 1$ it is necessary that

$$\omega_1 < \frac{N_{BV}}{N_{ac}} k_{ix} = \frac{2(\gamma - 1)^{\frac{1}{2}}}{\gamma} k_{ix}.$$

The region limiting $\omega_1 < N_{BV}/N_{ac}$ for larger k_{ix} and $\omega_1 < (2(\gamma - 1)^{\frac{1}{2}}/\gamma) k_{ix}$ for small k_{ix} defines the region of propagating gravity waves. These are the slow-mode waves. Acoustic waves are the fast-mode waves limited by $\omega_1 > 1$ and $\omega_1 > k_{ix}$.

A dispersion diagram for the acoustic-gravity waves is shown in Figure X-1. The propagating waves lie outside the interval between the curves. In a semi-infinite, isothermal atmosphere, the evanescent waves lying between the two curves do not propagate vertically. Their pressure and velocity oscillations are $90°$ out of phase. Tunneling effects, however, permit some vertical propagation of energy for the evanescent waves in a bounded atmosphere or in an atmosphere such as the solar atmosphere, where large temperature jumps occur.

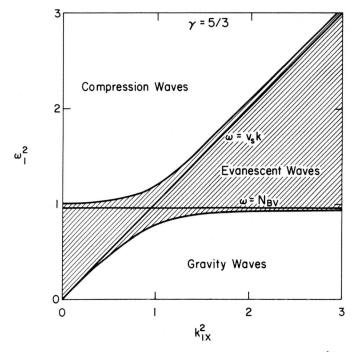

Fig. X-1.   A dispersion diagram for acoustic-gravity waves for $\lambda = \frac{5}{3}$.

The preceding restrictions on ω_1 and k_{ix} apply only if both k_{iz} and k_{ix} are not zero. In the case of purely horizontal propagation, k_{iz} and B are both zero. Equation (X-31) then becomes

$$\omega_1^4 - \omega_1^2 k_{ix}^2 + \frac{N_{BV}^2}{N_{ac}^2} k_{ix}^2 = 0 \,.$$

The solution for ω_1^2 is now

$$\omega_1^2 = \tfrac{1}{2} k_{ix}^2 \left[1 \pm \left(1 - 4 \frac{N_{BV}^2}{N_{ac}^2 k_{ix}^2} \right)^{1/2} \right] \,,$$

and we see that for ω_1 to be real

$$k_{ix} > 2 \frac{N_{BV}}{N_{ac}} = \frac{4(\gamma - 1)}{\gamma} \approx 2 \quad \text{for} \quad \gamma = 5/3 \,.$$

The restriction on ω_1 is simply that

$$\omega_1 > \frac{N_{BV}}{N_{ac}} = \frac{2(\gamma - 1)^{\frac{1}{2}}}{\gamma} \approx 1 \quad \text{for} \quad \gamma = 5/3 \,.$$

Both the acoustic (fast-mode) and gravity (slow-mode) waves are present. However, they now propagate in the same regions of the k_1, ω_1 plane.

For purely vertical propagation $k_{ix} = 0$ and $C = 0$. Thus, Equation (X-41) becomes

$$k_{iz} = -i \pm (\omega_1^2 - 1)^{1/2}$$

and the only restriction is that $\omega_1 > 1$. Only acoustic waves are present in this case.

The high frequency cut-off at $\omega_1 \lesssim k_{ix}$ for gravity waves arises from compressibility effects, and the low frequency cut-off at $\omega_1 > k_{ix}$ for sound waves arises from buoyancy effects. Also, the low frequency cut-off of sound waves at $\omega_1 > 1$ is caused by the density stratification. The high frequency cut-off of gravity waves at $\omega_1 \lesssim N_{BV}/N_{ac}$ appears to represent a natural resonance mode of the atmosphere (Worrall, 1972).

Acoustic waves may propagate in any direction. Their group velocity is less than or equal to the sound speed and their phase velocity is greater than the sound speed. The particle motions in the presence of gravity are elliptical.

Gravity waves have the curious property that they do not propagate in the purely vertical direction ($k_{ix} = 0$ and $\omega_1 < 1$), and they propagate energy and phase in opposite directions. Both the phase velocity and group velocity are less than the sound speed. Particle motions are elliptical in the gravity waves. Surface water waves are a particular case of gravity waves, but, because of the surface effects, they possess properties that are peculiar to surface waves.

Note that the plot in Figure X-1 for the isothermal case is normalized to be independent of temperature. In an ω, k_x plot, however, the critical frequencies N_{BV} and N_{ac} and the limiting frequency $\omega = v_x k_x$ are each functions of temperature.

D. MAGNETO-ACOUSTIC WAVES

If we neglect both the buoyancy forces and the density stratification of the atmosphere by setting $g = 0$, we obtain the pure magnetohydrodynamic case. The complex parts of the dispersion relation vanish ($B = 0$ and $D = 0$) as does the term containing N_{BV}^2/N_{ac}^2. Thus, the dispersion relation for the coupled wave modes becomes

$$\omega_1^4 - (1 + \beta^2)\omega_1^2 k_1^2 + \beta^2 \cos\theta_{kH} k_1^4 = 0. \tag{X-42}$$

The solution for ω_1^2 is

$$\omega_1^2 = \frac{1 + \beta^2}{2} k_1^2 \left[1 \pm \left(1 - 4\frac{\beta^2}{(1 + \beta^2)^2} \cos^2\theta_{kH} \right)^{1/2} \right] \tag{X-43}$$

and the solution for k_1^2 is

$$k_1^2 = \tfrac{1}{2} (1 + \beta^2)\omega_1^2 \left[1 \pm \left(1 - 4\frac{\beta^2 \cos\theta_{kH}}{(1 + \beta^2)^2} \right)^{1/2} \right] \tag{X-44}$$

We remind the reader that we have assumed β to be constant with height.

Since $\beta^2/(1 + \beta^2)^2$ has a maximum of $1/4$ that occurs when $\beta = 1$, the term under the radical sign is never negative. Thus, there are no excluded regions k_1 in the ω_1 plane. The phase velocity of the waves is given by

$$v_p^2 = \frac{\omega_1^2}{k_1^2} v_s^2 = \frac{1+\beta^2}{2} v_s^2 \left[1 \pm \left(1 - 4 \frac{\beta^2}{(1+\beta^2)^2} \cos \theta_{kH} \right)^{1/2} \right] \qquad (X\text{-}45)$$

and has a fast-mode (+ sign) and slow-mode (− sign). For the singular case where $\beta = 1$ and $\cos \theta_{kH} = 1$ (propagation along the field lines with the Alfvén speed equal to the sound speed, which occurs when $H = (4\pi\gamma\rho_0)^{1/2}$), the distinction between fast and slow-mode disappears. In the case of propagation at right angles to the field lines, ($\cos \theta_{kH} = 0$), the slow-mode vanishes. Thus, the slow-mode waves propagate mainly in the direction of the field whereas the fast-mode waves propagate in all directions.

The regular Alfvén-mode that propagates with group velocity v_a along the field lines is present in addition to the fast and slow-mode waves. In the limit $v_s \gg v_a$, the phase velocity is just v_s, and, in the limit $v_a \gg v_s$, the phase velocity is just v_a. Both of these limits correspond to fast-mode waves. In general, v_p exceeds both v_a and v_s for the fast-mode and is less than both v_a and v_s for the slow mode.

Propagation characteristics of the fast and slow-mode waves, as summarized by Stein and Leibacher (1974) are given in Table X-2. The acoustic wave, in the weak field case, corresponds to the fast-mode. However, the acoustic wave in the strong field case is a slow-mode wave and is constrained to move along the field lines.

TABLE X-2
Characteristics of magneto-acoustic and Alfvén waves

Mode	Phase velocity	Material velocity and wave type	
		Weak field	Strong field
Fast	Isotropic $v_p > v_s$ or v_a	Parallel to k, longitudinal, gas pressure	Perpendicular to H, mixed, magnetic pressure.
Slow	approximately parallel to H	Perpendicular to k, transverse, magnetic tension	Parallel to H, mixed, gas pressure
Alfvén	parallel to H	Perpendicular to k and H, transverse, magnetic tension	

Figure X-2 shows a plot of the relation between k_{ix} and ω_1 for fixed values of k_{iz}. In making the plot, it has been assumed that $\beta^2 = 0.1$ and that the field makes an angle of $45°$ with the vertical. Note that for horizontal propagation ($k_z = 0$) the curves are straight lines of slope unity in this log-log plot. The same is true for vertical propagation ($k_x = 0$), i.e., $\omega \propto k_z$, as may be deduced from the intercepts at the left hand border of the plot.

Although the plots in Figure X-2 are for a field inclined $45°$ to the vertical, the plots have a similar appearance for other field orientations.

E. MAGNETO-ACOUSTIC-GRAVITY WAVES

Under the combined forces of gravity and magnetic fields there are two unique directions

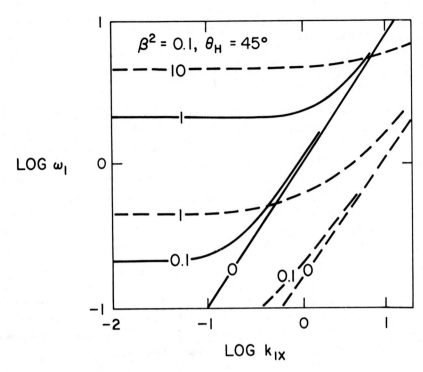

Fig. X-2.   A dispersion diagram for magneto-acoustic waves for the case $\beta^2 = 0.1$ and $\theta_H = 45°$ and for fixed values of k_{iz}. Values of k_{iz} are indicated on the curves. The fast-mode waves are shown as solid lines and slow-mode waves are shown as dashed lines.

associated with wave motions. The slow-mode Alfvén waves move preferentially along the field lines and the gravity waves move preferentially in the horizontal direction. Acoustic waves interact with both the gravitational and magnetic forces to add further complexity to the problem.

As in the preceding examples, we shall keep the problems as simple as possible by assuming that both v_a and v_s (hence β) are constant with depth. This undoubtedly oversimplifies the problem with respect to solar applications, but, nevertheless, provides useful insight to the nature of the wave motions to be expected. In order to further simplify the discussion we shall select special cases such that part of the B, C and D coefficients vanish.

It follows from Equation (X-37) that if $B = 0$, then the only cases for which ω_1 is real are those for which $D = 0$ also. By inspection of Equation (X-36), we note that $D = 0$ whenever the direction of propagation is at right angles to the magnetic field lines or whenever the magnetic field is horizontal. Purely horizontal wave propagation is possible, therefore, only when the field itself is horizontal or when the propagation is at right angles to the plane containing the **H** vector and the vertical.

As noted earlier in this section, the case in which he wave propagation is at right angles to the field lines is formally the same as the coustic-gravity case with the exception that the phase velocity is proportional to $v_s(1 + \beta^2)^{\frac{1}{2}}$ rather than to v_s. Thus, the simplest case to treat is one for which the wave propagation vector lies at right angles to the field lines and is in the horizontal plane:

Case i: Horizontal Propagation at Right Angles to Magnetic Field Lines
For this case, the coefficients B and D of the imaginary terms in Equation (X-37) vanish. In additions, the second term in C is zero. Thus, Equation (X-37) becomes

$$\omega_1^2 = \frac{1}{2}(1 + \beta^2)k_{ix}^2 \left[1 \pm \left(1 - 4\frac{N_{BV}^2}{N_{ac}^2(1 + \beta^2)^2 k_{ix}^2} \right)^{1/2} \right]$$

(X-46)

and the solution for k_{ix}^2 becomes

$$k_{ix}^2 = \frac{\omega_1^4}{\omega_1^2(1 + \beta^2) - N_{BV}^2\sqrt{N_{ac}^2}}$$

(X-47)

The condition for ω_1 to be real is

$$k_{ix}^2 > 4\frac{N_{BV}^2}{N_{ac}^2}\frac{1}{(1 + \beta^2)^2} = \frac{16(\gamma - 1)}{\gamma^2}\frac{1}{(1 + \beta^2)^2}.$$

(X-48)

and the condition for k_{ix} to be real is

$$\omega_1^2 > \frac{N_{BV}^2}{N_{ac}^2}\frac{1}{1 + \beta^2} = \frac{4(\gamma - 1)}{\gamma^2}\frac{1}{1 + \beta^2}.$$

(X-49)

Since $4(\gamma - 1)8\gamma^2 \approx 1$ for $\gamma = 5/3$, these limits on k_{ix} and ω_1 are approximately $k_{ix} > 2/(1 + \beta^2) \approx 1$, for $\beta = 1$, and $\omega_1 > (1 + \beta^2)^{\frac{1}{2}} \approx 0.7$, for $\beta = 1$. For large β, the limits on k_{ix} and ω_1 disappear. For $\beta = 0$, the limits become

$$k_{ix} > \frac{4(\gamma - 1)^{\frac{1}{2}}}{\gamma} \quad \text{and} \quad \omega_1 > \frac{2(\gamma - 1)^{\frac{1}{2}}}{\gamma}.$$

This latter case corresponds to horizontal propagation in the acoustic-gravity modes.

It is of interest to note that in the case where β and D are both zero Equation (X-38) reduces to $(A^2 - 4C)^2 - (A^2 - 4C) = 0$, which, in turn, is satisfied if $A^2 = 4C$. This is just the condition expressed by Equation (X-48).

The phase velocities given by Equation (X-46) for the two coupled waves present in this case are

$$v_p^2 = \frac{1}{2}(1 + \beta^2)v_s^2 \left[1 \pm \left(1 - \frac{16(\gamma - 1)}{\gamma^2(1 + \beta^2)^2 k_1^2} \right)^{1/2} \right].$$

(X-50)

The distinction between fast and slow-mode waves disappears at the low wave number cut-off

$$k_1 = \frac{4(\gamma - 1)^{1/2}}{\gamma(1 + \beta^2)}.$$

Both fast and slow-modes exist throughout the propagating region. It should be noted, again, that the results in this section are independent of the angle between the magnetic field and the vertical.

Case ii: Horizontal Propagation at an Acute Angle to a Horizontal Field

For this case we remove the restriction that the propagation be at right angles to the field but add the restriction that the field be horizontal. This retains the condition that B and D are both zero but removes the condition that the second term in C be zero.

The solutions, for ω_1^2 and k_{ix}^2 in this case are

$$\omega_1^2 = \frac{1}{2}(1 + \beta^2)k_{ix}^2 \left[1 \pm \left(1 - 4 \frac{\frac{N_{BV}^2}{N_{ac}^2} + \beta^2 k_{ik}^2 \cos^2 \theta_{kH}}{(1 + \beta^2)^2 k_{ix}^2} \right)^{1/2} \right] \qquad (X\text{-}51)$$

and

$$k_{ix}^2 = \frac{1}{2} \frac{\left[\omega_1^2(1 + \beta^2) - \frac{N_{BV}^2}{N_{ac}^2} \right]}{\beta^2 \cos^2 \theta_{kH}} \left[1 \pm \left(1 - 4 \frac{\omega_1^4 \beta^2 \cos^2 \theta_{kH}}{\left(\omega_1^2(1 + \beta^2) - \frac{N_{BV}^2}{N_{ac}^2} \right)^2} \right)^{1/2} \right] \qquad (X\text{-}52)$$

These equations impose the restrictions

$$k_{ix}^2 > \frac{16(\gamma - 1)/\gamma^2}{(1 + \beta^2)^2 - 4\beta^2 \cos^2 \theta_{kH}} \qquad (X\text{-}53)$$

and

$$\omega_1^2 < \frac{4(\gamma - 1)}{\gamma^2} \frac{1}{1 + \beta^2} \qquad (X\text{-}54)$$

It follows from the inequality (X-53) and the condition that k_{ix} be finite that

$$\cos \theta_{kH} \neq \frac{1 + \beta^2}{2\beta}. \qquad (X\text{-}55)$$

This is satisfied everywhere except $\cos \theta_{kH} = 1$ and $\beta = 1$. We have the interesting conclusion, therefore, that there is no stable horizontal propagation along field lines by acoustic and gravity waves when $v_a = v_s$ only the regular Alfvén mode propagates stably in this case. When $v_a \neq v_s$, the magneto-acoustic-gravity waves propagate preferentially across the field lines but show some propagation along the field as well. Both the fast and slow-modes are present.

For a given β and angle θ_{kH}, the restrictions on k_{ix} and ω_1 are similar to those for

Case i. Note that for $\beta = 0$ the restrictions on k_{ix} and ω_1 are the same as for the pure acoustic-gravity waves.

Since the only allowed stable propagation in the horizontal plane is covered by Cases i and ii, we turn next to a consideration of non-horizontal propagation.

Case iii: Propagation at Right Angles to Magnetic Field Lines

For this case the propagation vector has a vertical component and the field is at right angles. Since the terms in $\cos \theta_{kH}$ are zero, for this case, and since these are the only terms containing β other than the A term, this case is formally similar to the case of acoustic-gravity waves without a magnetic field.

We replace k_{iz}^2 and k_{ix}^2 in Equation (X-40) by $(1 + \beta^2)k_{iz}^2$ and $(1 + \beta^2)k_{ix}^2$ to obtain

$$(1 + \beta^2)k_{iz}^2 + 2ik_{iz} + \left[\omega_i^2(1 + \beta^2) - \frac{N_{BV}^2}{N_{ac}^2} \right] - \omega_i^2 = 0 \qquad (X-56)$$

Solving for k_{iz}, we obtain

$$k_{iz} = -i \pm \left\{ [\omega_i^2(1 + \beta^2) - 1] - (1 + \beta^2) \left[\omega_i^2(1 + \beta^2) - \frac{N_{BV}^2}{N_{ac}^2} \right] \frac{k_{ix}^2}{\omega_i^2} \right\}^{1/2}$$

$$(X-57)$$

The term under the radical sign is real ω_1 lies outside the interval

$$\omega_1 = (1 + \beta^2)^{-\frac{1}{2}} \quad \text{and} \quad \omega_1 = \frac{N_{BV}}{N_{ac}}(1 + \beta^2)^{-\frac{1}{2}} = \frac{2}{\gamma}\left(\frac{\gamma - 1}{1 + \beta^2}\right)^{1/2}.$$

Limits on k_{ix} are set by the requirement that, for

$$\omega_i^2(1 + \beta^2) \gg 1, \quad k_{ix} < \omega_1(1 + \beta^2)^{-\frac{1}{2}},$$

and by the requirement that, for

$$\omega_1(1 + \beta^2) \ll 1, \quad k_{ix} > \omega_1 \frac{\gamma}{2(\gamma - 1)^{\frac{1}{2}}(1 + \beta^2)^{\frac{1}{2}}}.$$

A dispersion diagram for this case is shown in Figure X-3 for $\beta = 1$ and $\gamma = 5/3$. It is essentially similar to Figure X-1 except for the changes in the limiting values of k_{ix} and ω_1. For $\beta \gg 1$, this case reverts to the pure magneto-acoustic fast mode case, and for $\beta \ll 1$ it reverts to the acoustic gravity case.

In the case of horizontal propagation across vertical magnetic field lines, Case iii reduces to Case i. For purely vertical propagation across a horizontal magnetic field, the only requirement is that $\omega_i^2 > 1/(1 + \beta^2)$. Only the fast-mode magneto-acoustic wave is present in this case.

In each of Cases i, ii and iii, the D term is zero. Thus, we next turn to cases where $D \neq 0$. Since we have already considered cases in which B and D are both zero, and cases

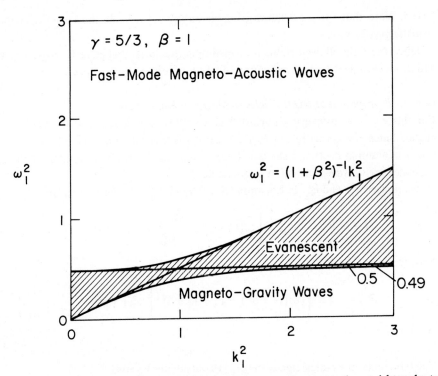

Fig. X-3.   A dispersion diagram for magneto-acoustic-gravity waves for propagation at right angles to the magnetic field.

for which $\cos \theta_{kH} = 0$, it follows that in the remaining cases none of the coefficients A, B, C or D is zero.

Case iv: Vertical Propagation at an Acute Angle to a Magnetic Field
In the case of vertical propagation, $k_{ix} = 0$ and $\cos \theta_{kh} = \cos \theta_H$. Stein and Leibacher (1974) give for this case the following relations between k_{iz} and ω_1:
    Acoustic Mode,

$$k_{iz}\left(\overset{\text{fast}}{\beta \ll 1}\right) = -i \pm \omega_1 \left[1 - \frac{1}{\omega_1^2} + \beta^2 \sin^2 \theta_H\right]^{1/2}, \tag{X-58}$$

$$k_{iz}\left(\overset{\text{slow}}{\beta \gg 1}\right) = -i \pm \frac{\omega_1}{\cos \theta_H}\left[1 - \frac{\cos^2 \theta_H}{\omega_1^2}\right]^{1/2}; \tag{X-59}$$

Magnetic Mode,

$$k_{iz}\left(\overset{\text{slow}}{\beta \ll 1}\right) = \pm \frac{\omega_1}{\beta \cos \theta_H} \tag{X-60}$$

$$k_{iz}\left(\begin{matrix}\text{fast}\\ \beta\gg1\ 1\\ \beta\gg\omega_1\end{matrix}\right)=-i\frac{\sin^2\theta_H}{\beta^2}\pm\frac{\omega_1}{\beta^2}\left[\beta^2-\frac{1}{\omega_1^2}+\sin^2\theta_H\right]^{1/2} \tag{X-61}$$

and

$$k_{iz}\left(\begin{matrix}\text{fast}\\ \omega_1^{-1}\gg\beta\gg1\end{matrix}\right)=\pm\frac{\omega_1}{\beta\cos\theta_H} \tag{X-62}$$

Propagation occurs when k_{iz} has a real component. Since this is true for all of the magnetic cases, there is no cut-off frequency for the magnetic modes. The acoustic modes have cut-off frequencies at

$$\omega_1=\frac{1}{(1+\beta^2\sin^2\theta_H)^{1/2}}, \qquad \beta\ll1, \tag{X-63}$$

and

$$\omega_1=\cos\theta_H, \qquad \beta\gg1. \tag{X-64}$$

Propagation occurs at higher frequencies. It may be seen by comparison to the acoustic-gravity case that the effect of the magnetic field is to decrease the acoustic cut-off frequency. This effect disappears for propagation in the direction of the field. In the strong field case, however, ω_1 is strongly dependent upon the field orientation and the acoustic cut-off frequency reduces to zero for propagation at right angles to the field lines.

2. Interpretation of 5 Minute Oscillations

A. LOCATION OF WAVES IN k_i, ω_1 PLANE

As is evident from the preceding section, wave motion under the combined forces of gravity, compression and magnetic fields is exceedingly complex and the identification of an observed wave phenomena with a particular wave mode is not easy. Since the magnetic fields in quiet solar regions seem to be confined largely to small knots of high field strength covering but a small fraction of the solar surface, it may be assumed safely that most of the observations of the 5 min oscillations refer to regions that are relatively free of magnetic field effects. At photospheric densities ($>10^{-8}$ g cm^{-3}), the Alfvén speed in a field of 30 G is less than 1 km s^{-1}. The sound speed, by comparison, is near 7 km s^{-1}. Field strengths as high as 30 G occur only in active centers and in the magnetic knots of the network. Thus, in so far as the current data on the 5 min oscillations are concerned, we need consider only the acoustic-gravity mode.

Observations of the 5 min oscillations permit us to locate the waves approximately in the ω_1, k_{ix} diagnostic diagram illustrated by Figure X-1. The acoustic cut-off frequency, N_{ac}, may be written in the form (Equation (X-27))

$$N_{ac} = \gamma^{1/2} \frac{g}{2} \left(\frac{\rho}{P}\right)^{1/2} = \gamma^{1/2} \frac{g}{2} \left(\frac{m_H}{kT}\right)^{1/2} = 1.51 \left(\frac{\gamma}{T}\right)^{1/2}. \qquad (X\text{-}65)$$

In order to estimate γ, we need to determine what fraction of the energy added by an increment in temperature ΔT is used in ionizing the gas. For a given volume, the change in ionization energy is given by $a\Delta n_p$, where $a = 13.56 \times 1.6 \times 10^{-12} = 2.17 \times 10^{-11}$ and Δn_p is the change in proton density. Since hydrogen is mostly in the ground state, we may determine Δn_p from the Saha equation for n_1, viz.,

$$n_1 = \left(\frac{h^2}{2\pi mk}\right)^{3/2} n_e n_p T^{-3/2} e^{\chi/T}.$$

Throughout most of the photosphere the metals are the main contributor to n_e and we may take n_e as being independent of T (the metals are mostly singly ionized). Thus, the derivative of the Saha equation gives

$$dn_1 = \left(\frac{h^2}{2\pi mk}\right)^{3/2} n_e e^{\chi/T} (T^{-3/2} \, dn_p - 3/2 T^{-5/2} n_p \, dT - \chi T^{-7/2} n_p \, dT).$$

It is readily shown that the term in dn_1 is small compared to the term in dn_p, in which case

$$dn_p = (3/2 + \chi/T)n_p \frac{dT}{T}.$$

It follows that the gain in ionization energy, dU, is given by

$$dU = 2.17 \times 10^{-11} \left(\frac{3}{2} + \frac{\chi}{T}\right) n_p \frac{dT}{T}. \qquad (X\text{-}66)$$

The corresponding change in kinetic energy,

$$dW = nk \, dT \approx n_1 k \, dT. \qquad (X\text{-}67)$$

Thus, the ratio dU/dW is given by

$$\frac{dU}{dW} = 2.17 \times 10^{-11} (3/2 + \chi/T) \frac{n_p}{n_1 kT}. \qquad (X\text{-}68)$$

Using the Saha equation for n_1/n_p, we find

$$\frac{dU}{dW} = \frac{2.17 \times 10^{-11} \left(\frac{3}{2} + \frac{\chi}{T}\right) T^{3/2} e^{-\chi/T}}{(h^2/2\pi mk)^{3/2} P_e}$$

$$= 5.2 \times 10^4 \frac{\left(\frac{3}{2} + \frac{\chi}{T}\right) T^{3/2} e^{-\chi/T}}{P_e}. \qquad (X\text{-}69)$$

Values of dU/dW for combinations of T and P_e found in the Harvard Smithsonian

Reference Atmosphere are given in Table X-3. It is clear from the results in this table that most of the increase in internal energy at photospheric levels between $\tau = 1$ and 10^{-3} is in the translational motion. For this case, we may treat the photospheric gas as an ideal, monatomic gas and set $\gamma = 5/3$.

TABLE X-3
Values of dU/dW, N_{ac}[a] and ω_1 for the HSRA

τ	10	1	0.1	10^{-2}	10^{-3}	10^{-5}
T	8880	6390	5160	4660	4380	5300
P_e	2600	56	4	0.9	0.24	0.07
dU/dW	6.6	0.25	0.009	0.002	0.006	1.2
N_{ac}	0.021(5/3) 0.016(1)	0.024	0.027	0.029	0.029	0.027(5/3) 0.021(1)
T_w	300(5/3) 390(1)	260	230	220	220	230(5/3) 300(1)
ω_1^2	1.0(5/3) 1.7(1)	0.76	0.59	0.52	0.52	0.59(5/3) 1.0(1)
$\dfrac{\Delta h}{\lambda}$	−0.027	0	0.07	0.14	0.21	0.42

[a]Values of N_{ac}, T_w and ω_1 are given for $\gamma = 5/3$ and $\gamma = 1$ at $\tau = 10$ and 10^{-5}.

Values of N_{ac}, the corresponding wave period, T_w, and the value of ω_1^2 for a 300 s wave are given in rows 5, 6 and 7 of Table X-3.

In order to estimate k_{ix}^2, we adopt a horizontal wave number of 2.5×10^{-4} to 5×10^{-4} km^{-1}, corresponding to distances of 4000 and 2000 km, respectively, for the size of the oscillating features. The scale height in the photosphere is approximately 180 km. Thus, $k_{ix} = 2Hk_x \approx 0.09 - 0.18$ and $k_{ix}^2 \approx 0.01 - 0.03$.

If we use the preceding estimates of ω_1^2 and k_{ix}^2 in the diagnostic diagram shown in Figure X-1, then the observed waves seem to lie clearly in the evanescent region to the lower left in the diagram. It should be noted, however, that Figure X-1 is based on several assumptions, including the assumptions that radiation losses are not important in the energy balance and that the atmosphere is more or less homogeneous over one wavelength. Both of these assumptions are invalid in the photosphere. A wave of 300 s period moving with the photospheric sound speed, for example, has a wavelength of approximately 2000 km. This is approximately 10 vertical scale heights and approximately 5 granule diameters. Thus, neither the vertical nor the horizontal structure of the photosphere is sensibly uniform over one wavelength (Ulrich, 1970).

At deep photospheric levels and in the middle chromosphere most of the heat energy required to bring about an increase in temperature ΔT is used up in ionizing the hydrogen gas, which is mostly neutral. Under these conditions γ will approach unity. The two values of N_{ac}, T_w and ω_1^2 given in Table X-3 represent the two cases $\gamma = 1$ and $\gamma = 5/3$. It should be noted that at $\tau = 10$ both $\gamma = 1$ and $\gamma = 5/3$ give values of N_{ac} that are equal to or greater than the observed wave frequency, i.e., $\omega_1 \geqslant 1$. Thus, if the regions at $\tau = 10$ and deeper are important in determining the wave frequency, then the waves might very well be propagating compressional waves. The last row in Table X-3 gives the ratio of the height interval between the particular depth and the depth at $\tau = 1$ to the approximate wavelength of 2000 km. Note that the depth $\tau = 10$ is only a small fraction of a wavelength from the depth $\tau = 0.1$. It is not unreasonable to suppose, therefore, that sound waves of 300 s period observed in the photosphere may very well have properties characteristic of sub-photospheric regions. By the same token, we should not exclude the possibility that the 300 s oscillations are propagating waves even though they fall in a region of the ω_1, k_{ix} plane which, at photospheric depths, corresponds to evanescent waves.

The influence of radiation on wave structure has been investigated by a number of authors, including Spiegel (1957), Souffrin (1966) and Stix (1970). For a small amplitude perturbation in an optically thin medium in local thermodynamic equilibrium the radiative cooling time is given by (Spiegel, 1957)

$$t_r = \frac{\rho C_v}{16 \bar{K} \sigma T^3} , \tag{X-70}$$

where $\bar{K}$ is the mean, linear absorption coefficient, and the other symbols have their usual meaning. We may rewrite this expression for t_r as

$$t_r = \frac{1.4 \times 10^{11}}{K' T^3} ,$$

where K' is the mean opacity g^{-1}. Values of K', T and t_r for the HSRA are given in Table X-4.

TABLE X-4
Estimated values of t_r for the HSRA

τ	1	0.1	0.01	0.001
T	6390	5160	4660	4380
k'	0.71	0.11	0.039	0.014
t_r(s)	0.78	90	35	120

Souffrin (1966) has shown that in order for gravity waves to exist the radiative cooling time must be sufficiently long that $t_r > 1/2\gamma N_{BV}$, or from Equation (X-29) $t_r > 1/4(\gamma-1)^{1/2} N_{ac}$. For $\gamma = 5/3$ and a mean photospheric value of $N_{ac} = 0.05$ (Table X-3), the required limit on t_r is $t_r > 12$ s. This appears to exclude gravity waves from depths, where $\tau > 0.03$ because of the large radiative damping.

In order for sound waves to propagate in the presence of radiation, Souffrin (1966) finds that ω_i^2 must exceed a value given by

$$\omega_i^2 > \tfrac{1}{2} \left\{ 1 - \frac{\gamma}{t_r^2 N_{ac}^2} + \left[\left(1 - \frac{\gamma}{t_r^2 N_{ac}^2}\right)^2 + \frac{4}{t_r^2 N_{ac}^2} \right]^{1/2} \right\}. \tag{X-71}$$

This inequality reduces to the following three cases

$$\omega_i^2 > \frac{1}{\gamma} = 0.6, \qquad \frac{\gamma}{t_r^2 N_{ac}^2} \gg 1;$$

$$\omega_i^2 > \frac{1}{\gamma^{1/2}} = 0.77, \qquad \frac{\gamma}{t_r^2 N_{ac}^2} = 1;$$

and

$$\omega_i^2 > 1, \qquad \frac{\gamma}{t_r^2 N_{ac}^2} \ll 1.$$

At $\tau = 1$, $\gamma/(t_r N_{ac})^2 \approx 5000$, and, at $\tau = 0.01$, $\gamma/(t_r N_{ac})^2 \approx 1.6$. Thus, it appears that the effect of radiation on the acoustic waves in the low photosphere is to reduce the propagating range to $\omega_i^2 \approx 0.6$, which is below the values given in Table X-3. On the other hand, the assumption underlying the expression used for t_r, that the atmosphere is optically thin begins to fail in the low photosphere, and we have probably underestimated t_r.

About all that can be concluded about the 5 min oscillations observed in the photosphere is that they are near the acoustic cut-off limit. The long wavelength of the acoustic waves relative to the density scale height and the tendency for both the critical value of ω_i^2 and N_{ac} to decrease as τ increases enhances the likelihood that the waves are propagating acoustic waves, but this conclusion is not without ambiguity. It does seem quite certain, however, that the 5 min oscillations are not gravity waves, and, for that matter, that no gravity waves exist in the deeper photosphere.

Some authors have attempted to use the phase relations between velocity and brightness for a fixed line and between velocity at two different depths to determine the nature of the waves. It seems clear from the observations that the velocity peak lags the brightness peak at a given depth and that the velocity peak occurs progressively later with increasing height.

The velocity phase lag with height has been most carefully (and presumably most accurately) considered by Canfield and Musman (1973) and Tanenbaum (1971) using lines whose depths of formation have been treated quantitatively and realistically. Canfield and Musman find for lines formed over a range of heights of some 800 km that the vertical phase velocity is of the order of 100 km s^{-1}. Tanenbaum (1971) found somewhat lower phase velocities of 50 to 80 km s^{-1}. In both cases, the phase velocity is much in excess of the sound velocity.

The wave group velocity, v_g, is related to the phase velocity, v_{ph}, through the equation

$$v_g = \frac{v_s^2}{v_{ph}} \, .$$

Using the preceding values for v_{ph} and $v_s = 7 \ km \ s^{-1}$, we find $v_g \approx 0.5 \ km \ s^{-1}$ for Canfield and Musman's results and $v_g \approx 1$ to $0.6 \ km \ s^{-1}$ for Tanenbaum's results. The combination of low group velocity and high phase velocity suggests that the waves contain a mixture of propagating and standing components, which is consistent with the conclusion that the waves are located near the acoustic cut-off frequency in the k_{ix}, ω_1 plane. Although some authors have reported somewhat different phase lags, hence different phase velocities, the conclusion that v_g is finite and less than v_s seems consistent with nearly all of the observations.

Velocity-brightness phase lags also shed some informaton on the nature of the waves, but the data are not easily interpreted. In a simplistic interpretation, the peak brightness leads the peak velocity by $90°$ for a pure standing wave, whereas for an upwards moving sound wave with group velocity equal to v_s the brightness and velocity peaks are in phase. For the case where v_g is less than v_s, the peak brightness should lead the peak velocity by something greater than $0°$ and less than $90°$.

Tanenbaum et al. (1969) found phase leads of $40 \pm 15°$ for a deep photospheric C I line and $98 \pm 8°$ for the D_1 line of Na I formed in the high photosphere – low chromosphere. They concluded from this that the waves are propagating in the deeper photosphere, but become standing waves in the high photosphere – low chromosphere. It must be remembered, however, that when one looks at first a brightness signal then a velocity signal in a spectral line the layers of the atmosphere producing the two signals do not necessarily coincide, i.e., the two phenomena may occur in different regions of the contribution function and may be displaced from each other by as much as a scale height. If this is the case, which is not unlikely, then the phase relationship observed is not the phase relationship that exists at a given depth. For this reason, the conclusion that the waves become standing waves in the low chromosphere should not be accepted without strong reservation. In fact, the study by Canfield and Musman (1973) covers approximately the same height range as the study by Tanenbaum et al. and the former authors find no clear evidence for a change of phase velocity with height.

Very little is known at the time of this writing concerning the nature of the horizontal motions associated with the oscillatory phenomena. Before the waves can be positively identified with respect to their mode and propagation characteristics, both the vertical phase relations and the horizontal propagation need to be studied in more detail.

Several models for generating the 5 min oscillations have been proposed. In the following, we review briefly those models that seem most pertinent.

B. MODELS FOR 5 MINUTE OSCILLATIONS

Stein and Leibacher (1974) have recently reviewed the proposed models for the 5 min oscillations. They classify the models into two types: (1) resonant 'ringing' of the photosphere in response to cellular convection associated with sub-photospheric granules, and (2) trapping of waves in specific layers of the atmosphere. This latter class includes both trapped acoustic and trapped gravity waves and the trapping regions range from the upper convection zone to the chromosphere.

The first class of models has been discussed by Schmidt and Zirker (1963), Meyer and Schmidt (1967), Stix (1970), and Schmidt and Stix (1973). Several objections can be made to this class of models. In such a model, it would be expected that the wave period would be characteristic of the photosphere and further that the waves would not be observed at the interface between the stable layers of the photosphere and the top of the convection zone. As we have seen, the acoustic cut-off frequency in the photosphere itself is too high to be in agreement with the observed 300 s period. Also, the waves are observed even in the deepest layers of the photosphere where the granulation occurs. Neither of these arguments are very compelling, however. The wavelength of the waves is much too great to restrict the identification of the wave frequency with the photospheric values of N_{ac}, and the contribution function for a given optical frequency is so broad that two phenomena such as granules and oscillations may be physically separated by more than one density scale height but still show in observation that appear to come from the same depth.

More convincing arguments against the interpretation of the 5 min oscillations as standing waves excited by granular convection are the lack of any marked correlation between the appearance of a granule and the onset of an oscillation (Frazier, 1968). Also, Kato (1966) has argued that the 5 min oscillations decay much too quickly to be a resonant mode induced by sub-photospheric convection.

Three regions favorable to trapping of waves are present near the visible layers of the Sun's atmosphere. These occur in the upper convection zone, the temperature minimum region and the middle chromosphere. Trapping in the upper convection zone has been discussed by Ulrich (1970), Leibacher and Stein (1971) and Wolff (1972). Trapping of acoustic waves in the temperature minimum region has been discussed by Kahn (1961, 1962), and trapping of gravity waves in the temperature minimum region has been discussed by Uchida (1965) and Thomas et al. (1971). Chromospheric trapping of acoustic waves has been proposed by Bahng and Schwarzschild (1963) and McKenzie (1971).

Perhaps the most pertinent observational aspect of the 5 min oscillation that has bearing on the possibility of the waves being trapped waves is the strong tendency for the energy density in the waves to decrease upwards. The energy density is given by $\frac{1}{2}\rho v_a^2$, where v_a is the velocity amplitude. Typical values of ρ and v_a in the low photosphere are 10^{-7} g cm^{-3} and 0.4 km s^{-1}. By comparison, the typical values in the low chromosphere are 10^{-5} g cm^{-3} and 1 km s^{-1}. Thus the energy density in the waves in the low photosphere is some ten times greater than the energy density in the waves in the low chromosphere. If the wave trapping occurred either in the chromosphere or in the temperature minimum region, the energy density in the waves should decrease with depth and have a lower value in the photosphere than in the trapping region. The observed opposite effect makes it unlikely that the photospheric waves result from trapping in the higher layers. A further argument given by Stein and Leibacher (1974) against the interpretation that the photospheric waves is a result of trapping at higher layers is that non-linear momentum transfer at the top of the trapping layers destroys the phase coherence within the trapping layer.

The preceding arguments against trapping in the higher layers apply equally as well to gravity and acoustic waves. Gravity waves were eliminated in the preceding section as a possible explanation for the deeper photospheric waves because of the short radiative cooling time. The same argument does not apply in the temperature minimum region and the chromosphere. However, internal gravity waves have the characteristic that their phase velocity and group velocity are oppositely directed. Observational evidence suggests that both the group velocity and phase velocity are directed upwards, which again argues against the gravity wave model.

Waves trapped below the photosphere where the acoustic cut-off period is 300 s or greater satisfy the two requirements that the dominant wave period be of the order of 300 s and that the wave energy density decrease upwards. Waves with periods longer than about 420 s are not trapped at the lower boundary and waves with periods less than about 220 s propagate on through the photosphere and temperature minimum region (Stein and Leibacher, 1974).

Waves trapped in sub-photospheric layers become evanescent waves in the photosphere, which appears, superficially, to argue against this mechanism as an explanation for the photospheric waves. However, Zhugzhda (1972) has pointed out that waves of 300 s period propagate in the chromosphere as well as in the sub-photosphere and, as a result, some of the wave energy will 'tunnel' through the photospheric barrier. Thus, one would expect to observe some propagation velocity.

Of the various ringing and trapped wave models proposed, the sub-photospheric trapping model seems to be in best overall agreement with observations. While this model is favored at the present time, however, the matter is by no means closed. The observations remain at a relatively crude stage, and the theoretical basis for the diagnostic tools is far too simplistic to permit a conclusive decision.

A third possibility for the 5 min oscillations is that they are simply sound waves generated by the compressional action of convection cells in the convection zone and are propagating through the photosphere. Stein (1967, 1968) has shown, however, that most of the sound energy generated in the convection zone is carried in waves whose periods lie between 20 and 60 s. Much less energy is present at 300 s, so one still has to appeal to trapping or selective filtering to account for the observed 300 second period. In this connection, Souffrin (1966) has suggested that the high acoustic cut-off frequency in the photosphere makes the photosphere act as a high-pass filter. The 300 s period is then interpreted as the filtered output of a broad band input. This model carries some of the same objections as the resonant 'ringing' models.

3. Generation of Waves by the Convection Zone

It is generally assumed that all of the surface phenomena of the Sun that are not predicted by a hydrostatic, radiative equilibrium model are traceable, one way or another, to the hydrogen convection zone. Both the 5 min oscillations and the heating of the chromosphere and corona fall within this category. It is tempting to relate the two phenomena by supposing that the 5 min oscillations are indeed the waves that provide the heating. That this is not totally implausible can be seen by estimating the energy flux in the waves. Thus, if one supposes that the oscillations propagate vertically with a group velocity equal to the sound speed of 7 km s^{-1} and have a velocity amplitude of 0.5 km s^{-1} at a depth where the matter density is 10^{-8}, then the energy flux, $\frac{1}{2}\rho v_a^2 v_s$, in the waves is of the order of $10^7 \text{ erg cm}^{-2} \text{ s}^{-1}$, which is ample to heat both the chromosphere and corona. On the other hand, the available evidence suggests that the group velocity is much less than v_s. Also, as we shall note in the following section of this chapter, waves whose periods are of the order of 300 s suffer very little dissipation in the low chromosphere. Thus, since most of the heating is required in the low chromosphere, the 5 min oscillations are not good candidates for the primary heating. Nevertheless, they may play an important role in heating the high chromosphere and the corona.

It is important that we understand the various mechanisms for producing wave energy in the Sun together with the nature and energy spectra of the waves that are produced. At present, however, neither the energy flux nor the energy spectrum of the waves is understood in more than crude outline. Existing theories of wave generation rely upon either generation within the convection zone itself or generation in the stable layers overlying the convection zone. The latter category of waves includes those generated by convective overshoot as convection cells boil over into the stable layers, those generated by overstability and those generated by the Eddington valve mechanism. Waves generated within the convection zone itself are primarily produced by the convective cells interacting with the ambient medium or with each other. Each of these mechanisms will be discussed briefly in the remainder of this section.

A. CONVECTIVE OVERSHOOT

A convectively stable layer overlying a convectively unstable layer is not totally shielded from the convection within the unstable layer. Statistical fluctuations within the unstable layer will lead to convective elements arriving at the base of the stable layer with some remaining momentum and some remaining bouyancy force. In such cases, the convective eddy will penetrate into the stable layer as a 'convective tongue' (cf. Lighthill, 1967). This phenomena is familiar to meteorologists as 'thermals' and 'plume' or 'chimney' clouds. The latter are relatively frequent cloud forms in tropical latitudes.

In the mixing length theory of convection one usually equates the bouyancy force, $ga\Delta T$, where a is the coefficient of expansion, to the stress due to eddy viscosity, $\nu v l^{-2}$, where l is the mixing length and ν is the eddy viscosity. One then sets $\nu = vl$ and obtains the result

$$v = (ga\Delta Tl)^{\frac{1}{2}}.$$ (X-72)

The temperature fluctuation ΔT is given by

$$\Delta T = -1 \left[\frac{\mathrm{d}T}{\mathrm{d}h} - \left(\frac{\mathrm{d}T}{\mathrm{d}h} \right)_{\mathrm{ad}} \right]$$

Thus, $v = 0$ when the temperature gradient equals the adiabatic gradient, and we see that in this formulation there is no convective overshoot. Moore (1967) notes that this formulation neglects terms of the type $\partial v/\partial t$ and $\partial T/\partial t$, and that it is the neglect of these terms that suppresses the overshoot.

No really satisfactory method of dealing with overshoot problems has been developed. In the case of thermals in the Earth's atmosphere one of the major effects is the intake of material from the ambient medium by the hot bubble, which acts to decelerate the bubble. Most of the intake occurs by entrainment in the wake of the bubble, and the importance of this effect depends upon the size of the bubble and its momentum. Magnetic field effects undoubtedly complicate the solar case.

Howe (1969) has suggested a model in which overshooting tongues of turbulence produce gravity waves which move upwards inducing torsional MHD waves in the magnetic field lines. He investigates the properties of a spherical vortex supposedly produced by this process but does not treat the generation of the vortex itself.

A somewhat similar problem has been treated by Musman (1972) in connection with the phenomena of 'exploding' granules. In Musman's model, the granule eddy is visualized as a circulating cell with upwards motion along its central axis and downward motion at its borders as in a typical Bernard Cell. The hot granule rises until it meets the stable layer. Conservation of angular momentum then causes the granule to spread into a vortex ring that expands laterally. Motion pictures of granule evolution lend some support to a model of this type.

Aside from these rather sketchy cases, the questions raised by convective overshoot at the top of the solar convection zone and its possible role in generating wave phenomena in the higher layers of the photosphere have not been investigated.

B. OVERSTABILITY

A fluid element moving vertically in the solar atmosphere under the combined influences of thermal instability, a restoring force and energy dissipation may oscillate about an equilibrium position with an amplitude that increases with time (Spiegel, 1964; Moore and Spiegel, 1966). Such a situation is said to be 'overstable'. Overstability is thus a particular manifestation of thermal instability applied to the case of oscillatory motion. However, Syrovatsky and Zhugzhda (1968) and Ulrich (1973) have shown that overstability may occur in some cases even when the temperature lapse rate is subadiabatic. The original discussion of overstable oscillations by Chandrasekhar (1953) was applied to convection in a rotating body. In this case, rotation supplies the necessary restoring force for the oscillation. More recent applications by Spiegel (1964) and Moore and Spiegel (1966) discuss the mechanism of overstability with respect to wave motions where pressure forces or magnetic forces provide the restoring force.

In the case of adiabatic oscillations in a convectively unstable atmosphere, a parcel (bubble) of fluid passing through the origin at the beginning of the upper half cycle of an oscillation cools adiabatically and expands during the upward phase of the cycle. Since the surrounding medium has a superadiabatic lapse rate, the bubble experiences an upward buoyancy force. In the downward path, the bubble is compressed and heated adiabatically, but it continues to experience a buoyancy force. At a given distance from the equilibrium position the buoyancy force is the same regardless of whether the bubble is moving upwards or downwards. All of the processes acting on the bubble are reversible. Thus, at each passage of the origin the bubble has the same velocity, and the oscillation is stable.

Radiative dissipation introduces an irreversible process and leads to an asymmetry in the buoyancy forces. During the upward half cycle of the oscillation, for example, the oscillating bubble is hotter than its surroundings and has an additional cooling due to loss of energy by radiation. This excess cooling continues throughout most of the upper half cycle with the result that on the downward part of the cycle the bubble is cooler and has less buoyancy force than it had at the same distance from the equilibrium position on the upward part of the cycle. Since the buoyancy force is directed upwards, the bubble has experienced a net decrease in the upward force during the upper half of the cycle. As a result the bubble arrives back at the origin with a downward velocity that exceeds the upwards velocity at the previous passage of the origin. Similarly, during the lower half cycle the bubble is heated by radiation from the surroundings and experiences a net increase in the buoyancy forces. Thus, the next passage of the origin is at a still higher velocity. In the idealized case treated by Moore and Spiegel (1966), the amplitude of the oscillation grows exponentially with time with a time constant of the order of the radiative cooling time, t_r.

Leibacher (1971) has suggested a second mechanism for producing overstable oscillations in the upper convection zone and low photosphere. The so-called, Eddington valve mechanism becomes effective when the opacity is an increasing function of temperature.

It was noted in the discussion of the line blanketing effect in Chapter IX that an increase in opacity at photospheric depths produced a local heating. Added continuum opacity produces an entirely similar effect. Thus, when an oscillating volume element with $d\kappa/dT > 0$ is heated by adiabatic compression during its descending phase it experiences an additional heating from its increased opacity. Conversely, during the upward expansion and cooling phase the volume element experiences additional cooling due to reduced opacity. The net effect, again, is a growth in the amplitude of the oscillation.

The photospheric opacity is mainly due to H^- and may be expressed in the form

$$\kappa = \alpha N_e T N_H, \qquad\qquad (X\text{-}73)$$

where α is a slow function of T, which we shall ignore. We are interested in the case where hydrogen is only partially ionized but in which most of the electrons are supplied by hydrogen. (Note that if the electrons come mainly from the metals both N_H and N_e will be essentially independent of temperature.) For this case, we may approximate the temperature dependence of κ by

$$\frac{d \ln \kappa}{dT} = \frac{d \ln N_e}{dT} + \frac{1}{T} \tag{X-74}$$

To evaluate $d \ln N_e/dT$, we note that the ground state population of neutral hydrogen is essentially independent of T when hydrogen is predominantly neutral and is given by

$$N_1 = \text{const } N_e N_p T^{-3/2} \exp (157\,000/T). \tag{X-75}$$

Since we are considering the case where $N_e \approx N_p$, the logarithmic derivative of Equation (X-75) gives

$$0 = 2 \frac{d \ln N_e}{dT} - \frac{3}{2T} - \frac{157\,000}{T^2}.$$

Solving this expression for $d \ln N_e/dT$ and substituting into Equation (X-74), we find

$$\frac{d \ln \kappa}{dT} = \frac{78\,500}{T^2} + \frac{1.75}{T}. \tag{X-76}$$

It follows that for the case we have assumed, viz., $N_e \approx N_p$ and hydrogen mainly neutral, κ is an increasing function of T. These conditions prevail in the low photosphere and the upper convection zone. At large depths hydrogen becomes strongly ionized and in the high photosphere hydrogen is so weakly ionized that the electrons come mainly from metals.

Overstable oscillations have the interesting property that they extract energy from the radiation and thermal reservoir directly into wave energy. Ulrich (1973) has shown that the evanescent waves trapped in the photosphere with periods of 200 to 400 s are overstable in the fundamental mode and first two or three overtones. He proposes this overstability as the source of the 300 s oscillations and, subsequently, as the source of energy for heating the corona. In a somewhat similar vein, Wolff (1972) has proposed that a combination of overstability and/or the Eddington-valve mechanism will drive non-radial oscillations of the entire Sun. Since the driving forces are near the solar surface only the high-order modes are excited. A number of authors have proposed overstability as a source of wave phenomena in sunspots where magnetic field strengths are high.

As noted by Ulrich (1970), the wavelength of the overstable modes in the photosphere is of the order of 1000 to 2000 km., whereas the photospheric layers in which overstability occurs are only a few hundred km thick. The mathematical treatment of the oscillations assumes that the wavelength is small compared to the characteristic scale of the atmosphere and, therefore, is only a crude approximation at best. Nevertheless, this mechanism of wave generation is very promising and deserves considerable attention.

C. CONVECTIVE COMPRESSION

In all cases of unsteady fluid flow, the fluid motions give rise to local pressure fluctuations. The pressure fluctuations that propagate away from their source region are sound waves. Those that do not propagate are referred to as pseudo sound. Our interest here is in the propagating sound waves produced by the turbulent eddies within the convection zone itself. The mechanism is simply the statistical fluctuations in pressure that are expected to occur in association with the convective eddies.

A reliable, semi-quantitative description of sound generation by a turbulent medium was first developed by Lighthill (1952) and was first applied to the problem of coronal heating by Proudman (1952). This mechanism for heating the corona and chromosphere has injoyed considerable popularity. Lighthill (1967) has expressed some pessimism concerning the quantititive aspects of the energy that could be made available in this way, but this pessimism was based on an incorrect notion of the amount of energy required. Osterbrock (1961) had estimated that the amount of mechanical energy required for the chromosphere and corona was about 10^{-3} of the solar constant, whereas Lighthill's (1967) estimate of the maximum acoustic energy available was of the same order. However, Lighthill estimated that only a small fraction of the total acoustic energy would be available for heating. As we have seen in Chapter IX, Osterbrock's (1961) estimate is too high by about an order of magnitude. The currently accepted energy requirements leave the distinct possibility that the sound generated by the Lighthill mechanism may indeed be sufficient to supply the necessary mechanical energy. Attempts to apply this theory of heating to the chromosphere and corona will be discussed in Part 4 of this chapter. Our discussion here is limited to a brief description of the nature of the sound generation. For more thorough discussion, the reader is referred to the review article by Ffowcs-WIlliams (1969), to a paper by Stein (1967) and to chapter on Acoustics in Moving Media in Morse and Ingard (1968). The following discussion is derived from these sources plus a review article by Leibacher (1974), which is not widely available.

In order to help the reader understand the nature of the Lighthill mechanism, we will comment on three modes of motion associated with turbulent eddies — the monopole, dipole and quadrapole modes. Only the latter of them is of major interest in the Lighthill mechanism, but it is important to understand why the first two are of lesser interest.

By a monopole in a turbulence medium one means a single, moving vortex element. A simple analogy is a piston moving into a stationary fluid. Associated with such a system is a local pressure fluctuation, P. Bernoulli's law requires that

$$P + \tfrac{1}{2}\,\rho v^2 + \Omega = \text{const}, \qquad\qquad (\text{X-77})$$

where P is the local pressure, ρ is the local density, v is the local velocity and Ω is the gravitational potential, which we assume to be constant. Since the velocity fluctuation is expected to be of the order of v, Bernoulli's law requires that the pressure fluctuation be of the order of ρv^2. If the pressure fluctuation occurs over a scale length d, the sound pressure at $R > d$ is proportional to

$$P_s \propto \rho v^2 \frac{d}{R} \tag{X-78}$$

and the sound energy flux is proportional to

$$F_1 \propto R^2 \frac{p_s^2}{\rho v_s} . \tag{X-79}$$

These two proportionalities, together with the definitions

$$T_w = \frac{d}{v}, \qquad W = \rho v^2 d^3$$

and

$$M = \frac{v}{v_s},$$

yield

$$F_1 \propto \frac{W}{T_w} M. \tag{X-80}$$

The quantity W is proportional to the total kinetic energy associated with the velocity amplitude of the sound wave and T_w is the characteristic time for the pressure fluctuation, i.e., the wave period. The important conclusion to be drawn from Equation (X-80), therefore, is that the efficiency with which energy is converted from the statistical pressure fluctuation of a monopole source to sound energy is proportional to the mach number M. In arriving at this conclusion, it has been assumed that the sound pressure does not react back on the fluid motion, which amounts to assuming that M is small compared to unity. The formalism breaks down when M approaches unity.

The solar convection zone is bordered by relatively quiescent media with no net exchange of mass or momentum. It follows that a volume of space that is large compared to the size of a typical convective eddy will contain a large number of monopoles whose phase distribution is random. Thus, positive and negative pressure fluctuations will be uncorrelated and will average to zero with the net effect that no sound energy is radiated.

Two nearby monopole sources moving out of phase can be considered acoustically as a dipole source. For such a source, it is well known that the pressure fluctuation at large distance from the source is reduced from the pressure fluctuation due to one of the monopoles making up the dipole by a factor d/λ, where d is the dipole length and λ is the wavelength of sound. Since

$$\lambda = v_s T_w = v_s \frac{d}{v} = \frac{d}{M},$$

and since the energy flux is proportional to the square of the pressure fluctuation, the energy flux for the dipole is proportional to

$$F_2 \propto \frac{W}{T_w} M^3 . \tag{X-81}$$

Again, we note that by assumption $M \ll 1$, and, hence, that the dipole is much less efficient than the monopole as a source of sound energy in turbulent flow. Also, as was the case for the monopole, a turbulent layer bordered by quiescent layers has no net dipole component. A net dipole can exist only if the boundary regions transmit momentum to the surrounding layers, which is not the case if the surroundings are quiescent.

Two adjacent dipole structures that are moving in antiphase with respect to each other constitute an acoustic quadrapole. At large distance from such a source, the sound pressure is reduced by an additional factor $d/\lambda = M^{-1}$ from the dipole source. Hence, the sound energy flux is now given by

$$F_4 = C_4 \frac{W}{T_w} M^5 , \tag{X-82}$$

where C_4 is a constant of proportionality. Unlike the cases for the monopole and dipole, however, a turbulent layer bordered by quiescent layers can have a net quadrapole moment. For this reason, sound energy generated by the solar convection zone is thought to be associated predominantly with quadrapole radiation.

Note that since $W \propto v^2$ and $T_w \propto v^{-1}$ the sound energy flux from a quadrapole is in reality proportional to v^8. Also, a more vigorous derivation of the acoustic energy flux gives a value for C_4 that is much different from unity.

In a detailed treatment of the acoustic energy flux resulting from a turbulent flow, the scale length, d, is replaced by a spatial spectrum of turbulent eddie sizes and the characteristic time, T, is replaced by a frequency spectrum. For each of these distribution functions and for the velocity distribution function as well, we are forced to rely upon the mixing length theory of turbulence. The result is that neither the velocity amplitude nor the spatial and frequency spectra are well known.

Figure X-4 shows two velocity distributions computed from mixing-length theory by Vitense (1953) for different assumed values of the mixing length, l_0. In the upper 500 km of the convection zone, which extends from $\log P_g \approx 5$ to 6, the ratio of velocities given by the two cases is about 1.4:1. The factor 1.4 raised to the eighth power equals 15. Since most of the acoustic flux can be expected to come from this upper region of the convection zone where v is largest, the acoustic power predicted for the two velocity distributions differ by a factor of about 15.

The peak values of v shown in Figure X-4 correspond to mach numbers of 0.23 and 0.16. These are large enough that some question exists as to whether the assumption that the acoustic energy does not react back on the turbulent energy is questionable (Stein, 1967).

The frequency and spatial spectra of turbulence are both uncertain. Stein (1967) considers several forms of the spatial spectrum, two of which are shown in Figure X-5. In

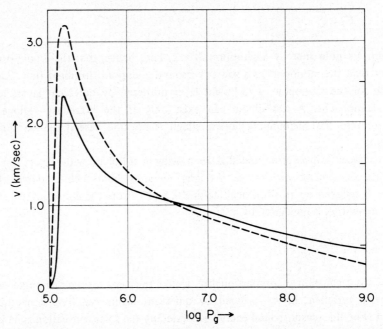

Fig. X-4.   Values of turbulent velocity predicted for the solar convection zone by Vitense (1953) for two values of the mixing length: ———, $l_0 = H$; – – – – –, $l_0 = 2H$. $P_g = 10^5$ occurs near $\tau_s \approx 0.5$, which is some 40 km above the surface $\tau_s = 1$. The depth where $P_g = 10^6$ lies about 500 km below the surface $\tau_s = 1$.

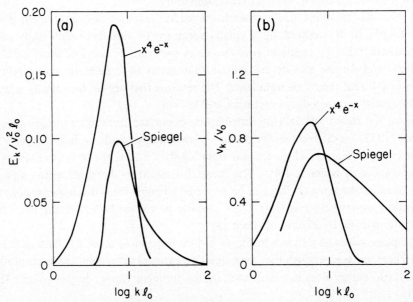

Fig. X-5.   Spatial energy (a) and spatial velocity (b) spectra of turbulence used for the computation of the acoustic flux in Figure X-6. For the curves labeled $x^4 e^{-x}$, $x = (2/\pi)l_0 k$. The curves labeled Spiegel represent a particular case of a family of spectra derived by Spiegel (1962). Both sets of curves are taken from Stein (1967).

both cases, the spectra satisfy the proper asymptotic conditions at large and small k and are properly normalized. Although the two cases have been chosen so that the maxima occur at nearly equal values of k, they differ quite markedly in their behavior at $k/l_0 \ll 10$ and $k/l_0 \gg 10$. Different spatial energy spectra are associated, of course, with different spatial velocity spectra. The spatial velocity spectra corresponding to the two energy spectra are shown in Part b of Figure X-5. This latter plot emphasizes the difference in the two spectra at high wave numbers.

For the frequency spectrum Stein (1968) adopts two cases: an exponential form

$$\phi_\omega = \frac{1}{k v_k} \exp - (|\omega|/k v_k) ; \tag{X-83}$$

and a gaussian form

$$\phi_\omega = \frac{2}{\pi^{1/2} k v_k} \exp - (\omega/k v_k)^2 . \tag{X-84}$$

The acoustic energy flux computed for three combinations of the spatial spectra in Figure X-5 and the two frequency spectra are shown in Figure X-6. The integrated power for each of the three cases is

$$F_{EE} = 9.9 \times 10^7 \text{ erg cm}^{-2} \text{ s}^{-1},$$

$$F_{EG} = 7.2 \times 10^6 \text{ erg cm}^{-2} \text{ s}^{-1},$$

and

$$F_{SE} = 5.5 \times 10^7 \text{ erg cm}^{-2} \text{ s}^{-1} .$$

Thus, the Gaussian frequency spectrum produces an order of magnitude less power than the simple exponential spectrum, and the Spiegel wave number spectrum produces about half the power of the $\chi^4 e^{-\chi}$ spectrum. All of the above cases are computed for the velocity distribution for $l_0 = 2H$ in Figure X-4.

Considering all of the different uncertainties in the computation of the acoustic flux, Stein (1968) adopts a value of 2×10^7 erg cm^{-2} s^{-1}, but he cautions that this figure is uncertain by about an order of magnitude. It is of interest to note that most of the energy falls between $\omega = 0.08$ to 0.3, corresponding to periods of approximately 80 to 20 s in the EE case and between $\omega = 0.1$ to 0.4, corresponding to periods of approximately 60 to 15 s, in the SE case. For the Gaussian case the maximum power occurs near a period of 80 s. As we shall later see, these results are in general agreement with the required frequency spectrum for heating the low chromosphere.

In an effort to find more efficient mechanisms for generating wave energy from the turbulent eddies of the convection zone, some authors have suggested that the force fields present in the convection zone will lead to dipole and monopole radiation. Unno (1964) showed that buoyancy forces derived from the gravitational potential will result in both

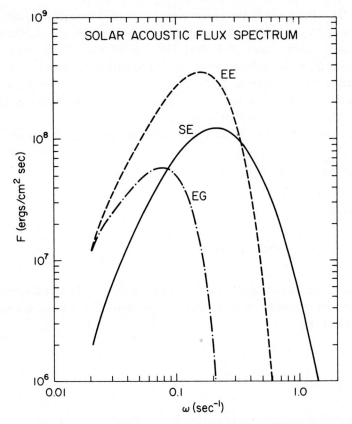

Fig. X-6.   Acoustic power output of the convection zone in the vertical direction computed by Stein (1968) for: the $x^4 e^{-x}$ wave number spectrum and exponential frequency spectrum (EE), the $x^4 e^{-x}$ wave number spectrum and Gaussian frequency spectrum (EG), and the Spiegel wave number spectrum and exponential frequency spectrum (SE) (courtesy *Astrophys. J.*).

dipole and monopole radiation. However, Stein (1967, 1968) found that after taking these effects into account the quadrapole radiation was still dominant.

Magnetic fields offer a second possibility for body forces that may couple with wave generation to produce monopole and dipole modes. The generation of sound waves is unaffected by the presence of a uniform magnetic field. However, Parker (1964) has shown that turbulent eddies impinging on concentrations of vertical magnetic field will produce magneto-acoustic waves via the monopole mechanism. This much more efficient mechanism for extracting wave energy from the turbulence energy provides the possibility for additional strong mechanical heating over sunspots and magnetic knots. Also, Kulsrud (1955) has shown that the presence of turbulent magnetic field structure leads to enhanced efficiency in the generation of sound waves. For the case in which the turbulence hydrodynamic energy density and the turbulence magnetic energy density are equal, the efficiency of producing sound energy flux increases by an order of magnitude. This latter mechanism may serve to increase the efficiency of the generation of acoustic flux where the measured vertical field strengths are low. However, this suggestion is purely speculative at this point.

In summary, we note that the suggestion that the chromosphere and corona are heated by sound waves generated in or by the convection zone appears to be very plausible. Whether the wave energy is actually generated by convective overshoot, overstability or convective compression within the convection zone remains an open question, however. Similarly, the role of magnetic fields remains somewhat ambiguous other than the conclusion that in regions of strong vertical fields both he energy flux in wave form and mechanical heating are undoubtedly enhanced.

4. Heating

The dissipation of compressional waves propagating upwards from the convection zone provides the most plausible and most widely accepted explanation for the mechanical heating in these layers. Magnetic fields clearly lead to enhanced efficiency in the heating process, and this added heating probably involves non-compressional wave modes. Nevertheless, it seems clear that chromospheric heating, at least, is not restricted to regions of strong magnetic field, and it follows that heating by either acoustic or gravity modes is of major importance. Since vertical propagation of energy in gravity wave modes appears to be inefficient, the remaining likely candidate is the acoustic mode. As we have shown in the preceding section, there is a strong possibility that ample energy is produced in acoustic modes in or near the convection zone to provide the necessary mechanical flux for heating the chromosphere and corona.

A complete theory of heating by the acoustic mode waves is non-existent. Such a theory would start with a fixed spectrum of energy flux generated in the convective zone and would follow the propagated flux to its regions of dissipation. This dissipated flux would be balanced against radiative, conductive and convective losses. Part of the acoustic flux of interest lies at frequencies that do not propagate in the photosphere. However, these frequencies tunnel through the evanescent region (Zhugzhda, 1972) in a non-linear fashion. Also, as the waves propagate upwards they reach shock strength where non-linear effects are again important. The steep temperature gradient in the transition region and the pronounced inhomogeneities existing in the upper chromosphere and corona lead to complicated reflection and refraction effects that must be taken into account. An additional complication arises from the loss of wave energy directly as radiation from the shock heated material behind the shock front (Ulmschneider, 1971a,b; Stein and Schwartz, 1972; Ulrich, 1973). An appreciable fraction of the wave energy can be lost by this process, but the exact amount is clearly dependent upon the nature of the shock wave itself. Since only a small fraction of the wave energy incident on the low chromosphere survives to reach the corona, any coronal model derived from waves followed through the chromosphere will be very sensitive to small errors in the dissipative and refractive phenomena.

The difficulties of following the wave energy through the photosphere and of properly treating the shock formation, radiation losses from the shocked material, refraction and dissipation are formidable problems by themselves. The other aspects of the problem, viz., the radiation losses from the heated chromosphere and corona, the thermal conduction through the transition region and into interplanetary space, the acceleration of the solar wind and the generation of spicules are each complicated problems in their own right. No one of these problems is sufficiently well understood in the presence of structural and magnetic inhomogeneities to be applied properly to a self-consistent mechanically heated model. Those models that have been constructed thus far have necessarily invoked great simplifications. The typical simplifications of the atmosphere itself commonly include: spherical symmetry, no magnetic field and negligible flow velocities. The acoustic flux is usually taken in some prescribed manner at the base of the chromosphere, or at the base of the transition region, and linear theories of shock formation and propagation are normally used. Radiation losses are commonly treated by the effectively thin approximation, which is sufficient for the transition region and the corona but not for the chromosphere.

According to Leibacher and Stein (1974), acoustic waves propagating upwards in a gravitationally stratified atmosphere reach shock strength after propagating an approximate distance

$$\Delta h = 2H \ln \left(1 + \frac{\pi \gamma g}{(\gamma + 1)\omega v_0} \right) , \qquad (\text{X-85})$$

where H is the density scale height. Using the relations

$$H = \frac{v_s^2}{\gamma g}$$

and

$$\omega = \frac{2\pi}{T_w} ,$$

we may rewrite Equation (X-85) as

$$\Delta h = 2H \ln \left(1 + \frac{1}{\gamma + 1} \frac{v_s}{2H} \frac{T_w}{M_0} \right) . \qquad (\text{X-86})$$

At the base of the chromosphere, we set $\gamma = 5/3$, $v_s = 6$ km s^{-1} and $H = 130$ km to obtain

$$\Delta h = 260 \ln \left(1 + 0.07 \frac{T_w}{M_0} \right) .$$

Thus, waves at $h = 0$ with strength $M_0 = 0.1$ and period 10 s will reach shock strength at about 500 km. Waves of the same strength but with a period 30 s will reach shock strength near 800 km. It is readily seen from these results that most of the heating of the low chromosphere is done by waves with periods of the order of 10 to 30 s (Ulmschneider, 1971a,b).

For the region where the Lyman continuum of hydrogen forms, viz., near $T = 8500°$, $\Delta h \approx 1500$ km, $v_s \approx 10$ km s^{-1} and $H \approx 250$ km. These values correspond to

$$\frac{1}{\gamma + 1} \frac{v_s}{2H} \frac{T_w}{M_0} \approx 20, \quad \text{or} \quad \frac{T_w}{M_0} \approx 2700. \text{ Thus, for } M_0 = 0.1, T_w \approx 270.$$

It is clear from this result that the waves with periods near 300 s will be effective in heating the high chromosphere and corona (Ulmschneider, 1971a,b).

A. CHROMOSPHERIC MODELS

The most notable attempts to construct chromospheric models from the dissipation of acoustical energy flux are those by Jordan (1969), Ulmschneider (1971a,b), and Kaplan *et al.* (1972).

Jordan (1969) assumes a monochromatic acoustic flux with a period of 300 s and an energy dissipation given by

$$\frac{dF_m}{dh} = 1.025 \times 10^5 \rho T \eta_s^3 ,$$

where η_s is the shock strength

$$\eta_s = \frac{\gamma + 1}{\gamma - 1} \left(1 + \frac{2}{\gamma - 1} \frac{1}{M^2} \right)^{-1} - 1.$$

For $\gamma = 5/3$, this becomes

$$\eta_s = \frac{4M^2}{M^2 + 3} - 1$$

Jordan assumes a functional form for η_s given by

$$\eta_s = 1 - \exp - (h/400),$$

where h is given in km. Thus, at $h = 0$, $\eta_s = 0$ and $M = 1$, at $h = 500$, $\eta_s = 0.71$ and $M = 1.5$, and, at $h = 1500$, $\eta_s = 0.98$ and $M = 1.7$. The only defense given for these choices for η_s and M is that they do not appear to be unreasonable. As we have just seen, however, waves with periods near 300 s are not expected to reach shock strength in the low and middle chromosphere.

Jordan treats the radiation losses by forming rather crude solutions of the transfer equations for lines of hydrogen, Si II, Ca II and Mg II for the middle and upper chromosphere. These crude solutions are inadequate to properly describe the radiation losses. This is particularly true for the Balmer lines of hydrogen, which should dominate the losses in the middle chromosphere but which are rendered relatively unimportant by the approximations used. In the low chromosphere, Jordan considers radiation losses due to H⁻ ions. Conductive energy losses are unimportant in the regions of the chromosphere treated by Jordan and the chromosphere is assumed to be at rest.

As might be expected from the type of approach used by Jordan (1969), the theoretical model shows a general temperature rise in the chromosphere but gives poor quantitative agreement with empirical models.

Ulmschneider (1971a,b) used a rather different approach in that he adopts the HSRA model and then compares the predicted dissipation rates for acoustic flux with the estimated radiation loss rates. For the acoustic energy flux, he uses Stein's (1968) SE and EE spectra shown in Figure X-6. The energy dissipation is computed using a distorted wave approach (Landau and Lifshitz, 1959) with allowance for radiative dissipation of the wave energy. The latter is computed using a reasonably valid approximation to the Balmer series losses. Ulmschneider concludes that most of the chromospheric heating is done by waves with periods of the order of 20 to 40 s and that the total energy dissipation by the waves agrees within a factor of three with the radiative losses estimated for the HSRA model in the height range 800 to 1200 km. Since he considers this degree of agreement to be acceptable, Ulmschneider makes no attempt to revise the model atmosphere.

Ulmschneider (1971a) treats the radiation losses in the chromosphere in a reasonably accurate way. However, the fact that he relies upon empirical values for some of the critical parameters removes his work from the class of purely theoretical models. The general agreement he finds between mechanical dissipation and radiation losses for the HSRA model is comforting, but this does not necessarily prove that the mechanical flux computed by Stein (1968) does, in fact, predict the HSRA model.

Kaplan *et al.*, (1972) developed a self-consistent model in which the energy dissipation by monochromatic waves is balanced against radiation losses. However, they treated only the optically thin case for a gray radiator, and their derived temperature models do not agree well with empirical models such as the HSRA.

B. TRANSITION REGION AND CORONAL MODELS

Many authors have constructed models of the transition region and the corona based upon mechanical heating. There exists among these models, however, a wide range in the treatment of the conductive, radiative and convective energy processes. All of the models prior to about 1970 used poor estimates for the radiative loss terms. Also, many of the models neglected thermal conduction, which is now known to be of major importance. Similarly, most models have neglected coronal expansion, although this neglect seems not to be of major consequence. No model to date has taken into account convective motions in the transition region associated with the network and spicule structure. This neglect is likely to be serious, but until such effects are taken into account in at least a semi-valid way we can do little more than speculate about their importance.

In the following, we will comment on only those models that consider thermal conduction and radiation losses in addition to mechanical heating. Four models of this type appeared prior to 1970, but each used a radiative loss rate that is too low by one or two orders of magnitude. Thus, we will not pay much attention to the specific models resulting from these earlier computations.

In all except two cases (Uchida, 1963; Ulmschneider, 1971a), the models referred to in this section start with a given acoustic energy flux at the base of the transition region. In other words, the problems associated with propagation of the acoustic flux through the photosphere and chromosphere are largely ignored.

De Jager and Kuperus (1961) showed that the acoustic flux predicted by the Lighthill-Proudman theory was capable of producing a corona somewhat resembling the solar corona when both thermal conduction and radiation losses were included. As mentioned above, however, they grossly underestimated the radiative losses. Kuperus (1965) extended this work including approximate treatments of reflection and refraction of the acoustic waves in the transition region and the effect of the acoustic waves in providing momentum to the atmosphere. Kuperus noted, also, that the energy spectrum of the acoustic flux was of major importance, and he treated several cases of differing wave periods. The major shortcoming of his work was the continued use of a radiative loss term that is too low by about two orders of magnitude.

Uchida (1963) treated a similar problem, and, in addition, attempted to include the chromosphere. Again, however, his estimates of the radiation losses were too crude to provide realistic results. One of the interesting effects Uchida included was that arising from a horizontal magnetic field component. The effect of this field was to delay the onset of shock formation with the result of reduced heating in the low corona and an increase in the coronal maximum temperature. The horizontal field also suppressed conduction effects and steepened the transition region.

Kopp (1968) extended the treatment of the acoustic flux to include damping of the shock wave flux by both the shocked material in the wake of the shock front and the inhomogeneities and motions in the pre-shocked matter ahead of the shock front. He allowed also for expansion of the corona to form the solar wind. Although Kopp used an improved radiative loss rate over that used by previous workers, he adopted a rate that was below the Cox and Tucker (1969) rate by a about an order of magnitude.

By the time of Kopp's work the properties of the transition region were known much more reliably than they were at the time of Kuperus' and Uchida's work. Thus, Kopp was able to construct more realistic models of the corona and Kopp's theoretical model gives a maximum coronal temperature of about 1.4×10 K, which is reasonably consistent with empirical models. He identifies five sub-regions within the transition region and corona where the nature of the energy balance differs. Within the transition region (Kopp's region III), the models show that the main energy balance is set by the constant conductive flux. Both radiation and mechanical dissipation are of little consequence. This is followed by region IV near the temperature maximum where mechanical heating and radiation both become important. In region V just beyond the temperature maximum radiative losses diminish again and expansion and outwards conduction become important. Still further out, in region VI, the mechanical heating drops out and conductive flux supplies the expansion energy. This region extends from $r \approx 2R_\odot$ to beyond 1 AU. Eventually, region VII, the coronal expansion becomes adiabatic.

Ulmschneider (1971a) discusses shock heated models of the transition region and corona, but, as he did for the chromosphere, he adopts a model and then asks whether the dissipated fluxes are consistent with the model. Consistency checks of this type are useful, but they cannot claim to have the status of a theoretical model. Also, they tend to suppress most of the physically interesting problems associated with complex energy balance in the transition region and corona.

In each of the preceding models, it is assumed that the wave energy is dissipated in the corona by the formation of shocks and the associated dissipation of the shocks. More recent authors have challenged this assumption.

Moore (1972) has studied the mechanism of wave dissipation in the corona for wave periods of the order of 10 to 300 s. He concludes that the main energy dissipation mechanism in the waves is by thermal conduction and Landau damping rather than by the formation of shocks. Landau damping was suggested by D'Angelo (1969) as an important dissipative mechanism in the corona. This type of damping occurs in longitudinal oscillations in a plasma where the ion (in this case protons) mean-free-path, λ_p, is comparable to or greater than $\lambda/2\pi$, where λ is the wavelength of the oscillation. The damping mechanism is the absorption of wave energy by the ions moving in the electric field produced by the wave itself (cf. Stix, 1962).

The proton mean-free-path in the corona is given by (Spitzer, 1962)

$$\lambda_p = 8.0 \times 10^3 \frac{T^2}{\eta_p} ,$$

and the sound wavelength is given by

$$\lambda = v_s T_w = 1.7 \times 10^4 T^{1/2} T_w .$$

Thus, Landau damping occurs when

$$\frac{\lambda_p}{\lambda} = \frac{0.47 T^{3/2}}{\eta_p T_w^*} > \frac{1}{2\pi} ,$$

or when

$$T_w^* < 3 \frac{T^{3/2}}{\eta_p} .$$

In the inner corona where $T = 1 \times 10^6$ K and $\eta_p = 6 \times 10^8$ cm^{-3}, T_w^* must be of the order 5 s or less. However, near the temperature maximum where $T \approx 2 \times 10^6$ K and $\eta_p \approx 3 \times 10^8$, T_w^* increases to about 30 s, and, still further out, where η_p has decreased to $\eta_p \approx 1 \times 10^8$, T_w^* has increased to near 90 s. Since waves with periods less than about 100 s are dissipated in the chromosphere, Landau damping appears to be of major importance beyond the temperature maximum.

Inside the temperature maximum, dissipation by ordinary thermal conduction appears to dominate. The compressed regions in the wave are hotter than their surroundings and lose energy by conduction to the adjacent regions.

Moore's (1972) main interest is in the transition region rather than the corona, and he assumes that the dissipation of wave energy can be ignored in the transition region. Thus, in his model, the waves are assumed to pass through the transition region and to be dissipated in the corona. Thermal conduction and radiation then determine the transition region model.

In a recent paper, McWhirter *et al.* (1974) consider coronal and transition region models in which the acoustic flux is dissipated by thermal conduction. The energy dissipated by the acoustical flux is lost by radiation from the corona and the lower transition region. However, most of the energy radiated from the lower transition region is actually dissipated in the corona and carried through the transition region by thermal conduction. The atmosphere is assumed to be static.

The radiation loss term used by McWhirter *et al.* is similar in magnitude to the Cox and Tucker (1969) results but differs in some details. Most of the differences arise from differences in the assumed abundances. For an assumed acoustic flux of 2.5×10^5 erg cm^{-1} s^{-1}, the maximum coronal temperature is 2.3×10^6 K and occurs at a height near 3×10^5 km. An eight-fold increase of the acoustic flux increases T_{max} to only 3.4×10^6 K. The excess energy is disposed of mainly by increased thermal conduction through the transition region which, in turn, is radiated away by an increased density in the lower transition region. The inclusion of Landau damping and shock dissipation in the higher layers would change both the value and location of T_{max}.

Although all models of the transition region and corona derived from dissipation of acoustical flux assume the transition region to be more or less static, McWhirter *et al.* (1974) have shown that the sound wave pressure given by

$$\frac{F_s}{v_s} \frac{1+R}{1-R},$$

where R is the coefficient of reflectivity within the transition zone, F_s is the energy flux incident on the transition region and v_s is the sound velocity in the upper chromosphere, actually exceeds the hydrostatic pressure in the transition region. Even with $R = 0$, the acoustic pressure for $F_s = 3 \times 10^5$ erg cm^{-2} s^{-1} and $v_s = 30$ km s^{-1} is 0.1 dyn cm^{-2}. By comparison, the hydrostatic pressure in the transition region is only 0.16 dyn cm^{-2}.

It seems clear, therefore, that hydrostatic models of the transition region are incompatible with an acoustic flux sufficiently large to heat the corona. There is ample evidence of motion within the transition region and little or no evidence to support the hypothesis that the region is in hydrostatic equilibrium. We may find, indeed, that the transition region is highly turbulent, and that its average properties are only statistical in nature.

5. Acoustic Production of Stellar Chromospheres and Coronas

The Sun is but a typical star of a spectral type that is expected to have a moderate convection zone. More strongly developed convection zones are expected for cooler stars in the sense that the convection carries a larger fraction of the total flux. Stars earlier than about F_0 have convection zones that carry a smaller fraction of the total flux. However, mixing length theory predicts that the maximum convective velocity, which occurs in the upper convection zone, increases towards earlier spectral type reaching a

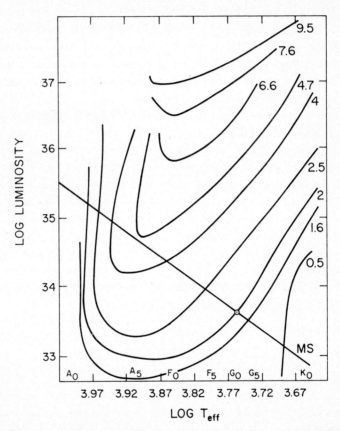

Fig. X-7.   A plot of the maximum convective velocities as a function of stellar luminosity and effective temperature. The curves are labeled by the velocity in km s^{-1}, and are computed from mixing length theory with $l = H_p$ (de Loore, 1970; courtesy *Astrophys. Space Sci.*).

maximum around spectral type A_5 on the main sequence. Maximum convective velocities computed by de Loore (1970) as a function of luminosity and spectral type (or effective temperature) are shown in Figure X-7. The curves are isopleths of constant velocity. Since the acoustic flux is strongly dependent upon the maximum velocity, the predicted acoustic power from the convection zone follows rather closely the pattern of the maximum velocity. This is illustrated in Figure X-8. The curves drawn in Figure X-8 are

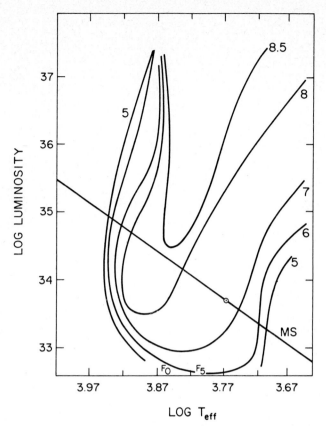

Fig. X-8.   Acoustic fluxes computed from the convective velocities illustrated in Figure X-7. The curves are labeled by the logarithm of the flux in erg cm^{-2} s^{-1} (de Loore 1970, courtesy *Astrophys. Space Sci.*).

isopleths of constant acoustic flux and are labeled by the logarithm of the flux. Note that the maximum predicted acoustic flux is for stars of type F_0 to F_s.

De Loore's computations are made with mixing length theory with the mixing length equated to the pressure scale height, H_p. As is well known from mixing length theory, however, the maximum turbulent velocity and, hence, the maximum acoustic flux is strongly dependent upon the assumed mixing length (see Figure X-4). Figure X-9. taken from the work of Castellani *et al.* (1971) shows a plot similar to Figure X-8 but with the mixing length set equal to 1.5 H_ρ, where H_ρ is the density scale height. Since $H_\rho \gg H_\rho$, this latter plot gives much larger acoustical fluxes. The isopleths in this plot are labeled by the ratio of the acoustic flux to the stellar radiation flux. Note that this ratio reaches a maximum of about 0.5. The maximum in Figure X-8, by comparison, is about 10^{-3} of the stellar radiation flux. Note that the plot in Figure X-9 represents the acoustic flux multiplied by the stellar surface area, and therefore contains the stellar radius.

The acoustic fluxes shown in Figure X-9 seem unrealistically large. However, they

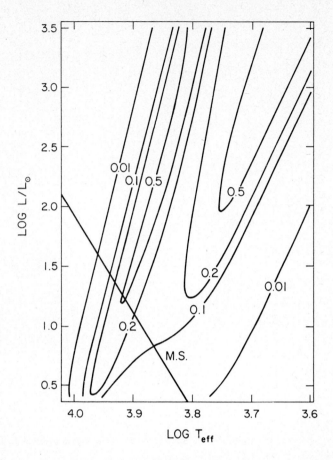

Fig. X-9.   Acoustic fluxes expressed in units of stellar luminosity computed from mixing length theory with $l = 1.5\,H_\rho$, where H_ρ is the density scale height. The flux was computed for a star of mass $M = 0.6\,M_0$. Note that the acoustic fluxes shown here are much larger than those shown in Figure X-8. This is a result of the larger value of l used here (Castellani *et al.*, 1971; courtesy *Astrophys. Space Sci.*)

serve, by comparison to the results in Figure X-8, to illustrate the uncertainty involved in predicting the acoustic fluxes for other stars. Castellani *et al.* (1971) suggest that the two tongues of maximum acoustic flux may be associated with variable star phenomena. They note that Cephied variables tend to lie in the left hand tongue and that RV Tauri, irregular and long period variables tend to lie in the right hand tongue.

De Loore (1970) has used the results in Figure X-8 to compute stellar coronas for stars with T_{eff} between 5000 and 8320 K and with $\log g$ between 4 and 5. He includes thermal conduction, radiation and solar wind losses and treats the wave dissipation as shock dissipation. For the solar case, he obtains a maximum temperature of 1.3×10^6 and an electron density at the temperature maximum of 6×10^9 cm^{-3}. The density is too high by about a factor of 20; and de Loore suggests that this may be due to an inhomogeneous energy supply with the average value being less than the adopted value.

A summary of the coronal maximum temperatures and densities for the models computed by de Loore is shown in Figure X-10.

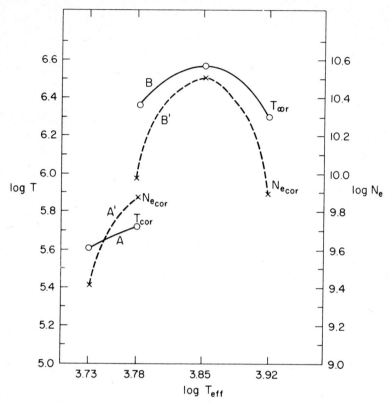

Fig. X-10.   Coronal maximum temperatures and electron densities at T_{max} computed by de Loore (1970) from the acoustic fluxes in Figure X-8 (courtesy *Astrophys. Space Sci.*).

An attempt to identify chromospheres and coronas on other stars (Praderie 1973) immediately encounters the dual difficulties of first knowing what to look for in the observational data and secondly of knowing how to define a stellar chromosphere and corona. In the solar case, the outwards increase in temperature and the dissipation of mechanical energy are closely coupled, and one could define the chromosphere either by the presence of the inversion in dT/dh or by the presence of mechanical energy dissipation. Similarly the existence of the corona, mass loss and mechanical dissipation are also closely coupled in the Sun.

The presence of the solar wind and the presence in the disk spectrum of the emission lines in the XUV spectrum, the high radio brightness temperature, both the width and the intensities of the central reversals in the Ca II and Mg II lines, the central intensities of strong lines of Fe I, Na I, Mg I, Ca I, the intensity and width of the Balmer-α line and the intensities and widths of the neutral helium lines, are valid indicators of the existence of the solar chromosphere and corona.

However, the lack of ambiguity in the solar case exists only because the outwards temperature increase is caused by mechanical dissipation of energy. Some stars may have a reversal in the temperature gradient that is purely radiative in character and has nothing to do with mechanical energy dissipation. The Cayrel Mechanism can lead to an outwards increase in temperature in stars whose spectrum is not exceedingly rich in lines. This mechanism seems to be particularly effective in early type stars and some radiative equilibrium models show temperature structures that are suggestive of chromospheres.

Some of the giant stars have extended atmospheres and undergo mass loss because of their low surface gravities. Such extended atmosphere stars often exhibit emission lines in addition to mass loss, but neither is indicative of mechanical energy dissipation. Also, mass exchange between binary systems and nearby nebulosities may produce observational effects that resemble chromospheric and coronal phenomena.

On the other side of the coin, it seems highly probable that acoustic energy flux is only one of several possibilities for transferring energy non-radiatively from the deeper layers of a star to its outer atmosphere. Thus, we should not restrict our consideration of stellar chromospheres and coronas to just those cases where acoustical energy flux is expected.

Still a third indicator of the solar chromosphere and corona is the outwards increase in the Doppler widths of lines, most of which is due to non-thermal motion within the chromosphere itself. The Wilson-Bappu effect in stars with K_2 emission is, in all probability, a good indicator of chromospheric structure. However, of the many solar lines with a strong chromospheric component, the Ca II lines are the only visual lines that show self-reversal. Thus, we should recognize that the absence of reversal in the H and K lines is not necessarily good evidence for the absence of a chromosphere. In particular, Ca II will be relatively less abundant in hotter stars and the emission reversals may disappear simply because the opacity in the H and K lines is diminished. Thus, the tendency for K_2 emission to disappear in stars earlier than type F may be more a result of the increased effective temperature than of the absence of mechanical heating.

It is generally concluded from the intensities of the H and K reversals that chromospheric activity increases for stars later than the Sun (cf. Praderie, 1970), whereas the results in Figures X-8 and X-9 suggest just the opposite. An assertion that this disproves the theory of heating by acoustic flux is not warranted, however. The predictions of acoustic flux are highly uncertain, and the absence of H and K reversals is not a guarantee that mechanical heating is not present.

In those stars that do exhibit reversals in the H and K lines, the width of the reversals is coupled to the Doppler broadening velocity in the chromospheric layers. Thus a proper understanding of the Wilson-Bappu effect together with a valid explanation for the source of the line broadening appear to be one of the more promising approaches to the study of stellar chromospheres.

References

Bahng, J. and Schwarzschild, M.: 1963, *Astrophys. J.* **137**, 901.
Bel, N. and Mein, P.: 1971, *Astron. Astrophys.* **11**, 234.
Biermann, L.: 1946, *Naturwissenschaften* **33**, 118.
Bray, R. J. and Loughhead, R. E.: 1974, *The Solar Chromosphere*. Chapman and Hall, London.
Canfield, R. C. and Musman, S.: 1973, *Astrophys. J.* **184**, L131.
Castellani, V., Puppi, L. and Renzini, A.: 1971, *Astrophys. Space Sci.* **10**, 136.
Chandrasekhar, S.: 1953, *Proc. Roy. Astron. Soc. London* **A217**, 306.
Chiu, Y. T.: 1971, *Phys. fluid* **14**, 1717.
Cox, D. P. and Tucker, W. H.: 1969, *Astrophys. J.* **157**, 1157.
D'Angelo, N.: 1969, *Solar Phys.* **7**, 321.
Ffowcs-Williams, J. E.: 1969, *Ann. Rev. Fluid Mech.* **1**, 197.
Frazier, E. N.: 1968, *Z. Astrophys.* **68**, 345.
Howe, M. S.: 1969, *Astrophys. J.* **156**, 27.
Jager, C. de and Kuperus, M.: 1961, *Bull. Astron. Inst. Neth.* **16**, 71.
Jordan, S. D.: 1969, *The Temperature Distribution in the Solar Chromosphere*, NASA Tech. Rep. TR/R–291.
Kahn, F. D.: 1961, *Astrophys. J.* **134**, 343.
Kahn, F. D.: 1969, *Astrophys. J.* **135**, 547.
Kaplan, S. A., Ostrovskii, L. A., Petrukhin, N. S. and Fridman, V. E.: 1972, *Astron. Zh.* **49**, 1267. (English, *Soviet Astron.* **16**, 1013.)
Kato, S.: 1966, *Astrophys. J.* **143**, 893.
Kopp, R. A.: 1968, *The Equilibrium Structure of a Shock-heated Corona*, Sci. Rep. No. 4, Air Force Cambridge Res. Lab. Bedford.
Kulsrud, R. M.: 1955, *Astrophys. J.* **121**, 461.
Kuperus, M.: 1965, *Rech. Astron. Obs. Utrecht* **17**, 1.
Landau, L. D. anu Lifshitz, E. M.: 1959, *Fluid Mechanics*, Pergamon, London.
Leibacher, J.: 1971, Thesis, Harvard University.
Leibacher, J.: 1974, *Generation of Waves in Stellar Atmospheres*, National Center for Atmospheric Research, Boulder.
Leibacher, J. and Stein, R. F.: 1971, *Astrophys. Letters* **7**, 191.
Leighton, R. B.: 1960, in R. N. Thomas (ed.), 4th Symp. of Cosmical Gas Dynamics, *IAU Symp.* **12**.
Lighthill, M.: 1952, *Proc. Roy. Astron. Soc. London* **A211**, 564.
Lighthill, M.: 1967, in R. N. Thomas (ed.), 'Aerodyanamic Phenomena in Stellar Atmospheres', *IAU Symp.* **28**.
Loore, C. De: 1970, *Astrophys. Space Sci.* **6**, 60.
McLellan, A. and Winterberg, F.: 1968, *Solar Phys.* **4**, 401.
McKenzie, J. F.: 1971, *Astron. Astrophys.* **15**, 450.
McWhirter, R. W. P., Thonemann, P. C. and Wilson, R.: 1975, *Astron. Astrophys.* **450**, 63.
Meyer, F. and Schmidt, H. U.: 1967, *Z. Astrophys.* **65**, 274.
Moore, D. W.: 1967, in R. N. Thomas (ed.), 'Aerodynamic Phenomena in Stellar Atmospheres', *IAU Symp.* **28**.
Moore, D. W. and Spiegel, E. A.: 1966, *Astrophys. J.* **143**, 871.
Moore, R. L.: 1972, *The Structure and Heating of the Chromosphere-Corona Transition Region*, SUIPR Report No. 463, Inst. for Plasma Res., Stanford Univ.
Morse, P. M. and Ingard, K. U.: 1968, *Theoretical Acoustics*, McGraw-Hill, New York.
Musman, S.: 1972, *Solar Phys.* **26**, 290.
Osterbrock, D. E.: 1961, *Astrophys. J.* **134**, 347.
Parker, E. N.: 1964, *Astrophys. J.* **140**, 1170.
Praderie, F.: 1970, in H. G. Groth and P. Wellman (eds.), *Extended Atmosphere Stars*, N.B.S. Spec. Pub. 332.
Praderie, F.: 1973, in S. D. Jordan and E. H. Avrett (eds.), *Stellar Chromospheres*, NASA SP–317.
Proudman, I.: 1952, *Proc. Roy. Astron. Soc. London* **A214**, 119.
Schmidt, H. U. and Stix, M.: 1973, *Mitt. Astron. Gesell.* **32**, 182.
Schmidt, H. U. and Zirker, J. B.: 1963, *Astrophys. J.* **138**, 1310.
Schwarzschild, M.: 1948, *Astrophys. J.* **105**, 1.
Souffrin, P.: 1966, *Ann. Astrophys.* **29**, 55.
Spiegel, E. A.: 1957, *Astrophys. J.* **126**, 202.
Spiegel, E. A.: 1962, *J. Geophys. Res.* **67**, 3063.

Spiegel, E. A., 1964, *Astrophys. J.* **139**, 959.
Spitzer, L.: 1962, *Physics of Fully Ionized Gases,* Interscience, New York.
Stein, R. F.: 1967, *Solar Phys.* **2**, 385.
Stein, R. F.: 1968, *AStrophys. J.* **154**, 297.
Stein, R. F. and Leibacher, J.: 1974, *Ann. Rev. Astron. Astrophys.* **12**, 407.
Stein, R. F. and Schwartz, R. A.: 1972, *Astrophys. J.* **177**, 807.
Stix, M.: 1970, *Astron. Astrophys.* **4**, 189.
Stix, T. H.: 1962, *The Theory of Plasma Waves*, McGraw-Hill, New York.
Syrovatsky, S. I. and Zhugzhda, Y. D.: 1968, in K. O. Kiepenheuer (ed.), 'Structure and Development
    of Solar Active Regions', *IAU Symp.* **35**, 127.
Tanenbaum, A. S.: 1971, *A Study of Five-minute Oscillations, Supergranulation, and Related
    Phenomena in the Solar Atmosphere*, Space Sciences Lab., Univ. of Calif., Berkeley.
Tanenbaum, A. S., Wilcox, J. M., Frazier, E. N. and Howard, R.: 1969, *Solar Phys.* **9**, 328.
Thomas, J. H., Clark, P. A. and Clark, A.: 1971, *Solar Phys.* **16**, 51.
Uchida, Y.: 1963, *Publ. Astron. Soc. Japan* **15**, 376.
Uchida, Y.: 1965, *Astrophys. J.* **142**, 335.
Ulmschneider, P.: 1971a, *Astron. Astrophys.* **12**, 297.
Ulmschneider, P.: 1971b, *Astron. Astrophys.* **14**, 275.
Ulrich, R. K.: 1973, in S. D. Jordan and E. H. Avrett (eds.), *Stellar Chromospheres*, NASA, SP–317.
Ulrich, R. K.: 1973, in S. D. Jordan and E. H. Avrett (eds.), NASA, SP–317.
Unno, W.: 1964, *Trans. IAU* **12B**, 555.
Vitense, E.: 1953, *Z. Astrophys.* **32**, 135.
Wolff, C. L.: 1972, *Astrophys. J.* **177**, L87.
Worrall, G.: 1972, *Astrophys. J.* **172**, 749.
Zhugzhda, Y. D.: 1972, *Solar Phys.* **25**, 329.

INDEX

ASTROPHYSICS AND SPACE SCIENCE LIBRARY

Edited by

J. E. Blamont, R. L. F. Boyd, L. Goldberg, C. de Jager, Z. Kopal, G. H. Ludwig, R. Lüst,
B. M. McCormac, H. E. Newell, L. I. Sedov, Z. Švestka, and W. de Graaff

1. C. de Jager (ed.), *The Solar Spectrum. Proceedings of the Symposium held at the University of Utrecht, 26–31 August, 1963.* 1965, XIV + 417 pp.
2. J. Ortner and H. Maseland (eds.), *Introduction to Solar Terrestrial Relations, Proceedings of the Summer School in Space Physics held in Alpbach, Austria, July 15–August 10, 1963 and Organized by the European Preparatory Commission for Space Research.* 1965, IX + 506 pp.
3. C. C. Chang and S. S. Huang (eds.), *Proceedings of the Plasma Space Science Symposium, held at the Catholic University of America, Washington, D.C., June 11–14, 1963.* 1965, IX + 377 pp.
4. Zdeněk Kopal, *An Introduction to the Study of the Moon.* 1966, XII + 464 pp.
5. B. M. McCormac (ed.), *Radiation Trapped in the Earth's Magnetic Field. Proceedings of the Advanced Study Institute, held at the Chr. Michelsen Institute, Bergen, Norway, August 16–September 3, 1965.* 1966, XII + 901 pp.
6. A. B. Underhill, *The Early Type Stars.* 1966, XII + 282 pp.
7. Jean Kovalevsky, *Introduction to Celestial Mechanics.* 1967, VIII + 427 pp.
8. Zdeněk Kopal and Constantine L. Goudas (eds.), *Measure of the Moon. Proceedings of the 2nd International Conference on Selenodesy and Lunar Topography, held in the University of Manchester, England, May 30–June 4, 1966.* 1967, XVIII + 479 pp.
9. J. G. Emming (ed.), *Electromagnetic Radiation in Space. Proceedings of the 3rd ESRO Summer School in Space Physics, held in Alpbach, Austria, from 19 July to 13 August, 1965.* 1968, VIII + 307 pp.
10. R. L. Carovillano, John, F. McClay, and Henry R. Radoski (eds.), *Physics of the Magnetosphere, Based upon the Proceedings of the Conference held at Boston College, June 19–28, 1967.* 1968, X + 686 pp.
11. Syun-Ichi Akasofu, *Polar and Magnetospheric Substorms.* 1968, XVIII + 280 pp.
12. Peter M. Millman (ed.), *Meteorite Research. Proceedings of a Symposium on Meteorite Research, held in Vienna, Austria, 7–13 August, 1968.* 1969, XV + 941 pp.
13. Margherita Hack (ed.), *Mass Loss from Stars. Proceedings of the 2nd Trieste Colloquium on Astrophysics, 12–17 September, 1968.* 1969, XII + 345 pp.
14. N. D'Angelo (ed.), *Low-Frequency Waves and Irregularities in the Ionosphere. Proceedings of the 2nd ESRIN-ESLAB Symposium, held in Frascati, Italy, 23–27 September, 1968.* 1969, VII + 218 pp.
15. G. A. Partel (ed.), *Space Engineering. Proceedings of the 2nd International Conference on Space Engineering, held at the Fondazione Giorgio Cini, Isola di San Giorgio, Venice, Italy, May 7–10, 1969.* 1970, XI + 728 pp.
16. S. Fred Singer (ed.), *Manned Laboratories in Space. Second International Orbital Laboratory Symposium.* 1969, XIII + 133 pp.
17. B. M. McCormac (ed.), *Particles and Fields in the Magnetosphere. Symposium Organized by the Summer Advanced Study Institute, held at the University of California, Santa Barbara, Calif., August 4–15, 1969.* 1970, XI + 450 pp.
18. Jean-Claude Pecker, *Experimental Astronomy.* 1970, X + 105 pp.
19. V. Manno and D. E. Page (eds.), *Intercorrelated Satellite Observations related to Solar Events. Proceedings of the 3rd ESLAB/ESRIN Symposium held in Noordwijk, The Netherlands, September 16–19, 1969.* 1970, XVI + 627 pp.
20. L. Mansinha, D. E. Smylie, and A. E. Beck, *Earthquake Displacement Fields and the Rotation of the Earth. A NATO Advanced Study Institute Conference Organized by the Department of Geophysics, University of Western Ontario, London, Canada, June 22–28, 1969.* 1970, XI + 308 pp.
21. Jean-Claude Pecker, *Space Observatories.* 1970, XI + 120 pp.
22. L. N. Mavridis (ed.), *Structure and Evolution of the Galaxy. Proceedings of the NATO Advanced Study Institute, held in Athens, September 8–19, 1969.* 1971, VII + 312 pp.
23. A. Muller (ed.), *The Magellanic Clouds. A European Southern Observatory Presentation: Principal Prospects, Current Observational and Theoretical Approaches, and Prospects for Future Research. Based*

on the Symposium on the Magellanic Clouds, held in Santiago de Chile, March 1969, on the Occasion of the Dedication of the European Southern Observatory. 1971, XII + 189 pp.

24. B. M. McCormac (ed.), *The Radiating Atmosphere*. Proceedings of a Symposium Organized by the Summer Advanced Study Institute, held at Queen's University, Kingston, Ontario, August 3–14, 1970. 1971, XI + 455 pp.

25. G. Fiocco (ed.), *Mesospheric Models and Related Experiments*. Proceedings of the 4th ESRIN-ESLAB Symposium, held at Frascati, Italy, July 6–10, 1970. 1971, VIII + 298 pp.

26. I. Atanasijević, *Selected Exercises in Galactic Astronomy*. 1971, XII + 144 pp.

27. C. J. Macris (ed.), *Physics of the Solar Corona*. Proceedings of the NATO Advanced Study Institute on Physics of the Solar Corona, held at Cavouri-Vouliagmeni, Athens, Greece, 6–17 September 1970. 1971, XII + 345 pp.

28. F. Delobeau, *The Environment of the Earth*. 1971, IX + 113 pp.

29. E. R. Dyer (general ed.), *Solar-Terrestrial Physics/1970*. Proceedings of the International Symposium on Solar-Terrestrial Physics, held in Leningrad, U.S.S.R., 12–19 May 1970. 1972, VIII + 938 pp.

30. V. Manno and J. Ring (eds.), *Infrared Detection Techniques for Space Research*. Proceedings of the 5th ESLAB-ESRIN Symposium, held in Noordwijk, The Netherlands, June 8–11, 1971. 1972, XII + 344 pp.

31. M. Lecar (ed.), *Gravitational N-Body Problem*. Proceedings of IAU Colloquium No. 10, held in Cambridge, England, August 12–15, 1970. 1972, XI + 441 pp.

32. B. M. McCormac (ed.), *Earth's Magnetospheric Processes*. Proceedings of a Symposium Organized by the Summer Advanced Study Institute and Ninth ESRO Summer School, held in Cortina, Italy, August 30–September 10, 1971. 1972, VIII + 417 pp.

33. Antonin Rükl, *Maps of Lunar Hemispheres*. 1972, V + 24 pp.

34. V. Kourganoff, *Introduction to the Physics of Stellar Interiors*. 1973, XI + 115 pp.

35. B. M. McCormac (ed.), *Physics and Chemistry of Upper Atmospheres*. Proceedings of a Symposium Organized by the Summer Advanced Study Institute, held at the University of Orléans, France, July 31–August 11, 1972. 1973, VIII + 389 pp.

36. J. D. Fernie (ed.), *Variable Stars in Globular Clusters and in Related Systems*. Proceedings of the IAU Colloquium No. 21, held at the University of Toronto, Toronto, Canada, August 29–31, 1972. 1973, IX + 234 pp.

37. R. J. L. Grard (ed.), *Photon and Particle Interaction with Surfaces in Space*. Proceedings of the 6th ESLAB Symposium, held at Noordwijk, The Netherlands, 26–29 September, 1972. 1973, XV + 577 pp.

38. Werner Israel (ed.), *Relativity, Astrophysics and Cosmology*. Proceedings of the Summer School, held 14–26 August, 1972, at the BANFF Centre, BANFF, Alberta, Canada. 1973, IX + 323 pp.

39. B. D. Tapley and V. Szebehely (eds.), *Recent Advances in Dynamical Astronomy*. Proceedings of the NATO Advanced Study Institute in Dynamical Astronomy, held in Cortina d'Ampezzo, Italy, August 9–12, 1972. 1973, XIII + 468 pp.

40. A. G. W. Cameron (ed.), *Cosmochemistry*. Proceedings of the Symposium on Cosmochemistry, held at the Smithsonian Astrophysical Observatory, Cambridge, Mass., August 14–16, 1972. 1973, X + 173 pp.

41. M. Golay, *Introduction to Astronomical Photometry*. 1974, IX + 364 pp.

42. D. E. Page (ed.), *Correlated Interplanetary and Magnetospheric Observations*. Proceedings of the 7th ESLAB Symposium, held at Saulgau, W. Germany, 22–25 May, 1973. 1974, XIV + 662 pp.

43. Riccardo Giacconi and Herbert Gursky (eds.), *X-Ray Astronomy*. 1974, X + 450 pp.

44. B. M. McCormac (ed.), *Magnetospheric Physics*. Proceedings of the Advanced Summer Institute, held in Sheffield, U.K., August 1973. 1974, VII + 399 pp.

45. C. B. Cosmovici (ed.), *Supernovae and Supernova Remnants*. Proceedings of the International Conference on Supernovae, held in Lecce, Italy, May 7–11, 1973. 1974, XVII + 387 pp.

46. A. P. Mitra, *Ionospheric Effects of Solar Flares*. 1974, XI + 294 pp.

49. Z. Švestka and P. Simon (eds.), *Catalog of Solar Particle Events 1955–1969*. Prepared under the Auspices of Working Group 2 of the Inter-Union Commission on Solar-Terrestrial Physics. 1975, IX + 428 pp.

50. Zdeněk Kopal and Robert W. Carder, *Mapping of the Moon*. 1974, VIII + 237 pp.

51. B. M. McCormac (ed.), *Atmospheres of Earth and the Planets*. Proceedings of the Summer Advanced Study Institute, held at the University of Liège, Belgium, July 29–August 9, 1974. 1975, VII + 454 pp.

52. V. Formisano (ed.), *The Magnetospheres of the Earth and Jupiter*. Proceedings of the Neil Brice Memorial Symposium, held in Frascati, May 28–June 1, 1974. 1975, XI + 485 pp.